JACARANDA PHYSICS 1

VCE UNITS 1 AND 2 | FIFTH EDITION

MICHAEL ROSENBROCK

DAN O'KEEFFE

CATHERINE BELLAIR

GRAEME LOFTS

ROSS PHILLIPS

PETER PENTLAND

JANE COYLE

BARBARA MCKINNON

CONTRIBUTING AUTHORS

John Whitehead

Samuel Watkins

jacaranda
A Wiley Brand

Fifth edition published 2023 by
John Wiley & Sons Australia, Ltd
155 Cremorne Street, Cremorne, Vic 3121

First edition published 1996
Second edition published 2003
Third edition published 2008
eBookPLUS edition published 2015
Fourth edition published 2020

Typeset in 10.5/13 pt TimesLTStd

ISBN: 978-1-119-88789-8

A catalogue record for this book is available from the National Library of Australia

Printed in Singapore
M WEP295522 110724

Contents

UNIT 2 HOW DOES PHYSICS HELP US TO UNDERSTAND THE WORLD?   319

10 Energy and motion

AREA OF STUDY 2 OPTIONS: HOW DOES PHYSICS INFORM CONTEMPORARY ISSUES AND APPLICATIONS IN SOCIETY?

online only

11 Socio-scientific issues

    11.1 How does physics explain climate change?

    11.2 How do fusion and fission compare as viable nuclear energy power sources?

    11.3 How do heavy things fly?

    11.4 How do forces act on structures and materials?

    11.5 How do forces act on the human body?

    11.6 How is radiation used to maintain human health?

    11.7 How does the human body use electricity?

    11.8 How can human vision be enhanced?

    11.9 How is physics used in photography?

    11.10 How do instruments make music?

    11.11 How can performance in ball sports be improved?

    11.12 How can AC electricity charge a DC device?

    11.13 How do astrophysicists investigate stars and black holes?

    11.14 How can we detect possible life beyond Earth's solar system?

    11.15 How can physics explain traditional artefacts, knowledge and techniques?

    11.16 How do particle accelerators work?

    11.17 How does physics explain the origins of matter?

    11.18 How is contemporary physics research being conducted in our region?

AREA OF STUDY 3 HOW DO PHYSICISTS INVESTIGATE QUESTIONS?

online only

12 Scientific investigations

    12.1 Overview

    12.2 Key science skills and concepts in physics

    12.3 Characteristics of scientific methodology and primary data generation

    12.4 Health, safety and ethical guidelines

    12.5 Accuracy, precision, reproducibility, repeatability and validity of measurements

    12.6 Ways of organising, analysing and evaluating primary data

    12.7 Challenging scientific models and theories

    12.8 The limitations of investigation methodology and conclusions

    12.9 Conventions of science communication

    12.10 Review

About this resource

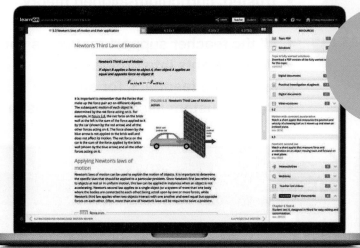

YEAR 11
NEW FOR 2023
YEAR 12 COMING FOR 2024

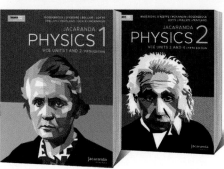

JACARANDA

PHYSICS 1

VCE UNITS 1 AND 2
FIFTH EDITION

Developed by expert Victorian teachers for VCE students

Tried, tested and trusted. The NEW Jacaranda VCE Physics series continues to deliver curriculum-aligned material that caters to students of all abilities.

Completely aligned to the VCE Physics Study Design

Our expert author team of practising teachers and assessors ensures 100% coverage of the new VCE Physics Study Design (2023–2027).

Everything you need for your students to succeed, including:

- **NEW!** Access targeted questions sets including exam-style questions and all relevant past VCAA exam questions since 2013. Ensure assessment preparedness with practice SACs.

- **NEW!** Enhanced practical investigation support including practical investigation videos, and eLogbook with fully customisable practical investigations — including teacher advice and risk assessments.

- **NEW!** Teacher-led videos to unpack challenging concepts, VCAA exam questions, exam-style questions, investigations and sample problems to fill learning gaps after COVID-19 disruptions.

Learn online with Australia's most

Everything you need for each of your lessons in one simple view

- Trusted, curriculum-aligned theory
- Engaging, rich multimedia
- All the teacher support resources you need
- Deep insights into progress
- Immediate feedback for students
- Create custom assignments in just a few clicks.

Practical teaching advice and ideas for each lesson provided in teachON

Each lesson linked to the Key Knowledge (and Key Science Skills) from the VCE Physics Study Design

Reading content and rich media including embedded videos and interactivities

---

learn on Jacaranda Physics 2 VCE Units 3 & 4 4e

📖 10.6 Interference using light   ≡   10.6 teachON

10.6 Interference using light

KEY CONCEPTS

- Explain the results of Young's double-slit experiment with reference to:
  - evidence for the wavelike nature of light
  - constructive and destructive interference of coherent waves in terms of path differences: $n\lambda$ and $\left(n - \frac{1}{2}\right)\lambda$ respectively
  - effect of wavelength, distance of screen and slit separation on interference patterns: $\Delta x = \frac{\lambda L}{d}$.

10.6.1 Young's double-slit experiment

Thomas Young (1773–1829) was keenly interested in many things. He has been called 'the last man who knew everything'. He was a practising surgeon as well as a very active scientist. He analysed the dynamics of blood flow, explained the accommodation mechanism for the human eye and proposed the three-receptor model for colour vision. He also made significant contributions to the study of elasticity and surface tension. His other interests included deciphering ancient Egyptian hieroglyphics, comparing the grammar and vocabulary of over 400 languages, and developing tunings for the twelve notes of the musical octave. Despite these many interests, the wave explanation of the nature of light was of continuing interest to him.

Young had already built a ripple tank to show that the water waves from two point sources with synchronised vibrations show evidence of interference.

‹ 10.5 DIFFRACTION OF LIGHT

powerful learning tool, learnON

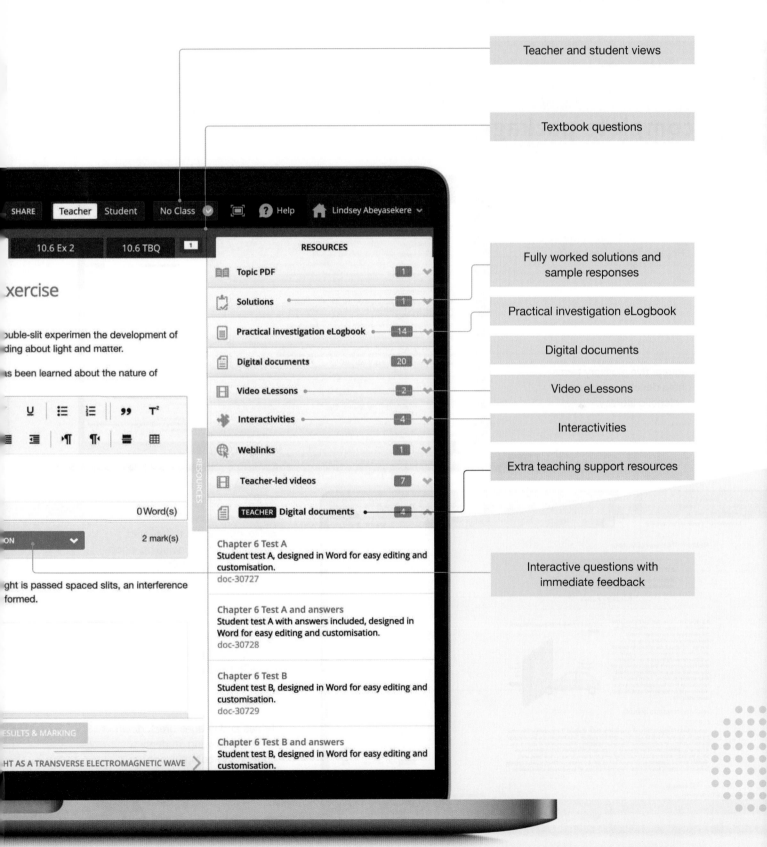

Teacher and student views

Textbook questions

Fully worked solutions and sample responses

Practical investigation eLogbook

Digital documents

Video eLessons

Interactivities

Extra teaching support resources

Interactive questions with immediate feedback

Get the most from your online resources

Online, these new editions are the **complete package**

Trusted Jacaranda theory, plus tools to support teaching and make learning more engaging, personalised and visible.

Each subtopic is linked to Key Knowledge (and Key Science Skills) from the VCE Physics Study Design.

Interactive glossary terms help develop and support scientific literacy.

onResources link to targeted digital resources including video eLessons and weblinks.

Pink highlight boxes summarise key information and provide tips for VCE Physics success.

Tables and images break down content, allowing students to understand complex concepts.

Sample problems break down the process of answering questions using a think/write format and a supporting teacher-led video.

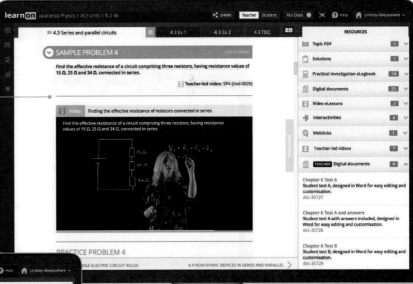

Practical investigations are highlighted throughout topics, and are supported by teacher-led videos and downloadable student and teacher version eLogbooks.

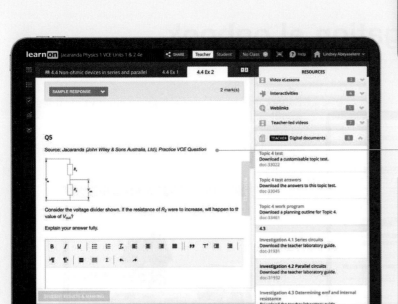

- Online and offline question sets contain practice questions and past VCAA exam questions with exemplary responses and marking guides.
- Every question has immediate, corrective feedback to help students to overcome misconceptions as they occur and to study independently — in class and at home.

Topic reviews

A summary flowchart shows the interrelationship between the main ideas of the topic. This includes links to both Key Knowledge and Key Science Skills.

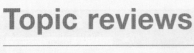
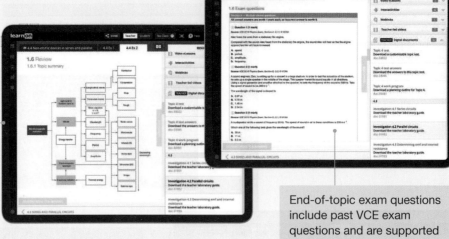

End-of-topic exam questions include past VCE exam questions and are supported by teacher-led videos.

Area of Study reviews

Areas of study reviews include practice examinations and practice SACs with worked solutions and sample responses. Teachers have access to customisable quarantined SACs with sample responses and marking rubrics.

Practical investigation eLogbook

Enhanced practical investigation support includes practical investigation videos and an eLogbook with fully customisable practical investigations — including teacher advice and risk assessments.

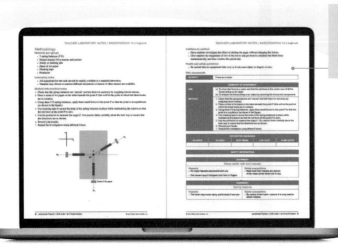

A wealth of teacher resources

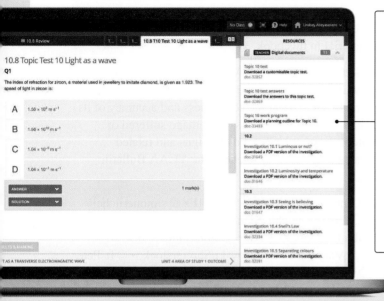

Enhanced teacher support resources, including:

- work programs and curriculum grids
- teaching advice
- additional activities
- teacher laboratory eLogbook, complete with solution and risk assessments
- quarantined topic tests (with solutions)
- quarantined SACs (with worked solutions and marking rubrics).

Customise and assign

A testmaker enables you to create custom tests from the complete bank of thousands of questions (including past VCAA exam questions).

Reports and results

Data analytics and instant reports provide data-driven insights into progress and performance within each lesson and across the entire course.

Show students (and their parents or carers) their own assessment data in fine detail. You can filter their results to identify areas of strength and weakness.

Acknowledgements

The authors and publisher would like to thank the following copyright holders, organisations and individuals for their assistance and for permission to reproduce copyright material in this book.

Selected extracts from the VCE Physics Study Design (2023–2027) are copyright Victorian Curriculum and Assessment Authority (VCAA), reproduced by permission. VCE® is a registered trademark of the VCAA. The VCAA does not endorse this product and makes no warranties regarding the correctness and accuracy of its content. To the extent permitted by law, the VCAA excludes all liability for any loss or damage suffered or incurred as a result of accessing, using or relying on the content. Current VCE Study Designs and related content can be accessed directly at www.vcaa.vic.edu.au. Teachers are advised to check the VCAA Bulletin for updates.

• © Andrew James/Alamy Stock Photo: **6** • © fStop Images GmbH/Alamy Stock Photo: **41** • © simoncritchell/Shutterstock: **66** • © 977_ReX_977/Shutterstock: **383** • © a. Juergen Faelchle/Shutterstock; © b. Ozgur Coskun/Shutterstock: **89** • © a. Lukasz Pawel Szczepanski/Shutterstock; © b. Alex Konon/Shutterstock: **61** • © a. Photoskills; © b. Photoskills; c. Photoskills: **43** • © a. Vladimir A Veljanovski/Shutterstock; • © b. Peter Pentland: **214** • © agefotostock/Alamy Stock Photo: **417** • © Alexandr Shevchenko/Shutterstock: **375** • © Allexxandar/Shutterstock: **39** • © AndrewJohnson/E+/Getty Images: **134** • © Andrey Armyagov/Shutterstock: **326** • © Antonio Guillem/Shutterstock: **296** • © Aphelleon/Shutterstock: **16** • © Artur_eM/Shutterstock: **144** • © Atomic Heritage Foundation: **158, 159** • © blickwinkel/Alamy Stock Photo: **162** • © Bork/Shutterstock: **149** • © Brett Jordan/Unsplash.com: **203** • © By Brady Holt - Own work: CC BY 3.0, https://commons.wikimedia.org/w/index.php?curid=11113684, **418** • © By Kebuk awan - Own work: CC BY-SA 4.0, https://commons.wikimedia.org/w/index.php?curid=110506871, **137** • © Caron Badkin/Shutterstock: **299** • © cassiede alain/Shutterstock: **465** • © ChameleonsEye/Shutterstock: **3** • © cherezoff/Shutterstock: **23** • © conrado/Shutterstock: **440** • © CP DC Press/Shutterstock: **329** • © Cris DiNoto/Unsplash.com: **24** • © D. Pimborough/Shutterstock: **97** • © DAMIEN LOVEGROVE/Science Photo Library: **66** • © Daniel Holland/Unsplash.com: **12** • © danpastia/Shutterstock: **186** • © David Gee 3/Alamy Stock Photo: **53** • © David Greene/Shutterstock: **62** • © Diego Barbieri/Shutterstock: **457** • © dikobraziy/Shutterstock: **104** • © DIZ Muenchen GmbH: Sueddeutsche Zeitung Photo/Alamy Stock Photo, **153** • © EECA/© Commonwealth of Australia 2019.: **290** • © Egor_Kulinich/Shutterstock: **295** • © Everett Collection/Shutterstock: **62** • © Filip Fuxa/Shutterstock: **62** • © Francesco Ocello/Shutterstock: **16** • © FS Stock/Shutterstock: **485** • © Georgios Kollidas/Shutterstock: **383** • © Germanskydiver/Shutterstock: **392** • © gualtiero boffi/Shutterstock: **18** • © hfzimages/Shutterstock: **322** • © Igor Prahin/Alamy Stock Photo: **374** • © IgorZh/Shutterstock: **189** • © Ilhamjb23/Shutterstock: **265** • © Ivan Smuk/Shutterstock: **398** • © Jan Babak/Shutterstock: **265** • © Jane Fownes/Alamy Stock Photo: **37** • © Joel_420/Shutterstock: **284** • © John Wiley & Sons Australia/Photo by Ron Ryan: **44** • © kaband/Shutterstock: **84** • © lightpoet/Shutterstock: **491** • © La Objetiva/Shutterstock: **283** • © leoleobobeo/1415 images/Pixabay: **4** • © manzrussali/Shutterstock: **416** • © MARGRIT HIRSCH/Shutterstock: **422** • © mariolav/Shutterstock: **373** • © Mark_Kostich/Shutterstock: **123** • © Martyn F. Chillmaid/Science Photo Library: **84** • © maxpro/Shutterstock: **416** • © mekoo/Getty Images: **121** • © melhijad/Shutterstock: **439** • © Mike Focus/Shutterstock: **293** • © mkrol0718/Shutterstock: **78** • © mooinblack/Shutterstock: **415** • © MPIX/Shutterstock: **296** • © MXW Stock/Shutterstock: **68** • © Nagy-Bagoly Arpad/Shutterstock: **88** • © Nasky/Shutterstock: **370** • © Neale Cousland/Shutterstock: **461** • © Newscom/Alamy Stock Photo: **124** • © Nils Versemann/Shutterstock: **234** • © Patricia F. Carvalho/Shutterstock: **197** • © petrroudny43/Shutterstock: **19** • © Phil Crosby/Alamy Stock Photo: **67** • © photastic/Shutterstock: **80** • © Phovoir/Shutterstock: **316** • © picturepartners/Shutterstock: **462** • © Polryaz/Shutterstock: **80** • © praphab louilarpprasert/Shutterstock: **100** • © Radu Razvan/Shutterstock: **374** • © raigvi/Shutterstock: **233** • © RalphHuijgen/Shutterstock: **267** • © Razor527/Shutterstock: **368** • © Rita_Kochmarjova/Shutterstock: **367** • © Robert Green/Moment/Getty Images: **205** • © rootstudio/Shutterstock: **321** • © Science History Images/Alamy Stock Photo: **213** • © Science Photo Library/Alamy Stock Photo: **59, 137**

• © sciencephotos/Alamy Stock Photo: **301** • © Sergio Bertino/Shutterstock: **52** • © SIPLEY/Alamy Stock Photo: **193** • © Skumer/Shutterstock: **324** • © Source: Visionlearning: Inc., **17** • © Stason4ik/Shutterstock: **401** • © Tanya Puntti/Shutterstock: **190** • © tuulijumala/Shutterstock: **38** • © udaix/Shutterstock: **29, 63** • © Vasilyev Alexandr/Shutterstock: **319** • © vldkont/Shutterstock: **288** • © wildestanimal/Shutterstock: **77** • © yerv/Shutterstock: **1** • © Yevhen Rychko/Alamy Stock Photo: **286** • © GIPhotoStock/Science Source: **5**

Every effort has been made to trace the ownership of copyright material. Information that will enable the publisher to rectify any error or omission in subsequent reprints will be welcome. In such cases, please contact the Permissions Section of John Wiley & Sons Australia, Ltd.

1 How is energy useful to society?

Source: VCE Physics Study Design (2023–2027) extracts © VCAA; reproduced by permission.

1 Electromagnetic radiation and waves

KEY KNOWLEDGE

In this topic, you will:
- identify all electromagnetic waves as transverse waves travelling at the same speed, c, in a vacuum as distinct from mechanical waves that require a medium to propagate
- identify the amplitude, wavelength, period and frequency of waves
- calculate the wavelength, frequency, period and speed of travel of waves using: $\lambda = \dfrac{v}{f} = vT$
- explain the wavelength of a wave as a result of the velocity (determined by the medium through which it travels) and the frequency (determined by the source)
- describe electromagnetic radiation emitted from the Sun as mainly ultraviolet, visible and infrared
- compare the wavelength and frequencies of different regions of the electromagnetic spectrum, including radio, microwave, infrared, visible, ultraviolet, x-ray and gamma, and compare the different uses each has in society
- calculate the peak wavelength of the radiated electromagnetic radiation using Wien's Law: $\lambda_{max} T = $ constant
- compare the total energy across the electromagnetic spectrum emitted by objects at different temperatures.

Source: VCE Physics Study Design (2023–2027) extracts © VCAA; reproduced by permission.

PRACTICAL WORK AND INVESTIGATIONS

Practical work is a central component of VCE Physics. Experiments and investigations, supported by a **practical investigation eLogbook** and **teacher-led videos**, are included in this topic to provide opportunities to undertake investigations and communicate findings.

EXAM PREPARATION

Access exam-style questions and their video solutions in every lesson, to ensure you are ready.

1.1 Overview

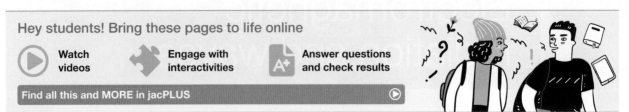
1.1.1 Introduction

We are familiar with many common phenomena related to the behaviour of waves. Consider the ripple of waves on the surface of water, waves on a string such as a guitar, the propagation of sound through a medium, and the properties of light. Waves can also transfer energy from one place to another. They seem to travel or propagate at a fixed speed in a medium regardless of the energy they contain. Waves have a tendency to spread out in all directions. Waves can pass through each other; they don't collide with each other in the same way particles collide — instead, they combine and interfere. In short, waves do things that particles cannot, and particles do things that waves cannot.

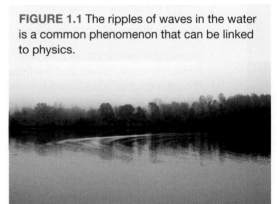

FIGURE 1.1 The ripples of waves in the water is a common phenomenon that can be linked to physics.

Waves can diffract, interfere, propagate through a medium, reflect and refract. All waves have the properties of wavelength, amplitude and frequency.

As you journey through your studies in Physics, you will discover that both a wave and a particle model are required to describe the properties of light and electromagnetic radiation. In Unit 1, you will investigate the properties of light as a particle modelled as a ray to understand reflection and refraction. In Unit 4, you will investigate in detail that both a wave and a particle model are needed to fully understand natural phenomena related to electromagnetic radiation and its interaction with matter.

LEARNING SEQUENCE

on Resources

☑ **Solutions**	Solutions — Topic 1 (sol-0787)
🔬 **Practical investigation eLogbook**	Practical investigation eLogbook — Topic 1 (elog-1570)
📄 **Digital documents**	Key science skills — VCE Physics Units 1–4 (doc-36950)
	Key terms glossary — Topic 1 (doc-36951)
	Key ideas summary — Topic 1 (doc-36952)
A+ **Exam question booklet**	Exam question booklet — Topic 1 (eqb-0069)

1.2 Explaining waves as the transmission of energy

1.2.1 Introducing waves

A **wave** is a disturbance that travels through a medium from the source to the detector without any movement of matter. There are two categories of waves that exist in nature: mechanical waves and electromagnetic waves. Electromagnetic waves do not require a medium to propagate; they can travel through a vacuum. Mechanical waves, however, do need a medium, such as air or water, to propagate. Particles of the matter vibrate up and down or back and forth about their rest position, transferring energy from one place to another. Waves therefore transfer energy without any net movement of particles. **Periodic waves** are a type of wave that transfers energy at regular intervals throughout a medium.

Looking at the examples in table 1.1, two different types of waves can be identified. For the **pulse** on the rope and the ripples on the water surface, the disturbance is at right angles to the direction the wave is travelling. These types of waves are called **transverse waves** and can be seen in solids.

wave the transfer of energy through a medium without any net movement of matter

periodic wave a disturbance that repeats itself at regular intervals

pulse a wave of short duration

transverse wave a wave for which the disturbance is at right angles to the direction of propagation

disturbance the movement of particles due to an energy wave passing through them

longitudinal wave a wave for which the disturbance is parallel to the direction of propagation

In the examples of the sound wave travelling through air and the compression moving along the spring, the **disturbance** is parallel to the direction the wave is travelling. These types of waves are called **longitudinal waves** and can be seen in solids as well as liquid and gases.

TABLE 1.1 Some examples of waves

Wave	Source	Medium	Detector	Disturbance	Type of wave
Sound	Push/pull of loudspeaker Compressions Speaker Sound waves	Air	Ear	Increase and decrease in air pressure	Longitudinal
Rope	Upward flick of hand Pulse on a rope	Rope	Person at other end of the rope	Section of rope moves up and then back down	Transverse

(continued)

TABLE 1.1 Some examples of waves *(continued)*

Wave	Source	Medium	Detector	Disturbance	Type of wave
Stretched spring	Push of hand Compressions Compressions moving along a stretched spring	Coils in the spring	Person at other end	Bunching of coils	Longitudinal
Water	Object dropped in water Ripples on water	Water	Bobbing cork placed in the water	Water surface at a point moves up then back down	Transverse

1.2 Activities

1.2 Quick quiz **on**	1.2 Exercise	1.2 Exam questions

1.2 Exercise

1. How is a periodic wave different from a single pulse moving along a rope?
2. How is a periodic longitudinal wave different from a transverse wave?
3. What is the speed of sound in air if it travels a distance of 996 m in 3.00 s?
4. Describe the motion of particles in a medium as a transverse wave passes through the medium.
5. Describe the motion of particles in a medium as a longitudinal wave passes through the medium.

1.2 Exam questions

▶ **Question 1 (2 marks)**

Select the correct options to complete the following sentence describing the characteristics of waves.

Waves transmit **matter/energy** without the net transfer of **matter/energy**.

▶ **Question 2 (1 mark)**

What is necessary for the transmission of mechanical waves?

Question 3 (1 mark)

Source: VCE 2018 Physics Exam, Section A, Q.10; © VCAA

MC A loudspeaker is producing a sound wave of constant frequency. Consider a tiny dust particle 1.0 m in front of the loudspeaker.

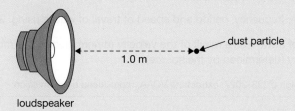

Which one of the following diagrams best describes the motion of the dust particle?

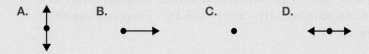

A.   B.   C.   D.

Question 4 (1 mark)

Source: VCE 2009 Physics Exam 2, Section B, Q.2; © VCAA

MC A stretched spring, attached to two fixed ends, is compressed on the right end and then released, as shown in Figure 1. The resulting wave travels back and forth between the two fixed ends until it comes to a stop.

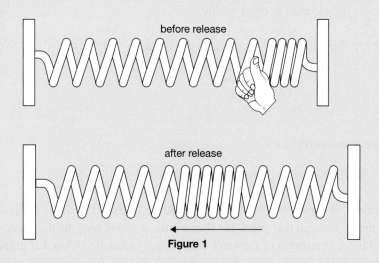

Figure 1

This wave is best seen as an example of
A. a transverse wave.
B. a longitudinal wave.
C. diffraction.
D. an electromagnetic wave.

Question 5 (3 marks)

a. What characterises a longitudinal wave?   (1 mark)
b. Identify one example of a wave that is longitudinal in its nature.   (1 mark)
c. Identify one example of a wave that is transverse in its nature.   (1 mark)

More exam questions are available in your learnON title.

1.3 Properties of waves

1.3.1 Wavelength, frequency and speed

The **frequency** of a periodic wave is the number of times that it repeats itself every second. Frequency is measured in hertz (Hz) and $1\,\text{Hz} = 1\,\text{s}^{-1}$. Frequency can be represented by the symbol f.

The **period** of a periodic wave is the time it takes a source to produce a complete wave. This is the same as the time taken for a complete wave to pass a given point. The period is measured in seconds and is represented by the symbol T.

The period of a wave is the reciprocal of its frequency. For example, if five complete waves pass every second, that is, $f = 5$ Hz, then the period (the time for one complete wavelength to pass) is $\dfrac{1}{5} = 0.2$ seconds. In other words:

frequency a measure of how many times per second an event happens, such as the number of times a wave repeats itself every second

period the amount of time, measured in seconds, that one cycle or event takes, such as the time taken for an object moving in a circular path and at a constant speed to complete one revolution

amplitude a periodic disturbance that is the maximum variation from zero

wavelength the distance between successive corresponding parts of a periodic wave

$$f = \frac{1}{T} \Rightarrow T = \frac{1}{f}$$

where:

f is the frequency of the wave, in Hz

T is the period of the wave, in s.

A displacement–time graph, as shown in figure 1.2, tracks the movement of a single point on a transverse wave over time as the wave moves through that point. In other words, it shows how the displacement of a single point on the wave varies over time. The period of the wave can be easily identified from this graph.

The **amplitude** of a wave is the size of the maximum disturbance of the medium from its normal state. The units of amplitude vary from wave type to wave type. For example, in sound waves the amplitude is measured in the units of pressure, whereas the amplitude of a water wave would normally be measured in centimetres or metres.

The **wavelength** is the distance between successive corresponding parts of a periodic wave. The wavelength is also the distance travelled by a periodic wave during a time interval of one period. For transverse periodic waves, the wavelength is equal to the distance between successive crests (or troughs). For longitudinal periodic waves, the wavelength is equal to the distance between two successive compressions (regions where particles are closest together) or rarefactions (regions where particles are furthest apart). Wavelength is represented by the symbol λ (lambda).

The displacement of all particles along the length of a transverse wave can be represented in a displacement–distance graph as shown in figure 1.3. A displacement–distance graph is like a snapshot of the wave at an instant in time. The amplitude and wavelength of the wave can be easily identified from this graph.

FIGURE 1.2 Displacement–time graph: the movement of a single point on a transverse wave over time

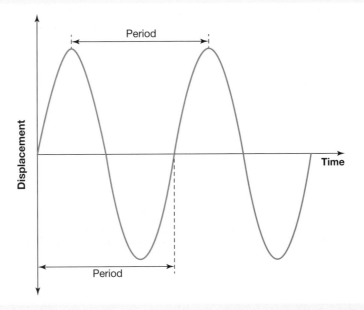

FIGURE 1.3 Displacement–distance graph: particle displacements along a transverse wave

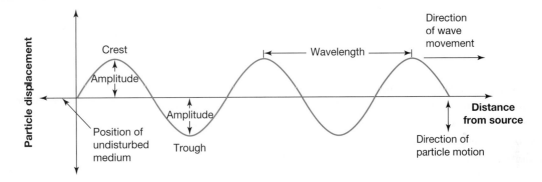

The **speed of a periodic wave**, v, is related to the frequency and period. In a time interval of one period, T, the wave travels a distance of one wavelength, λ. Thus:

> **speed of a periodic wave** the product of the wavelength of the wave multiplied by its frequency

$$\text{Speed} = \frac{\text{distance}}{\text{time}} = \frac{\lambda}{T} = \frac{\lambda}{\frac{1}{f}} = f\lambda$$

This relationship can be written as:

$$v = f\lambda = \frac{\lambda}{T}$$

where:

v is the speed of the wave, in m s^{-1}

f is the frequency of the wave, in Hz

λ is the wavelength of the wave, in m

T is the period of the wave, in s.

$$v = \lambda \times f$$

$$\lambda = \frac{v}{f} \qquad \lambda \times f \qquad f = \frac{v}{\lambda}$$

This relationship, $v = f\lambda$, is sometimes referred to as the universal wave equation.

The frequency of a periodic wave is determined by the source of the wave. The speed of a periodic wave is determined by the medium through which it is travelling. Because the wavelength is a measure of how far a wave travels during a period, if it can't be measured, it can be calculated using the formula $\lambda = \dfrac{v}{f}$.

In a longitudinal wave, as opposed to a transverse wave, the oscillations are parallel to the direction the wave is moving. Longitudinal waves can be set up in a slinky, as shown in figure 1.4a. Sound waves in air are also longitudinal waves, as shown in figure 1.4b. They are produced as a vibrating object (such as the arm of a tuning fork) first squashes the air, then pulls back to create a partial vacuum into which the air spreads.

FIGURE 1.4 Longitudinal waves in **a.** a slinky **b.** air

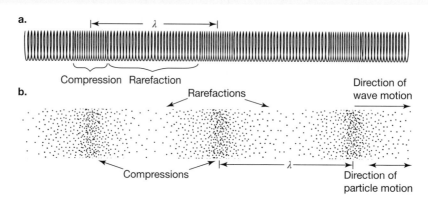

Longitudinal waves cause the medium to bunch up in places and to spread out in others. **Compressions** are regions in the medium where the particles are closer together. Referring to sound waves in air, compressions are regions where the air has a slightly increased pressure, as a result of the particles being closer together. **Rarefactions** are regions in the medium where the particles are spread out. This results in a slight decrease in air pressure in the case of sound waves.

compression a region of increased pressure in a medium during the transmission of a wave

rarefraction a region of reduced pressure in a medium during the transmission of a sound wave

The wavelength, λ, for longitudinal waves is the distance between the centres of adjacent compressions (or rarefactions, as shown in figure 1.4b). The amplitude of a sound wave in air is the maximum variation of air pressure from normal air pressure.

tlvd-3817

SAMPLE PROBLEM 1 Calculating the speed of a sound wave

What is the speed of a sound wave if it has a period of 2.0 ms and a wavelength of 68 cm?

THINK	WRITE
1. Note down the known variables in their appropriate units. Time must be expressed in seconds and length in metres.	$T = 2.0$ ms $= 2.0 \times 10^{-3}$ s $\lambda = 68$ cm $= 0.68$ m

2. Choose the appropriate formula.

$$v = f\lambda$$
$$\Rightarrow v = \frac{\lambda}{T}$$

3. Substitute values for the wavelength and period and then solve for v.

$$v = \frac{0.68 \text{ m}}{2.0 \times 10^{-3} \text{ s}}$$
$$= 340 \text{ m s}^{-1}$$

PRACTICE PROBLEM 1

What is the speed of a sound wave if it has a period of 1.5 ms and a wavelength of 51 cm?

tlvd-3818

SAMPLE PROBLEM 2 Calculating the wavelength of a sound wave

What is the wavelength of a sound of frequency 550 Hz if the speed of sound in air is 335 m s^{-1}?

THINK

1. Note down the known variables in their appropriate units. Frequency must be expressed in hertz and speed in m s^{-1}.

2. Choose the appropriate formula.

3. Substitute values for the frequency and speed and then solve for the wavelength.

WRITE

$f = 550$ Hz, $v = 335$ m s^{-1}

$$v = f\lambda$$
$$\Rightarrow \lambda = \frac{v}{f}$$

$$\lambda = \frac{335 \text{ m s}^{-1}}{550 \text{ Hz}}$$
$$= 0.609 \text{ m}$$

PRACTICE PROBLEM 2

A siren produces a sound wave with a frequency of 587 Hz. Calculate the speed of sound if the wavelength of the sound is 0.571 m.

elog-1596

INVESTIGATION 1.1

Investigating waves from a slinky spring

Aim

To observe and investigate the behaviour of waves (or pulses) travelling along a slinky spring

 Resources

▶ **Video eLesson** Properties of waves (eles-3213)

EXTENSION: Applying wavelength and frequency to the Doppler effect

The Doppler effect

Note that this content is useful if you are choosing to study option 2.14 (How can we detect possible life beyond Earth's solar system?) in Unit 2, Area of Study 2.

FIGURE 1.5 Both high-pitched and low-pitched sounds can be heard from the siren of an emergency vehicle.

Everyone is familiar with the change in pitch of sound made by a car when it passes them. This is most pronounced when an emergency vehicle races by. The sound always starts high but finishes low. This effect is called the Doppler effect, after Christian Andreas Doppler, who predicted it in 1842 before it had been observed. The Doppler effect is the result of a wave travelling at a constant speed through a medium while the source is in motion relative to the medium or if an observer is in motion relative to the medium. In either case, the frequency of the source will be different from the frequency as measured by an observer.

Consider a fire engine racing to attend a fire. While it is stuck in traffic with its siren blaring, a Physics student decides to measure the frequency and wavelength of the sound. The fire engine's siren alternates between a high-pitched sound and a low-pitched sound. The student measures the high-pitched sound to have a frequency of 500 Hz and the low-pitched sound to have a frequency of 200 Hz. After determining the speed of sound to be 340 m s^{-1}, and noticing that there is no wind, the student calculates the wavelengths using $v = f\lambda$:

$$\lambda = \frac{v}{f}$$
$$= \frac{340 \text{ m s}^{-1}}{500 \text{ Hz}}$$
$$= 0.680 \text{ m for the 500 Hz sound}$$

$$\lambda = \frac{v}{f}$$
$$= \frac{340 \text{ m s}^{-1}}{200 \text{ Hz}}$$
$$= 1.70 \text{ m for the 200 Hz sound}$$

FIGURE 1.6 The Doppler effect **a.** O1 and O2 both hear the same frequency sound. **b.** O1 hears a higher frequency than O2.

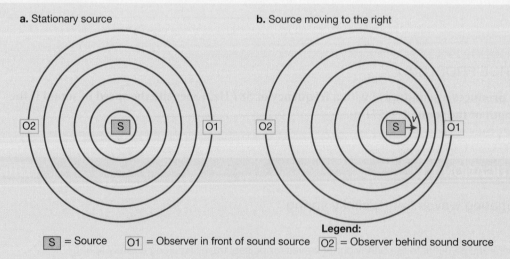

a. Stationary source

b. Source moving to the right

Legend:
S = Source   O1 = Observer in front of sound source   O2 = Observer behind sound source

Later, the traffic jam has cleared and another fire engine passes the Physics student. The fire engine travels at a velocity of 24 m s^{-1} relative to the road (and air). The speed of sound remains at 340 m s^{-1} through the air. The fire engine is identical to the first one, but now the student measures the frequencies to be 538 Hz and 215 Hz as the fire engine approaches, and 467 Hz and 187 Hz as the fire engine moves away. The student's frequency-measuring equipment is not faulty — the student could clearly hear the pitch drop as the fire engine passed.

When the air is still, something approaching a listener will sound higher in pitch than when it is at rest relative to the listener, and will sound lower in pitch when it is moving away. Doppler cleverly predicted this result before the advent of fast fire engines. His prediction was first confirmed experimentally by having a trumpeter play a note while passing on a 'relatively' fast-moving train.

The sound produced by the siren of the fire engine is a series of pressure variations in the air. When the fire engine produces a compression (region of higher than average air pressure) of the high-frequency sound, this compression moves forward at the speed of sound in air, 340 m s^{-1}. The next compression is produced T seconds later, where T is the period of the sound wave.

$$T = \frac{1}{f}$$

$$= \frac{1}{500 \text{ Hz}} = 0.002 \text{ s}$$

At 0.002 s, the first compression has travelled the following distance:

$$d = vt$$

$$= 340 \text{ m s}^{-1} \times 0.002 \text{ s} = 0.680 \text{ m}$$

In this time, the fire engine has moved:

$$d = vt$$

$$= 24 \text{ m s}^{-1} \times 0.002 \text{ s} = 0.0480 \text{ m}$$

The distance between compressions is therefore:

$$\lambda = 0.680 \text{ m} - 0.0480 \text{ m}$$

$$= 0.632 \text{ m}$$

For the fire engine that was stationary, the wavelength was 0.68 m. As sound is travelling at 340 m s^{-1} relative to the student on the roadside, and $v = f\lambda$, the shorter wavelength from the approaching fire engine will have a higher frequency than the stationary fire engine. In this case, the detected frequency measured by the student for the 500 Hz sound as the fire engine approached at 24 m s^{-1} would be:

$$f = \frac{v}{\lambda}$$

$$= \frac{340 \text{ m s}^{-1}}{0.632 \text{ m}} = 538 \text{ Hz}$$

In summary, when a source of waves is approaching an observer, the frequency appears to be greater than the source frequency. When the source of waves is receding from the observer, the frequency appears to be smaller than the source frequency. Likewise, if an observer is approaching a stationary source, the frequency appears higher and, if the observer is receding from the stationary source, the frequency appears lower. The equations for calculating these changes in frequency due to motion of a source and observer relative to a medium are not examinable.

Applying this to electromagnetic waves (light), if a source of light gets farther away from you, then it will appear to have a longer wavelength (shifting towards the red), while if it gets closer to you, it will appear to have a shorter wavelength (shifting towards the blue). Astronomers call this a redshift and a blueshift respectively, and use this to determine which stars and galaxies are moving away from us, and which are moving towards us.

 Resources

 Weblink Doppler effect applet

1.3 Activities

1.3 Quick quiz on	1.3 Exercise	1.3 Exam questions

1.3 Exercise

1. How far does a periodic wave travel in one period? Give your answer in terms of the wavelength of the periodic wave.
2. Do loud sounds travel faster than soft sounds? Justify your answer.
3. A marching band on the other side of a sports oval appears to be 'out of step' with the music. Explain why this might happen.
4. You arrive late to an outdoor concert and have to sit 500 m from the stage. Will you hear high-frequency sounds at the same time as low-frequency sounds if they are played simultaneously? Explain your answer.
5. A loudspeaker is producing a note of 256 Hz. How long does it take for 200 wavelengths to interact with your ear?
6. What is the wavelength of a sound that has a speed of 340 m s^{-1} and a period of 3.00 ms?
7. What is the speed of a sound if the wavelength is 1.32 m and the period is 4.00×10^{-3} s?
8. The speed of sound in air is 340 m s^{-1} and a note is produced that has a frequency of 256 Hz.
   a. What is its wavelength?
   b. This same note is now produced in water where the speed of sound is 1.50×10^3 m s^{-1}. What is the new wavelength of the note?
9. A stationary siren is producing a sound. The siren vibrates at 100 Hz to make the sound. One observer measures the sound to have a frequency of 110 Hz. A second observer measures the sound to have a frequency of only 90 Hz. Explain why this is the case.

1.3 Exam questions

▶ **Question 1 (1 mark)**
***Source:** VCE 2021 Physics Exam, Section A, Q.13; © VCAA*
MC The diagram below shows part of a travelling wave.

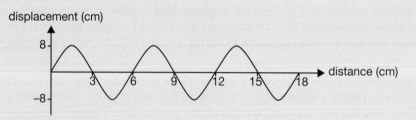

The wave propagates with a speed of 18 m s^{-1}.

Which of the following is closest to the amplitude and frequency of the wave?
A. 8 cm, 3.0 Hz
B. 16 cm, 3.0 Hz
C. 8 cm, 300 Hz
D. 16 cm, 300 Hz

Question 2 (3 marks)

Source: VCE 2019 Physics Exam, Section B, Q.12; © VCAA

A sinusoidal wave of wavelength 1.40 m is travelling along a stretched string with constant speed v, as shown in Figure 11. The time taken for point P on the string to move from maximum displacement to zero is 0.120 s.

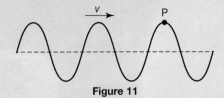

Figure 11

Calculate the speed of the wave, v. Give your answer correct to three significant figures. Show your working.

Question 3 (1 mark)

Green light has a wavelength of 510 nm and travels at the speed of light ($c = 3 \times 10^8$ m s^{-1}).

Calculate its period.

Question 4 (2 marks)

A student observes the motion of waves passing by a navigational marker at a fixed point on the surface of the ocean. They count 26 wave crests passing the marker in 2 minutes.

Calculate the frequency of the wave.

Question 5 (2 marks)

A sound wave for the note known as middle C has a frequency of 261.6 Hz.

Calculate its period.

More exam questions are available in your learnON title.

1.4 Energy from the Sun

KEY KNOWLEDGE

- Describe electromagnetic radiation emitted from the Sun as mainly ultraviolet, visible and infrared
- Calculate the peak wavelength of the radiated electromagnetic radiation using Wien's Law: $\lambda_{max}T =$ constant
- Compare the total energy across the electromagnetic spectrum emitted by objects at different temperatures

Source: VCE Physics Study Design (2023–2027) extracts © VCAA; reproduced by permission.

1.4.1 What is blackbody radiation?

The hottest object in our environment is our Sun. Nearly all of the energy available to Earth comes from the Sun, whose energy output is 3.86×10^{26} J s^{-1} (3.86×10^{26} W). This is known as its **luminosity**. A tiny portion of this energy hits Earth, heating and lighting it. When we examine the light from the Sun, we see the characteristic spectrum of light produced by hot objects, known as **blackbody radiation**.

luminosity the amount of radiated electromagnetic energy emitted by a light-emitting or luminous object

blackbody radiation the characteristic radiation emitted by a blackbody when heated

FIGURE 1.7 Nearly all the energy available to Earth comes from the Sun.

During the late nineteenth century, scientists conducted investigations into how much radiation was produced across the light spectrum and how this distribution changed with **temperature**. The interest was in part theoretical, but it was also a practical interest sparked by the needs of the growing public — and home — illumination industry. The eventual explanation of the shape of the blackbody radiation spectrum departed from classical physics and pointed the way to the development of the new model of the atom early in the twentieth century.

When a solid object is heated to several hundred degrees Celsius, it becomes **incandescent**; that is, it glows. A theoretically ideal object for producing incandescent light is called a **blackbody**.

At low temperatures, a theoretical blackbody will absorb all the **electromagnetic radiation** (sometimes abbreviated as EM radiation) that falls on it, as it is both a perfect absorber and a perfect radiator of electromagnetic radiation. Such an object could be an oven, or even a star like our Sun. When an oven is cold, light that passes from outside the oven is absorbed and not reflected back; it is dark inside. When the oven is very hot the electric coil can reach approximately 800 °C, and light is emitted by the oven that depends only on its temperature and not on its composition; the oven glows with a dull red colour, as shown in figure 1.8.

Figure 1.9 displays the solar intensity emitted by the Sun, which has a surface temperature of approximately 5500 °C, plotted against its electromagnetic emission. Note that the highest section of the peak is in the visible section of the electromagnetic spectrum; however, a significant amount of heat energy (infrared) is clearly illustrated, as well as a smaller but not insignificant amount of light in the ultraviolet section.

> **temperature** a measure of how hot or how cold something is
>
> **incandescent** refers to luminous objects that produce light because they are hot; the higher the temperature, the brighter the light, and the colour also changes
>
> **blackbody** an object that absorbs all radiation that falls on it
>
> **electromagnetic radiation** an electromagnetic wave or radiation that includes visible light, radio waves, gamma rays and x-rays

FIGURE 1.8 At very high temperatures, a dull red light is emitted from an oven.

Measurements of the spectra of blackbody radiation show that for any temperature, one wavelength has a greater intensity than all the others. In figure 1.9, which shows the electromagnetic emission from the Sun, the greatest intensity occurs in the green part of the visible spectrum. However, because the total intensity of light is highest overall in the yellow part of the visible spectrum, our Sun appears to be yellow in colour.

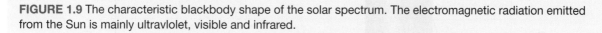

FIGURE 1.9 The characteristic blackbody shape of the solar spectrum. The electromagnetic radiation emitted from the Sun is mainly ultraviolet, visible and infrared.

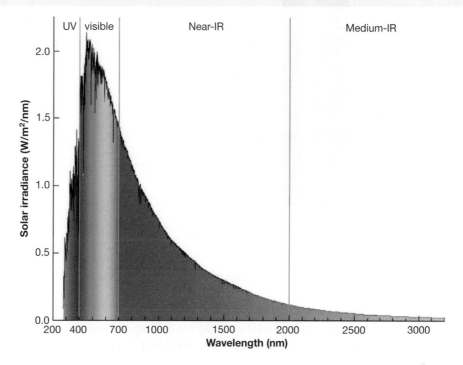

The results for blackbody spectra at different temperatures are displayed in figure 1.10. This plot of intensity versus wavelength shows the emission of objects with different temperatures, shown in degrees kelvin (K). While the nature of the kelvin temperature scale is discussed in more detail in subtopic 2.2, in simple terms, the conversion between degrees Celsius and kelvin is:

$$°C = K - 273$$

where:

°C is the temperature, in degrees Celsius

K is the temperature, in kelvin.

The graphs for the four different temperatures are generally the same shape. Starting from the right with long wavelengths, there is very little infrared radiation emitted. As the wavelength gets shorter, the radiation produced increases to a maximum; then finally, as the wavelength shortens even further, the amount of radiation drops quite quickly. The graphs for higher temperatures have a peak at a shorter wavelength and a much larger area under the graph, meaning a lot more energy is emitted (intensity). This shape is typical of blackbody radiation, and all heated objects exhibit these characteristics to some degree.

The intensity of emission at a temperature of 5000 K closely resembles the emission from the Sun based on its surface temperature ($\approx$ 5200 K). In contrast, the atmosphere and surface of Earth have temperatures ranging from approximately −50 °C to 50 °C (223 K to 323 K), which results in a peak blackbody emission in the non-visible infrared portion of the electromagnetic spectrum.

FIGURE 1.10 The variation in intensity versus wavelength produced for different hot objects. Note that not only does the intensity peak higher as temperature increases, but the spectrum moves to the left and becomes bluer (dotted line).

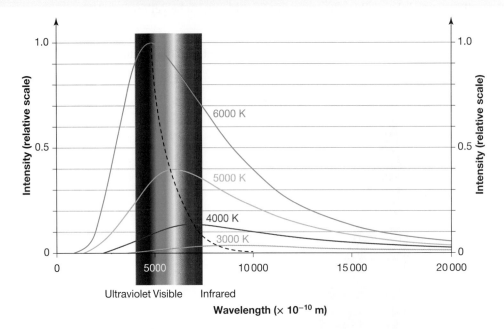

In summary, the most important features of this spectrum that occur as the temperature of an object (blackbody) increases are:
- the peak emission wavelength becomes shorter
- the peak (height of the emission curve) and total intensity (area under the emission curve) of electromagnetic emission both increase as an object heats up to higher temperatures.

The overall visible electromagnetic emission of an incandescent object becomes bluer as it heats up to high temperatures, as observed in figure 1.11.

FIGURE 1.11 The emission of objects becomes bluer (as shown to the right) as the temperature increases.

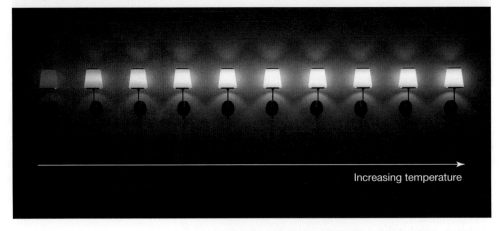

In everyday life, all objects are continually absorbing and emitting thermal energy by radiation.
- When an object absorbs more radiant energy than it emits, its temperature increases.
- When an object emits more radiant energy than it absorbs, its temperature decreases.

There are several physical factors that can affect the rate of emission and absorption of energy from an object. These include:

- *the temperature of an object and its surroundings.* The larger the difference between the temperature of an object and its surroundings, the greater the rate of energy transfer through the emission of radiation.
- *the surface area of an object.* The rate of radiant energy transfer is directly proportional to the surface area of an object.
- *the properties of the surface of an object.* Matte-black objects are almost perfect absorbers of radiant energy of all wavelengths, while white and shiny objects re-emit a significant proportion of radiant energy from the visible portion of the spectrum. Highly reflective surfaces, such as mirrors, reflect most radiant energy. A mylar thermal blanket for instance, with its reflective coating, can be used to to reduce heat loss and keep someone warm as it reflects back approximately 90 per cent of radiant heat.

FIGURE 1.12 Black surfaces absorb and emit radiant energy more easily than white surfaces.

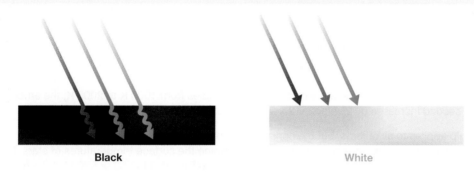

Black   White

In summary, rough or matte-black surfaces absorb and emit radiant energy more quickly than shiny, white surfaces. This means that matte-black surfaces heat up and cool down more quickly due to an increased rate of radiant energy transfer, more closely resembling the properties of a theoretical blackbody.

elog-1598

tlvd-3824

INVESTIGATION 1.2

Examining the Sun's spectrum

Aim

To show the different parts of the Sun's visible spectrum and compare this to the other sections of the electromagnetic spectrum the Sun emits

elog-1599

tlvd-0806

INVESTIGATION 1.3

The colour of temperature

Aim

To show that different colours have different thermal effects

1.4.2 Interaction of thermal energy and electromagnetic radiation

EXTENSION: Stefan–Boltzmann Law and the relationship between power and temperature

Note that this content is useful if you are choosing to study option 2.1 (How does physics explain climate change?) in Unit 2, Area of Study 2.

Early researchers such as Jozef Stefan were keen to find patterns and relationships in the data shown in figure 1.10 and to be able to explain their observations. In 1879, Stefan compared the area under the graph for different temperatures. This area is the total energy emitted every second across all wavelengths — in other words, the power.

He found that the **power radiated by a blackbody** was proportional to **absolute temperature** to the power of 4; that is, $P \propto T^4$.

Ludwig Boltzmann later proved this from a theoretical standpoint, and so the $P \propto T^4$ relationship is called the Stefan–Boltzmann Law.

power radiated by a blackbody the total energy radiated by a blackbody every second

absolute temperature the temperature of an object taken in the scale using absolute zero

Stefan–Boltzmann Law

$$P \propto T^4$$

This means that if the absolute temperature of a hot object doubles from 1000 K to 2000 K, the amount of energy emitted every second increases by 2^4 ($2 \times 2 \times 2 \times 2 = 16$ times).

This relationship applies to all objects, but the constant of proportionality depends on the size of the object and other factors. Using this law, astronomers were able to determine the surface temperatures of stars. For instance, Stefan was able to estimate the temperature of the surface of the Sun as 5430 °C or 5700 K, which is very close to the value known today of 5778 K.

The energy falling on Earth from the Sun is fairly constant. At the equator, an average of about 684 joules of energy from the Sun hits each square metre of Earth's surface every second; that is, 684 W m^{-2}, where 1 watt is a unit of power or the rate of energy delivery and equals 1 joule per second. This value varies from day to day by as much as 2 W m^{-2}, as well as having an approximate 11-year cycle of a similar magnitude.

With the Stefan–Boltzmann relationship, $P \propto T^4$, it is possible to use the amount of energy that Earth radiates into space to calculate the temperature of Earth as observed from space.

tlvd-0007

SAMPLE PROBLEM 3 Exploring the relationship between temperature and power using the Stefan–Boltzmann Law

a. When iron reaches about 480 °C it begins to glow with a red colour. How much more energy is emitted by the iron at this temperature, compared to when it is at a room temperature of 20 °C?
b. How much hotter than 20 °C would the iron need to be to emit 10 times as much energy?

THINK	WRITE
a. 1. Convert the temperature to kelvin.	a. Temperature of hot iron: $T_{(\text{kelvin})} = T_{(\text{Celsius})} + 273$ $= 480\,°C + 273$ $= 753\,K$ Temperature of cold iron: $T_{(\text{kelvin})} = T_{(\text{Celsius})} + 273$ $= 20\,°C + 273$ $= 293\,K$

2. Calculate the ratio.

Ratio of power (hot to cold) = ratio of temperatures to the power of 4

$$\frac{P_{\text{hot}}}{P_{\text{cold}}} = \left(\frac{T_{\text{hot}}}{T_{\text{cold}}}\right)^4$$

$$= \left(\frac{753}{293}\right)^4$$

$$\approx 44$$

3. State the solution.

The hot iron emits 44 times as much energy every second as it does when it is at room temperature.

b. 1. Convert the temperature to kelvin.

b. Temperature of cold iron:

$$T_{(\text{kelvin})} = T_{(\text{Celsius})} + 273$$
$$= 20\,^{\circ}\text{C} + 273$$
$$= 293\ \text{K}$$

2. Calculate the ratio.

Ratio of power (hot to cold) = ratio of temperatures to the power of 4

$$\frac{P_{\text{hot}}}{P_{\text{cold}}} = \left(\frac{T_{\text{hot}}}{T_{\text{cold}}}\right)^4$$

$$10 = \left(\frac{T_{\text{hot}}}{293}\right)^4$$

$$10^{\frac{1}{4}} = \frac{T_{\text{hot}}}{293}$$

$$\Rightarrow T_{\text{hot}} = 293 \times 10^{\frac{1}{4}}$$

3. Use your calculator to evaluate $10^{\frac{1}{4}}$, then use that value to calculate T_{hot}.

$$10^{\frac{1}{4}} \approx 1.778$$

$$T_{\text{hot}} = 293 \times 1.778 \approx 521\ \text{K}$$

4. Convert the temperature to Celsius.

$$T_{(\text{Celsius})} = 521\ \text{K} - 273$$
$$= 248\,^{\circ}\text{C}$$

5. State the solution.

The iron would need to be at 248 °C to emit 10 times the energy every second it does at 20 °C.

PRACTICE PROBLEM 3

The Sun has a surface temperature of 5778 K and radiates energy at a rate of 3.846×10^{26} W. How much energy would a star of similar size radiate if its surface temperature was 8000 K?

Wien's Law and the relationship between wavelength and temperature

In 1893, Wilhelm Wien (pronounced *Veen*) was able to show that as the temperature increased, the wavelength of maximum intensity of energy emitted decreased, and indeed the two quantities were inversely proportional. That is, the wavelength is proportional to the inverse of the temperature. This can be seen in figure 1.13.

FIGURE 1.13 The wavelength is proportional to the inverse of the temperature. These graphs illustrate that as the temperature increased, the wavelength of maximum intensity of energy emitted decreased.

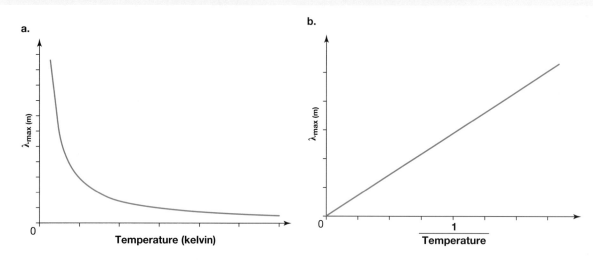

Wien's Law can be written as:

$$\lambda_{max} T = \text{constant}$$

The value of this constant is 2.90×10^{-3} mK (metre-kelvin).

This means that the shorter the peak wavelength, the hotter the temperature. For instance, a blue star with a peak wavelength of 246 nm has a temperature of 11 780 K, while a red star with a peak wavelength of 658 nm has a temperature of 4004 K. It has to be noted that the perception of colours differs from their actual colour temperature. Perceived 'warm colours', such as red and orange, have a lower colour temperature than perceived 'cool colours', such as blue and violet. 'Warm colours' in this case refers to colours with a longer wavelength, closer to infrared radiations, which make you feel warm.

tlvd-0008

SAMPLE PROBLEM 4 Determining the wavelength and spectrum of peak intensities of light from stars

a. **At what wavelength is the peak intensity of the light coming from a star whose surface temperature is 11 000 K (about twice as hot as the Sun)?**
b. **In what section of the spectrum is this wavelength?**

THINK

a. 1. State Wien's Law, and rearrange the equation to make λ_{max} the subject.

WRITE

a. $\lambda_{max} T = \text{constant}$

$\Rightarrow \lambda_{max} = \dfrac{\text{constant}}{T}$

2. Substitute the known values into the equation to determine λ_{max}.

$\text{Constant} = 2.90 \times 10^{-3} \text{ mK}, \ T = 11\,000 \text{ K}$

$$\lambda_{max} = \frac{\text{constant}}{T}$$
$$= \frac{2.90 \times 10^{-3} \text{ mK}}{11\,000 \text{ K}}$$
$$= 2.64 \times 10^{-7} \text{ m}$$

3. State the solution.

The light coming from a star whose surface temperature is 11 000 K has a peak intensity at a wavelength of 2.64×10^{-7} m.

b. Refer to figure 1.10.

b. The peak wavelength, 264 nm, is beyond the violet end of the visible spectrum, so it is in the ultraviolet section of the electromagnetic spectrum.

PRACTICE PROBLEM 4

Determine the surface temperature of a star that emits light at a maximum intensity of 450 nm.

1.4.3 How much energy does Earth get from the Sun?

The Sun is directly overhead the equator at midday on the equinox. At this place and time, the solar radiation is about 1368 watts per square metre of surface area (1368 W m^{-2}). But because Earth turns, producing night and day, this value has to be halved to 884 W m^{-2}. Also, Earth is curved, with the North and South Poles receiving much less light than the equator over a full year. This requires the number to be halved again. So the average solar radiation across Earth is 342 W m^{-2}, as shown in figure 1.14. About 100 W m^{-2} of this radiation is reflected straight back into space by the white surfaces of clouds and ice sheets. This leaves 242 W m^{-2} to heat up Earth. This radiation will be further explored in topic 2.

FIGURE 1.14 At Earth's equator at midday, the light intensity from the Sun is 1368 W m^{-2}. The average over night and day and from pole to pole is 342 W m^{-2}.

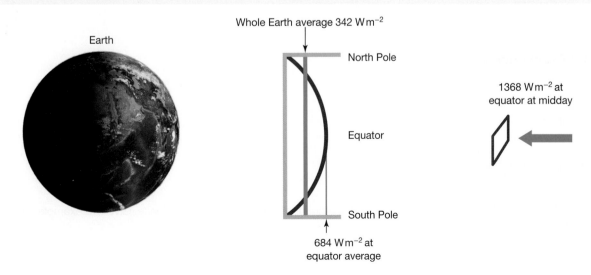

1.4 Quick quiz **on**	1.4 Exercise	1.4 Exam questions

1.4 Exercise

1. When electromagnetic radiation is shone on an object, which three interactions can occur between the radiation and the object?
2. A Thermos flask, which is designed to minimise heat loss from the liquid it contains, has a shiny and reflective inner and outer surface. Describe how this feature would assist in minimising heat loss.
3. Refer back to figure 1.10 to answer the following questions.
    a. A star has a surface temperature of 6000 K. What will the colour and wavelength of its peak radiation emission be?
    b. The star Spica is the 16th brightest star in the sky, and has a surface temperature of around 25 000 K. Extrapolating from the graph, what is the colour of this star likely to be? Justify your response.
4. All objects with a temperature above absolute zero (−273.15 °C) emit radiant thermal energy. As an object's temperature is increased, what happens to the *wavelength* and *frequency* of the radiation emitted by the surface of the object?
5. A metal filament is heated as a current passes through it, initially glowing a dull red colour as it begins to heat up. As the current is increased, the red colour of the filament becomes brighter, eventually turning a yellow colour. As the current allows the filament to reach its maximum temperature, it glows a white colour.

    With reference to the graph in figure 1.10, explain the reasons for the colour changes in the metal filament as it increases in temperature.

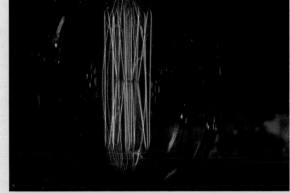

6. Earth's surface has an average temperature of 288 K. What is the wavelength of maximum emission from Earth's surface?
7. Our Sun gives off most of its light in the 'yellow' portion of the electromagnetic spectrum. λ_{max} is 510 nm. Calculate the average surface temperature of the Sun.

1.4 Exam questions

▶ Question 1 (1 mark)

MC Solar radiation incident on Earth is made up of _____ radiation.
A. infrared
B. visible light
C. ultraviolet
D. infrared, visible and ultraviolet

▶ Question 2 (2 marks)

Describe the effect of shiny white surfaces on the amount of energy reflected into space. Give two examples of these surfaces.

1.5 The electromagnetic spectrum

KEY KNOWLEDGE

- Compare the wavelength and frequencies of different regions of the electromagnetic spectrum, including radio, microwave, infrared, visible, ultraviolet, x-ray and gamma, and compare the different uses each has in society

Source: VCE Physics Study Design (2023–2027) extracts © VCAA; reproduced by permission.

1.5.1 Maxwell and electromagnetic waves

In 1864, James Clerk Maxwell (1831–1879) developed a theory predicting that an oscillating, and thus accelerating, electric charge would produce an oscillating electric field, together with a magnetic field oscillating at right angles to the electric field. These inseparable fields would travel together through a vacuum like a wave, and the speed of the wave would be the same, whether the oscillations were rapid (high frequency and a short wavelength) or very slow (low frequency and a long wavelength). Maxwell predicted their speed, using known electric and magnetic properties of a vacuum, to be 3×10^8 m s^{-1}: the speed of light!

Maxwell's model of an electromagnetic wave proposed that if a changing electric field is produced by a charged particle moving backwards and forwards, then this changing electric field will produce a changing magnetic field at right angles to it, as shown in figure 1.15. As the changing electric and magnetic fields — which vary in amplitude along the direction of the motion of the wave — self-propagate outwards, they can extend into space, with both electric and magnetic fields oscillating at the same frequency: the frequency of the electromagnetic radiation. This electromagnetic model of light indicated that electromagnetic radiation could be described as a transverse wave.

FIGURE 1.15 An electromagnetic wave. The electric (**E**) and magnetic (**B**) fields are uniform in each plane but vary in amplitude along the direction of the motion of the wave.

What Maxwell had produced was a theory that explained how light was produced and self-propagated through space as electromagnetic waves. This applied not only to visible light, but also to other radiation that people cannot see, such as gamma rays and x-rays, infrared and ultraviolet radiation, microwaves and radio waves.

> For light and electromagnetic radiation, due to the constant speed of light in a vacuum, the wave equation $v = f\lambda$ can be rewritten as:
>
> $$c = f\lambda$$
>
> where:
>
> f = frequency (Hz)
>
> λ = wavelength (m)
>
> c = speed of light = 3.0×10^8 m s^{-1}.
>
> The speed of light in a vacuum is the same for any wavelength/frequency.

The electromagnetic spectrum is the range of all electromagnetic radiation of different wavelengths and frequencies. Radiation is a form of energy that travels and spreads out in three dimensions as it moves away from its source. The electromagnetic spectrum is broken up into a range of different types based on its properties and uses. These sections of the electromagnetic spectrum include radio waves, microwaves, infrared radiation, visible light, ultraviolet radiation, x-rays and gamma rays.

1.5.2 The regions of the electromagnetic spectrum

All regions of the electromagnetic spectrum have proved to be useful, both for scientific observations and for a range of everyday technologies developed and used by human societies. Some types of electromagnetic radiation, particularly those with high frequencies and high energies, can also be hazardous to humans in certain circumstances.

For convenience, the electromagnetic spectrum is broken up into different regions based on the wavelength of the electromagnetic radiation. Most of the regions are named for historical purposes based on when they were discovered and/or the uses for each region. The range, uses and hazards of these regions are summarised in table 1.2.

TABLE 1.2 Properties of different regions of the electromagnetic spectrum

	Wavelength range	Uses	Hazards
Radio waves	1×10^{-2}–1000 m (1 cm–1 km)	• Radio transmission • Television transmission • Radar	None
Microwaves	2.5×10^{-2}–1.0×10^{-3} m (2.5 μm–1 cm)	• Cooking • Mobile phone signals	May cause internal heating of body tissues
Infrared (IR) radiation	7.5×10^{-7}–2.5×10^{-6} m (750 nm–2.5 μm)	• Optical fibre communication • Night vision/thermal vision equipment • TV remote controls	Radiation felt as heat — may cause the surface layers of skin/objects to burn or catch fire

(continued)

TABLE 1.2 Properties of different regions of the electromagnetic spectrum *(continued)*

	Wavelength range	Uses	Hazards
Visible light	4.0×10^{-7}–7.5×10^{-7} m (400 nm–750 nm)	• Human/animal vision • Optical fibres — transmitting information	Intense light may damage vision
Ultraviolet (UV) radiation	1.0×10^{-9}–4.0×10^{-7} m (1 nm–400 nm)	• Security marking on banknotes and other objects	Causes burning and damage to cells in deeper layers of skin; may cause cell damage, cell mutations (may lead to cancer) and cell death
X-rays	1.0×10^{-12}–1.0×10^{-9} m (1 pm–1 nm)	• Medical imaging of bones and other internal tissues • X-ray crystallography — determining atomic structure of crystals	Damages living cells, causing mutations (may lead to cancer) and cell death
Gamma rays	$<1.0 \times 10^{-12}$ m (<1 pm)	• Radiotherapy — killing cancer cells • Sterilising food and medical equipment	Damages living cells, causing mutations (may lead to cancer) and cell death

These wavelengths are further explored in figure 1.16.

FIGURE 1.16 Forms of radiation and their place in the electromagnetic spectrum. The visible portion of the spectrum is shown enlarged in the upper part of the diagram.

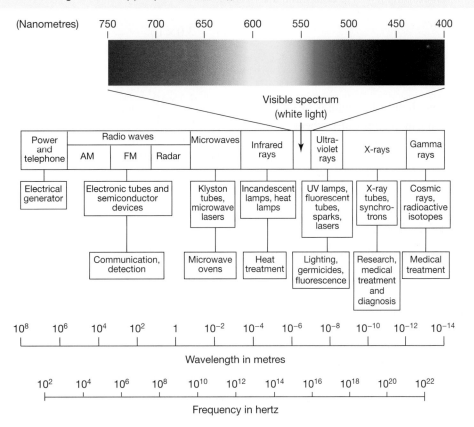

The frequency of a light ray is determined by the source (that is, the object producing the light). The speed of the light is determined by the material the light is passing through. This means that, when light passes from air into water, the frequency stays the same, the speed decreases and the wavelength must also decrease.

When you are under water and you look around, the objects you see still have the same colour. This means that your eye is responding to the frequency of the light ray and not to its wavelength. The world would be a strange place if the eye's response was the other way round.

tlvd-3830

SAMPLE PROBLEM 5 Calculating the wavelength and period of light when given frequency

When light with a frequency of 5.6×10^{14} Hz travels through a vacuum, what are the values of the following?
a. Its period
b. Its wavelength (in nanometres)
The speed of light in a vacuum is 3.0×10^8 m s^{-1}.

THINK	WRITE
a. The period of a wave is the reciprocal of its frequency.	a. $T = \dfrac{1}{f}$ $= \dfrac{1}{5.6 \times 10^{14}}$ $= 1.8 \times 10^{-15}$ s The period of the light is 1.8×10^{-15} s.
b. 1. Use the relationship $\lambda = \dfrac{c}{f}$ to determine the wavelength.	b. $\lambda = \dfrac{c}{f}$ $= \dfrac{3.0 \times 10^8}{5.6 \times 10^{14}}$ $= 5.4 \times 10^{-7}$ m
2. The wavelength of visible light is usually expressed in nanometres (nm), where 1.0 nm $= 1.0 \times 10^{-9}$ m.	$\lambda = \dfrac{5.4 \times 10^{-7}}{1.0 \times 10^{-9}}$ nm $= 5.4 \times 10^2$ nm The wavelength of the light is 540 nm.

PRACTICE PROBLEM 5

Calculate the frequency and period of light with a wavelength of 450 nm.

1.5.3 Visible light

Visible light is a form of radiation that can be modelled as transverse waves with colours differing in frequency. The light is visible to us because the frequency, wavelength and energy of visible light is 'just right' to interact with the cells on the rear surface (retina) of our eyes. Radiation with frequencies and wavelengths outside those of visible light (as explored in section 1.5.1) interact differently with our body and other objects on Earth and in the universe.

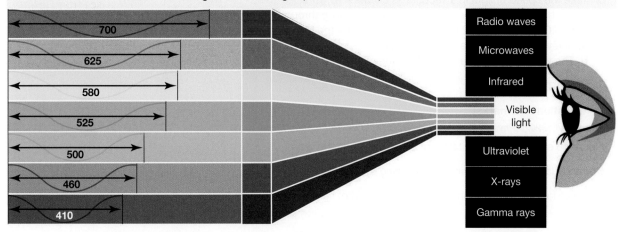

FIGURE 1.17 The different wavelengths of visible light (in nanometres)

700

625

580

525

500

460

410

Radio waves

Microwaves

Infrared

Visible light

Ultraviolet

X-rays

Gamma rays

TABLE 1.3 Frequency and wavelength of colours

	Red	Orange	Yellow	Green	Cyan	Blue	Violet
Frequency ($\times 10^{12}$ hertz)	430	480	520	570	600	650	730
Wavelength (nanometres)	700	625	580	525	500	460	410

1.5 Activities

learn on

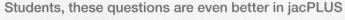

1.5 Quick quiz on	1.5 Exercise	1.5 Exam questions

1.5 Exercise

1. Take the speed of light in a vacuum as $c = 3.0 \times 10^8$ m s^{-1}.
   Calculate the period of orange light, which has a frequency of 4.8×10^{14} Hz.

2. Take the speed of light in a vacuum as $c = 3.0 \times 10^8$ m s^{-1}.
   Microwaves have a frequency ranging from 1.0×10^{10} through to 1.0×10^{12} Hz. Determine the range of wavelengths associated with microwaves.

3. Take the speed of light in a vacuum as $c = 3.0 \times 10^8$ m s^{-1}.
   X-rays used by dentists have a wavelength of 2.7×10^{-11} m. What is the frequency and hence the period of the x-rays produced?

4. Power lines that carry electrical energy use an AC current. These cables emit electromagnetic radiation with a period of 20 ms.
   a. What is the frequency of the radiation emitted by power lines?
   b. What is the wavelength of this radiation?

▶

5. Take the speed of light in a vacuum as $c = 3.0 \times 10^8$ m s^{-1}.

When blue light of frequency 6.5×10^{14} Hz travelling through the air meets a glass prism, its speed decreases from 3.0×10^8 m s^{-1} to 2.0×10^8 m s^{-1}. Calculate the following:

a. The wavelength of the blue light in the air
b. The wavelength of the blue light in the glass

1.5 Exam questions

Question 1 (3 marks)
Source: VCE 2020 Physics Exam, Section B, Q.14; © VCAA

Figure 13 shows a representation of an electromagnetic wave.

Correctly label Figure 13 using the following symbols.

E – electric field B – magnetic field c – speed of light λ – wavelength

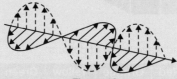

Figure 13

Question 2 (1 mark)
MC The range of frequencies of electromagnetic radiation is called the electromagnetic spectrum.

Which of the following is *not* part of the electromagnetic spectrum?
A. X-rays
B. Microwaves
C. Sound waves
D. Radio waves

Question 3 (1 mark)
MC Which of the following types of waves from the electromagnetic spectrum have wavelengths greater than 1 metre in length?
A. Gamma rays
B. Radio waves
C. X-rays
D. Ultraviolet rays

Question 4 (1 mark)
How fast do electromagnetic waves travel in a vacuum?

Question 5 (1 mark)
What is the wavelength of red light from the visible portion of the electromagnetic spectrum?

More exam questions are available in your learnON title.

1.6 Review

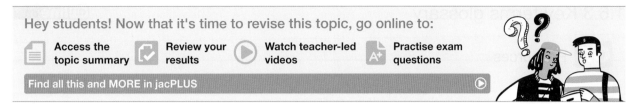
1.6.1 Topic summary

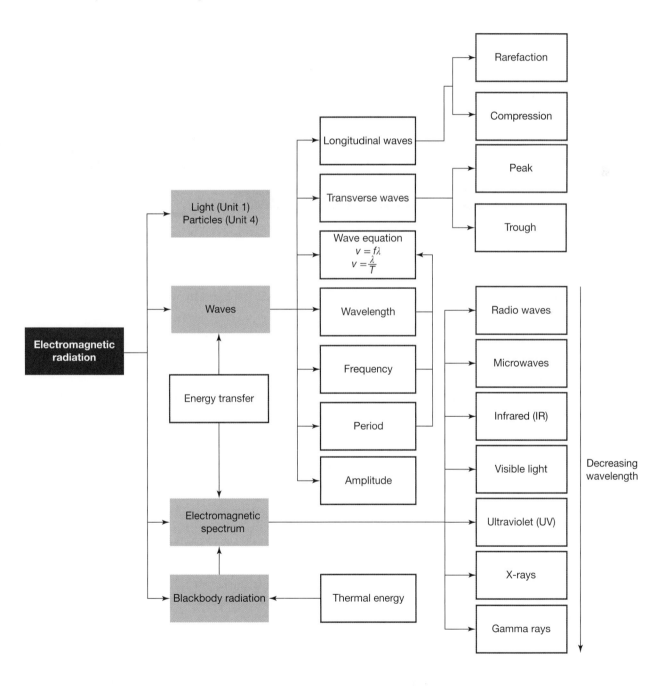

1.6.2 Key ideas summary

1.6.3 Key terms glossary

1.6 Activities

learn**on**

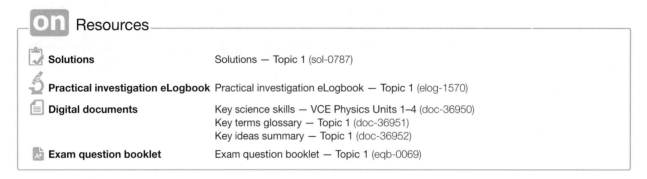

1.6 Review questions

1. Waves are capable of transferring energy from one place to another. Explain why waves require a medium to achieve this.

2. A student is studying surface waves using water at a local swimming pool. She makes waves at a rate of two every second and the ripples radiate away from the source.
   a. What is the period of the waves?
   b. The waves radiate away from the source at a speed of 2.5 m s⁻¹. What is the distance between two adjacent peaks; that is, the wavelength of the waves?
   c. If the student increases the rate at which she makes waves, what will happen to the wavelength of the waves? What will happen to the speed of the waves?

3. Sound produced by an opera singer has a frequency of 926 Hz.
   a. What is the period of the sound wave?
   b. Taking the speed of sound to be 340 m s⁻¹, what is the wavelength of this sound in air?

4. Humans can hear sounds ranging from approximately 20 Hz to 20 000 Hz. What is the wavelength associated with these frequencies in air?

5. Blue light has a frequency of 6.5×10^{14} Hz, and yellow light has a frequency of 5.2×10^{14} Hz. Determine the wavelengths of both blue and yellow light in air. Take the speed of light in air to be 3.0×10^8 m s⁻¹.

6. Lasers can be rapidly switched on and off to produce a pulse of light. A particular pulse of blue light (6.5×10^{14} Hz) consists of 1.0×10^6 complete cycles. What is the distance between the start and end of this particular pulse? *Hint:* How much time would it take to produce this pulse?

1.6 Exam questions

Question 1

Source: VCE 2018 Physics Exam, Section A, Q.11; © VCAA

Alex hears the siren from a stationary fire engine.

Compared with the sound Alex hears from the stationary fire engine, the sound Alex will hear as the fire engine approaches him will have increased

A. speed.
B. period.
C. amplitude.
D. frequency.

Question 2

Source: VCE 2016 Physics Exam, Section B, Q.2; © VCAA

A sound engineer, Dan, is setting up for a concert in a large stadium. In order to test the acoustics of the stadium, he sets up a single speaker in the middle of the stage. This speaker transmits sound equally in all directions. Using a signal generator and amplifier attached to the speaker, he sets the frequency of the sound to 500 Hz. Take the speed of sound to be $350 \, \text{m s}^{-1}$.

The wavelength of the signal is closest to

A. 0.07 m.
B. 0.70 m.
C. 1.43 m.
D. 2.34 m.

Question 3

Source: VCE 2015 Physics Exam, Section B, Q.1; © VCAA

A loudspeaker emits a sound of frequency 30 Hz. The speed of sound in air in these conditions is $330 \, \text{m s}^{-1}$.

Which one of the following best gives the wavelength of the sound?

A. 30 m
B. 11 m
C. 3.3 m
D. 0.091 m

Question 4

Source: VCE 2013 Physics Sample exam for Units 3 and 4, Q.2; © VCAA

A particle of dust is floating at rest 10 cm directly in front of a loudspeaker. The loudspeaker is not operating. The loudspeaker then emits sound of frequency of 10 Hz and speed of $330 \, \text{m s}^{-1}$. The loudspeaker now emits a sound of frequency 220 Hz.

dust particle

Which one of the following best gives the wavelength of the sound from the loudspeaker?

A. 0.67 m
B. 1.5 m
C. 220 m
D. 7.3×10^4 m

Question 5

Source: VCE 2008 Physics Exam 2, Section B, Q.2; © VCAA

A particle of dust is floating at rest 10 cm directly in front of a loudspeaker that is not operating.

The loudspeaker then emits sound of frequency of 10 Hz and spread of 330 m s^{-1}.

Which one of the following statements best describes the motion of the dust particle?

A. It vibrates vertically up and down at 10 Hz remaining on average 10 cm in front of the loudspeaker.

B. It vibrates horizontally backwards and forwards at 10 Hz remaining on average 10 cm in front of the loudspeaker.

C. It travels away from the loudspeaker at 330 m s^{-1} while moving horizontally backwards and forwards at 10 Hz.

D. It remains at rest.

Question 6

The time taken for a complete cycle of a wave is known as which of the following quantities?

A. Frequency

B. Period

C. Wavelength

D. Amplitude

Question 7

A student uses a high-speed camera to analyse the motion of waves in the strings of a guitar. They measure the maximum size of the disturbance produced by the wave.

Which of the following wave properties are they measuring?

A. Frequency

B. Period

C. Wavelength

D. Amplitude

Question 8

What would happen if Earth did not emit infrared radiation?

A. It would have no effect on Earth's climate.

B. Earth would be slightly warmer.

C. Earth would be too hot to support life.

D. Earth would emit visible light instead.

Question 9

Electromagnetic radiation from the Sun, corresponding with peak radiation, is primarily in the visible light range on the electromagnetic spectrum.

Electromagnetic radiation from Earth is primarily in which portion of the electromagnetic spectrum?

A. Infrared

B. Visible

C. Microwave

D. Ultraviolet

Question 10

Which of the following properties of sound is independent of the source producing the sound?

A. Frequency

B. Amplitude

C. Speed

D. None of the above

Section B — Short answer questions

Question 11 (5 marks)

a. In what way are longitudinal waves different from transverse waves? Give an example of each type of wave in your answer. **(2 marks)**

b. Jenny states that longitudinal and transverse waves have similarities. Discuss three ways in which both types of waves are similar. **(3 marks)**

Question 12 (3 marks)

Consider the wave pattern shown in the following diagram, which illustrates compressions and rarefactions of a sound wave at one instant progressing to the right.

Travelling wave moving to the right

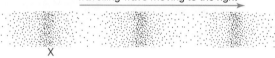

X

a. If the distance between a compression and adjacent rarefaction is measured to be 2.25 cm, determine the wavelength of the sound wave. **(1 mark)**

b. If the speed of the wave to the right is 38 cm s^{-1}, calculate the frequency of the source of the waves. **(1 mark)**

c. Consider the compression labelled X on the diagram.

Describe where the point X would be found 0.10 s later. **(1 mark)**

Question 13 (4 marks)

The following graph shows how λ_{max} (the wavelength of the peak of the radiation spectrum) for a range of stars varies with their surface temperatures.

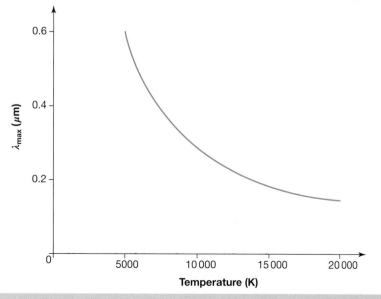

a. Estimate the surface temperature of a star whose intensity peaks at a wavelength of:
   i. 0.4 μm   **(1 mark)**
   ii. 0.25 μm.   **(1 mark)**

b. Estimate the peak wavelength for a star with a surface temperature of:
   i. 15 000 K   **(1 mark)**
   ii. 5555 K.   **(1 mark)**

▶ Question 14 (2 marks)

A sound wave with a frequency of 440 Hz travels at a speed of 4600 m s^{-1} in brass.

Calculate its wavelength.

▶ Question 15 (5 marks)

In the late nineteenth century, light was regarded as an electromagnetic wave because of the theoretical work done by James Maxwell.

a. Explain what is meant by the phrase *electromagnetic waves*.   **(1 mark)**

b. Consider four different types of electromagnetic waves: γ-rays, radio waves, UV light and microwaves.

   Match each of the four types of electromagnetic waves with the wavelengths listed in the following table.   **(4 marks)**

	Wavelength (m)	Electromagnetic wave type
i.	1×10^{3}	
ii.	1×10^{-2}	
iii.	1×10^{-12}	
iv.	1×10^{-8}	

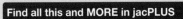 

2 Investigating light

KEY KNOWLEDGE

In this topic, you will:
- investigate and analyse theoretically and practically the behaviour of waves including:
  - refraction using Snell's Law: $n_1 \sin(\theta_1) = n_2 \sin(\theta_2)$ and $n_1 v_1 = n_2 v_2$
  - total internal reflection and critical angle including applications: $n_1 \sin(\theta_c) = n_2 \sin(90°)$
- investigate and explain theoretically and practically colour dispersion in prisms and lenses with reference to refraction of the components of white light as they pass from one medium to another
- explain the formation of optical phenomena: rainbows; mirages
- investigate light transmission through optical fibres for communication.

Source: VCE Physics Study Design (2023–2027) extracts © VCAA; reproduced by permission.

PRACTICAL WORK AND INVESTIGATIONS

Practical work is a central component of VCE Physics. Experiments and investigations, supported by a **practical investigation eLogbook** and **teacher-led videos**, are included in this topic to provide opportunities to undertake investigations and communicate findings.

EXAM PREPARATION

Access exam-style questions and their video solutions in every lesson, to ensure you are ready.

2.1 Overview

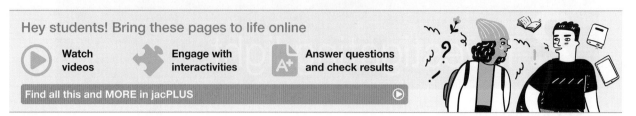
2.1.1 Introduction

Sight is the sense by which humans and a multitude of other animals get most of their information about the world. This sense responds to light. Light as electromagnetic radiation has already been introduced in topic 1. However, you might still have questions, such as where does light come from? What can it do? How can its properties be explained? What is colour?

Some obvious observations concerning light include the following:

- Sources of light are needed for a person to be able to see.
- Light travels very fast and appears to travel in a straight line.
- Light produces shadows.
- Light is sensed by people as exhibiting different colours.

FIGURE 2.1 Understanding light as a wave helps to explain many physical phenomena.

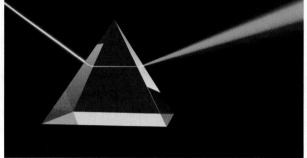

Such observations were made long before the development of scientific fields to study them. For instance, when fishing in rivers, Aboriginal and Torres Strait Islander fishermen knew to approach the banks of the river carefully to avoid being detected.

This topic will explore these early observations of light and develop your understanding that light acts as a wave.

on Resources

📋 **Solutions**	Solutions — Topic 2 (sol-0788)
🔬 **Practical investigation eLogbook**	Practical investigation eLogbook — Topic 2 (elog-1571)
📄 **Digital documents**	Key science skills — VCE Physics Units 1–4 (doc-36950)
	Key terms glossary — Topic 2 (doc-36953)
	Key ideas summary — Topic 2 (doc-36954)
📄 **Exam question booklet**	Exam question booklet — Topic 2 (eqb-0070)

2.2 Refraction using Snell's Law

2.2.1 BACKGROUND KNOWLEDGE: Light and its properties

Before a study of light can take place, some background information is necessary.

Sources of light

When you experience darkness at night or in an enclosed room, you know that a source of light, such as the Sun or a lamp, is needed to light up the darkness. Once a lamp is turned on, you can see features in the room because the light from the lamp shines on them and is then reflected into your eyes.

This means that objects can be classified into two groups. Objects seen because they give off their own light are called **luminous** objects; those seen because they reflect light are called non-luminous objects. The Sun, torches and candles are luminous objects. Tables, chairs, cats and dogs are non-luminous objects. Note that some living organisms, such as some jellyfish, glow-worms and fireflies, are bioluminescent, meaning they produce and emit their own light.

FIGURE 2.2 The Pleiades open star cluster in the constellation Taurus. All stars are incandescent sources of light.

As seen in topic 1 with blackbody radiation, some luminous objects produce light because they are hot. The Sun is one example. The higher the temperature, the brighter the light. This also causes a change in colour. These objects are called **incandescent**.

Other objects are cold and produce light in another way. This involves changes in the energy of electrons in the material brought about by either chemical or electrical processes.

luminous refers to objects seen because they give off their own light

incandescent refers to luminous objects that produce light because they are hot; the higher the temperature, the brighter the light, and the colour also changes

Propagation of light

The gap experienced between seeing lightning and hearing thunder shows that sound travels relatively slowly compared to light, which seems to travel so fast that its speed seems infinite; that is, events seem to be observed at the instant they happen. As seen in the previous topic, light is electromagnetic radiation travelling at the speed of light, c, in a vacuum.

SAMPLE PROBLEM 1 Calculating how long it takes for the light from the Sun to reach Earth

How long does light take to travel from the Sun to Earth, given that the speed of light is 3.0×10^8 m s^{-1} and the distance to the Sun is 1.50×10^{11} m?

THINK	WRITE
1. To calculate the time taken for light to travel from the Sun to Earth, the speed of light and the distance from the Sun to Earth need to be known.	Speed of light $= 3.0 \times 10^8$ m s^{-1} Distance from Sun to Earth $= 1.50 \times 10^{11}$ m
2. Substitute the values for the speed of light and the distance between the Sun and Earth into the following equation and solve for time: $\text{Average speed} = \dfrac{\text{distance travelled}}{\text{time taken}}$	$\text{Average speed} = \dfrac{\text{distance travelled}}{\text{time taken}}$ $\Rightarrow \text{Time taken} = \dfrac{\text{distance travelled}}{\text{average speed}}$ $= \dfrac{1.50 \times 10^{11} \text{ m}}{3.0 \times 10^8 \text{ m s}^{-1}}$ $= 0.50 \times 10^3$ s $= 500$ s $= 8$ minutes 20 seconds

PRACTICE PROBLEM 1

The Moon is approximately 380 000 km from Earth. How long would it take light to travel from a laser based on Earth to reflect off a mirror positioned on the Moon and then return back to Earth?

SAMPLE PROBLEM 2 Calculating the distance travelled by light in one year

How far does light travel in one year (one light-year) in a vacuum?

THINK	WRITE
1. To calculate the distance that light travels in one year, the speed of light and the number of seconds in a year need to be known.	Speed of light $= 3.0 \times 10^8$ m s^{-1} Seconds in a year $= 365.25 \times 24 \times 3600$ $= 31\,557\,600$ seconds
2. Substitute the values for the speed of light and the time taken into the following equation and solve for distance: Distance travelled $=$ average speed $\times$ time	Distance travelled $=$ average speed $\times$ time $= 3.0 \times 10^8 \times (31\,557\,600)$ $= 9.5 \times 10^{15}$ m $= 9.5 \times 10^{12}$ km

PRACTICE PROBLEM 2

In an optic fibre made from glass, it takes light $1.0\,\mu$s to travel a distance of 200 m. What is the speed of light in glass? Is it greater or less than the speed of light in a vacuum?

INVESTIGATION 2.1

online only

Luminous or not

Aim

To determine, from a list, which items are luminous and which items are non-luminous

Ray model for light

The need for sources of light, the great speed of light and the existence of sharp shadows can be described by a ray model. The model assumes that light travels in a straight-line path called a **light ray**. A light ray can be considered as an infinitely narrow beam of light and can be represented as a straight line, as shown in figure 2.3. In reality, light propagates away from its source in three dimensions. However, this idealised geometric model of light is useful in helping to predict the behaviour of light.

light ray an infinitely narrow beam of light, represented as a straight line

The bright Sun produces sharp shadows on the ground (see figure 2.4). The shape of the shadow is the same shape as the object blocking the light. This could happen only if light travels in a straight line.

FIGURE 2.3 Light rays leave a point on this pencil and travel in straight lines in all directions. The pencil is seen because of the 'bundle' of rays that enter the eye.

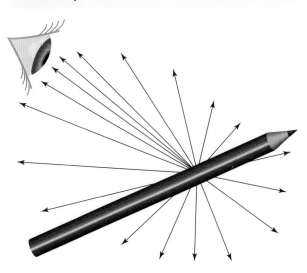

FIGURE 2.4 The straight rays passing the edge of a bird leave a sharp shadow on the ground.

Reflection: regular and diffuse

When you look at yourself in a plane mirror, some of the light rays from your nose, for example, travel in the direction of the mirror and reflect off in the direction of your eye (as seen in figure 2.5). What is happening at the surface of the mirror to produce such a perfect image?

FIGURE 2.5 Light rays from the tip of the nose reflect off the mirror and enter the eye. During reflection the frequency of the light is unchanged. Otherwise, your clothes might seem a different colour in the mirror!

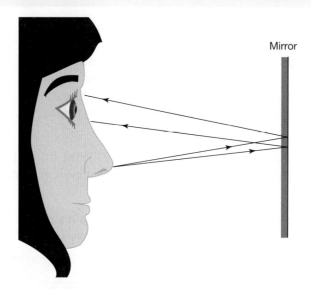

To investigate the reflection of light, the angles made by the rays need to be measured. Measurements of these angles show that, like a ball bouncing off a flat wall, the **angle of incidence** equals the **angle of reflection** (as observed in figure 2.6). The ray approaching the mirror is called the *incident ray*. The ray leaving the mirror is called the *reflected ray*. The **normal** is a line at right angles to the mirror. The angles are measured between each ray and the normal. It can be observed that the incident ray, the normal and the reflected ray all lie in the same plane.

angle of incidence the angle between an incident ray and the normal

angle of reflection the angle between a reflected ray and the normal

normal a line that is perpendicular to a surface or a boundary between two surfaces

FIGURE 2.6 The angle of incidence always equals the angle of reflection.

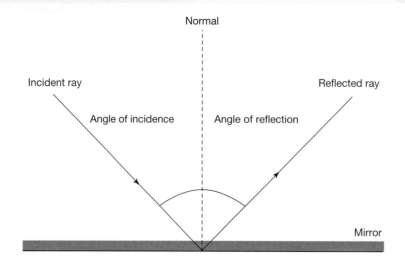

Reflection from a smooth surface is called **regular reflection** (also referred to as specular reflection). But what happens with an irregular surface, such as a page in a book? A page is not smooth like a mirror. At the microscopic level, 'hills and valleys' exist. As the light rays come down into these hills and valleys, they still reflect with the two angles the same but, because the surface is irregular, the reflected rays emerge in all directions. This is called **diffuse reflection**. Light rays from diffuse reflections — from the ground, trees and other objects — enter the eye and enable the brain to make sense of the world. Therefore, observers in all directions receive light from the surface.

regular reflection reflection from a smooth surface; also referred to as specular reflection
diffuse reflection reflection from a rough or irregular surface

FIGURE 2.7 The incident ray, the 'normal' to the surface of the mirror and the reflected ray all lie in the same plane, which is at right angles to the plane of the mirror.

FIGURE 2.8 In diffuse reflection, each of the incoming parallel rays meets the irregular surface at a different angle of incidence. The reflected rays will therefore go off in different directions.

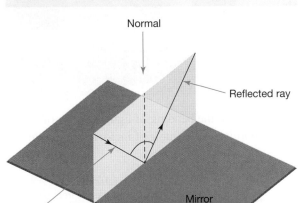

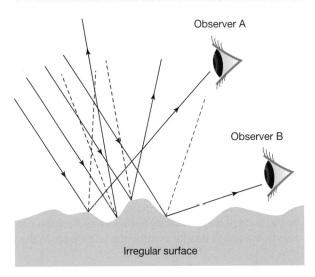

EXTENSION: What is colour?

Colours are an important part of humanity's language and environment. Colours can be peaceful to the eye or very stimulating. Colours are used in language to convey feelings and emotions (for example, *fiery red, warm orange* and *icy blue*).

At first, colour may seem to be a defining part of an object, like size, shape and texture. For example, you can say *green leaves, red earth* and *blue eyes*. It is only when experiments are done with light that you can observe that the colour or appearance of an object changes with the light that is shining on it, as shown in figure 2.9.

Here, it seems that colour — pure colour — is separate from any solid object. Colour is both a property of light and an aspect of human perception.

FIGURE 2.9 Changing the colour of the light on these flowers from white to red to blue changes one's perception of their colour.

a. White light   **b.** Red light   **c.** Blue light

2.2.2 Bending of light: Snell's Law

Experience shows that when you are spearing for fish in the shallows, you must aim the spear below where the fish appears to be in the water. At the beach or in a pool, people standing in the shallows appear to have shorter legs. Perception is distorted, but the reason is not apparent.

When a special situation is set up, such as in figure 2.10, where a straight rod is placed in a beaker of liquids that do not mix, the idea of the change of direction of the light is apparent. This change in direction is called **refraction**. Refraction occurs when light travels from one medium to another, such as when travelling through air and into water.

The ray model can help explain observations of light. If a fish seems closer to the surface of the water, the ray of light from the fish must have bent. To your eye, the ray seems to be coming from another direction. Given that light can travel both ways along a light path, the fish will see the spear-thrower further towards the vertical.

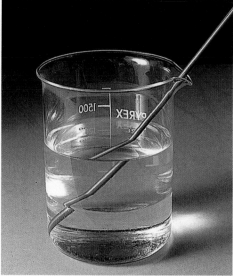

FIGURE 2.10 An example of refraction

The ray model not only provides a way of describing your observations of the bending of light, but also of taking measurements. The angle that a ray of light makes with the normal — the angle of incidence and **angle of refraction** — can be measured and investigated.

FIGURE 2.11 The rays from the fish bend when they enter the air. To the eye, the rays appear to come from a point closer to the surface. The fish appears closer to the surface and physically further away from the observer than it actually is. In reality, the fish is deeper in the water and closer to the observer than it appears.

> **refraction** the bending of light as it passes from one medium into another
>
> **angle of refraction** the angle between a refracted ray and the normal

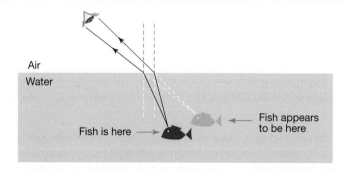

Air
Water
Fish is here →
Fish appears to be here ←

elog-1784

tlvd-3857

INVESTIGATION 2.2

online only

Seeing is believing

Aim

To observe the bending of light

on Resources

🔗 **Weblink** Bending light applet (web-7434)

Snell's Law

In 1621, the Dutch physicist Willebrord Snellius (1580–1626), known in the English-speaking world as Willebrand Snell, investigated the refraction of light and found that the ratio of the sines of the angles of incidence and refraction was constant for all angles of incidence.

$$\frac{\sin\theta_i}{\sin\theta_r} = \text{constant}$$

where:

θ_i is the angle of incidence

θ_r is the angle of refraction

Figure 2.12 shows how an incident ray is affected when it meets the boundary between air and water. The normal is a line at right angles to the boundary, and all angles are measured from the normal. Some of the light from the incident ray is reflected back into air, while the rest is transmitted into the water. The ratio shown in figure 2.12 is a constant for all angles of light travelling from air to water.

FIGURE 2.12 The ratio $\dfrac{\sin\theta_i}{\sin\theta_r}$ is constant for all angles for light travelling from air to water.

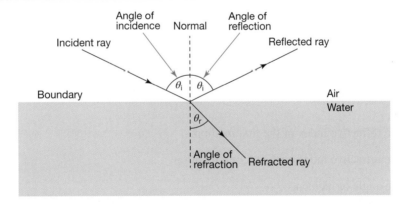

EXTENSION: Early history of refraction

Snell's Law was first discovered by Abū Sa'd al-'Alā' ibn Sahl (c. 940–c. 1000), a Muslim physicist in the court in Baghdad, in 984. He reported his findings in his book on burning mirrors and lenses. Ibn Sahl used the relationship to design a shape for lenses that overcame the problem of spherical aberration. Ptolemy (c. 100–c. 170), a Greco-Egyptian mathematician, had investigated refraction much earlier, compiling a table of angles for light travelling from air into water.

Snell repeated his experiments with different substances and found that the ratio between the sines of the angle of incidence was constant for a given substance. However, he found that each substance had a different value for this constant. This suggested that different substances bend light by different amounts. (Remember that some light is always reflected.)

In fact, there is a different ratio for each pair of substances (for example, air and glass, air and water). A different ratio is obtained for light travelling from water into glass. The value of the ratio is called the **relative refractive index** because it depends on the properties of two different substances. The relative refractive index represents the relative bending of light from one medium to another.

relative refractive index a measure of how much light bends when it travels from any one substance into any other substance

The bending of light always involves light travelling from one substance to another (and an angle of incidence different from zero in the first medium). It is not possible to find the effect of a particular substance on the deflection of light without adopting one substance as a reference standard. Once you have a standard, every substance can be compared with it. A natural standard is a vacuum — the absence of any substance. The **absolute refractive index** of a vacuum is given the value of one. From this, the absolute refractive index of all other substances can be determined. Some examples are given in table 2.1. (The word 'absolute' is commonly omitted and the term 'refractive index' usually refers to the absolute refractive index.) The absolute refractive index measures the bending ability of a material compared to that of a vacuum, whereby light travels in straight lines and is not bent.

TABLE 2.1 Values for absolute refractive index

Material	Value
Vacuum	1.000 00
Air at 20 °C and normal atmospheric pressure	1.000 28
Water	1.33
Perspex	1.49
Quartz	1.46
Crown glass	1.52
Flint glass	1.65
Carbon disulfide	1.63
Diamond	2.42

The refractive index is given the symbol n because it is a pure number without any units. This enables a more useful restatement of Snell's Law. For example:

$$n_{air}\sin\theta_{air} = n_{water}\sin\theta_{water}$$

absolute refractive index the relative refractive index for light travelling from a vacuum into a substance, commonly referred to as the refractive index

More generally this would be expressed as follows:

$$n_1\sin\theta_1 = n_2\sin\theta_2$$

where:

n_1 is the refractive index of the first medium

n_2 is the refractive index of the second medium

θ_1 is the angle of incidence

θ_2 is the angle of refraction.

FIGURE 2.13 A graphical depiction of Snell's Law for any two substances. Note that the light ray has no arrow, because the relationship is true for the ray travelling in either direction, and reflected light rays have been omitted for clarity.

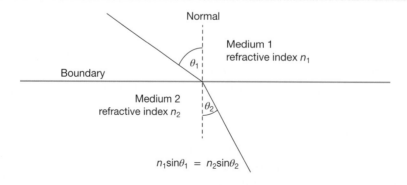

SAMPLE PROBLEM 3 Calculating the angle of refraction using Snell's Law

A ray of light moving through air strikes a glass block of refractive index 1.45 at an angle of incidence of 30°. What is the angle of refraction?

THINK	WRITE
1. List the known information.	$n_{air} = 1.0$; $\theta_{air} = 30°$; $n_{glass} = 1.45$; $\theta_{glass} = ?$
2. Use Snell's Law, $n_1 \sin\theta_1 = n_2 \sin\theta_2$, to determine the angle of refraction.	$1.0 \times \sin 30° = 1.45 \times \sin\theta_{glass}$
	$\sin\theta_{glass} = \dfrac{\sin 30°}{1.45}$
	$= 0.3448$
	$\Rightarrow \theta_{glass} = \sin^{-1}(0.3448)$
	$= 20.17°$
	$= 20°$

PRACTICE PROBLEM 3

A ray of light moving through air enters a plastic block at an angle of incidence of 40°. The angle of refraction is 30°. What is the refractive index of the plastic?

EXTENSION: Gravitational fields and bending light

Light can be bent by a strong gravitational field, such as that near the Sun. The gravitational field can act like a convex lens. Light from a distant star that is behind and blocked by the Sun bends around the Sun so that astronomers on Earth see an image of the star to the side of the Sun. This is quite different to how refraction occurs.

INVESTIGATION 2.3

on line only

Snell's Law

Aim

To observe the refraction of light and to use Snell's Law to determine the refractive index of a medium

on Resources

 Video eLesson Refraction of light and Snell's Law (eles-0037)

Interactivity Refraction of light and Snell's Law (int-0056)

INVESTIGATION 2.4   on line only

Refraction of waves

Aim

To observe the wavelength and refraction of waves travelling from deep to shallow water

Speed of light in a material

The results of the work of Jean Bernard Leon Foucault (1819–1868) and Hippolyte Fizeau (1819–1896), along with the work of Augustin-Jean Fresnel (1788–1827) (pronounced 'fray-NEL'), showed that the speed of light in water was less than the speed of light in air. This allowed scientists to determine the physical meaning of the refractive index:

$$\text{Absolute refractive index of water} = \frac{\text{speed of light in a vacuum}}{\text{speed of light in water}}$$

In general:

> The refractive index of a material is the ratio of the speed of light in a vacuum to the speed of light in the medium:
>
> $$n = \frac{c}{v}$$
>
> where:
>
> n is the refractive index of the medium
>
> c is the speed of light in a vacuum $\left(3.0 \times 10^8 \text{ m s}^{-1}\right)$
>
> v is the speed of light in the medium.

Hence, the higher the refractive index, the slower the speed of light in the medium.

The formula can be rearranged to give:

$$n_{\text{water}} \times v_{\text{water}} = c$$

where:

c = the speed of light in a vacuum

v_{water} = the speed of light in water.

Similarly, for glass, $n_{\text{glass}} \times v_{\text{glass}} = c$, which means $n_{\text{glass}} \times v_{\text{glass}} = n_{\text{water}} \times v_{\text{water}}$, or, as a general relationship for any two materials:

> $$n_1 v_1 = n_2 v_2$$
>
> where:
>
> n_1 and n_2 are the refractive indices of medium 1 and medium 2
>
> v_1 and v_2 are the speed of light in medium 1 and in medium 2.

tlvd-0197

SAMPLE PROBLEM 4 Calculating how fast light travels using the refractive index

a. The refractive index of glass is 1.5. How fast does light travel in glass?
b. Use the answer to part a to determine the speed of light in water ($n_{water} = 1.33$).

THINK	WRITE
a. Use the relationship $n_{glass} = \dfrac{c}{v_{glass}}$ to calculate the speed of light in glass.	a. $1.5 = \dfrac{3.0 \times 10^8}{(\text{speed of light in glass})}$ $\Rightarrow$ Speed of light in glass $= \dfrac{3.0 \times 10^8}{1.5}$ $= 2.0 \times 10^8 \, \text{m s}^{-1}$
b. Use the relationship $n_{glass} \times v_{glass} = n_{water} \times v_{water}$ to determine the speed of light in water.	b. $n_{glass} \times v_{glass} = n_{water} \times v_{water}$ $1.5 \times 2.0 \times 10^8 = 1.33 \times v_{water}$ $\Rightarrow v_{water} = \dfrac{1.5 \times 2.0 \times 10^8}{1.33}$ $= 2.3 \times 10^8 \, \text{m s}^{-1}$

PRACTICE PROBLEM 4

a. How fast does light travel in diamond ($n_{diamond} = 2.42$)?
b. Use the answer to part a to determine the speed of light in carbon disulfide ($n_{carbon\ disulfide} = 1.63$).

elog-1787

tlvd-0849

INVESTIGATION 2.5 online only

Using apparent depth to determine the refractive index

Aim

To determine the refractive index using apparent depth

2.2 Activities learn on

Students, these questions are even better in jacPLUS

 Receive immediate feedback and access sample responses

 Access additional questions

Track your results and progress

Find all this and MORE in jacPLUS

2.2 Quick quiz on	2.2 Exercise	2.2 Exam questions

2.2 Exercise

1. Calculate the longest and shortest time for a radio signal travelling at the speed of light (3.0×10^8 m s^{-1}) to go from Earth (the radius of Earth's orbit around the Sun is 1.50×10^{11} m) to a space probe when the space probe is:
   a. near Mars (the radius of Mars's orbit around the Sun is 2.28×10^{11} m)
   b. near Neptune (the radius of Neptune's orbit around the Sun is 4.50×10^{12} m).

2. A ray of light strikes a plane mirror with an angle of incidence of 25° and regular reflection takes place.
   a. What angle does the light ray make with the mirror surface?
   b. What is the angle of reflection?
3. A student arranges a beam of light from a laser to strike a plane mirror such that the incident beam and the reflected beam form a right angle. What is the angle of incidence that the light from the laser makes with the mirror?
4. A ray of light in air (n_{air} = 1.00) strikes water in a pond. What is the angle of refraction in water (n_{water} = 1.33) for an angle of incidence of 40°? If the angle of incidence is increased by 10°, by how much does the angle of refraction increase?
5. A ray of light enters a plastic block at an angle of incidence of 55° with an angle of refraction of 33°. What is the refractive index of the plastic?
6. A ray of light passes from air through a rectangular glass block (n_{glass} = 1.55). The angle of incidence as the ray enters the block is 65°.
   a. Calculate the angle of refraction at the first face of the block.
   b. Calculate the angle of refraction as the ray emerges on the other side of the rectangular block. Comment on your answers.
7. A glass block (n_{glass} = 1.55) is immersed in a container of water (n_{water} = 1.33). A light ray passes from the water into the glass block at an angle of incidence of 28.0°. Determine the angle of refraction in the glass block.
8. Determine the speed of light in the following materials:
   a. water (n_{water} = 1.33)
   b. glass (n_{glass} = 1.50)
   c. diamond ($n_{diamond}$ = 2.42).
9. Immiscible liquids are liquids that do not mix. Immiscible liquids will settle on top of one another, in the order of their density, with the densest liquid at the bottom. Some immiscible liquids are also transparent.
   a. Calculate the angles of refraction as a ray passes down through immiscible layers as shown in the diagram.
   b. If a plane mirror was placed at the bottom of the beaker, calculate the angles of refraction as the ray reflects back to the surface. Comment on your answers.

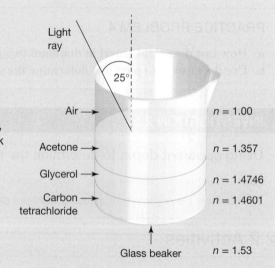

10. Calculate the sideways deflection as a ray of light goes through a parallel-sided plastic block ($n_{plastic}$ = 1.4) with sides 5.0 cm apart, as in the following figure.

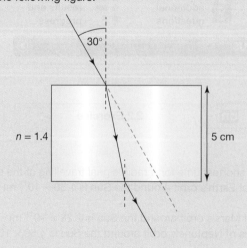

2.2 Exam questions

▶ Question 1 (1 mark)

Source: VCE 2019 Physics Exam, Section A, Q.9; © VCAA

MC A monochromatic light ray passes through three different media, as shown in the diagram below.

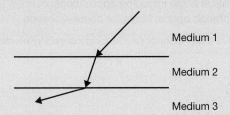

Assume that v_1 is the speed of light in Medium 1, v_2 is the speed of light in Medium 2 and v_3 is the speed of light in Medium 3.

Which one of the following would best represent the relative speeds in the media?

A. $v_1 > v_2 > v_3$
B. $v_1 > v_3 > v_2$
C. $v_3 > v_2 > v_1$
D. $v_3 > v_1 > v_2$

▶ Question 2 (1 mark)

MC Refraction occurs as light passes from one medium to another.

Refraction is associated with a change in which of the following pairs of properties of the light wave?

A. Frequency and direction
B. Speed and frequency
C. Speed and direction
D. Frequency and wavelength

▶ Question 3 (2 marks)

A light ray travelling in dry air ($n = 1.00$) enters glycerine ($n = 1.47$) at an angle of incidence of 17 degrees.

What is the angle of the refracted ray in the glycerine?

▶ Question 4 (2 marks)

A light ray travelling in crown glass ($n = 1.60$) enters an unknown material at an angle of incidence of 30 degrees. The refracted ray is found to be travelling at an angle of 42 degrees from the normal.

Calculate the refractive index of the unknown material.

▶ Question 5 (1 mark)

What is necessary for light to travel directly across the boundary between two materials without any change in direction?

More exam questions are available in your learnON title.

2.3 Total internal reflection and critical angle

KEY KNOWLEDGE

- Investigate and analyse theoretically and practically the behaviour of waves including:
  - total internal reflection and critical angle including applications: $n_1\sin(\theta_c) = n_2\sin(90°)$
- Investigate light transmission through optical fibres for communication

Source: VCE Physics Study Design (2023–2027) extracts © VCAA; reproduced by permission.

2.3.1 The critical angle

It has already been mentioned that some light is reflected off a transparent surface while the rest is transmitted into the next medium. This applies whether the refracted ray is bent towards or away from the normal.

However, a special situation applies when the refracted ray is bent away from the normal. This is illustrated in figure 2.15. As the angle of incidence increases, the angle of refraction also increases. Eventually the refracted ray becomes parallel to the surface and the angle of refraction reaches a maximum value of 90° (see figure 2.15b). When this condition is met, the corresponding angle of incidence is called the **critical angle**. If the angle of incidence is increased beyond the critical angle, all the light is reflected back into the water, with the angles being the same. This phenomenon is called **total internal reflection** (see figure 2.15c).

FIGURE 2.14 There are no mirrors in a fish tank but strange reflections can be seen. It appears that light is being reflected off the side of the fish tank and the water surface.

FIGURE 2.15 Three stages of refraction leading to total internal reflection

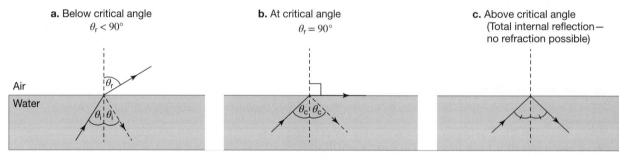

a. Below critical angle
$\theta_r < 90°$

b. At critical angle
$\theta_r = 90°$

c. Above critical angle
(Total internal reflection—
no refraction possible)

Air

Water

The critical angle can be calculated using Snell's Law:

$$n_1 \times \sin\theta_c = n_2 \times \sin90°$$

where:

n_1 is the refractive index of medium 1

n_2 is the refractive index of medium 2

θ_c is the critical angle for medium 1.

critical angle the angle of incidence for which the angle of refraction is 90°; the critical angle exists only when light passes from one substance into a second substance with a lower refractive index

total internal reflection the total reflection of light from a boundary between two substances; it occurs when the angle of incidence is greater than the critical angle

Note that when the second medium is air, the critical angle for medium one is simply: $\theta_c = \dfrac{1}{n_1}$.

SAMPLE PROBLEM 5 Determining the critical angle for water using the refractive index

What is the critical angle for water given that the refractive index of water is 1.3 and the refractive index of air is 1.0?

THINK	WRITE
1. List the known information.	$n_{air} = 1.0$; $\theta_{air} = 90°$; $n_{water} = 1.3$; $\theta_{water} = ?$
2. Use the relationship $n_{water} \times \sin\theta_c = n_{air} \times \sin90°$ to determine the critical angle for water.	$n_{water} \times \sin\theta_c = n_{air} \times \sin90°$

$$1.3 \times \sin\theta_c = 1.0 \times \sin90°$$
$$\Rightarrow \sin\theta_c = \frac{\sin90°}{1.3}$$
$$= 0.7692$$
$$\Rightarrow \theta_c = \sin^{-1}(0.7692)$$
$$= 50.28°$$
$$= 50°$$
The critical angle is 50°.

PRACTICE PROBLEM 5

Diamond has a refractive index of 2.42.
a. **What is the value of the critical angle for a diamond–air interface?**
b. **The diamond is now placed in water ($n = 1.33$). Would the critical angle increase, decrease or stay the same?**
c. **Determine the critical angle for a diamond–water interface.**

Total internal reflection is a relatively common atmospheric phenomenon (as in mirages) and it has technological uses (for example, in optical fibres).

optical fibre a thin tube of transparent material that allows light to pass through without being refracted into the air or another external medium

2.3.2 Internal reflection in optical fibres

Another example of total internal reflection, whereby light is completely reflected back into a medium, is in the important technological application of **optical fibres**. Optical fibres have a number of uses, such as in medical imaging, where a thin, flexible cable containing an optical fibre can be placed inside a person's body to transmit pictures of the condition of organs and arteries, without the need for invasive surgery. The same can be done in industry when there is a problem with complex machinery, or to investigate blocked pipes (see figure 2.16). Optical fibres are also the basis of the important telecommunications industry. They allow high-quality transmission of many channels of information in a small cable over very long distances, in both directions, and with negligible signal loss.

FIGURE 2.16 Fibre optic cable with a camera used to investigate a blockage in a pipe

An optical fibre is like a pipe with a light shone in one end that comes out the other end. An optical fibre is made of glass that is approximately 10 micrometres (10×10^{-6} m) thick. Light travels along it, as glass is transparent, but the fibre needs to be able to turn and bend around corners. The optical fibre is designed so that any ray meeting the outer surface of the glass fibre is totally internally reflected back into the glass. As shown in figure 2.18, the light ray meets the edge of the fibre at an angle of incidence greater than the critical angle and is reflected back into the fibre. In this way, nearly all the light that enters the fibre emerges at the other end.

FIGURE 2.17 A bundle of optical fibres. Each fibre in the bundle carries its signal along its length. If the individual fibres remain in the same arrangement, the bundle will emit an image of the original object.

FIGURE 2.18 A light ray travels along an optical fibre through total internal reflection.

Optical fibre

Light ray ⟶

FIGURE 2.19 Light rays entering the fibre at too sharp of an angle are refracted out of the fibre.

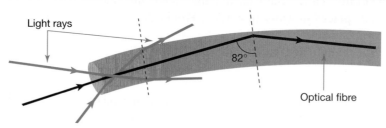

Light rays

82°

Optical fibre

If the glass fibre is exposed to the air, the critical angle for light travelling from glass to air is 42°, which is quite small. Any angle of incidence greater than this angle will produce total internal reflection. If the fibre is very narrow, this angle is easily achieved.

However, in both medical and telecommunications uses, fibres are joined in bundles with edges touching. The touching would enable light rays to pass from fibre to fibre, confusing the signal. To overcome this, a plastic coating or cladding is put around the glass to separate the glass fibres. The total internal reflection occurs between the glass and the plastic. The critical angle for light travelling from glass to plastic is 82°. This value presents a problem because light meeting the edge of the glass at any angle less than 82° will pass out of the fibre.

This has implications for the design of the optical fibre and the beam of light that enters the fibre. The fibre needs to be very narrow and the light entering the fibre has to be a thin beam with all the rays parallel.

SAMPLE PROBLEM 6 Explaining refraction and calculating critical angles in optical fibres

Optical fibres transport data using light. They are constructed using transparent materials with different refractive indices. The following diagram shows a typical fibre with a cylindrical core and surrounding cladding. Laser light is shone from air in the optical fibre.

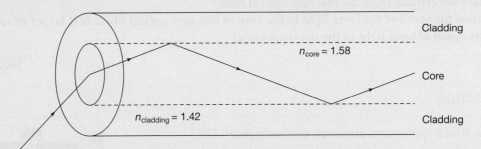

Cladding

$n_{core} = 1.58$

Core

$n_{cladding} = 1.42$

Cladding

a. Explain why it is necessary for the refractive index of the cladding to be smaller than the refractive index of the core.
b. Calculate the critical angle for the laser light at the core–cladding boundary.
c. Determine the speed of the laser light in the core of the optical fibre.

THINK	WRITE
a. Explain why it is necessary for the refractive index of the cladding to be smaller than the refractive index of the core.	**a.** If the refractive index of the cladding was larger than the refractive index of the core, the light would be refracted *towards the normal* and internal reflection would not be possible. When the refractive index of the cladding is less than that of the core, the light is refracted *away from the normal* so that total internal reflection is possible if the angle of incidence is larger than the critical angle.
b. Use the relationship $n_{core} \times \sin\theta_c = n_{cladding} \times \sin90°$ to determine the critical angle.	**b.** $n_{core} \times \sin\theta_c = n_{cladding} \times \sin90°$ $$1.58 \times \sin\theta_c = 1.42 \times \sin90°$$ $$\Rightarrow \sin\theta_c = \frac{1.42\sin90°}{1.58}$$ $$= \frac{1.42}{1.58}$$ $$\Rightarrow \theta_c = \sin^{-1}\left(\frac{1.42}{1.58}\right)$$ $$= 64.0°$$ The critical angle is 64.0°.
c. Analyse the refraction of the light at the entrance of the optical fibre, the air–core boundary. Use the relationship $n_{air}v_{air} = n_{core}v_{core}$.	**c.** $$n_{air}v_{air} = n_{core}v_{core}$$ $$1.00 \times (3.00 \times 10^8) = 1.58 \times v_{core}$$ $$\Rightarrow v_{core} = \frac{3.0 \times 10^8}{1.58}$$ $$= 1.90 \times 10^8 \text{ m s}^{-1}$$ The speed of the laser light in the core is 1.90×10^8 m s^{-1}.

PRACTICE PROBLEM 6

A different optical fibre is used with the same cladding ($n_{cladding} = 1.42$) as in sample problem 6, but with a core of refractive index 1.56.

a. Would the critical angle of this new optical fibre be larger or smaller than the previous fibre?

b. Calculate the critical angle for this new optical fibre.

c. Determine the speed of the laser light in the core of this new optical fibre. Is it larger or smaller than the speed of laser light in the previous fibre?

2.3 Activities

learn on

2.3 Quick quiz **on**	2.3 Exercise	2.3 Exam questions

2.3 Exercise

1. People enjoy the look of diamonds because they disperse white light.
   a. Calculate the critical angle for light travelling through a diamond ($n_{diamond} = 2.50$) towards its surface. You may assume that the diamond is in air ($n_{air} = 1.00$).
   b. The diamond is now placed in water ($n_{water} = 1.33$). Determine the critical angle now for light at a diamond–water interface.

2. Consider the triangular prism shown in the diagram.
   a. Calculate the refractive index of the glass triangular prism so that the light ray meets the faces at the critical angle. Is this value of the refractive index the minimum or maximum value for such a reflection?
   b. One ray enters the prism as shown (2 cm from the top of the prism). Another ray, parallel to the one shown, also enters the prism, 2 cm from the bottom of the prism. Describe the positions of the rays after refraction.
   c. Would the rays of light emerge at the same or different times?

3. Calculate the refractive index of the plastic coating surrounding an optical fibre if the critical angle for glass to plastic is 82.0° and the refractive index of glass is 1.50.

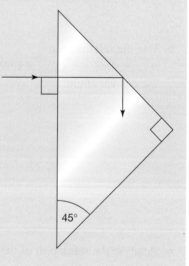

4. Light enters an optical fibre 1.0 μm in diameter, as shown in the following figure. Some light goes straight down the centre of the fibre. Another ray is angled, leaving the central line and meeting the outside edge at slightly more than the critical angle of 82°, then reflects back to the central line.

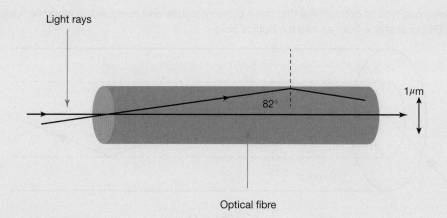

Light rays

82°

1μm

Optical fibre

    a. How much further did this second ray travel compared to the ray that travelled down the centre of the fibre?

    b. Calculate the speed of light in the glass and hence determine the time delay between the two rays after one internal reflection. Do you think this could be a problem when transmitting data in an optical fibre? If so, when? How could the problem be overcome?

5. A student conducts a series of measurements to determine the critical angle for light travelling from water ($n = 1.33$) into an unknown material. They determine that the critical angle is 70°. Use this result to determine the refractive index of the unknown material.

6. Calculate the critical angle for light travelling from diamond ($n = 2.40$) into air ($n = 1.00$).

2.3 Exam questions

▶ **Question 1 (1 mark)**

Source: VCE 2019 Physics Exam, Section A, Q.10; © VCAA

MC The horizontal face of a glass block is covered with a film of liquid, as shown below.

A monochromatic light ray is incident on the glass–liquid boundary with an angle of incidence of 62.0°.

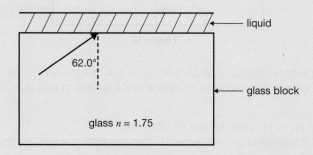

liquid

62.0°

glass block

glass $n = 1.75$

The minimum value of the liquid's refractive index, so that some light will just cross the interface into the liquid, is closest to

A. 1.33.

B. 1.55.

C. 1.88.

D. 1.98.

▶ **Question 2 (5 marks)**

Source: VCE 2018 Physics Exam, Section B, Q.12; © VCAA

Optical fibres are constructed using transparent materials with different refractive indices.

Figure 14 shows one type of optical fibre that has a cylindrical core and surrounding cladding. Laser light of wavelength 565 nm is shone from air into the optical fibre.

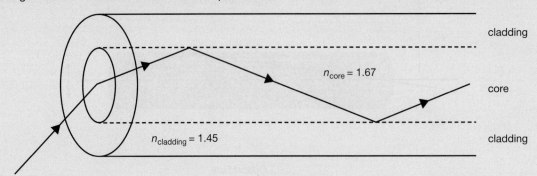

Figure 14

a. Calculate the frequency of the laser light before it enters the optical fibre. **(1 mark)**
b. Calculate the critical angle for the laser light at the cladding–core boundary. Show your working. **(2 marks)**
c. Calculate the speed of the laser light once it enters the core of the optical fibre. Give your answer correct to three significant figures. Show your working. **(2 marks)**

▶ **Question 3 (5 marks)**

Source: VCE 2017 Physics Exam, Section B, Q.14; © VCAA

A light ray from a laser passes from a glucose solution ($n = 1.44$) into the air ($n = 1.00$), as shown in Figure 12.

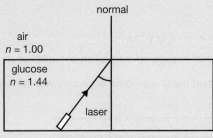

Figure 12

a. Calculate the critical angle (total internal reflection) from the glucose solution to the air. **(1 mark)**
b. The light ray strikes the surface at an angle of incidence to the normal of less than the critical angle calculated in **part a.**

On Figure 12, sketch the ray or rays that should be observed. **(2 marks)**
c. The angle to the normal is increased to a value greater than the critical angle. An observer at point X in Figure 13 says she cannot see the laser.

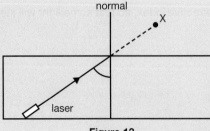

Figure 13

Explain why the observer says she cannot see the laser. **(2 marks)**

2.4 Dispersion

2.4.1 Producing colour from white light

White light can be separated into colours using a narrow beam of light and a glass triangular prism. This phenomenon is called **dispersion**. It was first analysed in this way by Isaac Newton in 1666, although René Descartes had sought an explanation for rainbows in 1637 by working with a spherical glass flask filled with water.

dispersion the separation of light into different colours as a result of refraction

FIGURE 2.20 The colours in white light disperse as they enter the glass and disperse even more when they leave. A continued spectrum of colours is produced when white light is passed through a prism. The red light is deflected the least and each colour in the visible part of the spectrum is deflected progressively more, with violet being the most deflected.

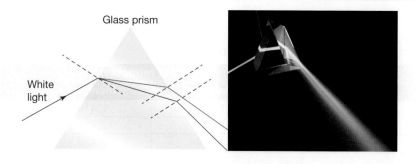

Glass prism

White light

As light enters a triangular glass prism, it is refracted towards the normal. It then travels through the prism to the other side, where it is refracted away from the normal, because the light is re-emerging into the air.

The colours spread as they enter the glass and travel on different paths through the triangular prism. They are spread even more as they leave the glass. Violet is bent the most and red the least. The order of the colours, from the colour that bends least to the colour that bends most, is red, orange, yellow, green, blue, indigo and violet. Different mnemonics exist to remember this, such as 'Rinse Out Your Granny's Boots In Vinegar', or 'Read Out Your Good Book In Verse'.

Each colour has a different angle of refraction. This means that the glass has a different refractive index for each colour. This can be expressed as a statement of Snell's Law, as follows:

$$n_{air}\sin\theta_i = n_{gl(red)}\sin\theta_{red} = n_{gl(violet)}\sin\theta_{violet}$$

For example, as shown in figure 2.21, violet light is bent more than red, so θ_{violet} is smaller than θ_{red}. This means that the refractive index of glass for violet light is greater than that for red light (and also that violet light propagates slower than red light; as seen previously, the greater the refractive index, the slower the light). This is also true for other materials (see table 2.2).

FIGURE 2.21 The angle of refraction for violet light is smaller than that for red light. *This means that the refractive index of the glass is different for different colours.* For violet it must be greater than that for red.

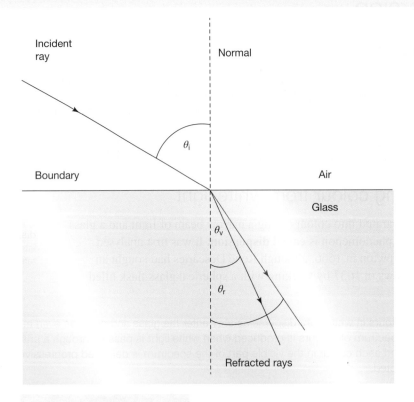

TABLE 2.2 Refractive index values vary for different coloured light

Colour	Index of refraction			
	Crown glass	Flint glass	Diamond	Water
Red	1.514	1.571	2.410	1.331
Yellow	1.517	1.575	2.418	1.333
Deep blue	1.528	1.594	2.450	1.340

INVESTIGATION 2.6

Separating colours

Aim

To determine which colour is bent the most by a prism

 Resources

📄 **Digital document** eModelling: Using a spreadsheet to explore refraction through a prism (doc-0058)

🧩 **Interactivity** Spreading the spectrum (int-6609)

2.4.2 EXTENSION: Scattering of light

There are certain observations in nature that have been universal for as long as man has looked up at the sky — the sky is blue during the day and red at sunset. Photographers are especially interested in this phenomenon and will often wait for different temperatures of light for their images.

It is tempting to assume that the air is entirely transparent and that energy like light can pass through the atmosphere without any interactions with the various molecules and particles that make up the air. This assumption, however, is incorrect. Although the light being emitted from the Sun has an intensity high enough to reach the surface of Earth, where it can be observed by the human eye, it must first pass through a reasonably dense atmosphere, where the probability of hitting something is reasonably high.

To understand the scattering effect of Earth's atmosphere, let's consider how different Earth's sky looks when observed from space. If you were to safely look towards the Sun from outside Earth's atmosphere, the Sun would look like a bright orb on a mostly black background, as shown in figure 2.22a. This is because the light emitted from the Sun would travel in a straight line to your eyes. (Never

FIGURE 2.22 Views from outside Earth's atmosphere **a.** Looking towards the Sun **b.** Looking away from the Sun

directly look at the Sun as this would damage your eyes.) If you were to turn around, not looking at the Sun, you would mainly observe darkness, as shown in figure 2.22b. This is because nothing would scatter the light emitted from the Sun, and hardly any light would travel to your eyes (you would still observe the light from more distant stars, and potentially from the Moon).

This is similar to what can be observed from the surface of the Moon, as shown in figure 2.23. On the surface of the Moon, where there is no atmosphere, there is nothing to scatter the light reflected from Earth and so it appears to be the only lit object visible.

FIGURE 2.23 Earthrise from the Moon

Compare this to what you are used to experiencing from inside Earth's atmosphere. How can we explain our perception of the sky being coloured?

FIGURE 2.24 A sky full of colours **a.** Ask anyone what colour the sky is, and blue is probably the answer you will get. **b.** How do we explain the colours observed at sunrise or sunset?

In subtopic 1.5 you have seen that the visible light from the Sun contains all the colours of the visible spectrum (from violet to red), with each colour corresponding to a specific wavelength range, such as 450 nm–495 nm for the colour blue. When this visible light travelling through the atmosphere hits a molecule, the light is scattered in several directions. The scattering is not equal, however; the shorter wavelengths of visible light are scattered more than the longer wavelengths, as shown in figure 2.25.

As the sunlight enters the atmosphere, the shorter wavelengths (violet and blue) are scattered the most, thus the sky should appear blueish-violet to humans. However, as the human eye is more sensitive to blue than to violet (we have three main colour receptors that respond largely to red, green and blue), and as the Sun emits more blue light than violet light, the sky appears blue to us. But what about at sunrise and sunset?

FIGURE 2.25 The scattering of visible light in relation to wavelength (in nanometres)

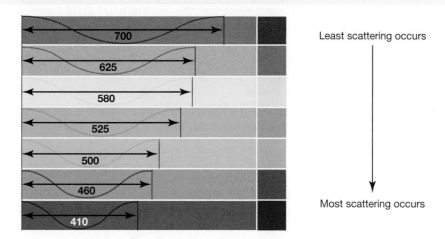

Least scattering occurs

Most scattering occurs

When the Sun is low enough on the horizon, the shorter wavelengths are scattered too much for us to perceive them. This is because the sunlight has to go through more air than during the day, encountering more molecules and so becoming more scattered. Only the longer wavelengths (yellow, orange and red) make it to our eyes, which explains the colours observed in figure 2.24b.

2.4 Activities

learnon

| 2.4 Quick quiz on | 2.4 Exercise | 2.4 Exam questions |

2.4 Exercise

1. **a.** White light enters a crown glass rectangular prism. Sketch the path of red and deep blue light through the glass and back into air. How does the direction of the emerging coloured rays compare with that of the incoming white ray?
   b. Suggest why a glass triangle is used to observe the visible spectrum, rather than a glass rectangle.
2. Which travels faster through crown glass — red light or violet light? What is the speed difference?
3. Green and violet light enter a triangular prism. Which is bent more?
4. A ray of white light in air strikes a rectangular glass block with an angle of incidence of 30°, as shown in the following diagram. For the glass block, the refractive index of red light is 1.514 and, for purple light, it is 1.530.
   a. On a copy of the diagram, carefully show the path taken by red light and by purple light.
   b. Use Snell's Law to find the angle of refraction for both the red light and the purple light.
   c. Determine the angle between the red light ray and the purple light ray in the rectangular glass.

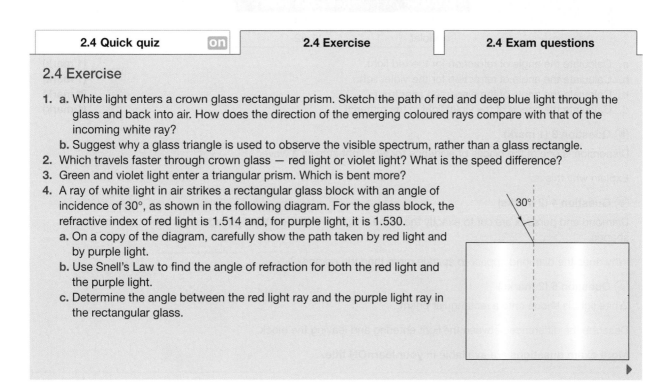

5. Red and violet beams of light enter a triangular prism at the same location and angle. The violet beam of light is refracted noticeably more than the red beam. What does this result indicate about the relative speeds of red and violet light in the material that the prism is made of?

2.4 Exam questions

▶ Question 1 (4 marks)

Source: VCE 2019 Physics Exam, Section B, Q.15; © VCAA

A student sets up an experiment involving a source of white light, a glass prism and a screen. The path of a single ray of white light when it travels through the prism and onto the screen is shown in Figure 14.

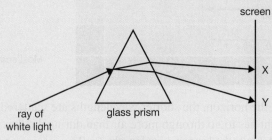

Figure 14

A spectrum of colours is observed by the student on the screen, which is positioned to the right of the prism.
a. Name and explain the effect observed by the student. **(3 marks)**
b. Points X and Y on Figure 14 represent either end of the visible spectrum observed by the student.

Identify the two visible colours observed at point X and at point Y. **(1 mark)**
Point X _____ Point Y _____

▶ Question 2 (4 marks)

A beam of white light is incident on a prism at an angle of 40.0°. The refractive indexes of the colours at each end of the visible light spectrum are shown in the following table.

Colour	Refractive index
Red	1.50
Violet	1.52

a. Calculate the angle of refraction for the red light. **(1 mark)**
b. Calculate the angle of refraction for the violet light. **(1 mark)**
c. Calculate the angle of the dispersed spectrum. **(1 mark)**
d. Calculate the speed of the violet light in the material. **(1 mark)**

▶ Question 3 (1 mark)

Dispersion is one of the pieces of evidence for why light acts as a wave.

Explain why this is.

▶ Question 4 (2 marks)

Diamond and perspex are cut to exactly the same shape. The same white light source is used to illuminate the shapes.

Why does the diamond appear to sparkle more than the perspex?

▶ Question 5 (2 marks)

White light is shone onto a rectangular prism.

Describe the difference between the light entering and leaving the block.

More exam questions are available in your learnON title.

2.5 Optical phenomena

KEY KNOWLEDGE

- Explain the formation of optical phenomena: rainbows; mirages

Source: VCE Physics Study Design (2023–2027) extracts © VCAA; reproduced by permission.

2.5.1 Rainbows

Rainbows are a common example of the dispersion of light. However, they are not only seen in the sky. You can also see a rainbow when you use a garden hose. Three conditions are necessary for a rainbow to be visible:

- The Sun needs to be visible.
- Some water droplets need to be airborne.
- An observer needs to be present.

For a rainbow to be seen, the usual arrangement of these three elements is to have the Sun oriented behind the observer and the airborne water to be present in front of the observer. The water droplets separate the colours in a similar way to that which occurs with the glass prism. The big difference is that, before the colours emerge from a water droplet, they are reflected from the opposite surface of the droplet.

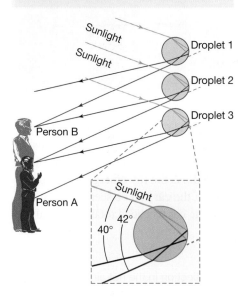

FIGURE 2.26 Each droplet of water in the air spreads the colours. Person A sees a rainbow between droplet 2 and droplet 3. Person B sees a rainbow between droplet 1 and droplet 2.

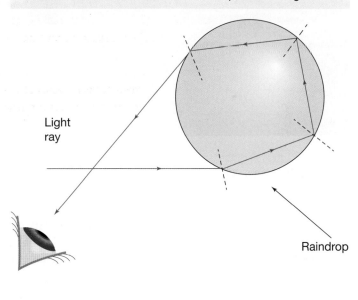

FIGURE 2.27 A light ray enters the bottom of the raindrop, is reflected twice off the wall of the raindrop, then emerges.

When you see a rainbow, each colour comes from a separate raindrop in the sky. If the red light from a raindrop enters your eye, then the violet light from that raindrop goes over your head to someone else. Each person sees his or her own personal rainbow. Your rainbow depends on raindrops in the sky being at a particular point so that the angle between you, the Sun and the raindrops is approximately 42°. The rainbow is not an image in the sky that everyone can see. In figure 2.27, the ray enters the eye at a higher angle than the primary rainbow. The colours are spread as they enter the raindrop and grow further apart the longer they are in the raindrop.

When the sky is very dark, a second, fainter rainbow may be visible on the outside of the bright one. This is due to the sunlight entering higher raindrops at the bottom and reflecting off the inside of the drop twice before emerging into the air.

2.5.2 Mirages

There are several types of mirages that can be seen when certain atmospheric conditions enable total internal reflection to occur. These mirages appear because the refractive index of air decreases with temperature.

A common type of mirage occurs in the desert or above a road on a sunny day. At ground level the air is hot with a refractive index close to 1. As height increases, the temperature of the air decreases and its refractive index increases.

FIGURE 2.28 A mirage on a country road

FIGURE 2.29 Mirages such as this are common on hot, sunny days.

Rays of light from a car, for example, go in all directions. The air above the ground can be considered as layers of air. The closer to the ground, the higher the temperature and the lower the refractive index. As a ray moves into hotter air, it bends away from the normal. After successive deflections, the angle of incidence exceeds the critical angle for air at that temperature and the ray is totally internally reflected. As the ray emerges, it follows a similar path, refracting towards the normal as it enters cooler air. An image of the car can be seen below street level, as shown in figure 2.30. The mirage is upside down because light from the car has been totally internally reflected by the hot air close to the road surface.

FIGURE 2.30 The mirage of the car appears upside down due to total internal reflection in the hot air close to the ground.

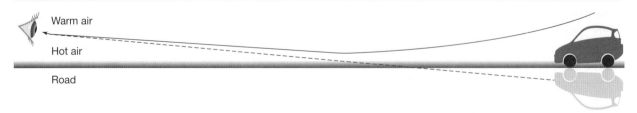

Another mirage that depends on layers of air at different temperatures is known as the 'Fata Morgana', in which vertical streaks, like towers or walls, appear. This occurs where there is a temperature inversion — very cold at ground level and warmer above — and very stable weather conditions.

The phenomenon is named after Morgan le Fay (Fata Morgana in Italian), who was a fairy and half-sister to King Arthur of the Celtic legend. She used mirages to show her powers and, in the Italian version of the legend, lived in a crystal palace under the sea. The mirage is often seen in the Strait of Messina and over arctic ice. As shown in figure 2.31, the light rays from a distant point are each refracted by the different layers of air, arriving at different angles to the eye. The effect is that the point source (P) becomes a vertically extended source, like a tower or wall. An example of the Fata Morgana, over an ice field in the Arctic Ocean off the coast of Svalbard, is shown in figure 2.32.

FIGURE 2.31 Ray paths for the Fata Morgana

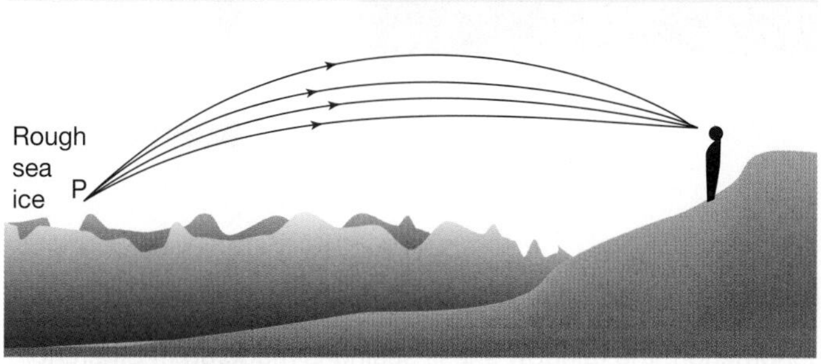

FIGURE 2.32 The Fata Morgana of Svalbard

2.5 Activities

learn on

Students, these questions are even better in jacPLUS

Receive immediate feedback and access sample responses

Access additional questions

Track your results and progress

Find all this and MORE in jacPLUS

| 2.5 Quick quiz | on | 2.5 Exercise | 2.5 Exam questions |

2.5 Exercise

1. In which direction would you need to look to see a rainbow early on a rainy morning, and from which direction should the rain be coming from?
2. **MC** When looking at a rainbow, what is the order of the colours from top to bottom?
   A. Red, orange, yellow, green, blue, indigo, violet
   B. Violet, indigo, blue, green, yellow, orange, red
   C. Blue, indigo, violet, green, red, orange, yellow
   D. Red, orange, yellow, blue, green, indigo, violet
3. The spectrum produced by a rainbow and by a prism look similar. Describe the differences between them.
4. **MC** A mirage occurs when
   A. the air is hot and the ground is cool.
   B. both the air and the ground are very hot.
   C. the ground is very hot and the air is cool.
   D. the air is slightly cooler than the ground.

5. Explain the optical phenomenon visible on the road through the bush that is shown in the following figure.

2.5 Exam questions

▶ Question 1 (3 marks)

Sophie and Emma are looking at a rainbow.

Sophie states that they are both seeing the same light from the same water droplets to form the rainbow.

Emma states that, while they are seeing the same phenomenon, the rainbow she is seeing is made up of different light from the one Sophie is seeing.

Determine who is correct. Justify your answer using theory.

▶ Question 2 (3 marks)

Dispersion is the phenomenon that explains the range of colours produced by a rainbow (whereas the natural formation of rainbows and their arc shape relates to total internal reflection).

Explain how a rainbow can be formed from a beam of white light.

▶ Question 3 (3 marks)

Give three conditions required to be able to observe a rainbow.

▶ Question 4 (2 marks)

After a storm, Anna walks along the Yarra River in the Royal Botanic Gardens. Looking to the west, she sees a rainbow that has formed over the National Gallery of Victoria.

Is it morning or evening? Explain how you know.

▶ Question 5 (1 mark)

MC A mirage is an example of
A. dispersion of light.
B. total internal reflection.
C. refraction of light and total internal reflection of light.
D. reflection of light and refraction of light.

More exam questions are available in your learnON title.

2.6 Review

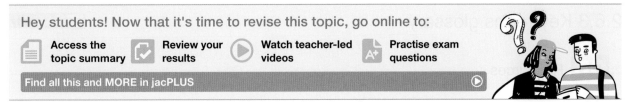

2.6.1 Topic summary

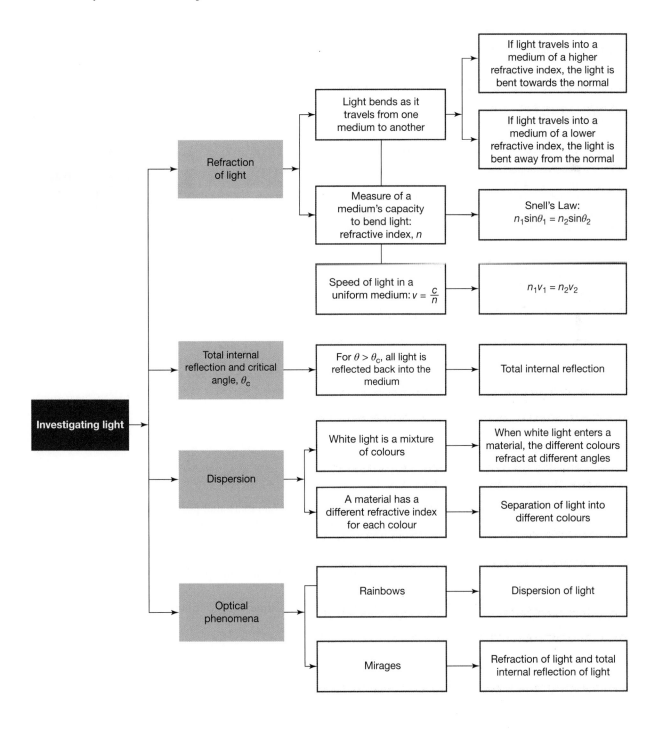

2.6.2 Key ideas summary

2.6.3 Key terms glossary

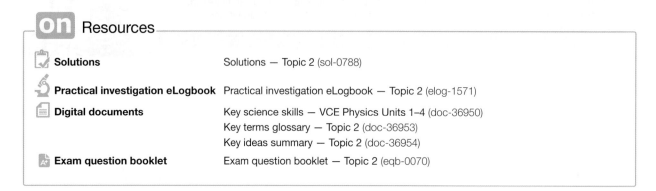

on Resources

☑ **Solutions**	Solutions — Topic 2 (sol-0788)
🔬 **Practical investigation eLogbook**	Practical investigation eLogbook — Topic 2 (elog-1571)
📄 **Digital documents**	Key science skills — VCE Physics Units 1–4 (doc-36950)
	Key terms glossary — Topic 2 (doc-36953)
	Key ideas summary — Topic 2 (doc-36954)
A⁺ **Exam question booklet**	Exam question booklet — Topic 2 (eqb-0070)

2.6 Activities

learn on

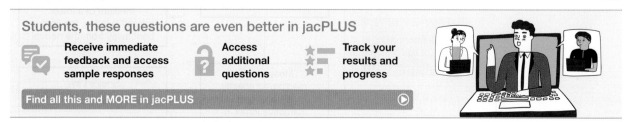

Students, these questions are even better in jacPLUS

Receive immediate feedback and access sample responses

Access additional questions

Track your results and progress

Find all this and MORE in jacPLUS ▶

2.6 Review questions

learn on

1. The following diagram shows a light source, S, placed in front of a plane mirror that is not shown. Two light rays, R_1 and R_2, that are reflected from the mirror are also shown.

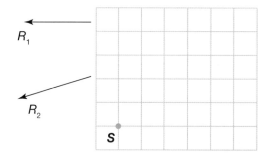

a. On a copy of the diagram, locate the position of the image S and mark that point with the letter I.
b. Draw in the location of the plane mirror, using a straight line to represent the mirror.

2. Blue light has a frequency of 6.5×10^{14} Hz, and yellow light has a frequency of 5.2×10^{14} Hz.

a. Determine the wavelength of both blue and yellow light in air. Take the speed of light in air to be 3.0×10^8 m s^{-1}.
b. Determine the wavelength of both blue and yellow light, this time with a refractive index of 1.50 for both colours.

3. Consider a ray of light incident on an air–perspex interface such that the angle of refraction is 16° when the angle of incidence is 22°.

   a. Calculate the refractive index of the perspex, assuming the refractive index of air is 1.00.
   b. If the angle of incidence is increased by 10°, what is the increase in the angle of refraction?
   c. Now consider a ray of light propagating from the perspex into the air with an angle of incidence of 22°. Calculate the angle of refraction and comment on your answer.

4. A group of students is investigating the refraction of light as it passes from a transparent block of carbon disulfide ($n = 1.63$) into very salty water ($n = 1.38$).

   a. Calculate the speed of light in carbon disulfide.

   Consider a ray of light passing from carbon disulfide into very salty water.

   b. Will the light speed up or slow down in the very salty water compared to its speed in carbon disulfide?
   c. Will the light bend towards the normal or away from the normal at the interface between carbon disulfide and the very salty water?
   d. If the angle of incidence is 45°, calculate the angle of refraction.

5. Consider a ray of white light incident on an air–flint glass interface with an angle of incidence of 25°. The refractive index for red light in flint glass is 1.571 and, for blue light, the refractive index is 1.594. The refracted beam disperses as it passes into the glass.

   a. Explain what is meant by the term *disperses* in this context.
   b. Determine the angle between the refracted red light and the refracted blue light in the flint glass.
   c. The angle of incidence is now increased from 25° to 50°. What is the angle between the red and blue refracted rays now?

6. A beam of white light is fired at a triangular prism as shown in the following diagram. The dotted line gives the direction of the incident beam as a reference line for your drawing.

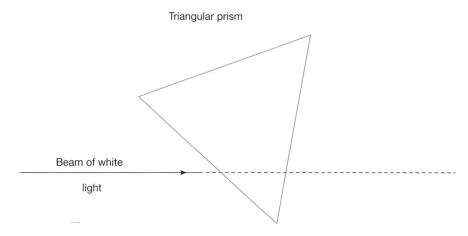

Triangular prism

Beam of white light

By drawing onto a copy of the diagram, show how the white light is dispersed as it passes through the prism. Clearly indicate on your diagram both the red and violet ends of the spectrum produced by white light passing through the prism.

7. Consider three different types of glass: X, Y and Z. Their refractive indices are $n_X = 1.53$, $n_Y = 1.55$ and $n_Z = 1.57$ respectively.

   a. Of the three different types of glass, state the glass with the smallest critical angle.
   b. Using your choice for part a, determine the critical angle for light.
   c. Now, consider glass Z making an optical interface with glass X. Determine the critical angle for this arrangement.

2.6 Exam questions

▶ Question 1

Source: VCE 2021 Physics Exam, Section A, Q.15; © VCAA

A Physics class is investigating the dispersion of white light using a triangular glass prism.

Which one of the following diagrams best shows the principle of dispersion?

A.

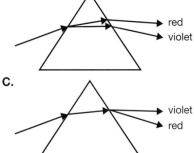

red
violet

B.

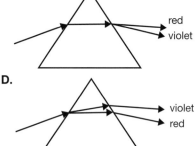

red
violet

C.

violet
red

D.

violet
red

▶ Question 2

A group of students is trying to measure the speed of light on a sports oval. They have placed a mirror 120 m from a laser source and light sensor, both of which are connected to a timer. The timer starts when the laser is switched on and stops when a pulse arrives at the sensor. Light from the laser strikes the mirror and is reflected directly back to the sensor.

Which one of the following do the students accurately measure the time to be?

A. 4.0×10^{-7} s

B. 1.6×10^{-6} s

C. 8.0×10^{-6} s

D. 8.0×10^{-7} s

▶ Question 3

In a practical session, students use a semicircular prism to investigate the refraction of light. When a narrow beam of light has an angle of incidence of 35.0°, the angle of refraction is 26.0°.

Which one of the following is the refractive index of the prism?

A. 1.31   **B.** 1.28   **C.** 0.75   **D.** 0.85

▶ Question 4

A ray of light inside a rectangular block of glass (n_{glass} = 1.53) makes an angle of incidence of 47.0°. The rectangular block is immersed in water (n_{water} = 1.33).

What is the angle of refraction for the ray of light as it exits into the water from the glass block?

A. 39.8°   **B.** 57.3°   **C.** 43.6°   **D.** 48.7°

Question 5

Consider two optical fibres each of length 10 m. Optical fibre X has a core refractive index of 1.54, whereas optical fibre Y has a core refractive index of 1.58.

Which one of the following statements is correct regarding the propagation of pulses of light sent down both fibres?

A. It will take a shorter amount of time for the light pulse to pass down optical fibre Y than optical fibre X.

B. It will take the same amount of time, as the fibres have the same length and the speed of light in both fibres is the same for all observers.

C. It will take a shorter amount of time for the light pulse to pass down optical fibre X than optical fibre Y.

D. It will take a different amount of time but one that is not possible to determine.

Question 6

Students are asked to calculate the critical angle for light propagating inside a transparent crystal. Their teacher tells them that the refractive index for the crystal is 1.387.

Which one of the following is the critical angle for this crystal when placed in air?

A. 47.3°   B. 45.4°   C. 48.5°   D. 46.1°

Question 7

White light is incident on an air–glass interface. Due to dispersion, the white light separates into colours.

Which one of the following statements is correct?

A. Blue light will have an angle of refraction that is greater than the angle of refraction of red light.

B. Red light will have an angle of refraction that is greater than the angle of refraction of green light.

C. Green light will have an angle of refraction that is greater than the angle of refraction of yellow light.

D. Orange light will have an angle of refraction that is greater than the angle of refraction of red light.

Question 8

If the refractive index of the plastic coating surrounding an optical fibre is 1.49 and the refractive index of glass is 1.52, then the closest value for the critical angle for glass to plastic is

A. 83.7°.

B. 11.4°.

C. 78.6°.

D. 84.3°.

Question 9

If the refractive index of an ethanol solution is 1.3425, then the closest value for the speed of light in this solution is

A. $2.2 \times 10^8 \text{ m s}^{-1}$.

B. $1.2 \times 10^8 \text{ m s}^{-1}$.

C. $2.0 \times 10^8 \text{ m s}^{-1}$.

D. $2.1 \times 10^8 \text{ m s}^{-1}$.

Question 10

A mirage occurs when

A. the air is slightly cooler than the ground.
B. the air is cool and the ground is very hot.
C. both the air and the ground are very hot.
D. the air is very hot and the ground is cool.

Section B — Short answer questions

Question 11 (2 marks)

A block of an unidentified clear material is completely immersed in glycerol, which has a refractive index of 1.467. A light ray passes from the glycerol into the block at an angle of incidence of 70.0° and the angle of refraction is measured to be 62.5°.

Determine the index of refraction of the clear material.

Question 12 (5 marks)

Source: VCE 2021 Physics Exam, Section B, Q.12; © VCAA

A Physics teacher is conducting a demonstration involving the transmission of light within an optical fibre. The optical fibre consists of an inner transparent core with a refractive index of 1.46 and an outer transparent cladding with a refractive index of 1.42. A single monochromatic light ray is incident on the optical fibre, as shown in Figure 12.

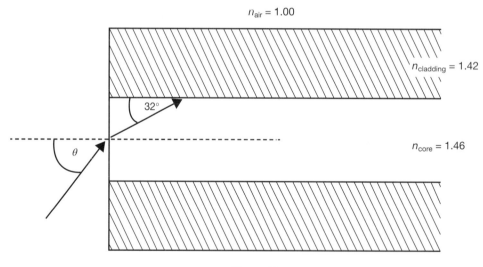

Figure 12

a. Determine the angle of incidence, θ, at the air–core boundary. Show your working. **(2 marks)**
b. Will any of the initial light ray be transmitted into the cladding? Explain your answer and show any supporting working. **(3 marks)**

Question 13 (7 marks)

A narrow beam of white light makes an angle of incidence with a rectangular prism having a width of 5.0 cm, as shown in the following diagram. The refractive index for red light in the prism is $n_{red} = 1.530$, and for blue light the refractive index is $n_{blue} = 1.548$.

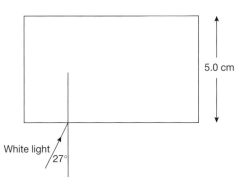

White light
27°
5.0 cm

a. On a copy of the diagram, carefully draw rays to indicate the propagation of red and blue light through the block of glass. For each path, indicate two points where refraction occurs and one where a reflection also occurs. **(2 marks)**

The narrow beam of white light makes an angle of incidence with the air–prism interface of 27°.

b. Determine the angle of refraction for red light. **(2 marks)**

c. Consider the time it takes for red light and blue light to propagate through the prism. **(3 marks)**

Which colour would take the least time? Or does it take the same time?

Question 14 (8 marks)

Optical fibres consist of transparent materials having an inner core and an outer cladding with different refractive indices. The following diagram shows a typical optical fibre. The refractive index of the core is 1.58 and the critical angle for light at the core–cladding boundary is 84°.

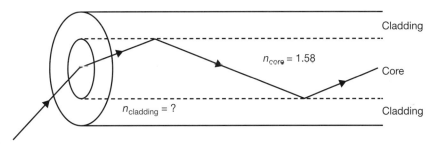

Cladding
$n_{core} = 1.58$
Core
$n_{cladding} = ?$
Cladding

a. For the fibre to operate correctly, would the refractive index of the cladding be greater than, the same size as, or smaller than the refractive index of the core? Explain your choice without a calculation. **(2 marks)**

b. Calculate the refractive index of the cladding, $n_{cladding}$. **(2 marks)**

c. Determine the speed of light in the core of the fibre optic. **(1 mark)**

d. In one instance, light of wavelength 400 nm is shone from air into the optical fibre.

Calculate the wavelength of the light in the core of the optical fibre. **(3 marks)**

Question 15 (3 marks)

A beam of light with an incident angle of 40.0° strikes an air–kerosene interface. The beam of light bends towards the normal with a direction change of 12.5°.

Determine the absolute refractive index of the oil.

3 Thermal energy and its interaction with electromagnetic radiation

KEY KNOWLEDGE

In this topic, you will:
- convert between Celsius and Kelvin scales
- describe temperature with reference to the average translational kinetic energy of the atoms and molecules within a system
  - distinguish between conduction, convection and radiation with reference to heat transfers within and between systems
  - explain why cooling results from evaporation using a simple kinetic energy model
- investigate and analyse theoretically and practically the energy required to:
  - raise the temperature of a substance: $Q = mc\Delta T$
  - change the state of a substance: $Q = mL$
- apply concepts of energy transfer, energy transformation, temperature change and change of state to climate change and global warming.

Source: VCE Physics Study Design (2023–2027) extracts © VCAA; reproduced by permission.

PRACTICAL WORK AND INVESTIGATIONS

Practical work is a central component of VCE Physics. Experiments and investigations, supported by a **practical investigation eLogbook** and **teacher-led videos**, are included in this topic to provide opportunities to undertake investigations and communicate findings.

EXAM PREPARATION

▶ Access exam-style questions and their video solutions in every lesson, to ensure you are ready.

3.1 Overview

3.1.1 Introduction

In this topic you will revisit your understanding of energy, utilise your knowledge of the particle theory and apply practical skills of measuring temperature. By the end of this topic you should be able to describe how the energy within a system is shared and transferred. You will be able to use the particle theory to explain internal energy, temperature and heat transfer. You will describe and explain internal energy as the random motion of particles in a substance. You will test and manipulate the relationships that dictate temperature changes in a substance (specific heat capacity) and changes of state (latent heat), and consider how these concepts are engineered to our benefit. You will also apply many of these concepts to your understanding of climate change and climate warming, and you will be able to explain how atmospheric gases create the greenhouse effect.

FIGURE 3.1 A cold day in two temperature scales. A balloon caught on a twig shrinks as the translational kinetic energy of the air inside is transferred to the surrounding air, losing the ability to stop the air pressure pushing it in.

LEARNING SEQUENCE

 Resources

3.2 Explaining heat using the kinetic theory

KEY KNOWLEDGE

- Convert between Celsius and Kelvin scales
- Describe temperature with reference to the average translational kinetic energy of the atoms and molecules within a system

Source: VCE Physics Study Design (2023–2027) extracts © VCAA; reproduced by permission.

3.2.1 What is heat?

It may surprise you that **heat** was not considered a form of energy until quite late in human history. Fire was one of the four elements in ancient times and it was clear that when substances combusted, heat was released. Thus, for centuries heat was considered a substance that could be transferred between objects and this was the starting point for scientific thought. In 1667 Johann Joachim Becher proposed that a substance he called *phlogiston* was released during combustion. Later, in 1789, the French scientist Antoine Lavoisier published a treatise on chemistry in which he described heat as an invisible, tasteless, odourless, weightless fluid that he called *calorific fluid*, which flowed from hot to cold objects. Although now defunct as an explanation of heat, the measure of the calorie is still used.

> **heat** the transfer of energy from one body to another due to a temperature difference
>
> **kinetic energy** the energy associated with the movement of objects; like all forms of energy, it is a scalar quantity
>
> **gravitational potential energy** the energy stored in an object as a result of its position relative to another object to which it is attracted by the force of gravity
>
> **joule** the SI unit of work or energy; 1 joule is the energy expended when a force of 1 newton acts through a distance of 1 metre

In 1798 Benjamin Thompson, later to be called Count Rumford, conducted an experiment on the nature of heat. The barrel of a cannon is made by drilling a cylindrical hole in a solid piece of metal. Rumford observed the piece of metal and the drill became quite hot. He devised an experiment to investigate the source of the heat and how much heat is produced. Rumford put the drill and the end of the cannon in a wooden box filled with water. He measured the mass of water and the rate at which the temperature rose. He showed that the amount of heat produced was not related to the amount of metal that was drilled out. He concluded that the amount of heat produced depended only on the work done against friction. He said that *heat was in fact a form of energy*, not an invisible substance that was transferred from hot objects to cold objects. Instead, a hot object had heat energy, in the same way as a moving object has **kinetic energy** or an object high off the ground has **gravitational potential energy**.

Rumford's ideas about heat were not adopted for a few decades, but in 1840 James Prescott Joule conducted a series of experiments to find a quantitative link between mechanical energy and heat. In other words, how much energy is required to increase the temperature of a mass by 1 °C?

Joule used different methods and compared the results.

- *Using gravity.* A falling mass spins a paddle wheel in an insulated barrel of water, raising the temperature of the water.
- *Using electricity.* Mechanical work is done turning a dynamo to produce an electric current in a wire, which heats the water.
- *Compressing a gas.* Mechanical work is used to compress a gas, which raises the gas's temperature.
- *Using a battery.* Chemical reactions at the battery terminals produce a current, which heats the water.
- *Using gravity.* Measure the temperature of water at the top and bottom of a waterfall.

Joule obtained approximately identical answers for all methods. This confirmed heat as a form of energy. To honour his achievement, the SI unit of energy is the **joule** (J).

The unit joule is used to measure the:
- kinetic energy of a runner
- light energy in a beam
- chemical energy stored in a battery
- electrical energy in a circuit
- potential energy in a lift on the top floor
- heat energy when water boils.

One joule is approximately the amount of energy needed to lift a 100 gram apple through a height of 1 metre.

The usual metric prefixes make the use of the unit joule more convenient. For example:

$$1\text{ kJ (kilojoule)} = 10^3 \text{ J} \quad 1\text{ MJ (megajoule)} = 10^6 \text{ J} \quad 1\text{ GJ (gigajoule)} = 10^9 \text{ J}$$

The chemical energy available from a bowl of breakfast cereal is usually hundreds of thousands of joules and is more likely to be listed on the packet in kilojoules. The amount of energy needed to boil an average kettle full of cold water is about 500 kJ.

Examples of 1 joule include the:
- kinetic energy of a tennis ball moving at about 6 m s^{-1}
- heat energy needed to raise the temperature of 1 gram of dry air by 1 °C
- heat energy needed to raise the temperature of 1 gram of water by 0.24 °C
- change in gravitational potential energy when an apple falls 1 metre to the ground
- amount of sunlight hitting a square centimetre every 10 seconds when the Sun is directly above
- amount of sound energy entering your eardrum at a loud concert over 3 hours
- amount of electrical energy used by an LED screen while on standby every second
- energy released by the combustion of 18 micrograms of methane.

3.2.2 Linking energy and heat: the kinetic theory of matter

Joule showed that a measure of energy was directly related to a change in temperature. But what happens to this energy within a material? The kinetic theory of matter enables us to explain the effect of energy on a material.

The kinetic theory of matter, which considers all objects as assemblies of particles in motion, is an old one, first described by Lucretius in 55 AD. The kinetic view of matter was developed over time by Hooke, Bernoulli, Boltzmann and Maxwell.

The evidence for the existence of particles includes the following:
- Gases and liquids diffuse; that is, a combination of two gases or two liquids quickly becomes a mixture — for example, a dye spreading in water. Even solids can diffuse; if a sheet of lead is clamped to a sheet of gold, over time the metals merge to a depth of a few millimetres.
- The mixing of two liquids gives a final volume that is less than the sum of their original volumes.
- A solid dissolves in a liquid.

The kinetic theory of matter assumes that:
- all matter is made up of particles in constant, random and rapid motion
- there is space between the particles.

FIGURE 3.2 Iodine crystals sublimate (turn directly into a gas) when heated.

This diagram shows a gas jar with iodine crystals.

As the crystals warm up, they produce a purple gas that diffuses throughout the jar.

After a long period of time, the crystals have completely sublimated.

The energy associated with the motion of the particles in an object is called the internal energy of the object. It is the sum of all kinetic energy of the particles that make up that object. The particles can move and interact in many ways, so there are a number of contributions to the internal energy.

Gases

In a gas made up of single atoms, such as helium, the atoms move around, randomly colliding with each other and the walls of the container. So, each atom has some **translational kinetic energy** (see figures 3.3 and 3.4). Note that the constant bombardment of the walls of the container by the gas particles is responsible for the pressure exerted by the gas in all directions. The collisions involving the gas particles are considered to be perfectly elastic (no net loss of kinetic energy involved).

translational kinetic energy the energy due to the motion of an object from one location to another

vibrational kinetic energy the energy due to the vibrational motion of objects

rotational kinetic energy the energy due to the rotational motion of objects

However, if the gas is made up of molecules with two or more atoms, the molecules can also stretch, contract and spin, so these molecules also have other types of kinetic energy called **vibrational kinetic energy** and **rotational kinetic energy** (see figure 3.4).

FIGURE 3.3 Moving single atoms have translational kinetic energy.

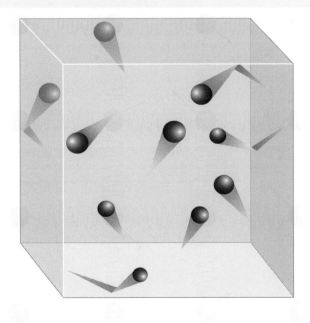

FIGURE 3.4 The movements of a molecule

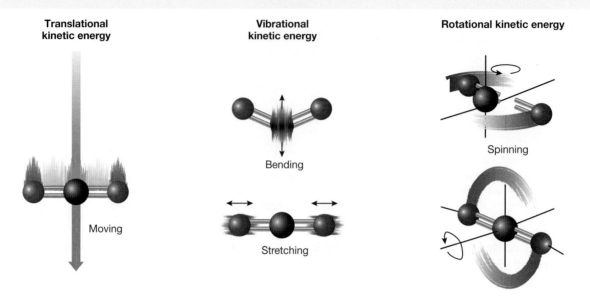

Translational kinetic energy

Moving

Vibrational kinetic energy

Bending

Stretching

Rotational kinetic energy

Spinning

Liquids

Like a gas, molecules in a liquid are free to move (they can slide past each other), but within the confines of the surface of the liquid. There is some attraction between molecules, which means there is some energy stored as molecules approach each other. Stored energy is called potential energy. It is the energy that must be overcome for a liquid to evaporate or boil.

Solids

In a solid, atoms jiggle rather than move around. They have kinetic energy, but they also have a lot of potential energy stored in the strong attractive **force** that holds the atoms together. This means that a lot of energy is required to melt a solid. This is seen in figure 3.5.

force an interaction between two objects that can cause a change

FIGURE 3.5 In a solid, atoms can jiggle around an essentially fixed location.

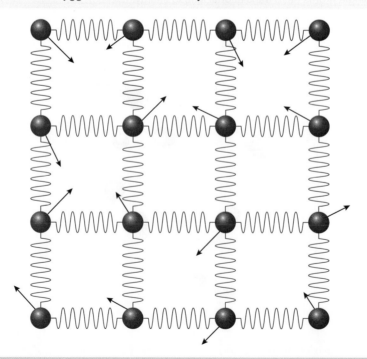

When the internal energy of a substance increases, this means the particle movement increases. Heating or cooling is reflected in a change in the particle movement. Heat energy will always transfer from a hotter substance to a colder substance as the collisions of the faster-moving particles transfer translational kinetic energy to the slower-moving particles.

TABLE 3.1 The internal energy of different objects

	Internal energy	
	Movement that is not related to temperature	**Movement that is related to temperature**
Atoms in a gas	None	Moving and colliding
Molecules in a gas	Spinning, stretching, compressing and bending	Moving and colliding
Molecules in a liquid	Spinning, stretching, compressing and bending	Moving and colliding
Atoms in a solid	Pulling and pushing	Jiggling
Energy types	Other types of kinetic energy, potential energy	Translational kinetic energy

3.2.3 Measuring and converting temperature

Measuring temperature

Temperature is generally thought of as a measure of how hot or how cold something is, expressed on a scale, and we have seen that heating or cooling something respectively increases or decreases particle movements. Thus, **temperature** is, more specifically, proportional to the average translational kinetic energy of particles.

Temperature is a measure of the average translational kinetic energy of particles.

The other contributions to the internal energy do not affect the temperature. This becomes important when materials melt or boil because the added heat must go somewhere, but the temperature does not change. In fact, the added heat is involved in breaking bonds between particles.

Measuring temperature is a relatively simple process. Our bodies tell us when it is hot or cold. Our fingers warn us when we touch a hot object.

Thermometers were designed as a way to measure temperature accurately. A good thermometer needs a material that changes in a regular and measurable way as its temperature changes. Many materials, including water, expand when heated, so the first thermometer, built in 1630, used water in a narrow tube with a filled bulb at the bottom. The water rose up the tube as the bulb was warmed.

German physicist Daniel Fahrenheit replaced the water with mercury in 1724. Liquid thermometers now use alcohol with a dye added. Fahrenheit developed a scale to measure the temperature, using the lowest temperature he could reach, an ice and salt mixture, as zero degrees, and the temperature of the human body as 100 degrees. Fahrenheit also showed that a particular liquid will always boil at the same temperature. Swedish astronomer Anders Celsius developed another temperature scale in 1742, which is the one we use today. Celsius used melting ice and steam from boiling water at atmospheric pressure to define 0 °C and 100 °C for his scale.

A third temperature scale was proposed in 1848 by William Thomson, later to be ennobled as Lord Kelvin. He proposed the scale based on the better understanding of heat and temperature that had developed by that time (this is discussed later in this section). This scale uses the symbol 'K' to stand for 'kelvin'.

temperature a measure of the average translational kinetic energy of particles

Other materials, including gases and metals, also expand with temperature and are used as thermometers. A bimetallic strip is two lengths of different metals, usually steel and copper, joined together. The two metals expand at different rates, so the strip bends one way as the temperature rises, or the other as it cools. A bimetallic strip can be used as a thermometer, a thermostat or as a compensating mechanism in clocks.

Properties that change with temperature and can be employed in designing a thermometer are:
- the electrical resistance of metals, which increases with temperature
- the electrical voltage from a thermocouple, which is two lengths of different metals with their ends joined; if one end is heated, a voltage is produced (see figure 3.6)
- colour change; liquid crystals change colour with temperature (see figure 3.7)
- the colour emitted by a hot object; in steelmaking, the temperature of hot steel is measured by its colour (see figure 3.8).

FIGURE 3.6 This thermocouple is connected to a voltmeter, which reads differing voltages as the thermocouple changes temperature.

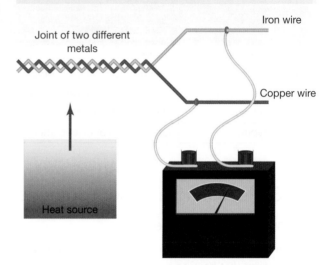

FIGURE 3.7 This liquid crystal thermometer indicates body temperature when the liquid crystals change colour. The thermometer is registering 37 °C.

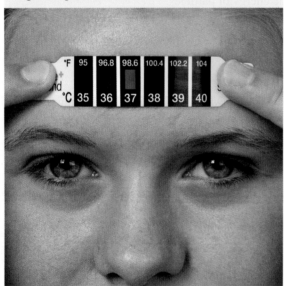

FIGURE 3.8 This steel is nearly 1000 °C and has recently been poured into a mould to shape it. The steel will continue to glow until it has cooled to about 400 °C.

The Kelvin scale

The kinetic theory of matter is the origin of the Kelvin temperature scale. If temperature depends on the movement of particles, then the slower they move, the lower the temperature. When the particles stop moving, the temperature will be the lowest that is physically possible. This temperature was adopted as **absolute zero**.

But how do we measure it and what is its value?

In the early 1800s gases were a good material to work with to explore the nature of matter. An amount of gas in a glass vessel could be heated and the variables of temperature, volume and pressure to keep the volume fixed could be easily measured. Joseph Gay-Lussac and Jacques Charles independently investigated how the volume of gases changed with temperature if they were kept under a constant pressure. They found that all gases kept at constant pressure expand or contract by $\frac{1}{273}$ of their volume at 0 °C for each Celsius degree rise or fall in temperature.

From that result you can conclude that if you cooled the gas, and it stayed as a gas and did not liquefy, you could cool it to a low enough temperature that its volume reduced to zero. The temperature would be absolute zero. According to their experiments, absolute zero was −273 °C. Nowadays more accurate experiments put the value at −273.15 °C.

absolute zero the lowest temperature that is physically possible, equal to 0 K or approximately −273 °C; at this temperature, particles cease to vibrate

In kelvin, absolute zero is 0 K. The increments in the Kelvin temperature scale are the same size as those in the Celsius scale, so if the temperature increased by 5 °C, it also increased by 5 K.

The conversion formula between the two temperature scales is:

$$\text{kelvin} = \text{Celsius} + 273$$

$$T_{(\text{kelvin})} = T_{(\text{Celsius})} + 273$$

FIGURE 3.9 By extrapolating his trend line back, Kelvin was able to establish absolute zero at −273 °C. With modern equipment, absolute zero is now determined to be −273.15 °C.

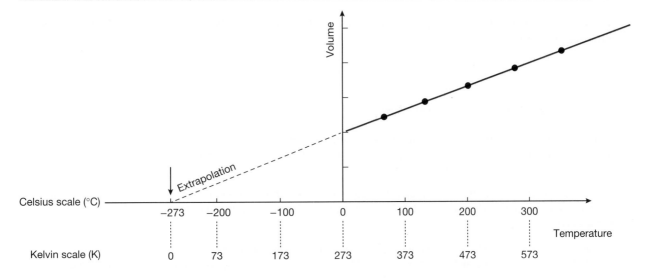

TABLE 3.2 Some temperatures on the Kelvin and Celsius scales

Event	Temperature	
	K	°C
Absolute zero	0	−273
Helium gas liquefies	4	−269
Lead becomes a superconductor	7	−266
Nitrogen gas liquefies	63	−210
Lowest recorded air temperature on Earth's surface (Vostok, Antarctica)	184	−89
Mercury freezes	234	−39
Water freezes	273	0
Normal human body temperature	310	37
Highest recorded air temperature on Earth's surface (Death Valley, USA)	330	57
Mercury boils	630	357
Iron melts	1535	1262
Surface of the Sun	5778	5505

Note: In 1968, the international General Conference on Weight and Measures decided that kelvin temperatures do not use the ° symbol, unlike Celsius and Fahrenheit temperatures.

tlvd-0003

SAMPLE PROBLEM 1 Converting from degrees Celsius to kelvin

What is the kelvin temperature at which ice melts?

THINK	WRITE
1. Recall the relationship between the two temperature scales.	$T_{(kelvin)} = T_{(Celsius)} + 273$
2. Ice melts at 0 °C. Substitute 0 °C into the conversion formula.	$T_{(kelvin)} = T_{(Celsius)} + 273$ $= 0 + 273$ $= 273 \text{ K}$
3. State the solution.	Ice melts at 273 K.

PRACTICE PROBLEM 1

The temperature of the universe predicted by the cosmic microwave background is 3 K. What is this in degrees Celsius?

EXTENSION: The lowest temperature

In October 2021 scientists in Germany, for a few seconds, created the coldest temperature ever recorded in a lab, reducing the temperature of a cloud of rubidium atoms to 38 trillionths of a degree above absolute zero. They did this by allowing the cloud of atoms to expand rapidly down a cooling tower.

There are no thermometers that can actually measure such a low temperature. In this case, the scientists used the kinetic energy of the atoms to calculate the temperature. The average speed of the atoms was less than 65 micrometres per second.

By comparison, the coldest place observed so far in the universe is the Boomerang Nebula, where the temperature is just 1 degree above absolute zero. At absolute zero no energy can be removed from particles.

3.2 Activities

3.2 Quick quiz on	3.2 Exercise	3.2 Exam questions

3.2 Exercise

1. Why is the Celsius scale of temperature commonly used rather than the Kelvin scale?
2. What is the main advantage of an absolute scale of temperature?
3. Estimate each of the following temperatures in kelvin.
   a. The maximum temperature in Melbourne on a hot summer's day
   b. The minimum temperature in Melbourne on a cold, frosty winter's morning
   c. The current room temperature
   d. The temperature of cold tap water
   e. The boiling point of water
4. The temperature of very cold water in a small test tube is measured with a large mercury-in-glass thermometer. The temperature measured is unexpectedly high. Suggest a reason for this.
5. Carbon dioxide sublimates, that is, goes directly from solid to gas, at −78.5 °C. What is this temperature in kelvin?
6. The temperature of the surface of Mars was measured by the Viking lander and ranged from 256 K to 166 K. What are the equivalent temperatures in degrees Celsius?
7. Use the particle theory to explain why you can smell what's for dinner from the front door of your house.
8. Explain the difference between translational kinetic energy and other types of internal energy.
9. Use the particle theory to explain what happens when your cup of tea cools down over time.
10. Cling film on a warm bowl of soup placed in the fridge gets sucked down when it cools. Use the particle theory to explain this.
11. James Joule showed that mechanical energy could be transformed into the internal energy of a substance or object. The temperature of a nail, for example, can be raised by hitting it with a hammer. List as many examples as you can of the use of mechanical energy to increase the temperature of a substance or object.
12. Explain, in terms of the kinetic particle model, why a red-hot pin dropped into a cup of water has less effect on the water's temperature than a red-hot nail dropped into the same cup of water.
13. Explain why energy is transferred from your body into the cold sea while swimming even though you have less internal energy than the surrounding cold water.

3.2 Exam questions

Question 1 (1 mark)

MC The boiling point of water is

A. 100 °F.
B. 100 K.
C. 0 °C.
D. 100 °C.

Question 2 (5 marks)

Temperature scales allow us to compare the average kinetic energy of different materials.
a. Describe what the Celsius and Kelvin scales are based on. **(2 marks)**
b. What is similar about both the Celsius and Kelvin scales? **(1 mark)**
c. Why is the Kelvin scale an absolute scale? **(1 mark)**
d. How are the units used in the Kelvin scale different to those used in the Celsius and Fahrenheit scales? **(1 mark)**

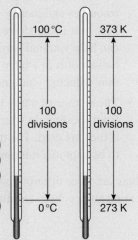

Question 3 (1 mark)

MC Absolute zero (0 K) is equivalent to (rounded off)

A. −273 °C.
B. 273 °C.
C. −373 °C.
D. 0 °F.

Question 4 (3 marks)

Explain, using the kinetic theory of matter, why a solid must be heated in order to turn it into a liquid.

Question 5 (2 marks)

When a tennis ball is hit it is momentarily squished by the contact with the racquet.

Explain what happens to the temperature of the air inside a tennis ball in this moment.

More exam questions are available in your learnON title.

3.3 Transferring heat

KEY KNOWLEDGE

- Distinguish between conduction, convection and radiation with reference to heat transfers within and between systems

Source: VCE Physics Study Design (2023–2027) extracts © VCAA; reproduced by permission.

From our understanding of the particle theory and its relationship to heat we know that energy is always transferred from a region of higher temperature to a region of lower temperature.

There are many situations in which it is necessary to control the rate at which the energy is transferred:

- Warm-blooded animals, including humans, need to maintain their body temperature in hot and cold conditions. Cooling of the body must be reduced in cold conditions. In hot conditions, it is important that cooling takes place to avoid an increase in body temperature.
- Keeping your home warm in winter and cool in summer can be costly, both in terms of energy resources and money. Applying knowledge of how heat is transferred from one place to another can help you find ways to reduce how much your house cools in winter and heats up in summer, thus reducing your energy bills.
- The storage of many foods in cold temperatures is necessary to keep them from spoiling. In warm climates most beverages are enjoyed more if they are cold. The transfer of heat from the warmer surroundings needs to be kept to a minimum.

There are three different processes through which energy can be transferred during heating and cooling: **conduction**, **convection** and **radiation** (see figure 3.11).

FIGURE 3.10 Reducing heat transfer to and from buildings saves precious energy resources, and reduces gas and electricity bills.

conduction the transfer of heat through a substance as a result of collisions between neighbouring vibrating particles

convection the transfer of heat in a fluid (a liquid or gas) as a result of the movement of particles within the fluid

radiation heat transfer without the presence of particles

FIGURE 3.11 Heat is transferred by conduction, convection and radiation.

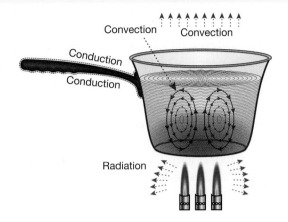

3.3.1 Conduction

Conduction is the transfer of heat through a substance as a result of collisions between neighbouring vibrating particles. The particles in the higher temperature region have more random kinetic energy than those in the lower temperature region. As shown in figure 3.12, the more energetic particles collide with the less energetic particles, giving up some of their kinetic energy. This transfer of kinetic energy from particle to particle continues until thermal equilibrium is reached. There is no net movement of particles during the process of conduction.

Solids are better **conductors** of heat than liquids and gases. In solids, the particles are more tightly bound and closer together than in liquids and gases. Thus, kinetic energy can be transferred more quickly. Metals are the best conductors of heat because free electrons are able to transfer kinetic energy more readily to other electrons and atoms.

conductor a material that contains charge carriers

insulator a substance that does not readily allow the passage of heat

FIGURE 3.12 Conduction is the transfer of thermal energy (heat) due to collisions between neighbouring particles.

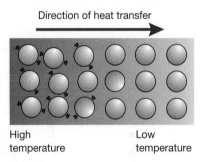

Materials that are poor conductors are called **insulators**. Materials such as polystyrene foam, wool and fibreglass batts are effective insulators because they contain pockets of still air. Air is a very poor conductor of heat. If air is free to move, however, heat can be transferred by a different method — convection.

FIGURE 3.13 a. Insulators like rock mineral wool have excellent thermal and acoustic properties. **b.** Fibreglass insulation is made from recycled glass bottles, sand and other materials.

3.3.2 Convection

Convection is the transfer of heat through a substance as a result of the movement of particles between regions of different temperatures. Convection takes place in liquids and gases, where particles are free to move around. In solids, the particles vibrate about a fixed position and convection does not generally occur, except under specific conditions (e.g. in the extreme temperature and pressure in Earth's mantle). Gases and liquids expand when heated, as their particles move further apart and thus become less dense. This reduction in density results in the warmer, less dense fluid rising.

The movement of particles during convection is called a **convection current**. Faster-moving particles in hot regions rise while slower-moving particles in cool regions fall. The particles in the warm water near the flame in figure 3.14 are moving faster and are further apart than those in the cooler water further from the flame. The cooler, denser water sinks, forcing the warm, less dense water upwards. This process continues as the warm water rises, gradually cools and eventually sinks again, replacing newly heated water.

FIGURE 3.14 Purple particles from a crystal of potassium permanganate carefully placed at the bottom of the beaker are forced around the beaker by convection currents in the heated water.

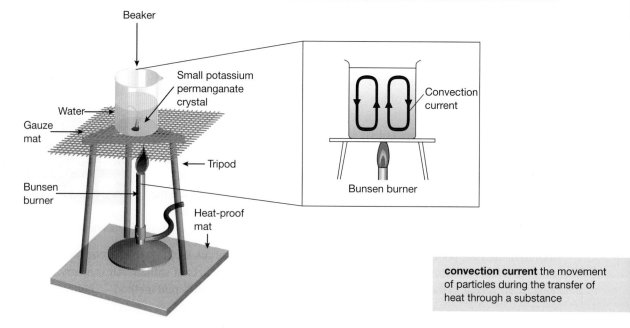

convection current the movement of particles during the transfer of heat through a substance

Convection currents are apparent in ovens that do not have fans. As the air circulates, the whole oven becomes hot. However, the top part of the oven always contains the hottest, least dense air. As the air cools, it sinks and is replaced by less dense hot air for as long as the energy source at the bottom of the oven remains on. Fans can be used to push air around the oven, providing a more even temperature.

Home-heating systems use convection to move warm air around. Ducted heating vents are, where possible, located in the floor. Without the aid of powerful fans, the warm air rises and then circulates around the room until it cools and sinks, being replaced with more warm air. In homes built on concrete slabs, ducted heating vents are in the ceiling. Fans are necessary to push the warm air downwards so that it can circulate more efficiently.

FIGURE 3.15 Convection currents circulate warm air pushed out by home-heating systems. The warm air rises and then circulates around the room until it cools and sinks, being replaced with more warm air.

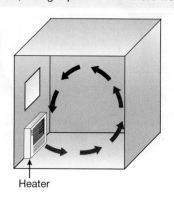

In summer, loose-fitting clothing is more comfortable because it allows air to circulate. Thus, heat can be transferred from your body by convection as the warm air near your skin rises and escapes upwards.

Convection, with its actual movement of particles, is more efficient than conduction to transfer heat, as with conduction, particles only vibrate at their position.

3.3.3 Radiation

Heat can be transferred without the presence of particles by the process of radiation. As outlined in subtopic 1.4, all objects with a temperature above absolute zero (0 K) emit **electromagnetic radiation** through the process of blackbody radiation. Visible light, microwaves, infrared radiation, ultraviolet radiation and x-rays are all examples of electromagnetic radiation. All electromagnetic radiation is transmitted through empty space at a speed of 3.0×10^8 m s^{-1}, which is most commonly known as the speed of light. Note that light in air is 1.0003 times slower than light in a vacuum. Thus, even in the atmosphere, the transfer of heat by radiation is extremely rapid and is in fact the fastest process of heat transfer.

Electromagnetic radiation can be absorbed by, reflected from or transmitted through substances. Scientists have used a wave model to explain much of the behaviour of electromagnetic waves. These electromagnetic waves transfer energy, and reflect and refract in ways that are similar to waves on water.

FIGURE 3.16 Radiant heat can be transmitted, absorbed or reflected.

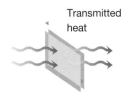

Transmitted radiant heat
Clear objects, like glass, allow light and radiant heat to pass through them. The temperature of these objects does not increase quickly when heat reaches them by radiation.

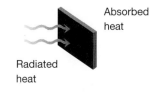

Absorbed radiant heat
Dark-coloured objects tend to absorb light and radiant heat. Their temperatures increase quickly when heat reaches them by radiation.

Reflected radiant heat
Shiny or light-coloured surfaces tend to reflect light and radiant heat away. The temperature of these objects does not change quickly when heat reaches them by radiation.

What distinguishes the different types of electromagnetic radiation from each other is:
- their wavelength (the distance the wave takes to repeat itself)
- their frequency (the number of wavelengths passing every second)
- the amount of energy they transfer.

These properties in turn determine their ability to be transmitted through transparent or opaque objects, their heating effect and their effect on living tissue.

Figure 3.17 shows the electromagnetic spectrum and demonstrates that higher-energy radiation corresponds to low wavelength.

electromagnetic radiation an electromagnetic wave or radiation that includes visible light, radio waves, gamma rays and x-rays

FIGURE 3.17 The electromagnetic spectrum. All objects emit some electromagnetic radiation.

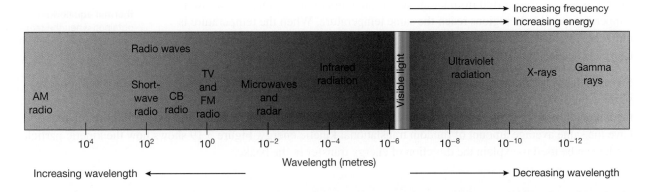

Why do hot objects emit electromagnetic radiation?

All matter is made up of atoms. At any temperature above absolute zero, these atoms are moving and colliding with each other. The atoms contain positive and negative charges. The motion of the atoms and their collisions with other atoms affect the motion of the electrons. Because they are charged and moving around, the electrons produce electromagnetic radiation. Electrons moving in an antenna produce a radio signal, but in a hot object the motion is more random with a range of speeds.

So, a hot object produces radiation across a broad range of wavelengths. If its temperature increases, the atoms move faster and have more frequent and more energetic collisions. These produce more intense radiation with higher frequencies and shorter wavelengths. However, this still occurs across a range of frequencies.

FIGURE 3.18 Electromagnetic radiation from a hot body

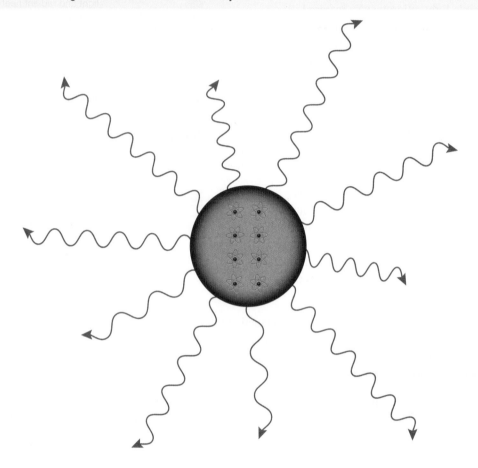

3.3.4 Thermal equilibrium

Energy is always transferred from a region of high temperature to a region of lower temperature until both regions reach the same temperature. When the temperature is uniform, a state of **thermal equilibrium** is said to exist and no net transfer of heat energy is occurring.

thermal equilibrium the state obtained when the temperature of two regions in contact is uniform

So, when a hot nail is dropped into a beaker of cold water, energy will be transferred from the hot nail into the water even though the hot nail has less total internal energy than the water. When thermal equilibrium is reached, the temperature of both the water and the nail is the same. The particles of water and the particles in the nail have the same average amount of random translational kinetic energy. Figure 3.19 shows how the kinetic particle model can be used to explain the direction of energy transfer in the beaker.

What is implicit in this discussion of thermal equilibrium and internal energy is the subtle, but important, point made by James Clerk Maxwell that 'All heat is of the same kind'.

FIGURE 3.19 The particles in the nail have more kinetic energy (on average) than those that make up the water. They collide with the particles of water, losing some of their kinetic energy and increasing the kinetic energy of individual particles of water. The temperature of the surrounding water increases.

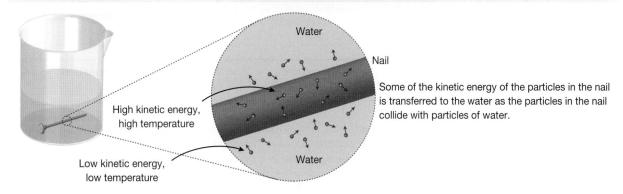

FIGURE 3.20 When you swim in a cold pool, energy is transferred from your body into the water. The water has much more total internal energy than your body because there is so much of it. However, the particles in your body have more random translational kinetic energy that can be transferred to the particles of cold water. Hopefully, you would not remain in the water long enough for thermal equilibrium to be reached.

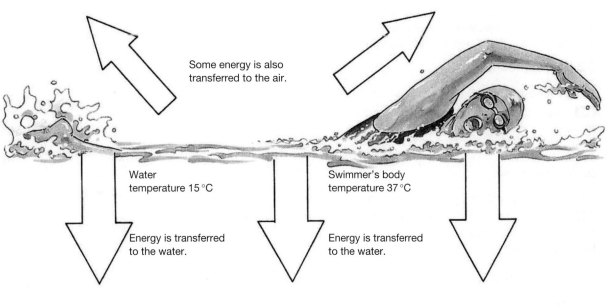

Resources

Interactivity Thermal equilibrium (int-6390)

3.3 Activities

| 3.3 Quick quiz `on` | 3.3 Exercise | 3.3 Exam questions |

3.3 Exercise

1. Explain, with the aid of a well-labelled diagram, how heat is transferred through a substance by conduction.
2. Why are liquids and gases generally poorer conductors of heat than solids?
3. Explain how convection occurs in water that is being heated in a test tube from below.
4. Why is it not possible for heat to be transferred through solids by convection?
5. At what speed does radiant energy move through space? What is significant about this speed?
6. When you swim in a still body of water on a hot afternoon there is a noticeable temperature difference between the water at the surface and the deeper water.
   a. Explain why this difference occurs.
   b. If the water is rough, the difference is less noticeable. Why?
7. Standing near the concrete wall of a city building after a hot day, you can instantly feel its warmth from a few metres away.
   a. How is the energy transferred to you?
   b. What caused the building to get hot during the day?
8. Why is it not practical to drink hot coffee in an aluminium picnic cup?
9. Why do ducts in the ceiling need more powerful fans than those in the floor?

3.3 Exam questions

▶ Question 1 (1 mark)

MC Conduction is a phenomenon that works best in
A. solids.
B. liquids.
C. gases.
D. all of the above.

▶ Question 2 (1 mark)

MC Which of the following sets of materials has the best conductors?
A. Iron, dry wood, water
B. Oil, steel, rock
C. Copper, petrol, air
D. Tin, copper, steel

▶ Question 3 (2 marks)

How is conduction affected by the temperature difference between materials?

▶ Question 4 (2 marks)

Why does heating water cause stirring of the water and formation of vortices (swirls)?

▶ Question 5 (3 marks)

Describe the motion of particles as a hot solid touches a colder solid.

More exam questions are available in your learnON title.

3.4 Specific heat capacity

3.4.1 Specific heat capacity

Once the temperature of materials could be accurately measured, it became apparent that, when heated, some materials increased in temperature more quickly than others. This phenomenon is called the 'specific heat capacity' of materials.

The 'specific heat capacity' of materials is defined as the amount of energy required to increase the temperature of 1 kilogram of the material by 1 °C (or K).

It takes more energy to increase the temperature of water by 1 °C than any other common substance. Water also needs to lose more energy to decrease in temperature. In simple terms, this means that water maintains its temperature well, cooling down and heating up more slowly than other materials. This explains, for instance, why it takes weeks for the average temperature of large bodies of water to increase or decrease by just 1 °C.

TABLE 3.3 Specific heat capacity of some common substances

Substance	Specific heat capacity (J kg^{-1} K^{-1})
Helium	5193
Water	4200
Human body (average)	3500
Cooking oil	2800
Ethylene glycol (used in car 'antifreeze')	2400
Ice	2100
Steam	2000
Fertile topsoil	1800
Neon	1030
Air	1003
Aluminium	897
Carbon dioxide	839
Desert sand	820
Glass (standard)	670
Argon	520
Iron and steel (average)	450
Zinc	387
Copper	385
Lead	129

Specific heat capacities differ because of two factors:
1. The different contributions to the internal energy by the forms of energy other than translational kinetic energy
2. The varying mass of individual atoms and molecules.

The internal energy of single-atom gases, such as helium, neon and argon, consists of only translational kinetic energy. So the specific heat capacities should be the same if you account for their difference in mass. Look up the atomic weight for each gas and multiply it by each gas's specific heat capacity and compare your answers.

The quantity of energy, Q, transferred to or from a substance in order to change its temperature is given as:

$$Q = mc\Delta T$$

The heat energy, Q, is directly proportional to three factors:
- The mass of the substance (m)
- The change in temperature (ΔT)
- The specific heat capacity of the substance (c).

tlvd-0005

SAMPLE PROBLEM 2 Calculations involving specific heat capacity

a. **How much energy is needed to heat 8.0 L (about 8.0 kg) of water from a room temperature of 15 °C to 85 °C?**
b. **A chef pours 200 g of cold water with a temperature of 15 °C into a hot aluminium saucepan with a mass of 250 g and a temperature of 120 °C. What will be the common temperature of the water and saucepan when thermal equilibrium is reached? Assume that no energy is transferred to or from the surroundings.**

THINK	WRITE
a. 1. Recall the relationship $Q = mc\Delta T$.	a. $Q = mc\Delta T$
2. Identify the values, using table 3.3 to determine the specific heat capacity and calculate the temperature difference in kelvin.	$m = 8.0 \, \text{kg}, \ c = 4200 \, \text{J kg}^{-1} \text{K}^{-1}$ $\Delta T = 85 \, °\text{C} - 15 \, °\text{C}$ $= 70 \, °\text{C}$ $= 70 \, \text{K (same change)}$
3. Substitute values into the formula and solve.	$Q = 8.0 \times 4200 \times 70$ $= 2\,352\,000$ $= 2.4 \times 10^3 \, \text{kJ}$
4. State the solution.	2.4×10^3 kJ of energy is needed to increase the temperature of 8.0 L of water from 15 °C to 85 °C.

b. The solution to this question relies on the following three factors:
- Energy is transferred from the saucepan into the water until both the saucepan and the water reach the same temperature ($T_f \ °\text{C}$).
- The amount of internal energy (Q_w) gained by the water will be the same as the amount of internal energy lost by the saucepan (Q_s).
- The internal energy gained or lost can be expressed as $mc\Delta T$. (ΔT can be expressed in K or °C since change in temperature is the same in both units.)

1. Recall the relationship $Q_w = Q_s$.

$$Q_w = Q_s$$
$$m_w c_w \Delta T_w = m_s c_s \Delta T_s$$

2. Identify the change in temperature of the water and saucepan.

$$\Delta T_w = T_f - 15°C$$
$$\Delta T_s = 120°C - T_f$$

3. Substitute values into the formula and solve.

$$0.200 \, \text{kg} \times 4200 \, \text{J kg}^{-1} \, °\text{C}^{-1} \times (T_f - 15 \, °\text{C})$$
$$= 0.250 \, \text{kg} \times 900 \, \text{J kg}^{-1} \, °\text{C}^{-1} \times (120 \, °\text{C} - T_f)$$
$$\Rightarrow 840 T_f - 12\,600 = 27\,000 - 225 T_f$$
$$1065 T_f = 39\,600$$
$$T_f = \frac{39\,600}{1065}$$
$$= 37.18 \, °\text{C}$$

4. State the solution (to the least number of significant figures provided in the question).

The saucepan and the water have reached a common temperature of 37 °C.

This example is a good illustration of the implications of a high specific heat capacity. Even though there was a smaller mass of water than aluminium, the final temperature was closer to the original water temperature than the original aluminium temperature.

PRACTICE PROBLEM 2

Estimate how much energy is needed to increase the temperature of your body by 1 °C.

EXTENSION: Caution! Contents may be hot

Eating a hot pie can be a health hazard! The temperature of the pastry and filling of a hot pie are the same. Thermal equilibrium has been reached. So why can you bite into a pie that seems cool enough to eat and be burnt by the filling?

The reason is that the filling is mostly water, while the pastry is mostly air. When your mouth surrounds that tasty pie, energy is transferred from the pie to your mouth. Each gram of water in the filling releases about 4 J of energy into your mouth for every 1 °C lost (since the specific heat of water is 4200 J kg^{-1} K^{-1}). Each gram of air in the pastry releases only about 1 J of energy into your mouth for every 1 °C lost (since the specific heat of air is 1000 J kg^{-1} K^{-1} Gram for gram, the filling transfers four times more energy into your mouth than the pastry.

3.4.2 Latent heat

In order for a substance to melt or evaporate, energy must be added. During the process of melting or evaporating, the temperature of the substance does not increase. The energy added while the state is changing is called **latent heat**. The word *latent* is used because it means 'hidden'. The usual evidence of heating, a change in temperature, is not observed.

> **latent heat** the heat added to a substance undergoing a change of state that does not increase the temperature

Similarly, when substances freeze or condense, energy must be released. However, during the process of changing state, there is no decrease in temperature accompanying the loss of internal energy.

In simple terms, the energy transferred to or from a substance during melting, evaporating, freezing or condensing is used to change the state rather than to change the temperature.

During a change of state, internal energy is gained or lost from the substance. Recall, however, that the internal energy includes the random kinetic and potential energy of the particles in the substance. The random translational kinetic energy of particles determines the temperature.

When a substance is being heated, the incoming energy increases the translational kinetic energy of the particles, increasing the temperature of the substance. When the substance being heated reaches its melting point, the incoming energy increases the potential energy of the particles rather than the random translational kinetic energy of the particles. The temperature of the substance does not change while it is melting. After the substance has melted completely, the incoming energy increases the kinetic energy of the particles again. When the substance is being cooled, the internal energy lost on reaching the melting (or freezing) point is potential energy. The temperature does not decrease until the substance has completely solidified.

The same process occurs at the boiling point of a substance. While evaporation or condensation takes place, the temperature of the substance does not change. The energy being gained or lost is latent heat, 'hidden' as changes in internal potential energy take place.

Specific latent heat of fusion

The **specific latent heat of fusion** is the quantity of energy required to change 1 kilogram of a substance from a solid to a liquid without a change in temperature. Note that the same quantity of energy is lost without a change in temperature during the change from a liquid to a solid. The specific latent heat of fusion of water is 334 kJ kg^{-1}. This value for the latent heat of fusion is also used when calculating the energy used during a phase change from liquid to solid. The difference is that changes of states from a more-ordered state to a less-ordered state (such as a solid to a liquid or a liquid to a gas) are endothermic reactions and *require* energy, while changes from a less-ordered state to a more-ordered state (such as a gas to a liquid or a liquid to a solid) are exothermic reactions and *release* energy.

Specific latent heat of vaporisation

The **specific latent heat of vaporisation** is the quantity of energy required to change 1 kilogram of a substance from a liquid to a gas without a change in temperature. Note that the same quantity of energy is lost without a change in temperature during the change from a gas to a liquid. The specific latent heat of vaporisation of water is $2.3 \times 10^3 \text{ kJ kg}^{-1}$. Note that this is approximately seven times greater that the value for the specific latent heat of fusion for water.

specific latent heat of fusion the quantity of energy required to change 1 kilogram of a substance from a solid to a liquid without a change in temperature

specific latent heat of vaporisation the quantity of energy required to change 1 kilogram of a substance from a liquid to a gas without a change in temperature

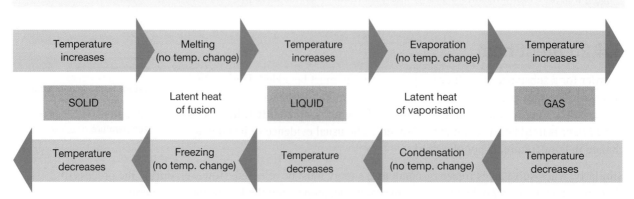

FIGURE 3.21 Changes of state and latent heat

Table 3.4 details both the specific latent heat of fusion and the specific latent heat of vaporisation of a number of common substances.

TABLE 3.4 Specific latent heat of some common substances

Substance	Specific latent heat of fusion (J kg⁻¹)	Specific latent heat of vaporisation (J kg⁻¹)
Water	3.3×10^5	2.3×10^6
Oxygen	6.9×10^3	1.1×10^5
Sodium chloride	4.9×10^5	2.9×10^6
Aluminium	2.2×10^3	1.7×10^4
Iron	2.8×10^5	6.3×10^6

The graph in figure 3.22 shows how the temperature of water increases as it is heated at a constant rate. During the interval BC, the temperature is not increasing. The water is changing state. The energy transferred to the water is not increasing the random translational kinetic energy of water particles. Note that the gradient of the section AB is considerably less than the gradient of the section CD. What difference in the properties of water and steam does this reflect?

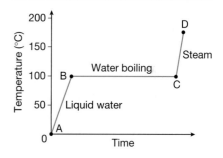

FIGURE 3.22 A heating curve for water being heated at a constant rate

Algebraically, the quantity of energy, Q, required to change the state of a substance without a change in temperature can be expressed as:

$$Q = mL$$

where:

m = mass of the substance

L = specific latent heat of fusion or vaporisation.

tlvd-0006

SAMPLE PROBLEM 3 The energy required to change the state of a substance

The graph in figure 3.22 might represent the melting of aluminium. The melting point of aluminium is 660.3 °C.
a. **How much energy must be transferred to completely melt 2.5 kg of aluminium?**
b. **What would be the temperature of the aluminium at the instant it has all melted?**

THINK

a. 1. Identify what type of change of state is occurring: fusion or vaporisation.

    2. Identify the values.

    3. Substitute values into the formula and solve.

    4. State the solution.

WRITE

a. Changing from solid to liquid is fusion.
$Q = mL$

$L = 2.2 \times 10^3 \text{ J kg}^{-1}$, $m = 2.5$ kg

$Q = 2.2 \times 10^3 \text{ J kg}^{-1} \times 2.5 \text{ kg}$

   $= 5.5 \times 10^3 \text{ J}$

5.5×10^3 J of energy is needed to completely melt 2.5 kg of aluminium.

b. All of the energy supplied is being used to change state, thus the temperature does not rise (see flat section of graph).

b. The temperature just after state change is still 660.3 °C.

PRACTICE PROBLEM 3

Determine how much energy is absorbed to evaporate 15 grams of water.

  Resources

Interactivity Changes of state (int-0222)

3.4.3 Evaporation

Your skin is not completely watertight, which allows water from the skin and tissues beneath it to evaporate. The latent heat of vaporisation required for the water to change state from liquid to gas is obtained from the body, reducing its temperature. Evaporation of water from the mouth and lungs also takes place during the process of breathing. Even without sweating, the energy used to evaporate water in the body accounts for about 17 per cent of the total heat transfer from the body to the environment.

Water evaporates even though its temperature is well below its boiling point. The temperature is dependent on the average translational kinetic energy of the water molecules. Those molecules with a kinetic energy greater than average will be moving faster than the others. Some of them will be moving fast enough to break the bonds holding them to the water and escape from the liquid surface. The fastest particles with the higher kinetic energies leave the surface of the water, thereby reducing the average kinetic energy of the remaining particles and thus the temperature of the water. This explains why you can keep a bottle and its contents cool on a hot day by wrapping a wet cloth around it.

FIGURE 3.23 Particle A experiences forces of attraction from the surrounding particles in all directions. Particle B does not experience as many forces, so it will need less kinetic energy to escape the forces of attraction and evaporate.

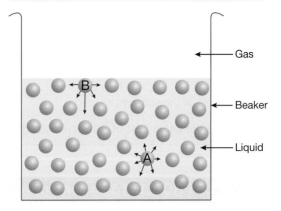

EXTENSION: Steam burns

A burn caused by steam at 100 °C is considerably more serious than a burn caused by the same mass of boiling water. Each gram of hot steam transfers 2600 J of energy to your skin as it condenses to water at 100 °C. Each gram of newly condensed steam then transfers 4.2 J for each degree drop in temperature as it cools to your body temperature of about 37 °C. That's about 265 J. The total quantity of energy transferred by each gram of steam is therefore about 2865 J, which is more than ten times the amount of energy a gram of boiling water would transfer to your skin as it cools down to your body temperature.

---

elog-1827

INVESTIGATION 3.1  online only

The good oil on heating

Aim

To show that different substances require different quantities of energy to change their temperatures by the same amount and to show that the quantity of energy required to change the temperature of a given substance is directly proportional to the mass of the substance

elog-1828

tlvd-0804

INVESTIGATION 3.2 online only

Cooling

Aim

To show that the internal energy of a substance can change without a subsequent change in temperature and to produce a cooling curve that illustrates the concept of latent heat

3.4 Activities

learn on

Students, these questions are even better in jacPLUS

 Receive immediate feedback and access sample responses

 Access additional questions

 Track your results and progress

Find all this and MORE in jacPLUS ▶

| 3.4 Quick quiz on | 3.4 Exercise | 3.4 Exam questions |

3.4 Exercise

1. The quantity of energy needed to increase the temperature of a substance is directly proportional to the mass, specific heat capacity and the change in temperature of the substance. If 200 kJ is used to increase the temperature of a particular quantity of a substance, how much energy would be needed to bring:
   a. twice as much of the substance through the same change in temperature
   b. three times as much of the substance through a temperature change twice as great?
2. Use table 3.3 to answer the following questions.
   a. Why is the specific heat capacity of the human body so high?
   b. Why is the specific heat capacity of deserts so much lower than that of fertile topsoil?
   c. When heating water to boiling point in a saucepan, some of the energy transferred from the hotplate is used to increase the temperature of the saucepan. Which would you expect to gain the most energy from the hotplate: an aluminium, copper or steel saucepan of similar size? (*Note:* Steel is nearly three times denser that aluminium.)
   d. Make some general comments about the order of substances listed in table 3.3.
3. Use the following data to determine the quantity of energy needed to evaporate 500 g of water without a change in temperature.

Substance	Specific latent heat of fusion (J kg^{-1})	Specific latent heat of vaporisation (J kg^{-1})
Water	3.3×10^5	2.3×10^6

4. The following graph shows the heating curve obtained when 500 g of candle wax in solid form was heated from room temperature in a beaker of boiling water.

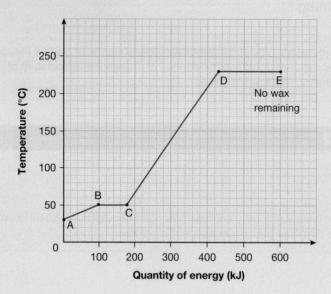

a. What is the boiling point of the candle wax?
b. During the interval BC there is no increase in temperature even though heating continued. What was the energy transferred to the candle wax being used for during this interval?
c. In which state of matter was the candle wax during the interval CD?
d. Use the heating curve to determine the latent heat of fusion of candle wax.
e. Which is higher: the specific heat capacity of solid candle wax or the specific heat capacity of liquid candle wax? Explain your answer.
f. Explain, in terms of the kinetic particle model, what is happening during the interval DE.
5. How much ice at 0 °C could be melted with 1 kilogram of steam at 100 °C, assuming no loss of energy to the surroundings? Use the specific latent heat values quoted in table 3.4. The specific heat capacity of water is 4200 J kg^{-1} K^{-1}.
6. Explain the importance of keeping a lid on a simmering saucepan of water in terms of latent heat of vaporisation.
7. How does the evaporation of water cause a reduction in the temperature of the remaining liquid water?

3.4 Exam questions

▶ Question 1 (1 mark)

MC The specific heat capacities of ice, water and steam are all different due to all three having
A. the same bonding strength.
B. different chemical compositions.
C. different amounts of kinetic and bonding energies.
D. the same structures.

▶ Question 2 (1 mark)

MC The relationship $Q = mc\Delta T$ can be used to calculate the
A. energy needed for a state change.
B. heat energy transfer.
C. internal energy change.
D. work done.

▶ Question 3 (3 marks)

Calculate the energy change, Q, if 550 g (mL) of water is heated from 34 °C to 89 °C. Use an appropriate scale unit in your answer.

3.5 Understanding climate change and global warming

KEY KNOWLEDGE

- Apply concepts of energy transfer, energy transformation, temperature change and change of state to climate change and global warming

Source: VCE Physics Study Design (2023–2027) extracts © VCAA; reproduced by permission.

Although the Sun is our primary source of energy on Earth, as seen in section 1.4, our secondary source of energy comes from within Earth. The heat energy within Earth comes from the **radioactive decay** of elements such as uranium. Gravity and Earth's rotation are other sources of energy. However, it is solar energy that drives the overall climate system.

radioactive decay the process by which an unstable atomic nucleus loses energy by radiation

3.5.1 Energy in balance

Every object attempts to achieve thermal equilibrium — that is, a balance between energy absorbed and energy emitted.

Imagine a steel ball placed in direct sunlight as shown in figure 3.24. When light energy heats and warms an object, like a steel ball, its particles start to get more excited and its translational kinetic energy starts to increase. As a result, the steel ball emits thermal radiant energy (usually in the infrared). The absorbed energy increases the temperature of the ball; the increased temperature means that the ball will radiate more thermal energy. If the amount of energy radiated is less than the amount absorbed, the temperature of the ball will continue to increase, leading to more energy being emitted, until a temperature is reached where energy in and energy out balance.

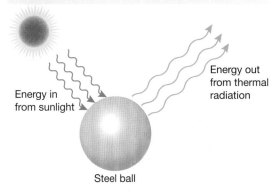

FIGURE 3.24 Achieving thermal equilibrium is a balance between energy absorbed and energy emitted.

Energy in from sunlight

Energy out from thermal radiation

Steel ball

Similarly, if a cloud moves in front of the Sun, the energy absorbed by the ball would suddenly decrease to less than the amount of energy being emitted. The temperature of the ball would drop and continue to fall until the amount of energy emitted matched the energy absorbed and a new equilibrium was reached.

Earth as viewed from space is like the steel ball. The energy falling on Earth from the Sun is fairly constant. At the equator, an average of about 684 joules of energy from the Sun hits each square metre of Earth's surface every second; that is, 684 W m^{-2}, where 1 watt is a unit of power or the rate of energy delivery and equals 1 joule per second. This value varies from day to day by as much as 2 W m^{-2}, as well as having an approximate 11-year cycle of a similar magnitude.

EXTENSION: Milankovitch cycles

Note that this content is useful if you are choosing to study option 2.1 (How does physics explain climate change?) in Unit 2, Area of Study 2.

While a prisoner of war in World War I, Milutin Milanković postulated several types of changes in Earth's movement around the Sun that could affect the amount of solar radiation Earth receives and its distribution. These changes can affect climate on a time span of many thousands of years and possibly explain the occurrence of ice ages. Some of the types of changes include:

- variation in the elliptical shape of Earth's orbit (**eccentricity**), with a cycle time of about 413 000 years
- **precession** of Earth's axis of rotation; like any spinning top, the axis itself rotates, once every 26 000 years
- the tilt of the axis (**obliquity**), which ranges from 22° to 24.5° every 41 000 years.

FIGURE 3.25 The variation in solar radiation that Earth is exposed to has changed the climate over tens and hundreds of thousands of years. These variations explain the ice ages that Earth has experienced in its past, but do not explain the current global warming.

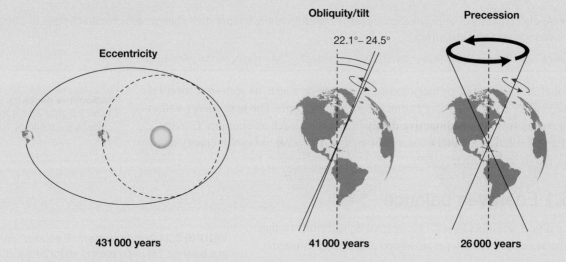

All these changes are due to gravitational interactions in the solar system between Earth, the Sun and other planets. It is thought that these factors may explain the long-term cooling trend that Earth has been in over the last 6000 years. From the slowness with which these changes occur, none can explain the unprecedented global warming in recent decades.

The law of conservation of energy states that energy can neither be created nor destroyed, but only converted from one form of energy to another. For instance, via **photosynthesis**, light energy can be converted to chemical energy.

We have seen that Earth's surface receives a certain amount of solar energy, depending on the time and place, and that this energy creates heat. If more heat than is received is radiated back out into space, Earth would cool down, while if less heat than received is dissipated back into space, Earth would warm up.

Some gases in the atmosphere, such as water vapour (H_2O) and carbon dioxide (CO_2), prevent some infrared wavelengths from being radiated back into space, thus trapping heat and warming our planet. This is the *greenhouse effect*.

The Stefan–Boltzmann relationship, $P \propto T^4$, enables the absolute temperature of an object to be determined if you know how much energy it is radiating. For the 242 W m^{-2} that Earth radiates, this gives a temperature of 255.6 K; that is, –18 °C. This seems incorrect; Earth is not that cold! What causes the difference between this calculated temperature (which is the temperature of Earth observed from space) and the temperature we observe at the surface? The explanation is that we ignored the greenhouse gases in the atmosphere.

eccentricity a measure of how elliptical an object's orbit is

precession a change in direction of the rotational axis of a spinning object

obliquity a measure of the angle tilt of a planet against its plane of orbit

photosynthesis a chemical reaction, converting light energy into chemical energy (sugars), that takes place in the chloroplasts of a plant cell consuming carbon dioxide and releasing oxygen as a byproduct

3.5.2 The greenhouse effect

The natural greenhouse effect

While the temperature of the upper atmosphere is -18 °C at an altitude of about 5 km above the surface, the average surface temperature of Earth nowadays is approximately 15 °C; this is 7 °C warmer than the average global temperature during the Last Glacial Maximum, which occurred around 20 000 years ago. Our atmosphere is a crucial protective layer, without which life would not be possible, and with the greenhouse gases it contains it acts as a 'warming blanket'.

FIGURE 3.26 Earth's surface is warmed by a 5 km thick blanket.

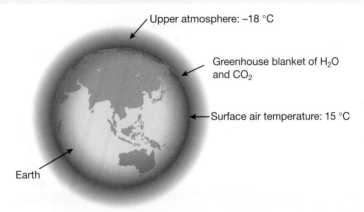

Upper atmosphere: -18 °C

Greenhouse blanket of H_2O and CO_2

Surface air temperature: 15 °C

Earth

How do water vapour and carbon dioxide act as a blanket to trap heat?

If you shine a broad spectrum of light through a gas, some colours will be absorbed by some molecules and then subsequently re-emitted, but in a random direction, while the light from the other parts of the spectrum passes through the gas without a change in direction.

Ultraviolet light has more energy than visible light, which has more energy than infrared light, but none of these types of radiation are energetic enough to break up molecules. They have only enough energy to stretch, twist and spin molecules. Molecules with two atoms such as nitrogen (N_2) and oxygen (O_2) have very strong bonds. When they absorb ultraviolet light, their bonds stretch. The energy of visible light and infrared radiation is too low to affect such molecules. This means the two gases are transparent to visible light and infrared.

H_2O and CO_2 have three atoms in each molecule, so they are more flexible than N_2 and O_2. These molecules can bend, whereas atoms with only two molecules cannot. When H_2O and CO_2 absorb infrared light, their bonds stretch and bend. Other molecules with more than two atoms, such as methane (CH_4), nitrous oxide (N_2O) or ozone (O_3), as well as industrial fluorinated gases, absorb infrared radiation in the same way.

EXTENSION: The ozone layer

Ozone, in the uppermost atmosphere, protects Earth from harmful ultraviolet radiation from the Sun. The protective benefits from the ozone layer far outweigh its contribution to the greenhouse effect.

Industrial gases such as chlorofluorocarbons (CFCs) and hydrofluorocarbons (HFCs), which are extremely potent greenhouse gases and can stay in the atmosphere for up to 100 years, were — until relatively recently — used in refrigerators, aerosols and air-conditioning. Once it became apparent that these gases were destroying the ozone layer, and in addition were contributing to global warming, they were banned; now there are safe, energy-efficient alternatives for industrial refrigerants.

Water vapour, carbon dioxide and methane all stretch and bend differently. This means each molecule will absorb different parts of the infrared spectrum and consequently, all these molecules contribute independently to the greenhouse effect.

Once a gas molecule has absorbed infrared radiation coming from Earth's surface, it re-emits the radiation but, importantly, in a random direction (see figure 3.27). So, for the gas as a whole, some radiation goes back down to Earth to increase its temperature and some is directed towards the top of the atmosphere and out into space. However, other molecules further up in the atmosphere can absorb this radiation and re-emit more back to Earth. The overall effect is that more than half the radiation emitted by the gas comes back to Earth's surface.

FIGURE 3.27 Infrared radiation from Earth's surface is absorbed by greenhouse gases and re-emitted in all directions.

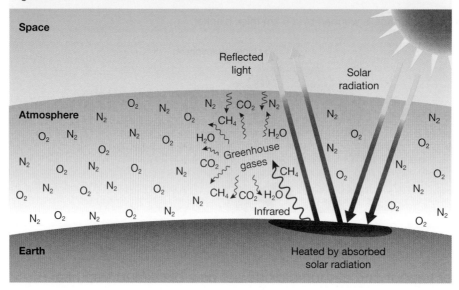

Enhanced greenhouse effect

Human activities, such as the burning of fossil fuels (coal, oil and natural gas), agriculture and land clearing, are increasing the concentrations of greenhouse gases in the atmosphere, especially since the Industrial Revolution. This increase is sometimes called the **enhanced greenhouse effect** (see figure 3.28).

Increased carbon dioxide concentration in the atmosphere means that more of the wavelengths that carbon dioxide absorbs will be re-emitted back to Earth, increasing the temperature of Earth.

enhanced greenhouse effect
the disruption to Earth's climate equilibrium caused by the release of greenhouse gases in the atmosphere due to human activities, which leads to an increase in global average surface temperatures

FIGURE 3.28 Comparison of the greenhouse effect and the enhanced greenhouse effect

Greenhouse effect

Sun

Earth

Enhanced greenhouse effect

Sun

Earth

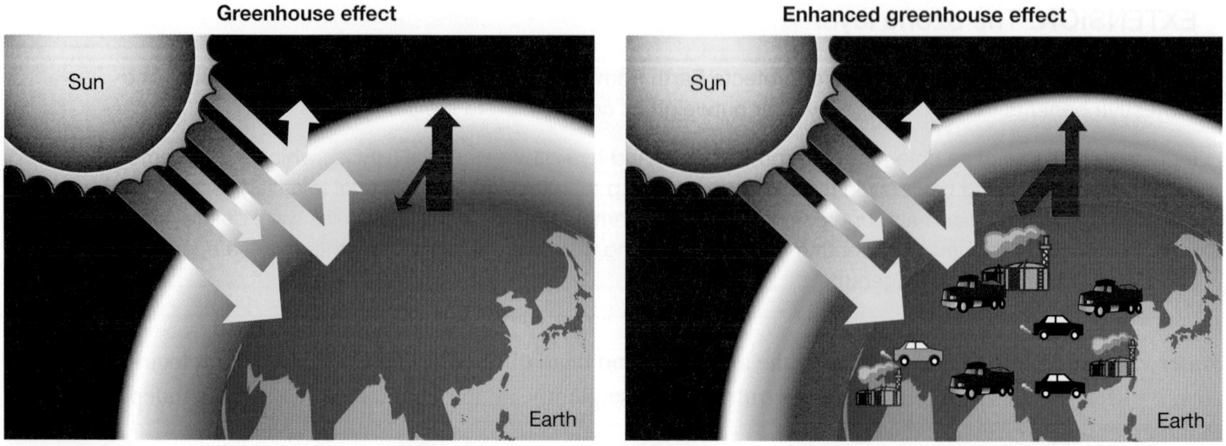

Our atmosphere is a mixture of gases, with CO_2, H_2O, N_2O, O_3 and CH_4 at low concentrations; thus, much of the radiation with the wavelengths that would be absorbed and re-emitted by these molecules has a good chance of passing through without ever hitting one of them. As long as the gases CO_2, H_2O, N_2O, O_3 and CH_4 make up a very small proportion of the atmosphere, the infrared radiation emitted from Earth's surface has a good chance of reaching outer space without being absorbed. However, with increased emissions of CO_2 and CH_4, interactions are more likely to occur, increasing both the level of radiation reaching Earth and its temperature.

3.5.3 Climate change

Climate change describes a long-term change in the typical weather patterns of Earth's global, regional and local climates. Earth's climates have changed drastically in the past. For instance, nowadays, the Sahara is the world's largest hot desert; but around 6000 years ago, the climate there was tropical, and the region was covered in grasslands and trees. Closer to home, the dried-out mega-lakes and inland seas in the interior of Australia are clues to a moister past, thousands of years before Australia became the second driest continent in the world (after Antarctica).

Climate changes — even abrupt changes — are not something new. However, it is now widely accepted that the recent changes observed in Earth's climate, and the rate of change, since the start of the Industrial Revolution are driven by human activities — particularly those increasing the levels of greenhouse gases in the atmosphere (such as fossil fuel burning) and thus, increasing Earth's global temperature. This long-term heating of Earth's climate as a result of human activities is called global warming.

EXTENSION: The evidence for rapid climate change

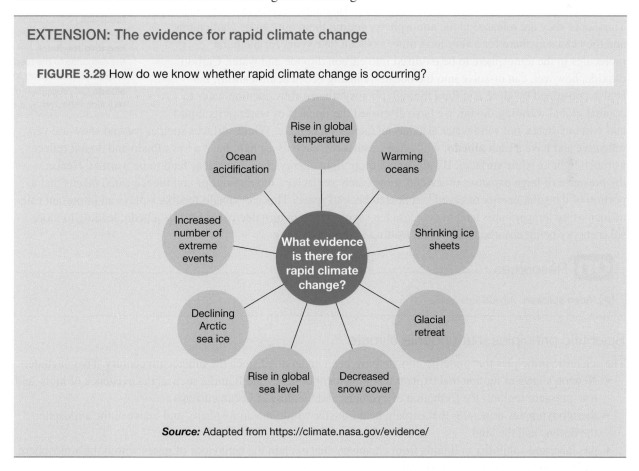

FIGURE 3.29 How do we know whether rapid climate change is occurring?

Source: Adapted from https://climate.nasa.gov/evidence/

Impact of water on the climate of Earth

As we have seen previously, water has a high specific heat capacity, meaning that a huge quantity of heat energy is stored in the water present on Earth. Moreover, liquid water and water vapour are fluids, with convective properties, and can transfer heat by convection. For instance, ocean convection is the driving force behind the Gulf Stream and other currents that impact climate patterns worldwide.

Additionally, water can be evaporated within the range of temperatures present on Earth, and the vaporisation of water requires heat, which is taken from the environment, cooling it down. However, water vapour is a greenhouse gas, and it is estimated that it accounts for around 60 per cent of Earth's greenhouse effect. The proportion of water vapour in the atmosphere is dependent upon the temperature of the atmosphere. As the atmosphere's temperature increases, so does the proportion of water vapour in it, and the more water vapour is present in the atmosphere, the more infrared radiation will end up being absorbed and re-emitted back to Earth. This in turn traps more heat and causes more water to evaporate, in a **feedback** loop.

The climate system is an extremely complex system made of nested and interlinked sub-systems, which are also complex. As a result, there are still a lot of unknowns. For instance, the increase of the proportion of water vapour in the atmosphere could lead to more cloud formation. As clouds reflect sunlight and therefore decrease the amount of solar energy reaching Earth's surface, this could produce a **negative feedback** effect — more water vapour being added to the atmosphere would lead to a decrease in temperatures and therefore more water vapour condensing, which in turn would lead to less heat being trapped in the atmosphere globally. However, locally, the presence of clouds would lead to a stronger greenhouse effect. To illustrate this, you have probably noticed that temperatures decrease less overnight if it is cloudy than if the night is clear. We still do not know if those two effects balance out and thus cancel each other out, or if one of them has a stronger effect than the other. What is known is that we, humans, do not control the water cycle. We are, however, actively increasing the proportion of other greenhouse gases such as carbon dioxide and methane in the atmosphere, and this can be controlled, albeit with a lot of effort from everyone.

Carbon dioxide and methane cannot condense at normal temperature and pressure. Thus, once they are released in the atmosphere, getting them out can be challenging and they can stay there for a very long time, as it can take years for methane molecules in the atmosphere to be oxidised into carbon dioxide and water. Carbon dioxide, however, can dissolve into the oceans (causing ocean acidification), but is also consumed by algae and flora through photosynthesis, thus opening ways to combat global warming. So far, we have discussed the impacts of water in its liquid and gaseous states, but solid water also has an impact on climate. White surfaces such as ice and snow are very reflective and have a high **albedo**, while darker surfaces, such as asphalt, have a low albedo and do not reflect as much light as white surfaces. By absorbing more light energy, darker surfaces tend to be warmer. Hence, the presence of large expanses of ice and snow (such as glaciers, mountain tops and the ice caps) means that a portion of the solar energy received is reflected back to space. This ice–albedo feedback plays an important role as increasing temperatures tend to decrease ice cover, which in turn decreases Earth's albedo, leading to more solar energy being absorbed and in turn leading to more warming.

> **feedback** refers to when a system's input is fed by its previous output
>
> **negative feedback** refers to when the response to the feedback is in the opposite direction to the input
>
> **albedo** the proportion of solar radiation reflected by a surface

 Resources

 Video eLesson Albedo (eles-2515)

Scientific principles that underlie climate

The scientific principles that underlie climate have been understood since the nineteenth century. They include:

- Newton's laws of motion that explain the movement of gases and liquids, such as the existence of high- and low-pressure regions, the formation of cyclones and the flow of ocean currents
- thermodynamics principles that explain heat transfer within the atmosphere, and between the atmosphere, the oceans and the land
- gas laws and solubility in liquids from chemistry that explain the behaviour of gases within the atmosphere and their interactions with the oceans.

These principles are known precisely with mathematical relationships, some of which you will come across in the VCE Physics and VCE Chemistry courses. The development of the computer meant these principles and their mathematical relationships could be applied to the biggest problem on Earth: Earth itself.

A climate model is an attempt to apply these relationships to Earth's whole atmosphere as well as the surface features of land, ice and sea. The model attempts to calculate aspects of the climate that are important to humans, such as rainfall and humidity, sea level rise, ocean acidity, wind strength and, of course, air temperature. Climate models are also able to calculate future trends in these aspects of climate and then investigate the effect on these trends of changes such as reducing greenhouse gas emissions or aerosol use.

3.5 Activities

learnon

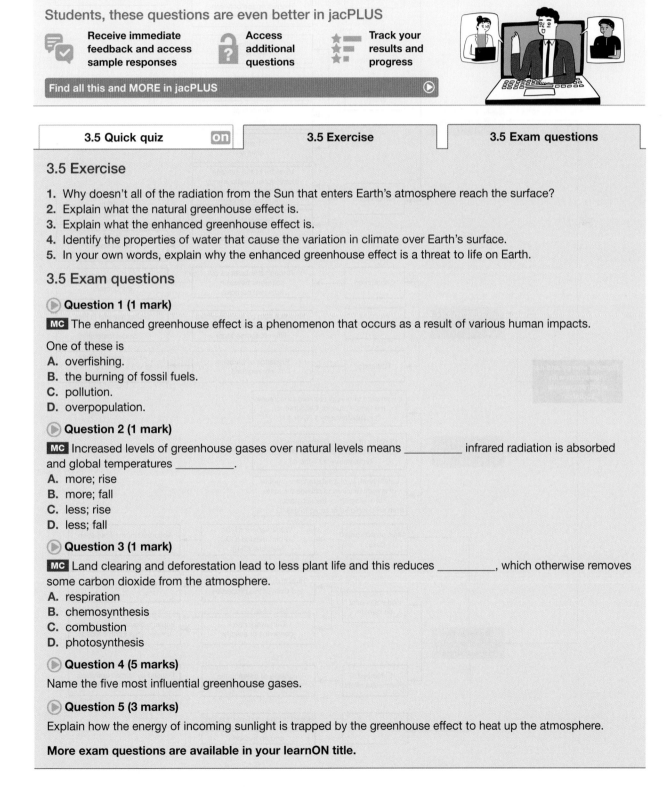

Students, these questions are even better in jacPLUS

Receive immediate feedback and access sample responses

Access additional questions

Track your results and progress

Find all this and MORE in jacPLUS

| 3.5 Quick quiz on | 3.5 Exercise | 3.5 Exam questions |

3.5 Exercise

1. Why doesn't all of the radiation from the Sun that enters Earth's atmosphere reach the surface?
2. Explain what the natural greenhouse effect is.
3. Explain what the enhanced greenhouse effect is.
4. Identify the properties of water that cause the variation in climate over Earth's surface.
5. In your own words, explain why the enhanced greenhouse effect is a threat to life on Earth.

3.5 Exam questions

Question 1 (1 mark)

MC The enhanced greenhouse effect is a phenomenon that occurs as a result of various human impacts.

One of these is
A. overfishing.
B. the burning of fossil fuels.
C. pollution.
D. overpopulation.

Question 2 (1 mark)

MC Increased levels of greenhouse gases over natural levels means _____ infrared radiation is absorbed and global temperatures _____.
A. more; rise
B. more; fall
C. less; rise
D. less; fall

Question 3 (1 mark)

MC Land clearing and deforestation lead to less plant life and this reduces _____, which otherwise removes some carbon dioxide from the atmosphere.
A. respiration
B. chemosynthesis
C. combustion
D. photosynthesis

Question 4 (5 marks)

Name the five most influential greenhouse gases.

Question 5 (3 marks)

Explain how the energy of incoming sunlight is trapped by the greenhouse effect to heat up the atmosphere.

More exam questions are available in your learnON title.

3.6 Review

3.6.1 Topic summary

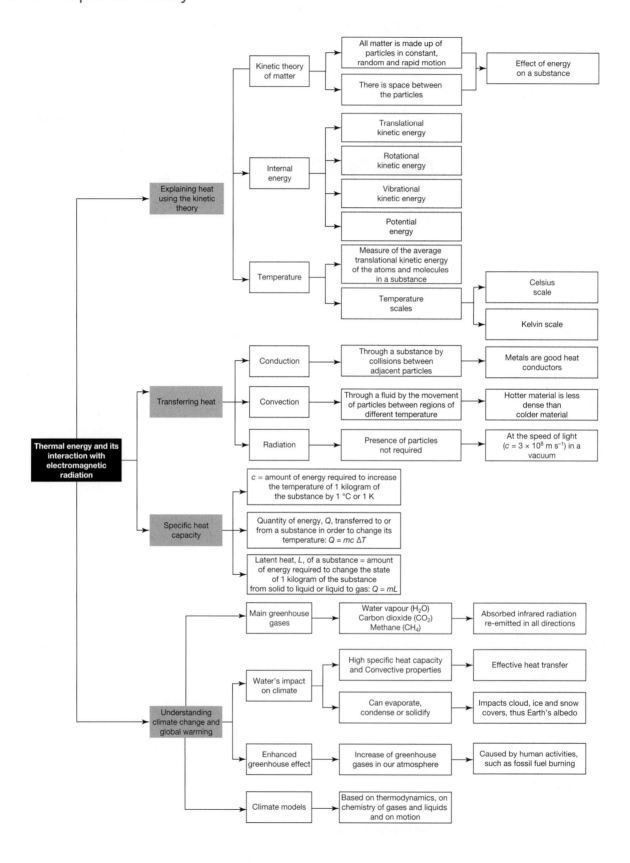

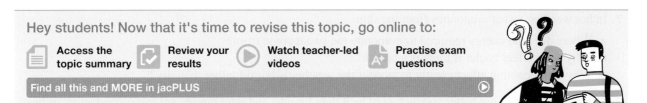

3.6.2 Key ideas summary

online only

3.6.3 Key terms glossary

online only

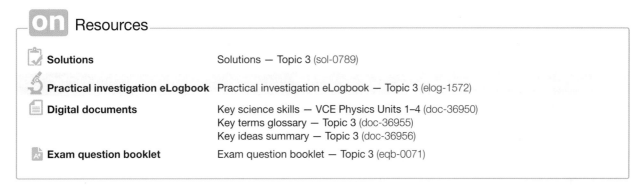

Resources

Solutions	Solutions — Topic 3 (sol-0789)
Practical investigation eLogbook	Practical investigation eLogbook — Topic 3 (elog-1572)
Digital documents	Key science skills — VCE Physics Units 1–4 (doc-36950)
	Key terms glossary — Topic 3 (doc-36955)
	Key ideas summary — Topic 3 (doc-36956)
Exam question booklet	Exam question booklet — Topic 3 (eqb-0071)

3.6 Activities

learn on

3.6 Review questions

1. Why can't you put your hand on your own forehead to estimate your body temperature?

2. If today's maximum temperature was 14 °C and tomorrow's maximum temperature is expected to be 28 °C, will tomorrow be twice as hot? Explain your answer. (*Hint:* Recall the relation between temperature and the average kinetic energy of particles.)

3. Explain, with the aid of a well-labelled diagram, how convection occurs in a liquid that is being heated from below.

4. Boiling water at 100 °C is poured into a ceramic coffee cup at 23 °C. After a while they are both at 77 °C.

   The two materials are at

   A. kinetic equilibrium.
   B. thermal equilibrium.
   C. potential equilibrium.
   D. internal energy balance.

5. Heat is a transfer of energy and is measured in

   A. kelvin (K).
   B. calories (C).
   C. joules (J).
   D. degrees Celsius (°C).

6. Why do conventional ovens without fans have heating elements at the bottom? What is the advantage of having an oven with a fan?

7. In hot weather, sweat evaporates from the skin.
    a. Where does the energy required to evaporate the sweat come from?
    b. Why do you feel cooler if there is a breeze?
    c. Why is the cooling effect greatly reduced if the weather is humid rather than dry?

8. Explain why metals are better conductors of heat than liquids, and why liquids are better conductors of heat than gases.

9. Give an example of positive feedback and an example of negative feedback in a system such as Earth's climate.

10. Explain what ice–albedo feedback is.

3.6 Exam questions

Section A — Multiple choice questions

All correct answers are worth 1 mark each; an incorrect answer is worth 0.

▶ Question 1

Which of the following is correct in relation to the kinetic theory of matter?

A. It states that all matter is made up of a large number of small particles.
B. It can be used to explain the process of convection and conduction.
C. It is needed to explain absolute zero.
D. All of the above

▶ Question 2

What is the internal energy of a substance?

A. The sum of all energy contained within the matter of that substance, including kinetic and potential energy
B. The sum of all kinetic energy in a substance
C. The sum of all the translational kinetic energy in a substance
D. The average of all the translational kinetic energy in a substance

▶ Question 3

What is temperature a measure of?

A. The average of all energy contained within the matter of that substance, including kinetic and potential energy
B. The sum of all kinetic energy in a substance
C. The sum of all the translational kinetic energy in a substance
D. The average of all the translational kinetic energy in a substance

▶ Question 4

The Kelvin scale is a temperature measure based on which of the following?

A. The temperature of the human body
B. The temperatures of when water boils and when it freezes
C. Absolute zero, when all particle movement stops
D. The temperature of the North Pole at midday in summer

▶ Question 5

How is heat best described?

- **A.** It is when a material gets hot.
- **B.** It is the energy transferred as a result of temperature difference.
- **C.** It is the energy inside matter.
- **D.** It is the translational kinetic energy of an object.

▶ Question 6

For two substances to reach thermal equilibrium, which of the following must they have?

- **A.** The same internal energy
- **B.** The same kinetic energy
- **C.** The same translational kinetic energy
- **D.** None of the above

▶ Question 7

What is the specific heat capacity a measure of?

- **A.** How much internal energy a substance can hold
- **B.** How much energy is required to raise the temperature of 1 kilogram of a substance by 1 °C
- **C.** How much energy is required to raise the temperature of 1 kilogram of a substance by 100 °C
- **D.** The temperature change for 1 kilogram of a substance when its internal energy has increased by 1 J

▶ Question 8

When the cooling curve of wax is examined, a flat section on the graph is found.

This section is due to which of the following?

- **A.** A pause in energy reaching the system, causing no temperature change
- **B.** The wax changing from a solid to a liquid and consuming extra energy without a temperature change
- **C.** A fault in the measuring equipment, as this type of flat section would never occur
- **D.** The wax changing state from liquid to solid, releasing extra energy without a temperature change

▶ Question 9

Which of the following is *not* a feature that influences climate change on Earth?

- **A.** The seasons
- **B.** The properties of water
- **C.** The tilt of Earth's axis
- **D.** The geological features (land mass and oceans)

▶ Question 10

On a warm sunny day, the Sun's radiation melts very little snow on the slopes of an alpine ski resort.

This is because

- **A.** snow, being a solid, is not affected by radiant heat.
- **B.** the snow particles are unable to transfer the translational kinetic energy and so cannot warm up.
- **C.** snow is a type of water and due to its high latent heat capacity does not melt easily.
- **D.** the white snow reflects up to 90 per cent of the radiant energy of the Sun.

Section B — Short answer questions

▶ Question 11 (3 marks)

An 800 g rubber hot-water bottle that has been stored at a room temperature of 15 °C is filled with 1.5 kg of water at a temperature of 80 °C. Before being placed in a cold bed, thermal equilibrium between the rubber and water is reached.

What is the common temperature of the rubber and water at this time? (Assume that no energy is lost to the surroundings. The specific heat capacity of rubber is 1700 J kg^{-1} K^{-1}. The specific heat capacity of water is 4200 J kg^{-1} K^{-1}.)

▶ Question 12 (3 marks)

How much energy does it take to completely convert 2.0 kg of ice at −5.0 °C into steam at 100 °C? (Assume that no energy is lost to the surroundings.)

▶ Question 13 (2 marks)

Explain, in terms of the kinetic particle model, why you can put your hand safely in a 300 °C oven for a few seconds, while if you touch a metal tray in the same oven your hand will be burned.

▶ Question 14 (5 marks)

The specific heat capacity of water is over four times that of sand.

a. What effect does this have on the heating and cooling of water and sand? **(1 mark)**
b. Explain why, on a hot day, sand is too hot to stand on in bare feet, while the water in the sea can be too cold for some people. **(2 marks)**
c. Why is the temperature of the sand of an inland desert almost always greater than that of the sand on a beach at the same latitude? **(2 marks)**

▶ Question 15 (2 marks)

In the tropical regions of Earth, more radiant energy is received from space than is lost. At the poles, more radiant energy is lost than is received. This would suggest that average temperatures in the tropics should be continually increasing, while the average temperatures at the poles should be continually decreasing.

Explain why this doesn't happen.

 Hey teachers! Create custom assignments for this topic

 Create and assign unique tests and exams 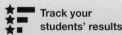 Access quarantined tests and assessments Track your students' results

Find all this and MORE in jacPLUS ▶

AREA OF STUDY 1 How are light and heat explained?

OUTCOME 1

Model, investigate and evaluate the wave-like nature of light, thermal energy and the emission and absorption of light by matter.

PRACTICE EXAMINATION

STRUCTURE OF PRACTICE EXAMINATION		
Section	Number of questions	Number of marks
A	20	20
B	6	20
	Total	40

Duration: 50 minutes

Information:

- This practice examination consists of two parts. You must answer all question sections.
- Pens, pencils, highlighters, erasers, rulers and a scientific calculator are permitted.
- You may use the VCAA Physics formula sheet for this task.

 Resources

🔗 **Weblink**   VCAA Physics formula sheet

Additional practice examination information	
Specific heat capacities	**Water** 4.2×10^3 J kg^{-1} K^{-1}
	Aluminium 0.92×10^3 J kg^{-1} K^{-1}
Specific latent heat of fusion	**Water** 3.3×10^5 J kg^{-1}
Specific latent heat of vaporisation	**Water** 2.3×10^6 J kg^{-1}
Refractive index of air	1.00
Refractive index of water	1.33
Speed of light in a vacuum	3.0×10^8 m s^{-1}
Wien's constant	2.90×10^{-3} m K

All correct answers are worth 1 mark each; an incorrect answer is worth 0.

1. The average kinetic energy of atoms and molecules are given by which of the following?
   A. Their temperature
   B. The heat input into them
   C. Their work output
   D. The heat output from them

2. The freezing point of water could be represented by which of the following?
   A. 0 K
   B. 0 °F
   C. 273.15 °C
   D. 273.15 K

3. The two graphs below record the motion of a wave in water.

Displacement–time graph

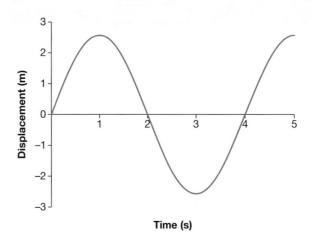

Displacement–distance graph

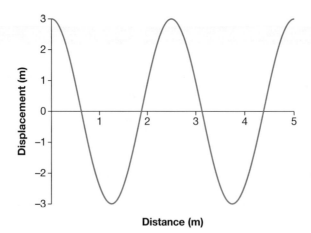

Using the information from the graphs, which of the following statements is correct?

A The period of the wave is 3 s.

B The wavelength of the wave is 3 m.

C The speed of propagation of the wave is 0.625 m s^{-1}.

D The frequency of the wave is 25 Hz.

4. A ray of light moving through water enters a plastic block at an angle of incidence of 40°. The angle of refraction is 30°.
   What is the refractive index of the plastic?

   A 1.03

   B 1.29

   C 1.71

   D 1.77

5. Which of the following is correct when comparing particles in a gas with those in a liquid?

   A. They are closer together and more energetic in a gas.

   B. They are further apart and less energetic in a gas.

   C. They are closer together and less energetic in a liquid.

   D. They are further apart and more energetic in a liquid.

6. Water and aluminium, each with a mass of 1 kg, are placed in their own insulated containers. Initially, they are both at the same temperature and are each heated by 1 kJ.
   Assuming the heat is evenly distributed across each mass, which of the following statements is correct about their final temperatures?

   A. The final temperatures of both water and aluminium are the same.

   B. The final temperature of the water is lower than that of the aluminium.

   C. The final temperature of the water is higher than that of the aluminium.

   D. There is insufficient information to conclude about their final temperatures.

7. The latent heat of fusion, denoted by the symbol L, is the amount of energy required to transform the state of 1 kg of a substance from solid to liquid, and vice versa.
   Which of the following could be used as a unit for L?

   A. J^{-1} kg

   B. J kg

   C. $J\,m^{-3}$

   D. $J\,kg^{-1}$

8. Meena would like to raise the temperature of a 2.2-kg bar of aluminium from 22 °C to 44 °C.
   What is the amount of energy, in kJ, required to achieve this (assuming 100% efficiency)?

   A. 4.45 kJ

   B. 8.90 kJ

   C. 44.5 kJ

   D. 89.0 kJ

9. An electric kettle supplies 67 kJ of thermal energy to 750 mL of water at an initial temperature of 18 °C. Determine the final temperature of the water, assuming that all the thermal energy is transferred into the water.

   A. 21.3 °C

   B. 34 °C

   C. 37.3 °C

   D. 39.3 °C

10. How much energy, in MJ, is liberated when 1.3 kg of steam condenses into water at 100 °C?

    A. 0.43 MJ

    B. 2.3 MJ

    C. 3 MJ

    D. 4.3 MJ

11. At the Mawson Antarctic Research station, an electric heater supplies 37 GJ of energy to melt ice to produce water for consumption by the scientists there.
How much water, in kg, at 0 °C would be produced by this amount of energy?
   A. 1.12×10^4 kg
   B. 1.12×10^5 kg
   C. 1.12×10^6 kg
   D. 1.12×10^7 kg

12. An electric hotplate with a temperature of 750 K emits 1200 W of thermal energy.
What is the expected power emitted if the filament temperature is doubled to 1500 K?
   A. 12 000 W
   B. 19 200 W
   C. 24 000 W
   D. 29 200 W

13. Thermite is a reaction between aluminium powder and iron oxide, which generates temperatures of about 2200 °C.
What is the wavelength of the radiation emitted by thermite?
   A. 1.32×10^{-5} m
   B. 1.17×10^{-5} m
   C. 1.32×10^{-6} m
   D. 1.17×10^{-6} m

14. A pyrometer is a device that senses temperature from the radiation emitted from a surface. One such device detected a peak radiation of 6.5 μm from an object.
What is the surface temperature of the object?
   A. 173 °C
   B. 273 °C
   C. 346 °C
   D. 446 °C

15. A ray of light moving through air enters a block of an unknown transparent material at an angle of incidence of 50.0°. The angle of refraction is measured to be 31.6°.
The speed of light in this material is closest to
   A. 1.90×10^8 m s^{-1}.
   B. 2.03×10^8 m s^{-1}.
   C. 1.54×10^8 m s^{-1}.
   D. 3.0×10^8 m s^{-1}.

16. Which group of atmospheric gases contains only greenhouse gases?
   A. Carbon dioxide, water vapour, methane, argon
   B. Oxygen, nitrogen, methane, carbon dioxide
   C. Nitrous oxide, ozone, chlorofluorocarbon
   D. Methane, water vapour, carbon dioxide, neon

17. A mirage happens when
   A. the air is hot and the ground is very cool.
   B. both the air and the ground are very hot.
   C. the ground is slightly cooler than the air.
   D. the ground is very hot and the air is cool.

18. Which of the following is an example of a positive feedback mechanism that enhances the effect of global warming on climate change?
    A. Rising ocean temperatures increase evaporation, which condenses into clouds to reflect the incoming solar radiation.
    B. Rising ocean temperatures increase evaporation, which increases the amount of water vapour, a greenhouse gas.
    C. Rising ocean temperatures disrupt the natural flow of ocean currents to transfer heat from the tropical regions to the polar regions, resulting in a new ice age.
    D. Rising ocean temperatures increase evaporation, which creates a humid greenhouse-like condition, which absorbs even more solar radiation.

19. What is the mechanism for heat transfer in convection?
    A. Vibrational transfer of kinetic energy
    B. Emission of electromagnetic radiation
    C. More energetic particles colliding with less energetic particles
    D. Movement of fluid due to differences in buoyancy

20. Greenhouse gases such as methane (CH_4) and carbon dioxide (CO_2) absorb infrared radiation emitted from the surface of Earth, unlike gases such as oxygen (O_2) or nitrogen (N_2).
    What is this absorption of infrared radiation due to?
    A. These compounds contain a carbon atom, hence the term 'carbon pollution'.
    B. Molecules with three or more atoms bend and stretch, and can better absorb infrared radiation.
    C. There is a higher number of atoms in those molecules.
    D. Compounds absorb infrared radiation better than pure elements such as oxygen or nitrogen.

SECTION B   Short answer questions

Question 21 (5 marks)

A small piece of beef and a small piece of chicken are placed in an oven, each with a meat thermometer stuck into it to measure the temperature at its centre. After 30 minutes in the oven the thermometer in the chicken indicated the temperature to be 100 °C. After a further 30 minutes, the thermometer in the beef indicated the temperature at its centre to also be 100 °C. They remained in the oven to cook for a further 60 minutes. During this period, the thermometers steadily indicated 100 °C.

a. The piece of beef and the piece of chicken could be said to be at thermal equilibrium after this time.
   By referring to the kinetic theory of matter, explain what is meant by 'thermal equilibrium'.   **(2 marks)**
b. Discuss why it may be possible to estimate the temperature of the air in the oven near these two masses of meat, even if the oven does not have a thermometer. Hence, give an estimate of the temperature of the air in the oven near these two masses of meat.   **(3 marks)**

Question 22 (3 marks)

a. The peak wavelength of our Sun is approximately $\lambda_{max} = 501.7$ nm.
   Determine its surface temperature to the nearest kelvin.   **(1 mark)**
b. Name the process by which heat from the Sun is transferred to Earth through space.   **(1 mark)**
c. The Sun's core is hotter than its surface and hot plasma from the core rises towards the surface.
   This is an illustration of which transfer of heat process?   **(1 mark)**

Question 23 (4 marks)

Anton is carrying out an experiment to determine the quantity of aluminium rivets with an initial temperature of 23 °C required to cool 250 grams of boiling water at 100 °C inside an insulated container. Assuming the experiment was carried out with negligible heat loss, both the aluminium rivets and the water will reach thermal equilibrium at 90 °C.

a. Calculate the magnitude of heat lost by the water to the aluminium as it cools from 100 °C to the equilibrium temperature of 90 °C. **(2 marks)**

b. The amount of heat lost by the water is the heat gained by the aluminium.
Determine the mass of aluminium rivets as it warmed up from the initial temperature of 23 °C to the equilibrium temperature of 90 °C. **(2 marks)**

Question 24 (2 marks)

Kym and Shan are observing light emitted from different light-emitting diodes (LED). One of the LEDs emitted light at twice the electromagnetic radiation frequency of another LED. Kym said: 'According to Wien's Law, if the electromagnetic radiation frequency of one LED is twice that of the other, it would also be at twice the absolute temperature of the other.'
Shan disagrees.
What explanation might Shan offer to show that Kym is incorrect?

Question 25 (2 marks)

Habib's electric barbeque radiates 2250 W of energy when the heating element is at a temperature of 850 °C.
What is the expected power output when the heating element is at a temperature of 900 °C?

Question 26 (4 marks)

Identify each of the four types of electromagnetic waves with the wavelengths listed in the table.

Wavelength (m)	Electromagnetic wave type
a. 1.0×10^{-14}	
b. 1.0×10^{-2}	
c. 1.0×10^{4}	
d. 1.0×10^{-8}	

PRACTICE SCHOOL-ASSESSED COURSEWORK

ASSESSMENT TASK — A REPORT ON A SELECTED SCIENTIFIC PHENOMENON

In this task, you will be required to report on the phenomenon of the greenhouse effect, linking this to thermodynamic principles.

- This practice SAC requires you to write a report; a structured set of questions is supplied to assist you to write your report.
- You may use the VCAA Physics formula sheet and a scientific calculator for this task.
- You may conduct research before commencing your write-up to assist you in this task.

Total time: 50 minutes (5 minutes reading and 45 minutes writing)

Total marks: 30 marks

A WORLDWIDE CRISIS — THE IMPACT OF THE GREENHOUSE EFFECT

Humans have a dramatic impact on the environment and unfortunately it is often a detrimental one. From deforestation to pollution, the environment is being continually affected, creating huge scientific and ecological problems.

Since the Industrial Revolution, human impact on increasing greenhouse gases has been particularly evident, and has lead to global warming and damage to ecosystems around the world. In Australia, the Great Barrier Reef has been significantly affected by coral bleaching caused by this global warming.

Write a report with reference to the following concepts. You should use subheadings to clearly frame your response.

1. Identify and describe the types of electromagnetic radiation emitted from the Sun.
2. Explain the roles of conduction, convection and radiation in moving heat around in Earth's mantle and surface, and its atmosphere. Identify which of these is most important in contributing to the greenhouse effect.
3. Provide background information on the greenhouse effect, with a clear link to global warming using thermodynamic principles.
4. Explain how greenhouse gases absorb and re-emit infrared radiation, and describe why different greenhouse gases are involved.
5. Clearly describe the evidence of the effect of human activity in creating an enhanced greenhouse effect.
6. Describe how you would collect evidence on the enhanced greenhouse effect, explaining how you can ensure reliability and validity while minimising uncertainty.
7. Apply thermodynamic principles to investigate one of the following issues (using evidence and data) related to the impacts of human activity on the enhanced greenhouse effect. In your response, you need to show a clear link to theory and describe how it may have an impact on the greenhouse effect. You should also suggest solutions that can help minimise the enhanced greenhouse effect for your given issue.
   - Proportion of national energy use due to heating and cooling of homes
   - Comparison of the operation and efficiencies of domestic heating and cooling systems: heat pumps, resistive heaters, reverse-cycle air conditioners, evaporative coolers, solar hot-water systems and/or electrical resistive hot-water systems
   - Possibility of homes being built that do not require any active heating or cooling at all
   - Use of thermal imaging and infrared thermography in locating heating losses in buildings and/or system malfunctions; cost savings implications

- Determination of the energy ratings of home appliances and fittings: insulation, double glazing, window size, light bulbs and/or electrical gadgets, appliances or machines
- Cooking alternatives: appliance options (microwave, convection, induction), fuel options (gas, electricity, solar, fossil fuel)
- Automobile efficiencies: fuel options (diesel petrol, LPG and electric), air delivery options (naturally aspirated, supercharged and turbocharged) and fuel delivery options (common rail, direct injection and fuel injection)

 Resources

Digital document School-assessed coursework (doc-38058)

4 Radiation from the nucleus and nuclear energy

KEY KNOWLEDGE

In this topic, you will:
- explain nuclear stability with reference to the forces in the nucleus including electrostatic forces, the strong nuclear force and the weak nuclear force
- model radioactive decay as random decay with a particular half-life, including mathematical modelling with reference to whole half-lives
- describe the properties of α, β^-, β^+ and γ radiation
- explain nuclear transformations using decay equations involving α, β^-, β^+ and γ radiation
- analyse decay series diagrams with reference to type of decay and stability of isotopes
- explain the effects of α, β and γ radiation on humans, including:
  - different capacities to cause cell damage
  - short- and long-term effects of low and high doses
  - ionising impacts of radioactive sources outside and inside the body
  - calculations of absorbed dose (gray), equivalent dose (sievert) and effective dose (sievert)
- evaluate the use of medical radioisotopes in therapy including the effects on healthy and damaged tissues and cells
- explain, qualitatively, nuclear energy as energy resulting from the conversion of mass
- explain fission chain reactions including:
  - the effect of mass and shape on criticality
  - neutron absorption and moderation
- compare the processes of nuclear fusion and nuclear fission
- explain, using a binding energy curve, why both fusion and fission are reactions that release energy
- investigate the viability of nuclear energy as an energy source for Australia.

Source: VCE Physics Study Design (2023–2027) extracts © VCAA; reproduced by permission.

PRACTICAL WORK AND INVESTIGATIONS

Practical work is a central component of VCE Physics. Experiments and investigations, supported by a **practical investigation eLogbook** and **teacher-led videos**, are included in this topic to provide opportunities to undertake investigations and communicate findings.

EXAM PREPARATION

Access exam-style questions and their video solutions in every lesson, to ensure you are ready.

4.1 Overview

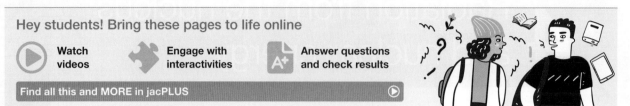
4.1.1 Introduction

The nucleus was first described in 1911 by Ernest Rutherford. Earlier discoveries, such as x-rays in 1895 by Wilhelm Röntgen, radioactivity in 1896 by Henri Becquerel, and new radioactive elements in 1898 by Marie and Pierre Curie, were both explained and made possible by the nuclear model of the atom developed by Rutherford.

By investigating how the energy from the nucleus is used, you will be better armed to identify some of the major challenges we face, whether they be scientific, economic, technological, environmental or medical. You will build on your understanding of how the processes of fusion and fission relate to energy production and learn about the effect of radiation on the human body.

FIGURE 4.1 The Super-Kamiokande neutrino observatory is situated 1000 metres underground to insulate it from other subatomic particles. Neutrinos are incredibly difficult to detect, with observatories such as Super-Kamiokande detecting only a handful each month.

LEARNING SEQUENCE

on Resources

Solutions   Solutions — Topic 4 (sol-0790)

Practical investigation eLogbook   Practical investigation eLogbook — Topic 4 (elog-1573)

Digital documents   Key science skills — VCE Physics Units 1–4 (doc-36950)
Key terms glossary — Topic 4 (doc-36957)
Key ideas summary — Topic 4 (doc-36958)

Exam question booklet   Exam question booklet — Topic 4 (eqb-0072)

4.2 Nuclear stability and nuclear radiation

4.2.1 Atoms and isotopes

Nuclear radiation, as the name suggests, is radiation emitted from the nucleus of an atom. In order to explain the mechanisms that release such radiation, it is important to understand a little about the structure of the atom.

Atoms

All matter is made up of atoms. Each atom consists of a tightly packed, positively charged centre called the **nucleus**, which is surrounded by a 'cloud' of negatively charged electrons. The particles in the nucleus are known collectively as **nucleons**, but there are two different types — protons and neutrons. Protons are positively charged, neutrons are chargeless, and both are about 2000 times heavier than the electrons that surround the nucleus, with the neutron having marginally more mass than the proton.

Scientists name atoms according to their atomic number, which is the number of protons in the nucleus. For example, all atoms with six protons are called carbon, all atoms with 11 protons are called sodium, and all atoms with 92 protons are called uranium. A substance consisting of atoms that all have the same name is called an **element**. Each element's name has its own shorthand symbol that scientists use. Carbon has the symbol 'C', sodium 'Na' and uranium 'U'. It is very important that the upper- or lowercase of the letters used in the symbols is kept the same. The names of all the elements, and their symbols, can be found in the periodic table in figure 4.3.

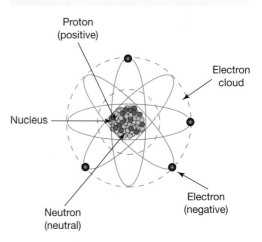

FIGURE 4.2 The structure of a typical atom

Proton (positive)

Electron cloud

Nucleus

Electron (negative)

Neutron (neutral)

Isotopes

Not all atoms of the same name (and therefore atoms having the same number of protons) exhibit the same number of neutrons. For instance, it is possible to find carbon atoms with six, seven and eight neutrons in the nucleus along with the six protons that make it carbon. These different forms of an element are called **isotopes**. To avoid confusion about which isotope is being referred to, scientists have a few standard ways of writing them. The number of nucleons, or **mass number**, of the particular isotope is used. The isotope of carbon with six protons and six neutrons is written as carbon-12 or ^{12}C, whereas the isotope of carbon with eight neutrons is written as carbon-14 or ^{14}C. Sometimes the number of protons, or **atomic number**, is written directly underneath the mass number, although this is not necessary (for example, $^{14}_{6}$C).

nucleus the solid centre of an atom where most of its mass is concentrated

nucleon a particle found in the nucleus — proton or neutron

element a substance that consists only of atoms of the same name

isotope an atom containing the same number of protons but a different numbers of neutrons

mass number the total number of nucleons in an atom

atomic number the number of protons in a nucleus

FIGURE 4.3 The periodic table

Key

2	Atomic number
Helium	Name
He	Symbol
4.0	Relative atomic mass

Period 1

1
Hydrogen
H
1.0

Group 1
Group 2
Group 3
Group 4
Group 5
Group 6
Group 7
Group 8
Group 9
Group 10
Group 11
Group 12
Group 13
Group 14
Group 15
Group 16
Group 17
Group 18

Period 1

1		2
Hydrogen		Helium
H		He
1.0		4.0

Period 2

3	4	5	6	7	8	9	10
Lithium	Beryllium	Boron	Carbon	Nitrogen	Oxygen	Fluorine	Neon
Li	Be	B	C	N	O	F	Ne
6.9	9.0	10.8	12.0	14.0	16.0	19.0	20.2

Period 3

11	12	13	14	15	16	17	18
Sodium	Magnesium	Aluminium	Silicon	Phosphorus	Sulfur	Chlorine	Argon
Na	Mg	Al	Si	P	S	Cl	Ar
23.0	24.3	27.0	28.1	31.0	32.1	35.5	39.9

Period 4

19	20	21	22	23	24	25	26	27	28	29	30	31	32	33	34	35	36
Potassium	Calcium	Scandium	Titanium	Vanadium	Chromium	Manganese	Iron	Cobalt	Nickel	Copper	Zinc	Gallium	Germanium	Arsenic	Selenium	Bromine	Krypton
K	Ca	Sc	Ti	V	Cr	Mn	Fe	Co	Ni	Cu	Zn	Ga	Ge	As	Se	Br	Kr
39.1	40.1	45.0	47.9	50.9	52.0	54.9	55.8	58.9	58.7	63.5	65.4	69.7	72.6	74.9	79.0	79.9	83.8

Period 5

37	38	39	40	41	42	43	44	45	46	47	48	49	50	51	52	53	54
Rubidium	Strontium	Yttrium	Zirconium	Niobium	Molybdenum	Technetium	Ruthenium	Rhodium	Palladium	Silver	Cadmium	Indium	Tin	Antimony	Tellurium	Iodine	Xenon
Rb	Sr	Y	Zr	Nb	Mo	Tc	Ru	Rh	Pd	Ag	Cd	In	Sn	Sb	Te	I	Xe
85.5	87.6	88.9	91.2	92.9	96.0	(98)	101.1	102.9	106.4	107.9	112.4	114.8	118.7	121.8	127.6	126.9	131.3

Period 6

55	56	57–71	72	73	74	75	76	77	78	79	80	81	82	83	84	85	86
Caesium	Barium	Lanthanoids	Hafnium	Tantalum	Tungsten	Rhenium	Osmium	Iridium	Platinum	Gold	Mercury	Thallium	Lead	Bismuth	Polonium	Astatine	Radon
Cs	Ba		Hf	Ta	W	Re	Os	Ir	Pt	Au	Hg	Tl	Pb	Bi	Po	At	Rn
132.9	137.3		178.5	180.9	183.8	186.2	190.2	192.2	195.1	197.0	200.6	204.4	207.2	209.0	(210)	(210)	(222)

Period 7

87	88	89–103	104	105	106	107	108	109	110	111	112	113	114	115	116	117	118
Francium	Radium	Actinoids	Rutherfordium	Dubnium	Seaborgium	Bohrium	Hassium	Meitnerium	Darmstadtium	Roentgenium	Copernicium	Nihonium	Flerovium	Moscovium	Livermorium	Tennessine	Oganesson
Fr	Ra		Rf	Db	Sg	Bh	Hs	Mt	Ds	Rg	Cn	Nh	Fl	Mc	Lv	Ts	Og
(223)	(226)		(261)	(262)	(266)	(264)	(267)	(268)	(271)	(272)	(285)	(280)	(289)	(289)	(292)	(294)	(294)

Lanthanoids

57	58	59	60	61	62	63	64	65	66	67	68	69	70	71
Lanthanum	Cerium	Praseodymium	Neodymium	Promethium	Samarium	Europium	Gadolinium	Terbium	Dysprosium	Holmium	Erbium	Thulium	Ytterbium	Lutetium
La	Ce	Pr	Nd	Pm	Sm	Eu	Gd	Tb	Dy	Ho	Er	Tm	Yb	Lu
138.9	140.1	140.9	144.2	(145)	150.4	152.0	157.3	158.9	162.5	164.9	167.3	168.9	173.1	175.0

Actinoids

89	90	91	92	93	94	95	96	97	98	99	100	101	102	103
Actinium	Thorium	Protactinium	Uranium	Neptunium	Plutonium	Americium	Curium	Berkelium	Californium	Einsteinium	Fermium	Mendelevium	Nobelium	Lawrencium
Ac	Th	Pa	U	Np	Pu	Am	Cm	Bk	Cf	Es	Fm	Md	No	Lr
(227)	232.0	231.0	238.0	(237)	(244)	(243)	(247)	(247)	(251)	(252)	(257)	(258)	(259)	(262)

Alkali metal
Alkaline earth metal
Transition metal
Lanthanoids
Actinoids
Unknown chemical properties
Post-transition metal
Metalloid
Reactive non-metal
Halide
Noble gas

tlvd-3742

SAMPLE PROBLEM 1 Identifying an isotope

Write the name of the isotope of an atom with 90 protons and 144 neutrons.

THINK	WRITE
1. What element has 90 protons?	Checking the periodic table, the element thorium (symbol Th, atomic number 90) has 90 protons.
2. What is the mass number of this isotope?	$90 + 144 = 234$
3. Write the name of the isotope.	The isotope of an atom with 90 protons and 144 neutrons is thorium-234 or ^{234}Th.

PRACTICE PROBLEM 1

What is the name of the isotope of an atom with 26 protons and 30 neutrons?

What holds the nucleus together?

The force that holds electrons around a nucleus is called an **electrostatic force**. Electrostatic forces increase as charges move closer together. Electrostatic attraction exists between unlike charges; electrostatic repulsion exists between like charges. So it seems strange that the positive charges inside a nucleus don't repel each other so strongly that the nucleus splits apart. In fact, two protons do repel each other when they are brought together, but in the nucleus they are so close to each other that the force of repulsion is overcome by an even stronger force — the **strong nuclear force**. This is an attractive force binding the protons and neutrons in the nucleus of an atom. While the strong nuclear force is a very strong force, it can act over incredibly small distances only (10^{-15} m). Inside a nucleus, the nucleons are sufficiently close that the pull of the strong nuclear force is much greater than the push of the protons repelling each other, and therefore the nucleus remains intact. Figure 4.4 shows how the strength and sense of the strong nuclear force changes with the separation of the nucleons. Its ability to hold two nucleons together is strongest at around 1×10^{-15} metres but is virtually zero at about 2.5 times this distance. Thus, this very strong force of attraction occurs over a relatively short range within the nucleus.

electrostatic force the force between two stationary charged objects

strong nuclear force the attractive force that holds nucleons together in a nucleus of an atom, acting over only very short distances

FIGURE 4.4 How the strong nuclear force between two nucleons varies with the separation of the nucleons

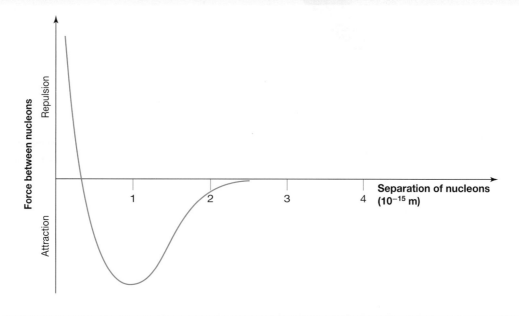

Unstable nuclei

Some nuclei are unstable. Isotopes that contain unstable nuclei are called **radioisotopes**. Unstable isotopes can emit various types of radiations. The reason that some nuclei are unstable can be explained by considering the interaction of the nucleons with the fundamental forces.

Though the strong nuclear force acts over very short distances only, the electromagnetic force continues to act with decreasing strength as the distance between charged particles increases. This is one reason that many nuclei are unstable. In a large nucleus, the net effect of all the protons in the nucleus results in an electrostatic repulsion on some protons greater than the strong force holding those protons in the nucleus. This nucleus is unstable and cannot remain like this forever. This explains why the number of neutrons in large atoms is greater than the number of protons; neutrons are chargeless, so do not contribute to the electromagnetic repulsion. This trend is apparent in figure 4.5.

FIGURE 4.5 This graph shows which nuclei are stable (green) and which are unstable (other).

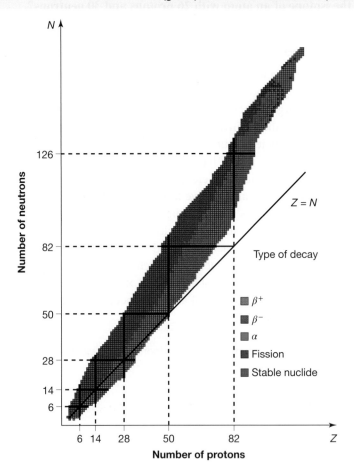

The **weak nuclear force** and the gravitational force are not significant in determining the stability of nuclei as they are comparatively weak over the distance of nucleon separation.

In order to become more stable, the nucleus in a radioisotope emits different types of nuclear radiation that increase the stability of the nucleus. The different types of radiation will be explained in subtopic 4.3.

radioisotope an unstable isotope
weak nuclear force the force that explains the transformation of neutrons into protons, and vice versa

TABLE 4.1 Types of forces, their relative strengths and ranges

Force	Strength	Range	Type
Strong nuclear	1	10^{-15} m (diameter of a nucleus)	Attractive
Electric and magnetic	$\dfrac{1}{137}$	Infinite	Attractive or repulsive
Weak nuclear	$\dfrac{1}{1\ 000\ 000}$	10^{-18} m (fraction of proton diameter)	Attractive
Gravity	$\dfrac{1}{10^{38}}$	Infinite	Attractive

tlvd-3743

SAMPLE PROBLEM 2 Explaining nuclear stability

The existence of nuclei larger than a single proton depends on forces holding neutrons and protons together. Describe the nature of this force.

THINK	WRITE
1. A nucleus larger than a proton includes protons and neutrons that are somehow held together.	There is an attractive force called the strong nuclear force that acts between protons and neutrons.
2. Protons repel each other as they have like charge.	The strong nuclear force can only hold a nucleus together when it is stronger than the electromagnetic force pushing the protons apart.
3. Describe the nature of the force.	The force holding protons and neutrons together to form nuclei must be an attractive force that works between neutrons and protons and is stronger than the electromagnetic force at close range.

PRACTICE PROBLEM 2

A helium-3 nucleus has two protons and one neutron. Describe the forces acting within it.

4.2.2 Half-life

It is not possible to predict exactly when a given unstable nucleus will decay, as this is a spontaneous event unaffected by any physical or chemical influences. However, we can predict what proportion of a certain number of nuclei will decay in a given time. It is rather like tossing a coin. We can't be sure that a given toss will result in a tail but we can predict that from 1000 tosses about 500 will result in tails. Scientists know that it will take 24 days for half of a group of unstable thorium-234 nuclei to decay to protactinium-234. The time taken for half a group of unstable nuclei to decay is called the **half-life**. Half-lives vary according to the isotope that is decaying. They range from trillionths of a nanosecond to thousands of millions of years, as illustrated in table 4.2.

Mathematicians and scientists often use graphs with the same basic shape as the one in figure 4.6. It shows what is known as a **decay curve**. This type of curve often appears in science. It is called exponential decay.

half-life the time taken for half of a group of unstable nuclei to decay

decay curve a graph of the number of nuclei remaining in a substance versus the time elapsed; the half-life of a substance can be determined by looking at the time that corresponds to half of the substance remaining

TABLE 4.2 Table of half-lives of some radioisotopes

Element	Half-life
Tellurium-128	2.2×10^{24} years
Indium-115	6×10^{14} years
Potassium-40	1.3×10^{9} years
Uranium-235	7.1×10^{8} years
Plutonium-239	24 110 years
Carbon-14	5730 years
Radium-226	1600 years
Actinium-227	22 years
Thorium-227	18 days
Carbon-11	20 minutes
Thallium-207	4.8 minutes
Magnesium-23	11 seconds
Polonium-212	3×10^{-7} seconds
Lithium-5	304×10^{-24} seconds

Looking at figure 4.6, in the first few days the number of atoms decaying every day is quite high but towards the end the number is quite low. If there was a Geiger counter near the source at the beginning, it would be clicking quickly, but near the end you would wait days for the next click. In fact, a graph of the count rate (the number of clicks per second) will have exactly the same shape and will show the same half-life. The number of decays per second of a radioactive source is a measure of its activity and is measured in becquerels (Bq).

TABLE 4.3 Activities of some everyday items

Item	Activity (Bq)
1 adult human	3000
1 kg coffee	1000
1 kg granite	1000
1 kg coal ash (used in cement)	2000
1 kg superphosphate fertiliser	5000

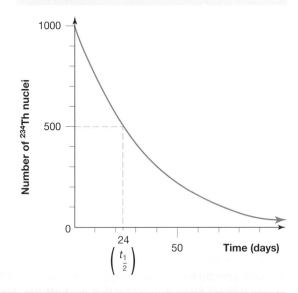

FIGURE 4.6 Graph showing decay of 1000 thorium-234 nuclei. The time taken for half of the original nuclei (500 nuclei) to remain is called the half-life. It can be seen from this graph that the half-life is 24 days. After the fourth half-life (at time = 96 days) it can be predicted that one-sixteenth of the thorium-234 (about 62 or 63 nuclei) would remain undecayed.

tlvd-0056

SAMPLE PROBLEM 3 Calculations involving half-lives

Technetium-99 is often used for medical diagnosis. It has a half-life of 6 hours. A patient has a small amount of the isotope injected into the bloodstream. What fraction of the original amount will remain after:

a. 12 hours

b. 48 hours?

THINK	WRITE
a. 1. How many half-lives have passed?	**a.** $\dfrac{12}{6} = 2$ half-lives
2. How much would be left after halving it that many times?	$\dfrac{1}{2} \times \dfrac{1}{2} = \left(\dfrac{1}{2}\right)^2 = \dfrac{1}{4}$
3. State the solution.	One-quarter would be left after 12 hours.
b. 1. How many half-lives have passed?	**b.** $\dfrac{48}{6} = 8$ half-lives
2. How much would be left after halving it that many times?	$\left(\dfrac{1}{2}\right)^8 = \dfrac{1}{256}$
3. State the solution.	$\dfrac{1}{256}$ would be left after 48 hours.

PRACTICE PROBLEM 3

Cobalt-60 is radioactive with a half-life of 5.27 years. It is produced from cobalt-59 by bombardment with neutrons at a nuclear reactor. It emits a low-energy beta particle followed by two high-energy gamma rays. It is used in the sterilisation of medical equipment, in radiotherapy and in industrial applications. A cobalt-60 source will need to be replaced when its activity decreases to $\frac{1}{16}$ of its initial value. How long will this take?

4.2 Activities

4.2 Quick quiz	4.2 Exercise	4.2 Exam questions

4.2 Exercise

1. Name the isotope that has an atomic number of 11 and contains 12 neutrons.
2. How many protons and neutrons are there in one atom of each of the following isotopes?
   a. Hydrogen-2, also known as deuterium
   b. Americium-241
   c. Europium-164
3. What force acts between a proton and a neutron and under what conditions does it act?
4. The half-life of caesium-134 is 2.06 years. What fraction of caesium is left after 10.3 years?
5. What is meant when an isotope is described to be stable?
6. If the strong nuclear force acts between all nucleons and is much stronger than the electromagnetic force at the distance of two neighbouring nucleons in a nucleus, explain how a nucleus might be unstable.
7. 100 grams of a radioactive isotope is delivered to a hospital. Three hours later there is only 6.25 grams left. What is the half-life of the isotope?
8. Sketch a graph of the decay of the isotope in question **7**.
9. Why are half-lives used for radioactive isotopes rather than just stating how long they take to decay?

10. The activity (decays per second) of a radioactive isotope halves with each half-life. Explain why that would be the case.

11. You need 20 grams of a radioactive isotope with a half-life of 3 hours. How much would you need to buy if it takes 24 hours to deliver it?

4.2 Exam questions

▶ Question 1 (1 mark)

MC What is the atomic number (Z) of an element?
A. The number of neutrons in the nucleus
B. The number of nucleons in the nucleus
C. The number of protons in the nucleus
D. The total number of nucleons and electrons in the atom

▶ Question 2 (1 mark)

MC Which of the following best describes all the isotopes of an element?
A. Same number of protons; different number of neutrons
B. Same number of neutrons; different number of protons
C. Same number of nucleons; different number of protons
D. Same number of nucleons; different number of neutrons

▶ Question 3 (1 mark)

MC A radioactive element has a half-life of 6.0 days. A sample of this material initially contains 12 400 nuclei.

How many radioactive nuclei are remaining after 12 days?
A. 6200
B. 3100
C. 1550
D. 775

▶ Question 4 (3 marks)

A radioactive source has a half-life of 50 years. It currently contains 10 000 of the unstable nuclei.

How many unstable nuclei were there in the source 200 years ago? Show your reasoning.

▶ Question 5 (3 marks)

The initial activity of a radioactive source is 8000 decays per second. After 1 day it has decreased to 1000 decays per second.

What is the half-life of this source (in hours)? Show your reasoning.

More exam questions are available in your learnON title.

4.3 Types of nuclear radiation

KEY KNOWLEDGE

- Describe the properties of α, β^-, β^+ and γ radiation
- Explain nuclear transformations using decay equations involving α, β^-, β^+ and γ radiation
- Analyse decay series diagrams with reference to type of decay and stability of isotopes

Source: VCE Physics Study Design (2023–2027) extracts © VCAA; reproduced by permission.

Unstable isotopes can emit various types of radiation while 'striving' to become more stable. There are three naturally occurring forms of nuclear radiation: α, β and γ (pronounced *alpha, beta* and *gamma*). Each type of radiation was named with a different Greek letter because, when the different types of radiation were discovered, scientists did not know what they consisted of. The emissions are described as decay processes because the nucleus changes into a different nucleus and the change is irreversible.

4.3.1 Alpha (α) decay

During α decay an unstable nucleus ejects a relatively large particle known as an **α particle**. This actually consists of two protons and two neutrons, and so may be called a helium nucleus. The remainder of the original nucleus, known as the **daughter nucleus**, is now more stable.

FIGURE 4.7 α decay: a nucleus ejects a helium nucleus

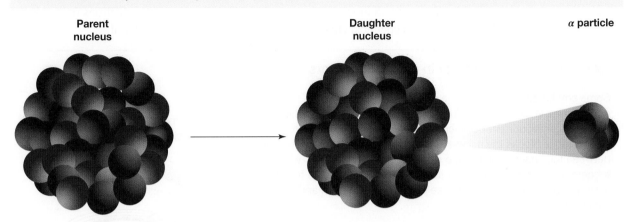

Parent nucleus

Daughter nucleus

α particle

The number of protons in the nucleus determines the elemental name of the atom. The daughter nucleus is therefore of a different element. For example, uranium-238 decays by emitting an α particle. The uranium-238 atom contains 92 protons and 146 neutrons (238 – 92 = 146). It emits an α particle, with two protons and two neutrons. The original nucleus is left with four less nucleons: 90 protons (92 – 2 = 90) and 144 neutrons (146 – 2 = 144). As the daughter nucleus now has 90 protons, it is called thorium and has the symbol Th. This particular isotope of thorium has 234 nucleons (90 protons and 144 neutrons) and is more correctly written as thorium-234.

The information in the previous paragraph can be written much more effectively in symbols. This is called the **decay equation**:

$$^{238}_{92}U \rightarrow\ ^{234}_{90}Th + ^{4}_{2}He + energy$$

$$or\ ^{238}_{92}U \rightarrow\ ^{234}_{90}Th + \alpha + energy$$

Note that the total number of nucleons (mass number) is conserved: in this case, there are 92 nucleons before and after (90 + 2) the decay.

This is the law of conservation of nucleon number, and it applies for any type of radioactive decay.

Properties of alpha radiation

The ejected α particle is relatively slow and heavy compared to other forms of nuclear radiation. The particle travels at 5–7 per cent of the speed of light: roughly 2×10^7 metres each second. Every object that moves has a form of energy known as kinetic energy, or energy of motion. Because the α particle is moving, it has kinetic energy. That energy is written into the decay equation.

In addition to having energy, the α particle has an overall charge of +2 because it contains two protons. This charge means the particle can be deflected by electric and magnetic fields — properties that helped scientists determine what an α particle consisted of.

α particle a relatively slow-moving decay product consisting of two protons and two neutrons, equivalent to a helium nucleus and carrying a positive charge

daughter nucleus the nucleus remaining after an atom undergoes radioactive decay; it is more stable than the original nucleus

decay equation a representation of a decay reaction; it shows the changes occurring in nuclei and lists the products of the decay reaction

The mass and charge of the particle ensure that it cannot travel more than ten centimetres through the air, and during this distance it interacts with other matter, including atoms and molecules. A sheet of paper will stop most α particles. For the same reason, an α particle's ability to penetrate the skin of animals and humans is extremely limited. However, if an α-emitting source decays inside the body, it can do significant damage. As a result, α emitters are rarely used in nuclear medicine, although their lethal properties are being put to work in a targeted way to destroy cancer tissue.

α particles are a strongly ionising form of radiation. As charged particles they can remove electrons from other atoms they approach, which can result in chemical changes in body tissue and genetic material.

EXTENSION: Radioactivity in smoke detectors

Some smoke detectors include a sample of the isotope americium-241, an α emitter. The α particles ionise the air, which sustains a small current between charged plates in the detector. When smoke blocks the access of the α particles to the air molecules between the charged plates, the current can no longer flow, which triggers an alarm to go off. It is estimated that an α particle travelling through the air for 10 cm will ionise up to 10 000 oxygen and nitrogen molecules.

FIGURE 4.8 Smoke alarms are set off when smoke blocks the path of α particles.

Discovering the nucleus

Ernest Rutherford was one of the central players in the investigation of radioactivity. By 1908, Rutherford and his team had determined that α particles were doubly charged helium ions and that they were moving very fast — at about five per cent of the speed of light. They quickly realised that these particles would be ideal 'bullets' to investigate the structure of the atom through a series of 'scattering' experiments involving various metals.

Two of Rutherford's younger colleagues, Hans Geiger and Ernest Marsden, fired α particles at a very thin foil of gold, about 400 atoms thick, and measured their different angles of deflection. Nearly all of the particles either went straight through or suffered a very small deflection, but about 1 in every 8000 rebounded from the gold's surface.

FIGURE 4.9 Geiger and Marsden's gold foil experiment

a.

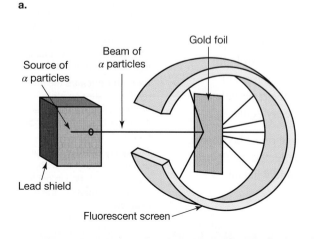

b.

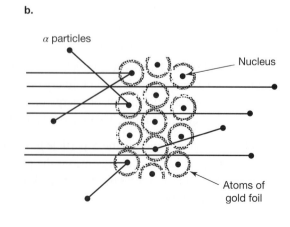

The positively charged α particle was repelled and deflected by the electrostatic interaction with the positive charges in the atom. Rutherford calculated that for an α particle to be turned around, these positive charges would need to be concentrated in a very small volume, which he called the *nucleus*. He calculated that the radius of such a nucleus would be about 10^{-14} metres, and the radius of an atom was about 10^{-10} metres.

Rutherford's nuclear model of the atom had nearly all the mass of the atom in the central nucleus and the much lighter electrons 'orbiting' around it.

However, this model was incomplete because it did not fully explain the mass of an atom. The neutron was subsequently discovered by James Chadwick in 1932.

4.3.2 Beta (β) decay

Two types of β decay are possible. The β^- particle is a fast-moving electron that is ejected from an unstable nucleus. The β^+ particle is a positively charged particle with the same mass as an electron and is the electron's antiparticle, the **positron**. Positrons are mostly produced in the atmosphere by **cosmic radiation**, but some nuclei do decay by β^+ emission.

FIGURE 4.10 β^- decay: a nucleus ejects an electron

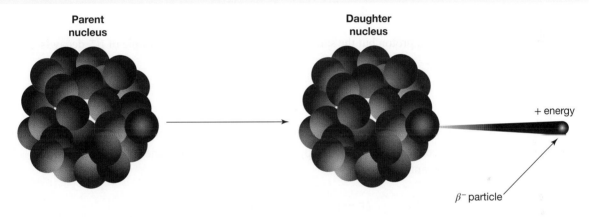

In β^- decay, an electron is emitted from inside the nucleus. Since nuclei do not contain any electrons this might seem strange, but it is in fact true! There is no change whatsoever to the electrons in the shells surrounding the nucleus.

Some very interesting changes take place inside a nucleus to enable it to emit an electron. One of the neutrons in the nucleus transforms into a proton and an electron. The proton remains in the nucleus and the electron is emitted and called a β^- particle:

positron a positively charged particle with the same mass as an electron, formed when a proton disintegrates to form a neutron and a positron

cosmic radiation very energetic charged particles, mainly protons, originating from beyond the solar system

$$\begin{smallmatrix}1\\0\end{smallmatrix}n \rightarrow \begin{smallmatrix}1\\1\end{smallmatrix}p + \begin{smallmatrix}0\\-1\end{smallmatrix}e$$

Note that the total charge is conserved: it is 0 before and after $(1 - 1)$ the decay.

The resulting daughter nucleus has the same number of nucleons as the parent, but one less neutron and one more proton. Thus, the daughter nucleus has an atomic number that has increased by one and constitutes a different element.

An example of β^- decay is the decay of thorium-234. This isotope is the result of the α decay of uranium-238. The nucleus is more stable than it was before the emission of the α particle but could become more stable by emitting a β^- particle. During this second decay, the mass number of the nucleus is unchanged (234). The number of protons, however, increases by one when a neutron changes into a proton and an electron. There are now 91 (90 + 1) protons in the nucleus, so the atom must be called protactinium-234.

β^- decay is accompanied by the emission of another particle called an antineutrino. β^+ decay is accompanied by the emission of a neutrino. Neutrinos are particles with no charge and nearly zero mass. They are non-ionising particles.

The decay equation is written as:

$$^{234}_{90}\text{Th} \rightarrow {}^{234}_{91}\text{Pa} + \beta^- + \bar{\nu} + \text{energy}$$

$$\text{or } {}^{234}_{90}\text{Th} \rightarrow {}^{234}_{91}\text{Pa} + {}^{0}_{-1}\text{e} + \bar{\nu} + \text{energy}$$

In β^+ decay, also known as positron emission, the positron is also emitted from inside the nucleus. In this case, strange as it may seem, the proton changes into a neutron and a positron, with the neutron staying in the nucleus:

$$^{1}_{1}\text{p} \rightarrow {}^{1}_{0}\text{n} + {}^{0}_{+1}\text{e} + \nu + \text{energy}$$

The resulting nucleus has one less proton but the same number of nucleons. An example of β^+ decay is sodium-22 decaying to neon-22:

$$^{22}_{11}\text{Na} \rightarrow {}^{22}_{10}\text{Ne} + {}^{0}_{+1}\text{e} + \nu + \text{energy}$$

β^- decay: A neutron, $^{1}_{0}\text{n}$, is converted to a proton, $^{1}_{1}\text{p}$, and the process releases energy and creates an electron, $^{0}_{-1}\text{e}$, and an electron antineutrino, $\bar{\nu}$.

$$^{1}_{0}\text{n} \rightarrow {}^{1}_{1}\text{p} + {}^{0}_{-1}\text{e} + \bar{\nu} + \text{energy}$$

β^+ decay: A proton, $^{1}_{1}\text{p}$, is converted to a neutron, $^{1}_{0}\text{n}$, and the process releases energy and creates a positron, $^{0}_{+1}\text{e}$, and an electron neutrino, ν.

$$^{1}_{1}\text{p} \rightarrow {}^{1}_{0}\text{n} + {}^{0}_{+1}\text{e} + \nu + \text{energy}$$

Properties of beta radiation

β particles are very light when compared to α particles. They travel at a large range of speeds — from that of an α particle up to 99 per cent of the speed of light. Just like α particles, β particles are deflected by electric and magnetic fields.

The lower charge, mass and higher speed of the β particle ensures that it can penetrate much further through air and body tissues than α particles. β particles can travel through centimetres of human skin and a metre or two of air. They can be stopped by a few centimetres of water or a few millimetres of aluminium. β radiation is useful for precisely measuring the thickness of paper in production and for treating some types of cancers. It is also a form of ionising radiation. However, it is less ionising than α radiation. In air, β particles travel further — up to 20 cm — and have a haphazard pathway due to their smaller mass, compared with the track of an α particle.

EXTENSION: Detecting radiation

We cannot see ionising radiation with the naked eye, but we have tools to detect it.

FIGURE 4.11 a. A cloud chamber and **b.** a Geiger counter

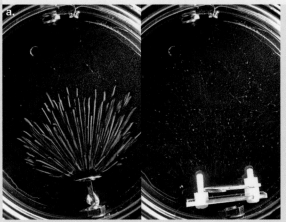

A cloud chamber (see figure 4.11a) detects ionising radiation including α and β particles, which momentarily produce distinctive tracks as they pass through the low-pressure supersaturated water vapour or alcohol in the chamber. α particles produce shorter, more distinct and straighter paths than β particles, which are longer, thinner and deviate in their paths.

A Geiger counter (see figure 4.11b) is an electronic device that detects and measures the frequency of ionising radiation including α and β particles, and gamma radiation. Radiation entering the chamber of the tube ionises the low-pressure gas within and creates a current within the device.

tlvd-0057

SAMPLE PROBLEM 4 Writing complete decay equations

Write down the complete decay equation in each of the following.

a. $^{234}_{92}\text{U} \rightarrow {}^{230}_{90}\text{Th} + ? + \text{energy}$

b. $^{210}_{82}\text{Pb} \rightarrow {}^{210}_{83}\text{Bi} + ? + \text{energy}$

c. $^{11}_{6}\text{C} \rightarrow {}^{11}_{5}\text{B} + ? + \text{energy}$

THINK	WRITE
a. 1. What is the mass number of the missing particle?	**a.** $234 - 230 = 4$
2. What is the atomic number of the missing particle?	$92 - 90 = 2$
3. Determine the missing particles from this information.	α particle
4. Write the equation.	The number of particles in the nucleus has decreased by 4, while the number of protons has decreased by 2. This implies that an α particle, or helium nucleus, has been released. The full equation is: $^{234}_{92}\text{U} \rightarrow {}^{230}_{90}\text{Th} + {}^{4}_{2}\text{He} + \text{energy}$

b. 1. What is the mass number of the missing particle?

2. What is the atomic number of the missing particle?

3. Determine the missing particles from this information.

4. Write the equation.

b. $210 - 210 = 0$

$82 - 83 = -1$

β^- particle

This equation cannot show an α emission, as the mass number remains constant. The atomic number has increased, indicating that a proton has been formed, and therefore β^- decay has occurred. The equation becomes:
$$^{210}_{82}\text{Pb} \rightarrow {}^{210}_{83}\text{Bi} + {}^{0}_{-1}\text{e} + \text{energy}$$

c. 1. What is the mass number of the missing particle?

2. What is the atomic number of the missing particle?

3. Determine the missing particles from this information.

4. Write the equation.

c. $11 - 11 = 0$

$6 - 5 = 1$

β^+ particle

The mass number stays the same, but there is one less proton, so it must be β^+ decay. The equation becomes:
$$^{11}_{6}\text{C} \rightarrow {}^{11}_{5}\text{B} + {}^{0}_{+1}\text{e} + \text{energy}$$

PRACTICE PROBLEM 4

Write the equations for:
a. the α decay of americium-241
b. the β^- decay of platinum-197
c. the β^+ decay of magnesium-23.

4.3.3 Gamma (γ) decay

This form of radioactive decay is quite different from either α or β decay. During γ decay, a small packet of electromagnetic energy called a **γ ray**, or photon, is emitted from the nucleus, rather than a particle. γ emission occurs after another form of nuclear decay has taken place. Following a decay, the arrangement of protons and neutrons in the nucleus may not be ideal, and the nucleus may need to release some extra energy to become more stable. Before the release of this energy, the nucleus is known as 'excited'. An **excited nucleus** is denoted by an asterisk (*) after the symbol for the element. The excess energy is emitted as a γ ray.

γ ray a packet of electromagnetic energy released when a nucleus remains unstable after α or β decay; γ rays travel at the speed of light and carry no charge

excited nucleus a nucleus that does not have an ideal arrangement of protons and neutrons within it; an excited nucleus emits γ radiation to become more stable

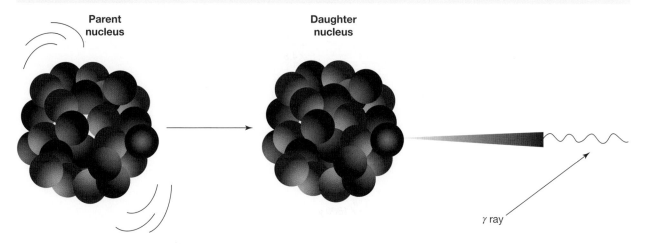

FIGURE 4.12 γ decay: an excited nucleus ejects a photon

Parent nucleus

Daughter nucleus

γ ray

One example of γ decay occurs after lead-210 emits a β^- particle and becomes bismuth-210. The excited daughter nucleus goes on to emit a γ ray:

$$^{210}_{83}\text{Bi}^* \rightarrow {}^{210}_{83}\text{Bi} + \gamma$$

Properties of gamma radiation

This γ ray is a packet of excess energy. It has no mass and no charge and is not deflected by electric or magnetic fields. Because it is a photon, or packet of electromagnetic energy, it travels at the speed of light.

γ radiation is extremely penetrating. A lead shield of a few centimetres thickness absorbs about 90 per cent of γ rays. γ radiation is ionising like α and β radiation. These properties of γ radiation make if very useful in medicine. It can be used to trace physiological processes by injecting or swallowing a chemical compound that contains γ emitters. The movement of the compound through the body can be traced with cameras outside the body detecting the γ rays. A newer technique involves using positron emitters. When the positrons are emitted they meet electrons and both are annihilated, producing γ rays that can be detected outside the body and an image can be constructed by computer. This is called positron emission tomography, or PET scans. γ rays are also a powerful means of killing cancerous tissue.

4.3.4 Decay series

In its 'quest' to become stable, an isotope may have to pass through many stages. As a radioactive isotope decays, the daughter nucleus is often radioactive itself. When this isotope decays, the resulting nucleus may also be radioactive. This sequence of radioisotopes is called a **decay chain** or decay series.

Uranium-238 undergoes 14 radioactive decays before it finally becomes the stable isotope lead-206. Two other decay chains, one starting with uranium-235 and another with thorium-232, also end with a stable isotope of lead. Another decay chain once passed through uranium-233, but this chain is almost extinct in nature now due to its shorter half-lives.

Not all naturally occurring radioisotopes are part of a decay series. There are about 10 with atomic numbers less than that of lead; for example, potassium-40. How can there be radioisotopes isolated in the periodic table? A look at the half-lives of potassium-40 and indium-115 reveals the answer (see table 4.2). Their half-lives are so long — greater than the age of Earth — that they are still decaying from when they were formed in a supernova explosion billions of years ago.

decay chain the sequence of stages that a radioisotope passes through to become more stable; at each stage, a more stable isotope forms, and the chain ends when a stable isotope forms; also known as the decay series

FIGURE 4.13 Radioactive series of uranium-238. The half-life is given beside each decay.

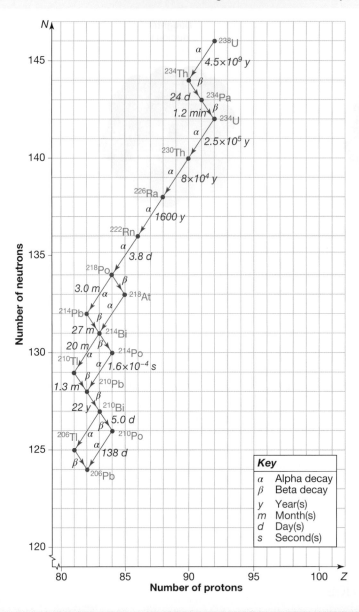

tlvd-3744

SAMPLE PROBLEM 5 Using the decay series to reflect on the stability of a daughter nucleus

Using the decay series, comment on whether the daughter nucleus is always more stable than its parent.

THINK	WRITE
1. How can I tell how stable a nucleus is?	The longer the half-life, the more stable the nucleus.
2. Do the daughter nuclei always have longer half-lives than the parent nucleus?	Often, in a decay chain, the daughter nucleus has a shorter half-life than the parent. For example, uranium-238 has a much longer half-life than its daughter thorium-234.
3. State the solution.	In a decay chain, a nucleus may decay to a less stable nucleus. This, however, is a step towards increased stability as the final nucleus is always stable.

140 Jacaranda Physics 1 VCE Units 1 & 2 Fifth Edition

PRACTICE PROBLEM 5

a. Apart from lead-206, which is stable, determine which nucleus in the uranium-238 decay chain is the most stable.

b. Why do you think the decay chain starts with a relatively stable nucleus?

EXTENSION: Nuclear transformations

α and β decay are natural examples of nuclear transformations. The numbers of protons and neutrons in the nucleus change during these processes. Artificial nuclear transformations are also possible. These are done either to investigate the structure of the nucleus or to produce specific radioisotopes for use in medicine or industry. The first artificial transformation was made by Ernest Rutherford, who fired α particles at nitrogen atoms to produce an isotope of oxygen.

$$^{14}_{7}N + {}^{4}_{2}He \rightarrow {}^{17}_{8}O + {}^{1}_{1}H$$

This result raised the intriguing possibility of achieving the alchemist's dream of changing lead into gold. Although prohibitively expensive, it appears to be theoretically possible.

The building of particle accelerators in the early 1930s enabled charged particles such as protons and α particles to be fired at atoms as well as α particles, but with the advantage that their energy could be pre-set. The limitation of both these particles is that since they are positively charged, they have to be travelling at very high speed to overcome the repulsion of the positively charged nucleus. This problem was overcome with the discovery of the neutron in 1932. The neutron, which has no net charge, can enter the nucleus at any speed. Both protons and neutrons are used today to produce radioisotopes. Particle accelerators firing positive ions produce neutron-deficient radioisotopes such as thallium-201 ($t_{\frac{1}{2}}$ = 73 days), which is used to show damaged heart tissue, and zinc-65 ($t_{\frac{1}{2}}$ = 244 days), which is used as a tracer to monitor the flow of heavy metals in mining effluent. Neutrons from nuclear reactors produce neutron-rich radioisotopes such as iridium-192 ($t_{\frac{1}{2}}$ = 74 days), which is used to locate weaknesses in metal pipes, and iodine-131 ($t_{\frac{1}{2}}$ = 8 days), which is used in the diagnosis and treatment of thyroid conditions.

Particle accelerators are also used to produce new elements with atomic numbers greater than that of uranium. The hunt is on for a new stable element. In 2007, calcium-48 ions were fired at californium-249 atoms to produce the element with atomic number 118. Only three atoms were produced and, as the half-life of this isotope is 0.89 milliseconds, they don't exist anymore.

INVESTIGATION 4.1

online only

elog-1831

tlvd-0817

Radioactive decay

Aim

To analyse the radioactive decay of a source

INVESTIGATION 4.2

online only

elog-1832

Background radiation

Aim

To measure the background radiation in the classroom

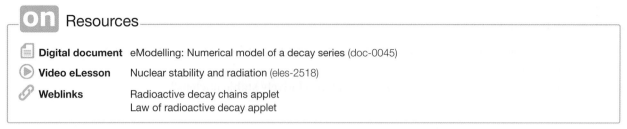

Resources

📄 **Digital document** eModelling: Numerical model of a decay series (doc-0045)

▶ **Video eLesson** Nuclear stability and radiation (eles-2518)

🔗 **Weblinks** Radioactive decay chains applet
Law of radioactive decay applet

4.3 Activities

learn**on**

Students, these questions are even better in jacPLUS

💬 Receive immediate feedback and access sample responses

🔓 Access additional questions

⭐ Track your results and progress

Find all this and MORE in jacPLUS

 4.3 Quick quiz on **4.3 Exercise**  **4.3 Exam questions**

4.3 Exercise

1. Write the nuclear equation for the α decay of radon-222.
2. Write the nuclear equation for the β decay of lead-214.
3. Write the nuclear equation for the γ decay of cobalt-60.
4. **a.** A particle is missing from the right-hand side of a nuclear equation. The atomic numbers on the left-hand side of a nuclear equation add to 20. The sum of the atomic numbers on the right-hand side of the equation add to 18. What is the atomic number of the missing particle?
   b. The mass numbers of the equation in part **a** add to 50 on the left-hand side. The sum of the mass numbers on the right-hand side of the equation add to 46. What is the mass number of the missing particle?
   c. What is the missing particle?
5. Identify two particles or emissions in nuclear equations that have a mass number and atomic number of zero.
6. Energy is released in nuclear decays other than in the form of γ rays. What form does this energy take?
7. The atomic number counts the number of protons in the nucleus of an atom. In nuclear equations some particles have atomic numbers but no protons. Given an example and explain what it means.

4.3 Exam questions

▶ **Question 1 (1 mark)**

MC The nucleus $^{236}_{92}U$ alpha decays into the nucleus $^{b}_{a}X$.

What are the numbers a and b?
A. $a = 94, b = 240$
B. $a = 90, b = 232$
C. $a = 91, b = 236$
D. $a = 93, b = 238$

▶ **Question 2 (1 mark)**

MC The radioactive isotope $^{230}_{90}Th$ undergoes three consecutive α particle decays before becoming an isotope of the element X.

What is the mass number of this isotope of element X?
A. 218
B. 221
C. 224
D. 227

4.4 Radiation and the human body

KEY KNOWLEDGE

- Explain the effects of α, β and γ radiation on humans, including:
  - different capacities to cause cell damage
  - short- and long-term effects of low and high doses
  - ionising impacts of radioactive sources outside and inside the body
  - calculations of absorbed dose (gray), equivalent dose (sievert) and effective dose (sievert)
- Evaluate the use of medical radioisotopes in therapy including the effects on healthy and damaged tissues and cells

Source: VCE Physics Study Design (2023–2027) extracts © VCAA; reproduced by permission.

4.4.1 Electromagnetic radiation and particle radiation

Ionising radiation is the collective name given to α and β particles, neutrons that have been released from the nucleus, γ rays and x-rays. These forms of radiation are grouped together because they have high energies and therefore similar effects on matter. Other forms of electromagnetic radiation, such as radio waves, microwaves and visible light, have lower energies and do not interact with matter in the same way. They are non-ionising radiations.

> **ionising radiation** high-energy radiation that can change atoms by removing electrons, thus giving the atom an overall charge
>
> **free radical** an uncharged fragment of a molecule resulting from a covalent bond being broken

X-rays are the only type of ionising radiation that are not formed by changes in the nucleus. They result from large energy losses by electrons when they are subjected to sudden deceleration.

Ionising radiation has sufficient energy to knock an electron from its orbit around a nucleus. Once an electron has been knocked away from the nucleus, the atom has more positive charges (protons) than negative charges (electrons), giving the atom an overall positive charge. Atoms that have an overall charge are called ions; hence, the name *ionising radiation*.

Sometimes the electron that is knocked from the atom is part of a bond between one atom and another. This causes the bond to be broken, and can result in the molecule being split in two. Each piece of the molecule would then have a charge. The charged pieces are called **free radicals**.

Both ions and free radicals are chemically very reactive. This may result in new chemical reactions taking place inside the substance that was exposed to the ionising radiation.

X-rays are electromagnetic waves of very high frequency and very short wavelength, in the range 0.001 nm to 10 nm. Because of their high frequency, and hence high energy, they can penetrate flesh and may cause ionisation of atoms they encounter on the way through.

When x-rays pass through the body, the body tissue absorbs energy and the intensity of the beam is reduced. The more dense material, such as bone, absorbs more x-radiation.

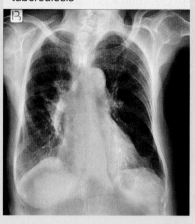

FIGURE 4.14 An x-ray of the lungs showing damage due to tuberculosis

4.4.2 Ionising radiation and living things

The chemical changes resulting from the production of ions and free radicals in living cells can have a range of effects. The cytoplasm (the part of the cell that surrounds the nucleus) has a high water content; therefore, it is often water molecules that are broken. This results in the production of H^+ and OH^- ions, which are chemically very reactive. The ions may react with important molecules, causing damage to DNA, the mechanisms for controlling cell division, or the production of molecules necessary for the life of the cell. Cells undergoing division when they are irradiated are particularly at risk.

x-rays electromagnetic waves of very high frequency and very short wavelength

Often the cell is able to repair itself, but sometimes the chemical changes cause the cell to die. This is not a problem for a living thing unless a large number of cells are damaged.

If the mechanism for cell division is damaged, the cell may begin to reproduce uncontrollably, forming a cancer. If DNA in the ovaries, the testes or in an unborn foetus is damaged, genetic mutations may be passed on. Usually these mutations are recessive and are not exhibited.

4.4.3 The effects of α, β and γ radiation on humans

How much radiation is safe? This seemingly simple question has a very complicated answer. The effect of radiation exposure can range from nausea to death. The damage caused by radiation depends on the type of radiation, the rate at which it is received, the part of the body exposed, the general health of the individual and many other factors.

FIGURE 4.15 Visualisation of the damage caused by ionising radiations of different penetration powers in a block of carbon

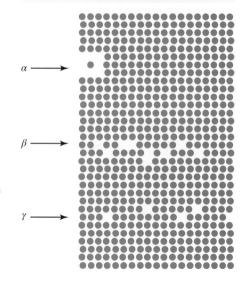

Absorbed dose

One measure of the amount of radiation received is the **absorbed dose**. This is the amount of energy absorbed by each kilogram of the tissue being irradiated. The unit of absorbed dose is the gray, which is given the symbol Gy (1 Gy = 1 J kg^{-1}).

$$\text{Absorbed dose} = \frac{\text{energy absorbed}}{\text{mass}}$$

Unfortunately, the number of grays absorbed by a person does not provide much information about the extent of the damage to that person. The penetrating power of the type of radiation is important. For example, α particles are stopped in a short distance. They pass on all their energy in a short space, causing a great deal of localised damage. β particles are more penetrating, so the damage they cause is less severe in any one area but is more widespread. Neutrons, γ rays and x-rays are far more penetrating than either α or β particles. They spread their energy over a large range.

Equivalent dose

To take into account the different styles of damage caused by the various forms of ionising radiation, another measure of the amount of radiation, **equivalent dose**, was developed. The units for equivalent dose are sieverts (Sv).

$$\text{Equivalent dose (Sv)} = \text{absorbed dose (Gy)} \times \text{quality factor}$$

The quality factor is determined by the type of radiation that delivered the energy (see table 4.4).

One sievert of radiation causes the same amount of biological damage, no matter what type of radiation is used.

TABLE 4.4 Quality factors for different types of radiation

Type of radiation	Approximate quality factor
γ rays and x-rays	1
β particles	1
Slow neutrons	3
Fast neutrons	10
α particles	10–20

tlvd-3745

SAMPLE PROBLEM 6 Calculating absorbed dose and equivalent dose

A 60-kilogram person absorbs 0.054 J of energy due to ionising radiation.
a. **Calculate the absorbed dose.**
b. **What would be the equivalent dose if the energy was delivered by γ rays?**
c. **What would be the equivalent dose if the energy was delivered by α particles? (Take the quality factor to be 20.)**
d. **Which would cause more biological damage to the person — the equivalent dose delivered by the γ rays or the α particles?**

THINK	WRITE
a. 1. Recall the formula for the absorbed dose.	**a.** Absorbed dose $= \dfrac{\text{energy absorbed}}{\text{mass}}$ $= \dfrac{0.054\,\text{J}}{60\,\text{kg}}$ $= 9 \times 10^{-4}\,\text{Gy}$
2. State the solution.	The absorbed dose is 9×10^{-4} Gy.
b. 1. The quality factor of γ radiation is 1.	**b.** Equivalent dose $=$ absorbed dose $\times$ quality factor $= 9 \times 10^{-4}\,\text{Gy} \times 1$ $= 9 \times 10^{-4}\,\text{Sv}$ $= 0.9\,\text{mSv}$
2. State the solution.	The equivalent dose if the energy was delivered by γ rays is 0.9 mSv.
c. 1. The quality factor of α radiation is 20.	**c.** Equivalent dose $=$ absorbed dose $\times$ quality factor $= 9 \times 10^{-4}\,\text{Gy} \times 20$ $= 1.8 \times 10^{-2}\,\text{Sv}$ $= 18\,\text{mSv}$
2. State the solution.	The equivalent dose if the energy was delivered by α particles is 18 mSv.
d. 1. Recall that damage is measured by equivalent dose, in sieverts.	**d.** Equivalent dose of γ rays $= 0.9$ mSv Equivalent dose of α particles $= 18$ mSv $\dfrac{18}{0.9} = 20$
2. State the solution.	The equivalent dose delivered by the α particles would cause about 20 times more damage than that delivered by the γ rays.

PRACTICE PROBLEM 6

On average, each crew member on the Apollo space missions received a dose of 12 mSv while in space. The exposure was mainly electrons and γ rays. Estimate an astronaut's mass and determine how much energy each astronaut absorbed.

Effective dose

Radiation affects different parts of the body in different ways. Each organ or tissue in the body has a different sensibility to radiation doses. For example, the head is less sensitive than the chest, while the gonads, the bone marrow or the digestive track are highly radiosensitive.

The **effective dose** is a number that is calculated for an individual patient. This number takes into account the absorbed dose, the quality factor (relative harm level) of the different types of radiation, and the sensitivity of each organ or tissue type to the different types of radiation.

effective dose the sum of the tissue-equivalent doses weighted by the weighting factors

The relative contribution of each tissue or organ to the total health detriment resulting from uniform irradiation of the body is represented by its **tissue weighting factor** (see table 4.5). The effective dose also takes into account the fact that different parts of the body will not receive the same amount of radiation when undergoing a medical procedure.

The calculation of the effective dose helps to estimate the risk to a patient from a procedure. The actual risk to an individual patient will also depend on such factors as the size and age of the patient.

tissue weighting factor a dimensionless factor reflecting the variable radiosensitivity of specific tissues and organs that is used to calculate the equivalent dose

The effective dose for a patient is the total of the equivalent doses for all the different parts of the body multiplied by their respective weighting factor. Effective dose is measured in sieverts.

TABLE 4.5 Tissue weighting factors for different organs and tissues

Organ or tissue	Approximate tissue weighting factor
Red bone marrow	0.12
Colon	0.12
Stomach	0.12
Breast	0.12
Lungs	0.12
Gonads (testes, ovaries)	0.08
Bladder	0.04
Liver	0.04
Oesophagus	0.04
Thyroid	0.04
Skin	0.01
Bone surfaces	0.01
Brain	0.01
Salivary glands	0.01
Remainder (spleen, cervix/prostate, heart, muscles, etc.)	0.12

Note: The sum of the weighting factors of the whole body is 1.
Source: Based on data from International Commission on Radiological Protection (2007); ICRP Publication 103, *Ann. ICRP* 37 (2–4), p.182.

As the sum of the weighting factors of all the body parts is 1, when the body is uniformly irradiated the effective dose is equal to the equivalent dose. However, this is not the case when a person is not uniformly irradiated, as generally happens with radiotherapy treatments focusing on a specific tissue or organ.

For example, if a person is exposed to an equivalent dose of 2 mSv to the whole body, and an additional 10 mSv to the lungs and 5 mSv to the liver, the effective dose is $2 + 0.12 \times 10 + 0.04 \times 5 = 3.4$ mSv. Thus, in terms of risk, this is the same as a whole-body dose of 3.4 mSv.

EXTENSION: Background radiation

The naturally occurring background radiation, which is a measure of the level of ionising radiation present in the environment at a particular location, is estimated to be 1.5–2 mSv per year (or around 4.11–5.48 μSv per day) on average in Australia.

The typical effective dose for a mammogram is 0.7 mSv, which correspond to roughly 24 weeks' worth of background radiation, while the typical effective dose for a CT scan of the head is 2 mSv, which corresponds to roughly 16 months' worth of background radiation.

▶

For comparison, the effective dose for a long-haul flight can range from 10 μSv to 80 μSv. In terms of risks from irradiation, one mammogram is roughly equivalent to seven return flights from Melbourne to London.

However, an effective dose of less than 20 mSv per year, averaged over five years with no more than 50 mSv for any given year, is considered safe.

tlvd-3746

SAMPLE PROBLEM 7 Calculating the effective dose of a breast cancer radiotherapy session and comparing it to background radiation

During a breast cancer radiotherapy session, Jo's breast is exposed to an equivalent dose of 2.5 mSv per day. The rest of her body is not irradiated.
a. Determine the corresponding effective dose for Jo's whole body, per day.
b. Jo lives in Australia, where the background radiation is around 1.5 mSv per year. Determine how long it would take for Jo to be exposed to the same effective dose from the background radiation.

THINK	WRITE
a. 1. Use table 4.5 to determine the weighting factor for the breast.	**a.** The weighting factor for the breast is 0.12.
2. Calculate the effective dose for the breast.	0.12×2.5 mSv $= 0.3$ mSv
3. Calculate the effective dose for the rest of Jo's body.	The rest of Jo's body is not irradiated, thus the effective dose for the rest of the body is 0 Sv.
4. Add the effective doses for each body part.	0.3 mSv $+ 0$ Sv $= 0.3$ mSv
5. State the answer.	The effective dose is 0.3 mSv.
b. 1. Calculate in how many days the exposure from the background radiation will amount to 0.3 mSv.	**b.** $\dfrac{0.3 \times 365}{1.5} = 73$ days
2. State the answer.	It would take 73 days for Jo to be exposed to an effective dose of 0.3 mSv from the background radiation.

PRACTICE PROBLEM 7

Bill's thyroid and brain are exposed to an equivalent dose of 10 mSv and 30 mS respectively. The rest of his body is not exposed to any additional radiation. Determine the effective dose.

4.4.4 The use of radiation in diagnosis and treatment

Radioactivity as a diagnostic tool

The best-known use of radioactive materials in medicine is probably in the 'radiotherapy' treatment of cancer. Less well known is the use of a radioactive material inside the body to diagnose disease. This use of radioactive material in the body may seem very risky because of the danger associated with radioactivity. In fact, the use of radioisotopes and, more recently, positron emission tomography (PET) to image organs and study their function has become a very common, effective and safe means of diagnosis.

For the purposes of medical diagnosis, radioactive substances may be introduced into the body and used to target areas of interest. The radiation produced is measured and used to determine the health of the organ or section of the body under investigation.

The effectiveness of nuclear medicine for diagnosis of disease has depended on the ability to:
- produce radioisotopes
- detect the γ radiation produced.

The production of radioisotopes became possible with the development of the cyclotron by E.O. Lawrence in 1931. From the mid-1940s, a range of radioisotopes from the United States and later from the United Kingdom was available.

The radioisotopes needed in nuclear medicine in Australia are made at the nuclear research reactor based at Lucas Heights in the south of Sydney, or in a cyclotron under the control of ANSTO (Australian Nuclear Science and Technology Organisation). Cyclotrons are needed to make radioisotopes for positron emission tomography (PET), a diagnostic technique discussed later in this topic. PET facilities are presently found in hospitals in Brisbane, Melbourne and Sydney.

FIGURE 4.16 A cyclotron for radioisotope synthesis in a clinical medical centre

Choosing the right medical radioisotope

When a radioisotope is introduced into the body, other factors in addition to the half-life of the radioisotope need to be considered. The radioisotope is removed from the patient's body by processes such as respiration, urination and defecation. However, some patients metabolise the chemical to which the radioisotope is attached more quickly than others, so it is important that the characteristics of the particular patient are considered when dosages are being determined.

The half-life of the radioisotope must be long enough to allow useful readings to be taken after it has been taken up by the targeted organ. Generally, if the radioisotope remains in the patient's body for a long period of time, its half-life should be comparable to the time taken to carry out the investigation, to minimise the dose to the patient. When the radioisotope is excreted in about the same time as is needed for the investigation, a longer half-life radioisotope can be safely used.

Radioisotopes that emit α particles are not used in the diagnosis of disease because the α particles cause damaging ionisation inside the body. β particles travel further than α particles before they are absorbed but their ionisation damage is much less. They are used in therapy but not in diagnosis of disease.

The most useful radioisotopes for nuclear medicine are those that emit γ radiation only. Technetium-99m and iodine-123 are two such isotopes. A γ-emitting radioisotope inside the body can be detected outside the body because γ radiation is very penetrating. Common radioisotopes used in medical diagnosis are listed in table 4.6.

TABLE 4.6 Radioisotopes used in medical diagnosis

Radioisotope	Production site	Half-life	Function
Chromium-51	Nuclear reactor	27.70 days	Used to label red blood cells and measure gastro-intestinal protein loss
Iodine-131	Nuclear reactor	8 days	Used to diagnose and treat various diseases associated with the thyroid gland; used in the diagnosis of the adrenal medullary; used for imaging some endocrine tumours
Iodine-123	Cyclotron	13 hours	Used to monitor thyroid function, evaluate thyroid gland size and detect dysfunction of the adrenal gland; used to assess stroke damage

(continued)

TABLE 4.6 Radioisotopes used in medical diagnosis *(continued)*

Radioisotope	Production site	Half-life	Function
Molybdenum-99	Nuclear reactor	65.94 hours	Used as the 'parent' in a generator to produce technetium-99m, which is the most widely used isotope in nuclear medicine
Technetium-99m	'Milked' from molybdenum-99	6 hours	Used to investigate bone metabolism and locate bone disease; to assess thyroid function; to study liver disease and disorders of its blood supply; to monitor cardiac output, blood volume and circulation clots; to monitor blood flow in lungs; to assess blood and urine flow in the kidneys and bladder; to investigate brain blood flow and function; to estimate total body plasma and blood count
Thallium-201	Cyclotron	3.05 days	Used to detect the location of damaged heart muscles

tlvd-3747

SAMPLE PROBLEM 8 Calculating the amount of remaining radioisotope in the body after a length of time

A 20-milligram sample of iodine-123 is to be used as a radioactive tracer in the body. The half-life of the iodine-123 is 13 hours.

a. How long will it take for 17.5 milligrams to decay?

b. Calculate how much iodine-123 will remain after 26 hours.

THINK	WRITE
a. 1. Determine the amount that decays in each half-life until the total amount has reached 17.5 mg.	**a.** In one half-life, 10 mg of iodine-123 will decay. This will leave 10 mg of iodine-123. In the second half-life, 5 mg of iodine-123 will decay, leaving 5 mg of iodine-123. In the third half-life, 2.5 mg of iodine-123 will decay.
2. State the solution.	17.5 mg (10 + 5 + 2.5) of iodine will have decayed in three half-lives, or 39 hours.
b. 1. Recall that 26 hours is two half-lives (2 × 13 hours).	**b.** After one half-life, 10 mg of iodine-123 will decay, leaving 10 mg of iodine-123. After two half-lives, 5 mg of iodine-123 will decay, leaving 5 mg of iodine-123.
2. State the solution.	5 mg of iodine-123 will remain after 26 hours.

PRACTICE PROBLEM 8

A radioisotope with a half-life of 13 hours is used in the diagnosis of a patient. A check 52 hours later reveals that 1 milligram of the radioisotope remains.

a. What mass of the radioisotope was used in the diagnosis?

b. How much of the radioisotope will remain after a further 52 hours?

tlvd-3748

SAMPLE PROBLEM 9 Calculating how long it takes for the activity of a radioisotope to drop to a certain level

A sample of a radioisotope has a half-life of 10 minutes.
a. Calculate the time it will take the radioisotope's activity to drop from 8 MBq (megabecquerels) to 4 MBq.
b. Calculate the time it will take for its activity to be 1 MBq.

THINK	WRITE
a. When half the sample has decayed, the activity will also halve. This assumes that the atoms formed are not radioactive.	a. The time needed to reduce the activity to 4 MBq is one half-life, or 10 minutes.
b. Halving the activity each half-life means that three half-lives have passed before the activity is 1 MBq.	b. The time taken is 30 minutes.

PRACTICE PROBLEM 9

A sample of a radioisotope with a half-life of 8 days has an activity of 8 MBq 16 days after it is placed in safe storage.
a. What was the activity of the sample when it was placed in safe storage?
b. What is the activity of the sample after a further 16 days?
c. How long will it take after the sample is placed in safe storage for its activity to decrease to 1 MBq?

4.4 Activities

learnon

Students, these questions are even better in jacPLUS

 Receive immediate feedback and access sample responses

 Access additional questions

 Track your results and progress

Find all this and MORE in jacPLUS

4.4 Quick quiz on	4.4 Exercise	4.4 Exam questions

4.4 Exercise

1. A 30-kilogram child receives 3 mGy of radiation. How much energy did the child absorb?
2. An adult (60 kg) absorbs 0.09 J of radiation. What is the adult's absorbed dose?
3. What is the equivalent dose of 3 mGy of radiation, assuming the energy was delivered by γ radiation?
4. What is the equivalent dose of a 60-kilogram adult who receives 3 mGy of radiation, assuming the energy was delivered by α radiation? Assume a quality factor of 18.
5. Why is α radiation given a higher quality factor than γ radiation?
6. Why is equivalent dose often a more useful measure than absorbed dose?
7. It is more dangerous for pregnant women to be exposed to high radiation levels than for other people. Why?
8. Australians receive on average 2 mSv of radiation each year. Assuming this radiation is all β particles with energy of 1 MeV, calculate how many β particles pass in or out of your body every second if your mass is 60 kg.
9. Ionising radiation can cause cancer, yet it can also cure cancer. Explain this contradictory statement.

4.4 Exam questions

▶ **Question 1 (1 mark)**

MC A 60-kilogram person absorbs 0.009 J of ionising radiation in the form of α particles.

What equivalent dose of radiation do they receive? (Take the quality factor to be 20.)
A. 15 mGy
B. 0.15 mGy
C. 3 mSv
D. 9 mSv

▶ **Question 2 (2 marks)**

An 85-kilogram person absorbs 16 mJ of ionising radiation in the form of α particles. Their equivalent dose is 2.45 mSv.

Determine the quality factor of the radiation.

▶ **Question 3 (1 mark)**

MC Which of the following radioisotopes is suitable for a 45-minute diagnosis appointment?
A. Nitrogen-16 (half-life: 7.13 seconds)
B. Carbon-11 (half-life: 20.3 minutes)
C. Fermium-252 (half-life: 72 hours)
D. Gold-196 (half-life: 148 hours)

▶ **Question 4 (2 marks)**

A hospital radiographer calculates the equivalent dose of radiation absorbed by a patient. During a scan of the patient's brain, the absorbed dose is measured as 1.5 mGy. The mass of the brain is 1.4 kg.

Calculate the energy absorbed by the brain during the scan.

▶ **Question 5 (4 marks)**

In a nuclear plant a worker's hand comes into contact with an object that emits α particles. The worker's hand has a mass of 0.50 kg and absorbs 6 μJ of energy.

Calculate:
a. the absorbed dose received by the worker's hand   **(2 marks)**
b. the equivalent dose received by the worker's hand if the quality factor is 20.   **(2 marks)**

More exam questions are available in your learnON title.

4.5 Energy from mass

KEY KNOWLEDGE

- Explain, qualitatively, nuclear energy as energy resulting from the conversion of mass
- Explain fission chain reactions including:
  - the effect of mass and shape on criticality
  - neutron absorption and moderation

Source: VCE Physics Study Design (2023–2027) extracts © VCAA; reproduced by permission.

4.5.1 Equivalence of mass and energy

The most famous physics equation of all is $E = mc^2$. This is an equation that Albert Einstein derived as a result of his work on relativity. We will now explore what this equation means and how it can be used.

We have learnt that some nuclei are unstable. They eventually decay by releasing particles, including β particles, neutrinos, α particles and γ rays. Which particles are released depends on the particular **isotope** involved. Before 1905, scientists would have said that these decays must obey two conservation laws: conservation of mass and conservation of energy.

isotope an atom containing the same number of protons but a different numbers of neutrons

Conservation of mass states that there is always the same amount of mass in a closed system — one where no mass is entering or leaving. This resulted from early understandings of chemical reactions, where the mass of the products is always the same as the mass of the reactants. Conservation of energy states that there is always the same amount of energy in an isolated system. Energy is never created or destroyed; it is transformed to different forms or transferred from one system to another.

However, in 1905, Einstein stunned the scientific world by showing that these two conservation laws are not strictly correct. He demonstrated that only one conservation law was needed because mass and energy are essentially the same thing. What is conserved is the combination of mass and energy, which is sometimes called mass–energy.

Einstein's equation was a result of theory. Like the prediction of particles such as the neutrino, positron and neutron, there was no experimental evidence for it at the time, and obtaining that evidence was not going to be easy.

FIGURE 4.17 Albert Einstein

$$E = mc^2$$

where:

E is the energy in joules

m is the mass in kilograms

c is the symbol for the speed of light ($c = 2.997\ 924\ 58 \times 10^8$ m s^{-1}).

Note: The speed of light is usually rounded to 3.00×10^8 m s^{-1} in calculations.

This equation is a statement that energy and mass are equivalent. If we need to know how much energy a certain mass is equivalent to, we use the equation.

For example, for a mass of 1 kilogram of any substance:

$$E = mc^2$$
$$= 1.0 \times \left(3.00 \times 10^8\right)^2$$
$$= 9.0 \times 10^{16} \text{ J}$$

That is a tremendous amount of energy. However, the Sun produces four billion times that every second. Einstein's $E = mc^2$ equation calculates that the Sun is losing energy at a rate that is equivalent to 4 billion kilograms every second!

Why do we not see this loss of mass in everyday life when a body radiates energy? Let's look at the example of an object cooling down, like a hot water bottle in your bed on a winter's night. The hot water bottle would transfer approximately 100 000 J of energy to your bed as it cools. Using $E = mc^2$:

$$E = mc^2$$

$$\Rightarrow m = \frac{E}{c^2}$$

$$= \frac{100\,000}{\left(3.00 \times 10^8\right)^2}$$

$$= 1.1 \times 10^{-12} \text{ kg}$$

That is one-millionth of one-millionth of a kilogram, far too tiny to notice or measure. As far as typical events are concerned, energy is conserved and mass is conserved. Even the mass lost by the Sun per second is such a tiny percentage of the Sun's mass that the Sun can continue losing mass at this rate for billions of years.

However, Einstein's equation has been tested in many ways. In 2008, researchers confirmed that $E = mc^2$ explained the mass of protons and neutrons.

FIGURE 4.18 α decay of uranium-238

Uranium-238

EXTENSION: Components of subatomic particles

Unlike electrons, which are fundamental particles without any substructure, protons and neutrons are made up of quarks, which are elementary particles with an electric charge. The quarks in protons and neutrons are held together by the strong nuclear force. Most of the mass of the proton and neutron derives from the energies of the quarks within each particle. To account for a proton having a mass of 1.67×10^{-27} kilograms we need to add up the mass of the quarks it contains and the mass equivalent of the energy involved by using $E = mc^2$.

Nuclear decays bring a nucleus to a state of lower energy and greater stability. The nucleus prior to the decay, or decay series, has a greater mass than the nucleus after decay, even when the masses of the particles emitted in the decay are accounted for. In the decay, mass is not conserved. Similarly, if the energies prior to the decay were compared with the energies of all of the particles after the decay, we find some energy that appears to have come from nowhere. The difference in mass (m) is equivalent to the difference in energy (E) as predicted by $E = mc^2$.

For example, consider the alpha decay of uranium-238, shown in table 4.7.

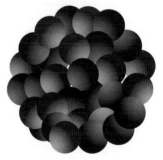

Thorium-234

TABLE 4.7 The decay of uranium-238 to thorium-234 and helium-4

	Nucleus	Mass (kg)
Parent	Uranium-238	$3.952\,93 \times 10^{-25}$
Daughter	Thorium-234	$3.886\,39 \times 10^{-25}$
Decay particle	Helium-4	$6.645\,758 \times 10^{-27}$
Mass after decay	He-4 + Th-234	$3.952\,85 \times 10^{-25}$
Change of mass resulting from decay	Mass defect	8.242×10^{-30}

Helium-4

The thorium nucleus and the α particle produced by the decay of uranium-238 have a slightly lower mass than the parent uranium nucleus. This change in mass is called the **mass defect** (or mass deficit). Mass is not conserved here.

Where is the missing mass?

The mass can be accounted for by applying $E = mc^2$. The mass is equivalent to 7.42×10^{-13} J. This energy is the kinetic energy of the remaining particles, most of it going to the α particle, which moves away from the thorium nucleus at approximately 10 000 km s^{-1}.

mass defect the difference in the mass of the products and reactants in a nuclear reaction; also known as the mass deficit

Energy is often better expressed in a unit called the electronvolt (eV) when dealing with nuclear physics, due to the small amounts of energy involved.

1 eV is the kinetic energy an electron or proton would gain if accelerated across a potential difference of 1 V. 1 eV = 1.6×10^{-19} J. The energy produced in the α decay of uranium-238 is then 7.42×10^{-13} J = 4.63 MeV.

tlvd-0065

SAMPLE PROBLEM 10 Calculating the energy released during alpha decay

Calculate the energy released when thorium-232 undergoes alpha decay. Thorium-232 has a mass of $3.853\,08 \times 10^{-25}$ kg and radium-228 has a mass of $3.786\,55 \times 10^{-25}$ kg. An α particle has a mass of $6.64\,648 \times 10^{-27}$ kg. Give your answer in MeV.

THINK	WRITE
1. Determine the mass of the parent nucleus.	$m_{\text{parent}} = 3.853\,08 \times 10^{-25}$ kg
2. Determine the mass of the daughter nucleus and the α particle.	$m_{\text{products}} = 3.786\,55 \times 10^{-25}$ kg $+ 6.646\,48 \times 10^{-27}$ kg $= 3.853\,01 \times 10^{-25}$ kg
3. Determine the change in the mass due to the decay.	$m_{\text{parent}} - m_{\text{products}} = 3.853\,08 \times 10^{-25}$ kg $- 3.853\,01 \times 10^{-25}$ kg $= 7 \times 10^{-30}$ kg
4. Determine the equivalent energy.	$E = mc^2$ $= 7 \times 10^{-30} \times \left(3.00 \times 10^8\right)^2$ $= 6.3 \times 10^{-13}$ J
5. Convert to MeV.	$E = \dfrac{6.3 \times 10^{-13}}{1.602\,176 \times 10^{-19}}$ MeV ≈ 3.9 MeV
6. State the solution.	3.9 MeV of energy is released when thorium-232 undergoes alpha decay.

PRACTICE PROBLEM 10

An α particle has a kinetic energy of 4.2 MeV as it leaves a nucleus. Calculate the mass defect of this reaction. (Assume that all the energy released went into the kinetic energy of the α particle.)

4.5 Activities

| 4.5 Quick quiz | 4.5 Exercise | 4.5 Exam questions |

4.5 Exercise

1. What does the equation $E = mc^2$ express about mass and energy?
2. What is the energy equivalent of 100 grams of mass in joules?
3. A positron and an electron (each of mass 9.1×10^{-31} kg) annihilate each other. How much energy is released?
4. What is the energy content of a 4500 kJ meal in eV?
5. Calculate the mass equivalent of 4500 kJ.

4.5 Exam questions

Question 1 (1 mark)

MC The difference between the mass of an alpha particle and the total mass of two protons and two neutrons is 5.040×10^{-29} kg.

Which of the following best describes the binding energy of an alpha particle? (*Hint:* Take $c = 3.00 \times 10^8$ m s^{-1}.)
A. 1.54×10^{-12} J   **B.** 3.54×10^{-12} J   **C.** 4.54×10^{-12} J   **D.** 7.54×10^{-12} J

Question 2 (1 mark)

MC An experimental fusion reaction results in each product nucleus undergoing a mass loss of 1.09×10^{-28} kg.

What is the energy released for each product nucleus?
A. 3.3×10^{-20} J   **B.** 3.3×10^{-12} J   **C.** 9.8×10^{-12} J   **D.** 9.8×10^{-4} J

Question 3 (1 mark)

MC A uranium-235 nucleus is hit by a neutron and undergoes nuclear fission, resulting in an energy release of 2.83×10^{-11} J due entirely to mass deficit.

What is the mass deficit of the fission products?
A. 3.1×10^{-31} kg   **B.** 3.1×10^{-28} kg   **C.** 9.4×10^{-28} kg   **D.** 9.4×10^{-20} kg

Question 4 (2 marks)

In the Sun, each fusion event liberates 2.81×10^{-12} J of energy with a corresponding loss of mass.

Calculate the mass lost in each fusion event.

Question 5 (2 marks)

When two helium nuclei combine to form a heavier nucleus, there is a mass deficit of 2.29×10^{-29} kg.

How much energy is liberated from this mass deficit?

More exam questions are available in your learnON title.

4.6 Energy from the nucleus

4.6.1 Binding energy

The amount of energy needed to overcome the strong nuclear force and pull apart a nucleus is known as the **binding energy**. This is the amount of energy that has to be added to a nucleus to split it into its individual nucleons — that is, to reverse the binding process. For example, it would take 2.23 MeV of energy to split a 'heavy' hydrogen nucleus into its separate proton and neutron.

binding energy the energy required to split a nucleus into individual nucleons

Each isotope has its own specific binding energy. Nuclei with high binding energies are very stable, as it takes a lot of energy to split them into separate protons and neutrons. Nuclei with lower binding energies are easier to split. Of course, it is difficult to supply sufficient energy to cause a nucleus to split totally apart. It is much more common for a nucleus to eject a small fragment, such as an α or β particle, to become more stable.

To compare the binding energies of various nuclei, and therefore their stability, it is more useful to compare the average binding energy per nucleon. The average binding energy per nucleon is calculated by dividing the total binding energy of a nucleus by the number of nucleons in the nucleus.

It can be seen from figure 4.19 that iron-56 has the highest binding energy per nucleon. In order to become more stable, other nuclei tend to release some of their energy. Releasing this energy would decrease the amount of energy they contained, and therefore increase the amount of energy that must be added to split them apart. Thus, the nucleons in more stable nuclei have less mass per nucleon than nucleons in less stable nuclei, as the greater binding energy per nucleon has its mass equivalent.

FIGURE 4.19 This graph of binding energy versus mass number peaks at nickel-62, although the much more common iron-56 is very close behind. Note that the graph takes a dip as the nucleus becomes larger. This is due to the ease with which these nuclei break apart.

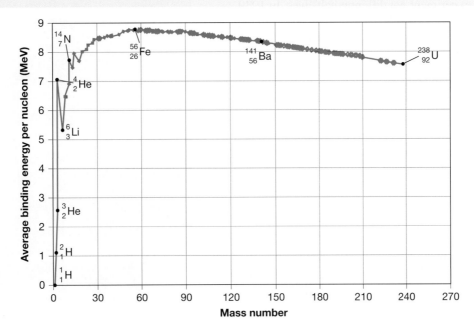

The binding energy is not only the amount of energy required to separate a nucleus into its component parts, but also the amount of energy released when those parts are brought together to form the nucleus; that is, when a proton and a neutron collide to form a 'heavy' hydrogen nucleus, 2.23 MeV of energy is released (twice the binding energy per nucleon).

nuclear fusion the process of joining together two nuclei to form a larger, more stable nucleus

nuclear fission the process of splitting a large nucleus to form two smaller, more stable nuclei

fission fragments the products from a nucleus that undergoes fission; they are smaller than the original nucleus

The curve of the graph in figure 4.19 indicates that if two nuclei with low mass numbers could be joined together to produce a single nucleus, then a lot of energy would be released. Similarly, if a nucleus with a very high mass number could split into two fragments with greater binding energy per nucleon, then once again a lot of energy would be released. These two possibilities were realised in the 1930s. The released energy can be calculated from Einstein's equation $E = mc^2$, where m is the difference between the total nuclear mass before and after the event, and c is the speed of light.

The joining of two nuclei is **nuclear fusion** and the splitting of a single nucleus is **nuclear fission**.

 Resources

 Interactivity Making nuclei (int-6393)

4.6.2 Nuclear fission

In 1934, Enrico Fermi investigated the effect of firing neutrons at uranium. The products had half-lives different from that of uranium. He thought that he had made new elements with atomic numbers greater than 92. Others repeating the experiment got different half-lives. In 1939, Otto Hahn and Fritz Strassmann chemically analysed the samples and found barium, which has atomic number 56, indicating that the nuclei had split.

Lise Meitner and Otto Frisch called this process 'fission' and showed that neutrons could also initiate fission in thorium and protactinium. Further chemical analysis revealed a range of possible fission reactions, each with a different combination of **fission fragments**, including bromine, molybdenum or rubidium (which have atomic numbers around 40) and antimony, caesium or iodine (which have atomic numbers in the 50s). Cloud chamber photographs showing two heavy nuclei flying off in opposite directions confirmed that fission had occurred. Meitner and Frisch also calculated from typical binding energies that the fission of one uranium-235 nucleus would produce about 200 MeV of energy, mainly as kinetic energy of the fission fragments.

FIGURE 4.20 The first chain reaction in a nuclear reactor was on a squash court at the University of Chicago during World War II

This is a huge amount of energy to be released by one nucleus, as can be seen when it is compared to the burning of coal in power plants. Each atom of carbon used in coal burning releases only 10 eV of energy — about 20 million times less than the energy released in the fission of uranium-235.

FIGURE 4.21 Nuclear fission reactions involve a large nucleus splitting into two smaller nuclei and several neutrons.

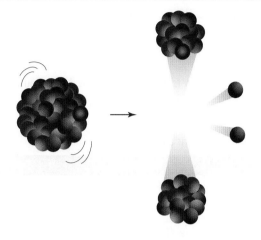

EXTENSION: Lise Meitner (1878–1968)

Lise Meitner was the physicist who coined the term 'fission' and, along with her nephew Otto Frisch, explained the splitting of uranium nuclei into barium and lanthanum.

FIGURE 4.22 Lise Meitner (1878–1968) was the first to describe how a nucleus could undergo fission.

Born in Vienna, Lise was fascinated by the world around her from an early age. A talented student, she wanted to understand the things she observed in nature. Having decided that she would like to pursue her interest in physics and mathematics, Lise engaged a private tutor to prepare her for the university entrance exams, as schools that taught such subjects would not accept girls at that time. She was the second woman to be granted a Doctorate in Physics from the University of Vienna, conferred in 1906.

Lise then moved to the Institute of Experimental Physics in Berlin to work with Otto Hahn. Initially, this proved difficult. Lise was forced to work in a converted workshop, as females were not permitted to use the facilities available to male students. As the place of women in the institute became more accepted, Lise was given positions of responsibility, finally being made a professor in 1926. During her time at the institute, Lise made many important contributions to atomic and particle physics, including the co-discovery with Otto Hahn of the radioactive element protactinium.

In 1938 Berlin became a dangerous place for Jews, and Lise moved to Sweden. It was there that she and Otto Frisch interpreted the results of experiments conducted by Otto Hahn and Fritz Strassman to come up with the first explanation of the fission process. In doing so, Lise was the first person to use Einstein's theory of mass–energy equivalence to calculate the energy released during fission.

Her international reputation led to an invitation to join the Manhattan Project in 1941 and work on the development of the atomic bomb. Lise objected to the project and declined the offer. She continued to work in Sweden until moving to England in 1960, finally retiring at the age of 82.

Also in 1939, Frédéric Joliot, Irène Joliot-Curie and their team confirmed that two or three fast neutrons were emitted with each fission reaction. This allowed for the possibility of a **chain reaction**, where one fission triggers further fissions, which could potentially release enormous amounts of energy.

Some possible equations for the fission of uranium-235 set off by the absorption of a neutron are:

$$^{235}_{92}U + ^{1}_{0}n \rightarrow ^{236}_{92}U \rightarrow ^{148}_{57}La + ^{85}_{35}Br + 3^{1}_{0}n + energy$$

$$^{235}_{92}U + ^{1}_{0}n \rightarrow ^{236}_{92}U \rightarrow ^{141}_{56}Ba + ^{92}_{36}Kr + 3^{1}_{0}n + energy$$

$$^{235}_{92}U + ^{1}_{0}n \rightarrow ^{236}_{92}U \rightarrow ^{140}_{54}La + ^{94}_{38}Br + 2^{1}_{0}n + energy$$

chain reaction a reaction occurring when neutrons, emitted from the decay of one atom, are free to initiate fission in surrounding nuclei

The data in figure 4.19 and Einstein's equation $E = mc^2$ can be used to calculate the amount of energy released in each of these fission reactions.

TABLE 4.8 Masses and binding energies of atoms

Nucleus	Symbol	Mass (kg)	Total binding energy (MeV)
Uranium-235	$^{235}_{92}U$	$3.902\ 9989 \times 10^{-25}$	1783.870 285
Uranium-236	$^{236}_{92}U$	$3.919\ 629 \times 10^{-25}$	1790.415 039
Lanthanum-148	$^{148}_{57}La$	$2.456\ 472 \times 10^{-25}$	1213.125 122
Bromine-85	$^{85}_{35}Br$	$1.410\ 057 \times 10^{-25}$	737.290 649
Barium-141	$^{141}_{56}Ba$	$2.339\ 939 \times 10^{-25}$	1173.974 609
Krypton-92	$^{92}_{36}Kr$	$1.526\ 470 \times 10^{-25}$	783.185 242
Xenon-140	$^{140}_{54}Xe$	$2.323\ 453 \times 10^{-25}$	1160.734 009
Strontium-94	$^{94}_{38}Sr$	$1.559\ 501 \times 10^{-25}$	807.816 711
Neutron	$^{1}_{0}n$	$1.674\ 746 \times 10^{-27}$	

Speed of light, $c = 2.997\ 924\ 58 \times 10^8$ m s^{-1}; 1 MeV = $1.602\ 176 \times 10^{-13}$ J

FIGURE 4.23 Graph of energies in a fission reaction. The sum of the binding energies of La-148 and Br-85 is greater than the binding energy of U-236. The difference is released as kinetic energy of the neutrons and the fission fragments. The total energy after fission is the same as the energy before. Mass–energy is thus conserved.

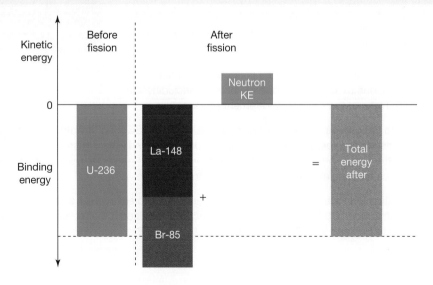

SAMPLE PROBLEM 11 Exploring the fission of uranium-236 to lanthanum-148 and bromine-85

Answer the following questions about the fission of uranium-236 producing lanthanum-148 and bromine-85. Use table 4.8 for data on masses and binding energies.

a. What is the difference between the binding energy of the uranium-236 nucleus and the sum of the binding energies of the two fission fragments?

b. What is the difference between the mass of the uranium-236 nucleus and the sum of the masses of all the fission fragments, including neutrons?

c. Use $E = mc^2$ to calculate the energy equivalent of this mass difference in joules and MeV.

THINK	WRITE
a. 1. Write out the fission equation.	a. $^{238}_{92}U \rightarrow \, ^{148}_{57}La + \, ^{85}_{35}Br + 3 \, ^{1}_{0}n + energy$
2. Use table 4.8 to find the binding energy of uranium-236.	Binding energy of uranium-236 = 1790.415 039 MeV
3. Use table 4.8 to calculate the sum of the binding energies of the fragments.	Binding energy of lanthanum-148 = 1213.125 122 MeV Binding energy of bromine-85 = 737.290 649 MeV 1213.125 122 + 737.290 649 = 1950.415 771 MeV
4. Calculate the difference between the binding energy of the uranium nucleus and its fission fragments.	Energy difference = 1950.415 771 − 1790.415 039 $\qquad\qquad\qquad$ = 160.000 732 MeV
5. State the solution.	The difference between the binding energy of the uranium-236 nucleus and the sum of the binding energies of the two fission fragments is 160.001 MeV.
b. 1. Use table 4.8 to find the mass of uranium-236.	b. Mass of uranium-236 = 3.919 629 × 10⁻²⁵ kg
2. Use table 4.8 to find the sum of the masses of the fragments.	Mass of lanthanum-148 = 2.456 472 × 10⁻²⁵ Mass of bromine-85 = 1.410 057 × 10⁻²⁵ Mass of a neutron = 1.674 746 × 10⁻²⁷ 2.456 472 × 10⁻²⁵ + 1.410 057 × 10⁻²⁵ + 3 × $\quad$ 1.674 746 × 10⁻²⁷ = 3.916 771 × 10⁻²⁵ kg
3. Calculate the difference between the mass of the uranium nucleus and its fission fragments.	Mass difference = 3.919 629 × 10⁻²⁵ − 3.916 771 × 10⁻²⁵ $\qquad\qquad\qquad$ = 2.857 62 × 10⁻²⁸ kg
4. State the solution.	The difference between the mass of the uranium-236 nucleus and the sum of the masses of all the fission fragments is 2.857 62 × 10⁻²⁸ kg.
c. 1. Use $E = mc^2$ to calculate the energy difference in joules. (The full value for c is used here, but you may use the rounded value.)	c. $E = mc^2$ $\quad = 2.857\,62 \times 10^{-28} \times \left(2.997\,924\,58 \times 10^8\right)^2$ $\quad = 2.568\,301 \times 10^{-11} \, J$
2. Convert the energy to MeV.	$\dfrac{2.568\,301 \times 10^{-11}}{1.602\,176 \times 10^{-13}} = 160.300\,789 \, MeV$
3. State the solution.	The energy equivalent of this mass difference is 2.568 301 × 10⁻¹¹ J or 160.300 789 MeV.

PRACTICE PROBLEM 11

Answer the following questions about the fission of uranium-236 to barium-141 and krypton-92. Use table 4.8 for data on masses and binding energies.

a. **What is the difference between the binding energy of the uranium-236 nucleus and the sum of the binding energies of the two fission fragments?**

b. **What is the difference between the mass of the uranium-236 nucleus and the sum of the masses of all the fission fragments, including neutrons?**

c. **Use $E = mc^2$ to calculate the energy equivalent of this mass difference in joules and MeV.**

 Resources

🔗 **Weblink** Fission animation

4.6.3 Nuclear fusion

Nuclear fusion is the process of joining two smaller nuclei together to form a larger, more stable nucleus. This was first observed by Australian physicist Mark Oliphant in 1932 when he was working with Ernest Rutherford. Oliphant was searching for other isotopes of hydrogen and helium. Heavy hydrogen (one proton and one neutron) was already known, but he discovered tritium (one proton and two neutrons) and helium-3 (with only one neutron). In his investigation he fired a fast heavy hydrogen nucleus at a heavy hydrogen target to produce a nucleus of tritium plus an extra neutron. This fusion reaction was to become the basis of the hydrogen bomb (in fact, most atomic bombs are a mix of both fusion and fission reactions), but Oliphant was interested only in the structure of the nucleus and did not realise the energy implications. For fusion to occur more extensively, high temperatures and pressures are needed, such as those that exist inside the Sun or in a fission bomb explosion.

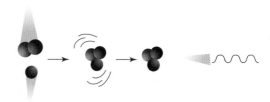

FIGURE 4.24 Nuclear fusion of hydrogen-2 and a proton. The resulting nucleus is in an excited state, which releases a γ ray.

The Sun's core has a temperature of more than 15 million K, just perfect for fusion to occur! Inside the Sun, hydrogen nuclei fuse together to form helium. As helium is more stable than hydrogen, the excess nuclear energy is released. This energy is emitted from the nuclei as γ radiation, and is eventually received on Earth as light and heat.

Fusion reactions also take place in other stars. Stars that are bigger than the Sun have such severe conditions that larger, more stable nuclei such as silicon and magnesium can be produced from the fusion of smaller nuclei. A star about 30 times more massive than the Sun would be needed to produce conditions that would enable the formation of iron by fusing smaller nuclei.

Our Sun

The chain of events occurring in the Sun is quite complex. The major component of the Sun is ${}^{1}_{1}$H; that is, nuclei consisting of only one proton and no neutrons. When collisions occur between ${}^{1}_{1}$H nuclei they have a chance of fusing together, but a change needs to occur for this pairing of positively charged particles to become a stable nucleus. One of the protons is changed into a neutron (in much the same way as a neutron is changed into a proton and an electron during β^- decay). This forms a ${}^{2}_{1}$H nucleus. The by-products of this process are a positron and a neutrino.

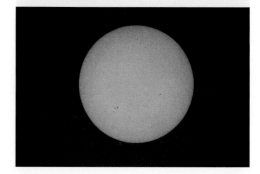

FIGURE 4.25 Our Sun is powered by fusion of small nuclei.

Positrons are also produced when some artificially produced isotopes undergo radioactive decay. As we learnt previously, positrons are the antiparticles of the electrons; they have the same mass but carry a positive charge. When a positron and an electron collide, they immediately annihilate each other. The only thing that remains of either is a γ ray. Neutrinos are produced when protons change into neutrons, and vice versa. They have no charge, are considered massless and travel at close to the speed of light. Fifty trillion neutrinos from the Sun pass through the human body every second. Neutrinos are non-ionising particles, unlike free neutrons.

When a $^1_1$H nucleus and a $^2_1$H nucleus collide, in the conditions found in the centre of the Sun, they form a more stable $^3_2$He nucleus and release the extra energy as a γ ray. If two $^3_2$He nuclei collide, they complete the process of turning hydrogen into helium. The collision results in the formation of a $^4_2$He nucleus and two $^1_1$H nuclei. Again, energy is released. The energy released during nuclear reactions inside the Sun provides energy for life on Earth.

EXTENSION: Formation of larger atoms

Most of the atoms that make up your body (and the rest of the atoms on Earth) were originally produced in a star. Fusion in stars produced most of the atoms other than hydrogen and helium. Those nuclei with atomic numbers up to that of iron were produced in regular stars. However, when large stars stop producing energy from fusion of elements up to iron, they implode, or collapse in on themselves. This causes conditions in which even larger atoms will fuse to produce very heavy elements such as gold, lead and uranium. These stars explode as supernovas, spreading the elements they have made out into space. Earth was formed from a cloud consisting of remnants of old supernovas. Other atoms form from radioactive decay processes and from high-energy collisions of cosmic rays.

The main sequence of nuclear equations that occur in the Sun to convert hydrogen to helium is:

$$^1_1\text{H} + ^1_1\text{H} \rightarrow ^2_1\text{H} + ^0_1\text{e} + \nu$$

$$^2_1\text{H} + ^1_1\text{H} \rightarrow ^3_2\text{He} + \gamma$$

$$^3_2\text{He} + ^3_2\text{He} \rightarrow ^4_2\text{He} + ^1_1\text{H} + ^1_1\text{H}$$

The binding energy can be used to determine the amount of energy released in a fusion reaction.

The difference between the binding energies of the reactants and the products is equivalent to the mass defect.

tlvd-3749

SAMPLE PROBLEM 12 Using the binding energy of helium isotopes to calculate the equivalent energy released by a fusion reaction

In the final nuclear equations that occur in the Sun, shown previously, two helium-3 nuclei collide to produce a helium-4 nucleus and two hydrogen-1 nuclei; that is, two protons.
Use the data in the following table to calculate the:
a. difference between the binding energy of the helium-4 nucleus and sum of the binding energies of the two helium-3 nuclei
b. difference between the sum of masses of the helium-4 nucleus and the two protons, and mass of two helium-3 nuclei
c. energy equivalent of this mass difference in joules and MeV.

Particle	Symbol	Mass (kg)	Total binding energy (MeV)
Helium-3	$^3_2$He	$5.006\ 942 \times 10^{-27}$	7.864 501
Helium-4	$^4_2$He	$6.645\ 758 \times 10^{-27}$	28.295 673
Proton	$^1_1$p or $^1_1$H	$1.673\ 351 \times 10^{-27}$	

THINK

a. 1. Use the table to find the binding energy of the helium-4 nucleus.

2. Use the table to find the binding energy of the helium-3 nuclei.

3. Calculate the difference in binding energies.

4. State the solution.

b. 1. Use the table to find the mass of the reactants.

2. Use the table to find the mass of the products.

3. Calculate the difference in mass.

4. State the solution.

c. 1. Use $E = mc^2$ to calculate the energy difference in joules.

2. Convert the energy to MeV.

3. State the solution.

WRITE

a. Binding energy of helium-4 nucleus = 28.295 673 MeV

Binding energy of two helium-3 nuclei = $2 \times 7.864\,501$
$= 15.729\,002$ MeV

Difference = $28.295\,673 - 15.729\,002$
$= 12.566\,671$ MeV

The difference between the binding energy of the helium-4 nucleus and sum of the binding energies of the two helium-3 nuclei is 12.566 671 MeV.

b. Mass before fusion = $2 \times 5.006\,942 \times 10^{-27}$
$= 1.001\,388 \times 10^{-26}$ kg

Mass after fusion = $6.645\,758 \times 10^{-27} +$
$\left(2 \times 1.674\,746 \times 10^{-27}\right)$
$= 9.995\,25 \times 10^{-27}$ kg

Mass difference = $10.013\,88 \times 10^{-27} - 9.995\,25 \times 10^{-27}$
$= 1.863 \times 10^{-29}$ kg

The difference between the sum of masses of the helium-4 nucleus and the two protons, and mass of two helium-3 nuclei is 1.863×10^{-29} kg.

c. $E = mc^2$
$= 1.863 \times 10^{-29} \times (2.997\,924\,58 \times 10^8)^2$
$= 1.674\,381 \times 10^{-12}$ J

$\dfrac{1.674\,381 \times 10^{-12}}{1.602\,176 \times 10^{-13}} = 10.45$ MeV

The energy equivalent of this mass difference is 1.674×10^{-12} J or 10.45 MeV.

PRACTICE PROBLEM 12

A fusion reaction in the Sun of hydrogen-1 and hydrogen-2 produces helium-3. Use the data in the following table to calculate the:
a. difference between the binding energy of the helium-3 nucleus and sum of the binding energies of the hydrogen-1 and hydrogen-2 nuclei
b. difference between the masses of the helium-3 nucleus and the sum of the masses of the hydrogen-1 and hydrogen-2 nuclei
c. energy equivalent of this mass difference in joules and MeV.

Particle	Symbol	Mass (kg)	Total binding energy (MeV)
Helium-3	$^3_2$He	$5.006\,942 \times 10^{-27}$	7.864 501
Hydrogen-2	$^2_1$H	$3.344\,132 \times 10^{-27}$	2.224 573
Proton	$^1_1$p or $^1_1$H	$1.673\,351 \times 10^{-27}$	

4.6 Activities

learn

| 4.6 Quick quiz **on** | 4.6 Exercise | 4.6 Exam questions |

4.6 Exercise

1. Define the terms *fusion* and *fission*.
2. Which of these reactions occurs in our Sun?
3. Explain why splitting uranium-235 nuclei releases energy, but joining hydrogen atoms also releases energy.
4. Use the graph of binding energy per nucleon (see figure 4.19) to estimate the amount of energy released when a uranium-235 nucleus is split into barium-141 and krypton-92. Think carefully about the number of significant figures in your answer. How well does your answer agree with the measured value of 200 MeV?
5. Why is energy released in the process of fusing two small nuclei together?
6. What aspect of the graph of binding energy per nucleon shows that fusion of light elements releases more energy than fission of heavy ones?
7. When two light elements fuse to form a single heavier one, is the product lighter or heavier than the sum of the masses of the two light atoms? Explain.
8. When uranium-236 undergoes fission to form two lighter nuclei plus some neutrons, is the mass of the uranium-236 greater than, equal to or less than the sum of the masses of the two nuclei plus the neutrons? Explain.

4.6 Exam questions

▶ **Question 1 (1 mark)**

MC The isotope Fe-56 sits at the highest point of the graph of average binding energy per nucleon versus mass number.

What is the implication of this fact?
A. Fe-56 is the most stable nucleus.
B. Fe-56 can be split into any pair of other nuclei with the release of energy.
C. Fe-56 is the most difficult nucleus to form.
D. Fe-56 has the smallest difference between its own mass and the total original mass of its constituents.

▶ **Question 2 (1 mark)**

MC If two smaller nuclei can be combined to form a larger nucleus, higher on the binding energy curve, then which of the following statements is correct?
A. Energy needs to be supplied, to provide the extra binding energy.
B. Mass needs to be supplied, to provide the extra mass.
C. Energy would be released, equal to the gain in binding energy.
D. Energy would be released, equal to the increase in mass of the product nucleus.

Question 3 (1 mark)

MC Which of the following equations represents a fusion reaction?

A. $^{235}_{92}U + ^{1}_{0}n \rightarrow ^{147}_{57}La + ^{86}_{35}Br + 3^{1}_{0}n$

B. $^{2}_{1}H + ^{1}_{1}H \rightarrow ^{3}_{2}He$

C. $^{27}_{13}Al + ^{1}_{1}H \rightarrow ^{24}_{12}Mg + ^{4}_{2}He$

D. $^{59}_{27}Co + ^{1}_{0}n \rightarrow ^{60}_{27}Co$

Question 4 (3 marks)

Consider the following fusion reaction:

$$^{2}_{1}H + ^{6}_{3}Li \rightarrow ^{4}_{2}He + ^{4}_{2}He$$

Calculate the energy released in this reaction, in MeV, using the binding energies (b. e.) given below.

$$^{2}_{1}H = 2.2246 \text{ MeV}$$

$$^{6}_{3}Li = 31.9946 \text{ MeV}$$

$$^{4}_{2}He = 28.2957 \text{ MeV}$$

Question 5 (3 marks)

Consider the following rare spontaneous fission reaction:

$$^{235}_{92}U \rightarrow ^{148}_{57}La + ^{87}_{35}Br$$

Calculate the energy released in this reaction (in J) using the data given below.

$$\text{Mass of Br-87} = 1.443\,318\,4 \times 10^{-25} \text{ kg}$$

$$\text{Mass of La-148} = 2.456\,414\,0 \times 10^{-25} \text{ kg}$$

$$\text{Mass of U-235} = 3.902\,904\,4 \times 10^{-25} \text{ kg}$$

$$c = 2.997\,92 \times 10^{8} \text{ m s}^{-1}$$

More exam questions are available in your learnON title.

4.7 Fission chain reactions

KEY KNOWLEDGE

- Explain fission chain reactions including:
  - the effect of mass and shape on criticality
  - neutron absorption and moderation

Source: VCE Physics Study Design (2023–2027) extracts © VCAA; reproduced by permission.

4.7.1 Fission fragments

In the previous subtopic, we have described three different fission reactions of uranium-236. The reactants in each case are a neutron and a uranium-235 nucleus, but there are many combinations of products. You saw three different reactions with six different products, but about 30 different elements can be produced, two at a time, with about 100 isotopes among those 30 elements.

FIGURE 4.26 Fission of uranium-236; distribution of fission fragments by mass number

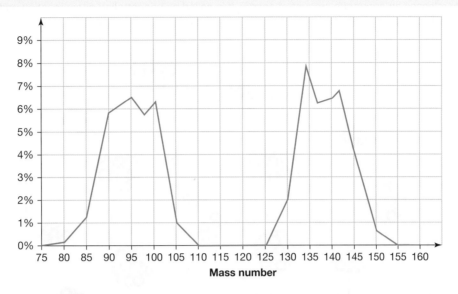

tlvd-3750

SAMPLE PROBLEM 13 Identifying the products of the fission of uranium-236 from the atomic mass number of one of the fragments

One of the fragments of the fission of uranium-236 has an atomic mass number of about 137. What is the species produced?

THINK	WRITE
Use the periodic table to determine which element has a mass number of 137.	The element with the mass number of 137 is barium, which has a relative atomic mass of 137.3.

PRACTICE PROBLEM 13

a. From the graph in figure 4.26, what is the mass number of the most common fragment of uranium-236 fission?
b. Look up the periodic table to find the most likely atomic number of this fragment.
c. Uranium-236 has 92 protons. Determine the atomic number of the other fragment in the most common fission reaction.

4.7.2 Fission chain reactions

In the three different fission reactions of uranium-236 we have seen, there were also two or three neutrons emitted with each fission reaction. This allows the possibility of a **chain reaction**, with one fission producing two or three others and so on.

> **chain reaction** a reaction occurring when neutrons, emitted from the decay of one atom, are free to initiate fission in surrounding nuclei

FIGURE 4.27 This is what happens if every free neutron goes on to produce another fission. A situation such as this quickly releases an enormous amount of energy. It is called an uncontrolled chain reaction and is what happens in a nuclear bomb.

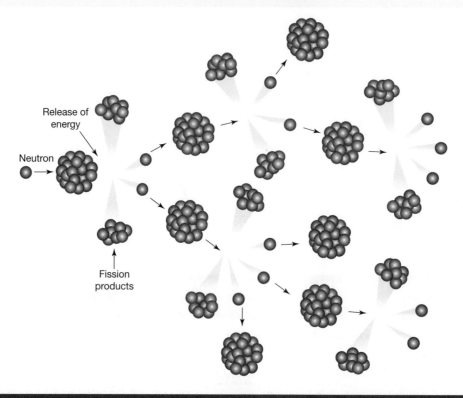

INVESTIGATION 4.3   online only

Chain reaction of dominoes

Aim

To model controlled and uncontrolled chain reactions in nuclear fission using dominoes

4.7.3 Achieving a chain reaction

There are three factors that make achieving a chain reaction very difficult:
1. Uranium has two main naturally occurring isotopes: uranium-235 (0.7%) and uranium-238 (99.3%). The isotopes' responses to an incoming neutron are different and depend on the speed of the neutron (see table 4.9).
2. The high speed of the ejected neutrons — they have kinetic energies of about 1 MeV, equivalent to speeds of 1.4×10^7 m s^{-1}.
3. If a chain reaction could be achieved, there needs to be a way of controlling or stopping it.

TABLE 4.9 Reactions of uranium isotopes to incoming neutrons

Neutron speed and energy	^{235}U	^{238}U
Very fast neutrons (5 MeV)	Few nuclei fission	Most nuclei fission
Fast neutrons (1 MeV)	Many nuclei fission	Very few nuclei fission
Slow neutrons (200 eV)	Most nuclei fission	Nearly all nuclei absorb neutrons
Very slow neutrons (0.03 eV)	All nuclei fission	All nuclei absorb neutrons

A high-speed neutron from the fission of a uranium-235 nucleus is travelling too slowly to cause a uranium-238 nucleus to split. By the time successive collisions slow the neutrons down to a speed to cause most of the uranium-235 to split, the neutrons will have been gobbled up by the uranium-238 nuclei, which outnumber the uranium-235 nuclei by about 140 to 1 in naturally occurring uranium. Therefore, a chain reaction cannot occur in a block of pure natural uranium.

Solutions have been developed for each of the three difficulties mentioned:
1. Increase the proportion of the uranium-235 isotope. This process is called **enrichment**. This in an important process because naturally occurring uranium does not have a high enough percentage of uranium-235 to sustain a chain reaction. The percentage of uranium-235 needs to be increased to 1–4 per cent for use in nuclear reactors.
2. Slow down the neutrons very quickly. This is done using a **moderator**.
3. To control the chain reaction, use a material that readily absorbs neutrons and takes them out of the reaction, and that can be quickly inserted at a moment's notice. This is done with **control rods**.

enrichment the process of increasing the percentage of uranium-235 in a sample of uranium to enable it to sustain a chain reaction

moderator a material that slows down the speed of a neutron

control rods rods of a neutron-absorbing material used in a nuclear reactor to regulate the rate of nuclear fission

4.7.4 Enriched uranium

Natural uranium cannot be used for nuclear bombs or power plants because it contains only small amounts of fissionable uranium-235. The uranium-238 absorbs free neutrons and prevents a sustainable chain reaction from occurring. Enrichment must be carried out to increase the percentage of uranium-235 in the ore. Because uranium-235 and uranium-238 isotopes are chemically identical, the process of separating them is difficult. A number of enrichment methods have been developed, but all are complex and costly. Enriched uranium for nuclear power plants must contain between 1 per cent and 4 per cent uranium-235. For nuclear bombs the percentage of uranium-235 must be closer to 97 per cent, because an uncontrolled chain reaction is required.

One enrichment method, called the centrifuge system, uses a rotating cylinder that sends the heavier isotope (uranium-238) in liquid uranium hexafluoride to the outside of the cylinder, where it can be drawn off, while the uranium-235 diffuses to the centre of the cylinder. To effectively enrich the uranium-235, thousands of centrifuges are connected in series.

FIGURE 4.28 One centrifuge cylinder

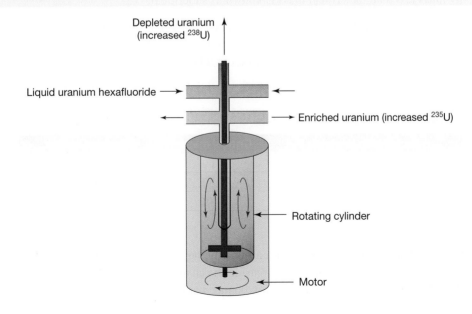

Depleted uranium (increased ^{238}U)

Liquid uranium hexafluoride →

← Enriched uranium (increased ^{235}U)

Rotating cylinder

Motor

Moderators

A fast neutron bouncing off a uranium nucleus is like a golf ball bouncing off a basketball. The large difference in mass means that the neutron does not lose much kinetic energy. To slow down quickly, the neutron needs to collide with something of similar mass, like one billiard ball hitting another billiard ball. Also, the moderating material should be a liquid or solid at room temperature, reasonably inexpensive, and not chemically reactive. These constraints mean that ordinary water is commonly used as a moderator. Other possible moderators are carbon (in the form of graphite) and 'heavy' water (water in which the hydrogen atoms are the deuterium isotope, which has one proton and one neutron).

Control rods

The elements cadmium and boron are deficient in neutrons and readily absorb them. Cadmium and boron are usually encased within steel control rods that can be rapidly raised or lowered into the reactor.

The first controlled fission reactor was constructed by Enrico Fermi at the University of Chicago in December 1942. He used graphite as the moderator and cadmium for the control rods.

The chain reaction

To create an uncontrollable chain reaction, a large proportion of fissionable nuclei is necessary. If uranium is the energy source, weapons-grade enriched ore (about 97 per cent uranium-235) is required. This means that there are very few nuclei present that can absorb the free neutrons without undergoing fission themselves. Therefore, more of the available neutrons cause an energy release.

Criticality

A large ball of weapons-grade material makes a large nuclear bomb, but a small ball of weapons-grade material may not make a small nuclear bomb! The difference lies in the volume-to-surface-area ratio. A small ball has a small ratio of volume to surface area. This means a low proportion of the free neutrons stay inside the ball, where they can initiate a fission. Large balls of weapons-grade material will have a higher ratio of volume to surface area; therefore, a higher proportion of the free neutrons stay inside the ball to produce further fissions.

Critical mass is the smallest mass of a fissionable substance which, when formed into a sphere, will sustain an uncontrolled chain reaction. This mass or criticality varies according to the percentage of fissionable nuclei in the material and the fissionable isotope used. Any mass less than the critical mass is known as subcritical.

critical mass the smallest spherical mass of a fissionable substance that will sustain an uncontrolled chain reaction

If the shape of the fissionable material is changed from a sphere to a brick shape, then there is more surface area available for neutrons to escape the material before initiating another fission reaction. A brick shape therefore needs to be larger and heavier before it becomes critical. The sphere is the shape with the smallest surface area for a given volume.

TABLE 4.10 Critical mass and diameter of a sphere of fissionable material

Fissionable nucleus	Critical mass	Critical diameter
Uranium-235	52 kg	17 cm
Plutonium-239	10 kg	10 cm
Uranium-233	15 kg	11 cm

The critical mass can be reduced by using beryllium around the outside as a neutron reflector. Escaping neutrons are 'reflected' back into the fissionable material, increasing the chances of a fission reaction. The neutrons collide with the beryllium nuclei and a high proportion bounce back.

FIGURE 4.29 An uncontrolled chain reaction

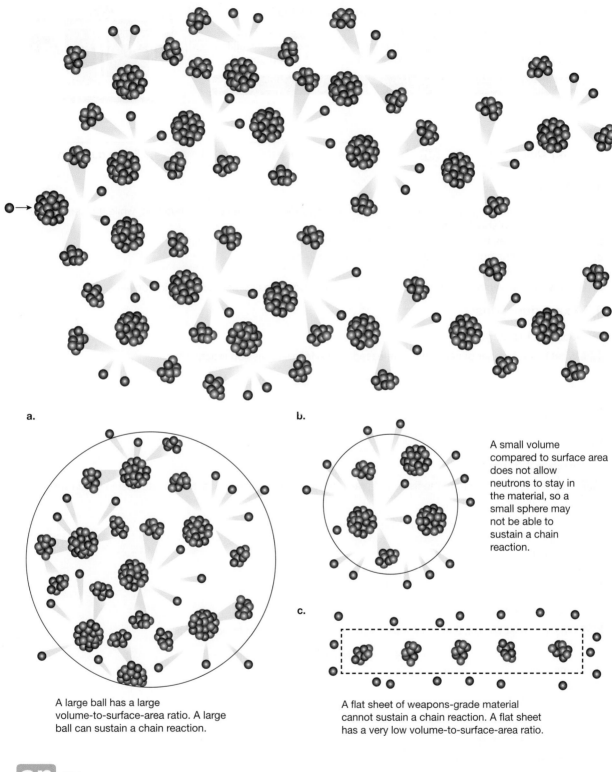

a.

A large ball has a large volume-to-surface-area ratio. A large ball can sustain a chain reaction.

b.

A small volume compared to surface area does not allow neutrons to stay in the material, so a small sphere may not be able to sustain a chain reaction.

c.

A flat sheet of weapons-grade material cannot sustain a chain reaction. A flat sheet has a very low volume-to-surface-area ratio.

 Resources

🔗 **Weblink** Nuclear power plant simulation

4.7 Activities

4.7 Quick quiz on	**4.7 Exercise**	**4.7 Exam questions**

4.7 Exercise

1. Explain why a large spherical mass of uranium may be able to sustain a chain reaction, while the same mass spread into a flat sheet could not.
2. In what form does the released energy from a nuclear fission reaction appear?
3. Why are neutrons good at initiating nuclear reactions?
4. Describe a chain reaction.
5. What are the advantages and disadvantages of fusion power as compared to fission power?

4.7 Exam questions

Question 1 (1 mark)

MC What is the number of neutrons, X, released by the following fission reaction?

$$^{235}_{92}U + {}^{1}_{0}n \rightarrow {}^{140}_{54}Xe + {}^{94}_{38}Sr + X{}^{1}_{0}n + energy$$

A. 1
B. 2
C. 3
D. 4

Question 2 (1 mark)

MC The neutrons ejected from the fission of uranium-235 are travelling too fast to initiate fission in surrounding uranium-235 nuclei.

This problem is overcome in nuclear fission reactors by using
A. enrichment.
B. a moderator.
C. control rods.
D. chain reactions.

Question 3 (1 mark)

MC What is the isotope, z, that must have been used in the following fission reaction?

$$z + {}^{1}_{0}n \rightarrow {}^{100}_{42}Mo + {}^{134}_{52}Te + 6{}^{1}_{0}n + energy$$

A. Uranium-235
B. Uranium-238
C. Plutonium-239
D. Plutonium-240

Question 4 (1 mark)

Explain what a fission chain reaction is.

Question 5 (2 marks)

Explain how control rods make it possible to control chain reactions in nuclear fission reactors.

More exam questions are available in your learnON title.

4.8 Review

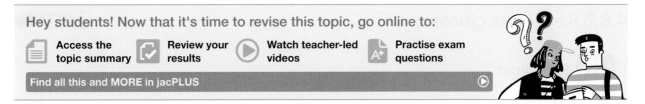

4.8.1 Topic summary

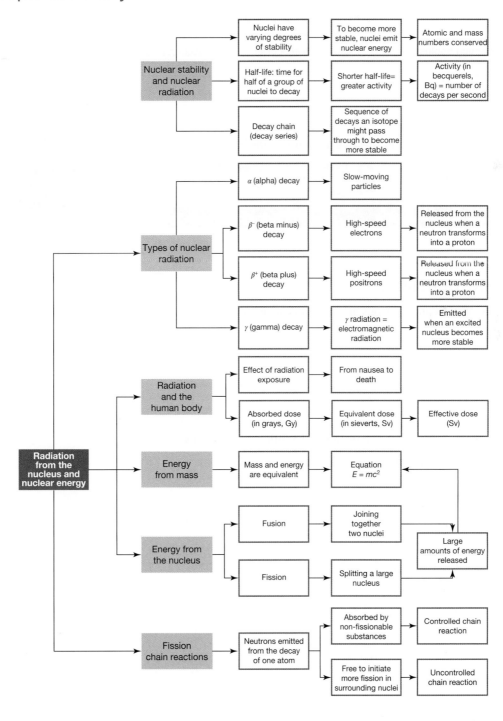

4.8.2 Key ideas summary

4.8.3 Key terms glossary

 Resources

Solutions	Solutions — Topic 4 (sol-0790)
Practical investigation eLogbook	Practical investigation eLogbook — Topic 4 (elog-1573)
Digital documents	Key science skills — VCE Physics Units 1–4 (doc-36950)
	Key terms glossary — Topic 4 (doc-36957)
	Key ideas summary — Topic 4 (doc-36958)
Exam question booklet	Exam question booklet — Topic 4 (eqb-0072)

4.8 Activities

learn on

Students, these questions are even better in jacPLUS

Receive immediate feedback and access sample responses

Access additional questions

Track your results and progress

Find all this and MORE in jacPLUS

4.8 Review questions

1. How many protons and neutrons are in the following atoms?

   a. $^{66}_{30}\text{Zn}$

   b. $^{230}_{90}\text{Th}$

   c. $^{45}_{20}\text{Ca}$

   d. $^{31}_{14}\text{Si}$

2. Write the symbols for isotopes containing the following nucleons.

   a. 2 neutrons and 2 protons
   b. 7 protons and 13 nucleons
   c. 91 protons and 143 neutrons

3. Determine the particle, X, that has been released in each of the following decay equations.

   a. $^{14}_{6}\text{C} \rightarrow {}^{14}_{7}\text{N} + X + \text{energy}$

   b. $^{90}_{38}\text{Sr} \rightarrow {}^{90}_{39}\text{Y} + X + \text{energy}$

   c. $^{238}_{92}\text{U} \rightarrow {}^{234}_{90}\text{Th} + X + \text{energy}$

4. Write a decay equation to show the following.

   a. α decay of:

      i. radium-226
      ii. polonium-214
      iii. americium-241.

   b. β decay of:

      i. cobalt-60
      ii. strontium-90
      iii. phosphorus-32.

5. What is the half-life of the substance represented in the following graph?

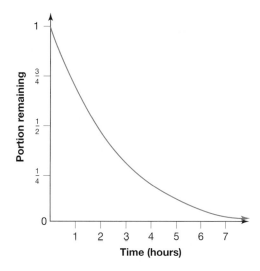

6. Assume the half-life of carbon-14 is 5730 years. If you had one gram of carbon-14, how many years would it take for one-eighth of it to remain?

7. What is the definition of *absorbed dose*?

8. What conditions are required before two protons will fuse to form a single nucleus? Why are these conditions required?

9. What is the critical mass of a fissionable substance?

10. Given the general equation A + B → C + D and using the following table, calculate the:

    a. difference between the binding energy of products and reactants
    b. difference between the sum of masses of the products and reactants
    c. energy equivalent of this mass difference in joules and MeV.

Symbol	Mass (kg)	Total binding energy (MeV)
A	$3.610\,422 \times 10^{-25}$	1467.164 791
B	$1.956\,117 \times 10^{-27}$	0
C	$2.307\,189 \times 10^{-25}$	890.065 063
D	$1.321\,596 \times 10^{-25}$	644.312 162

4.8 Exam questions

All correct answers are worth 1 mark each; an incorrect answer is worth 0.

▶ Question 1

There are four fundamental forces:

i. Gravity
ii. Electromagnetic force
iii. Weak nuclear force
iv. Strong nuclear force.

What is the correct order, from strongest to weakest, at the distance between two nucleons in a nucleus?

A. i, ii, iii, iv
B. iv, iii, ii, i
C. iv, ii, iii, i
D. iv, ii, i, iii

▶ Question 2

How is the half-life of an isotope best described?

A. It is half of the time it would take for the whole sample to decay.
B. It is the time it would take for half of the existing sample to decay.
C. It is the time it would take for the stability of the nucleus to halve.
D. It is a period during which the isotope is not very active.

▶ Question 3

In a series of decays, the mass number drops by eight and the atomic number drops by two.

Which of the following sequences of decays could result in this transmutation?

A. An alpha, a beta minus, a beta minus and another alpha
B. An alpha, a beta plus, a beta minus and an alpha
C. A beta plus, an alpha, a beta minus and an alpha
D. An alpha, an alpha, a beta minus and another alpha

▶ Question 4

Which of the following is the most penetrating radiation?

A. Alpha
B. Beta
C. X-rays
D. Gamma

Question 5

A radioactive isotope has a half-life of 8 days.

What percentage of substance is left after 24 days?

A. 50%
B. 12.5%
C. 25%
D. 87.5%

Question 6

Which of the following is a source of the Sun's energy?

A. Fission of uranium-235
B. Fusion of carbon
C. Exothermic chemical reactions
D. Fusion of protons

Question 7

Under what conditions can fusion occur?

A. Electric field
B. Bright light
C. High temperature
D. In water

Question 8

When uranium-235 collides with a neutron, which of the following is a possible nuclear equation?

A. $^{235}_{92}U + ^1_0n \rightarrow ^{141}_{46}Pd + ^{92}_{36}Kr + 3^1_0n$
B. $^{235}_{92}U + ^1_0n \rightarrow ^{87}_{35}Br + ^{146}_{67}La + 3^1_0n$
C. $^{235}_{92}U + ^1_0n \rightarrow ^{141}_{56}Br + ^{92}_{36}Kr + 3^1_0n$
D. $^{235}_{92}U + ^1_0n \rightarrow ^{147}_{58}Ce + ^{89}_{36}Kr + 3^1_0n$

Question 9

One of the fragments of fission of uranium-236 has an atomic mass number of about 96.

What is the most likely species produced?

A. Gold
B. Barium
C. Caesium
D. Rubidium

Question 10

The process to slow down neutrons very quickly is done using which of the following?

A. A net
B. A moderator
C. Water
D. Heat

Section B — Short answer questions

Question 11 (6 marks)

The following table shows the masses and binding energies of some atoms.

Nucleus	Symbol	Mass (kg)	Total binding energy (MeV)
Plutonium-240	$^{240}_{94}\text{Pu}$	$3.985\ 755 \times 10^{-25}$	1813.454 956
Strontium-90	$^{90}_{38}\text{Sr}$	$1.492\ 791 \times 10^{-25}$	782.631 470
Barium-147	$^{147}_{56}\text{Ba}$	$2.439\ 632 \times 10^{-25}$	1204.158 203
Uranium-234	$^{234}_{92}\text{U}$	$3.885\ 920 \times 10^{-25}$	1778.572 388
Zirconium-95	$^{95}_{40}\text{Zr}$	$1.575\ 810 \times 10^{-25}$	821.139 160
Tellurium-136	$^{136}_{52}\text{Te}$	$2.256\ 760 \times 10^{-25}$	1131.440 918
Neutron	$^{1}_{0}\text{n}$	$1.674\ 746 \times 10^{-27}$	
Proton	$^{1}_{1}\text{p}$ or $^{1}_{1}\text{H}$	$1.673\ 351 \times 10^{-27}$	
Hydrogen-2	$^{2}_{1}\text{H}$ or $^{2}_{1}\text{D}$	$3.344\ 132 \times 10^{-27}$	2.224 573
Hydrogen-3	$^{3}_{1}\text{H}$ or $^{3}_{1}\text{T}$	$5.007\ 725 \times 10^{-27}$	8.481 821
Helium-4	$^{4}_{2}\text{He}$	$6.645\ 758 \times 10^{-27}$	28.295 673
Lithium-6	$^{6}_{3}\text{Li}$	$9.987\ 263 \times 10^{-27}$	31.994 564

A plutonium-239 nucleus absorbs a neutron to become plutonium-240, which splits to form strontium-90, barium-147 and three neutrons.

a. What is the difference between the binding energy of the plutonium-240 nucleus and the sum of the binding energies of the two fission fragments? **(2 marks)**
b. What is the difference between the mass of the plutonium-240 nucleus and the sum of the masses of all the fission fragments, including neutrons? **(2 marks)**
c. Use $E = mc^2$ to calculate the energy equivalent of this mass difference in joules and MeV. **(2 marks)**

Question 12 (4 marks)

A 75-kilogram person absorbs 0.068 J of energy due to ionising radiation.

a. Calculate the absorbed dose. **(1 mark)**
b. What would be the equivalent dose if the energy was delivered by γ rays? **(1 mark)**
c. What would be the equivalent dose if the energy was delivered by α particles? (Quality factor of 20) **(1 mark)**
d. Which would cause more biological damage to the person? **(1 mark)**

Question 13 (6 marks)

A 35-milligram sample of technetium-99m is used for medical diagnosis in the body. The half-life of technetium-99m is 6 hours.

a. How long will it take for 26.25 milligrams to decay? **(3 marks)**

b. Calculate how much technetium-99m will remain after 18 hours. **(3 marks)**

Question 14 (6 marks)

Consider the graph of nuclear stability in figure 4.5 (in subtopic 4.2). Patterns can be seen where β and α decay occur in relation to the line of stability.

a. Would adding a proton or a neutron to a stable nucleus make it more likely to undergo β^- decay? **(1 mark)**

b. Explain why β^- decay is likely to occur in this scenario. **(2 marks)**

c. Explain why α emission is restricted to very large nuclei by referring to the forces involved in the nucleus. **(3 marks)**

Question 15 (9 marks)

Consider the fission reaction $^{236}_{92}\text{U} \rightarrow \, ^{137}_{52}\text{Te} + \, ^{97}_{40}\text{Zr} + 2\,^{1}_{0}\text{n}$.

Use the following values.

Nucleus	Symbol	Mass (kg)	Total binding energy (MeV)
Uranium-236	$^{236}_{92}\text{U}$	$3.919\,629 \times 10^{-25}$	1790.415 039
Tellurium-137	$^{137}_{52}\text{Te}$	$2.273\,507 \times 10^{-25}$	1139.760 337
Zirconium-97	$^{97}_{40}\text{Zr}$	$1.609\,304 \times 10^{-25}$	837.065 893
Neutron	$^{1}_{0}\text{n}$	$1.674\,746 \times 10^{-27}$	

Calculate the:

a. difference between the binding energy of the uranium-236 nucleus and the sum of the binding energies of the fission fragments **(2 marks)**

b. difference between the mass of the uranium-236 nucleus and the sum of the masses of the fission fragments, including neutrons **(2 marks)**

c. energy equivalent of this mass difference in joules and in MeV **(2 marks)**

d. energy released per nucleon in MeV **(1 mark)**

e. percentage of mass transformed into energy. **(2 marks)**

AREA OF STUDY 2 How is energy from the nucleus utilised?

OUTCOME 2

Explain, apply and evaluate nuclear radiation, radioactive decay and nuclear energy.

PRACTICE EXAMINATION

STRUCTURE OF PRACTICE EXAMINATION		
Section	Number of questions	Number of marks
A	20	20
B	4	20
	Total	40

Duration: 50 minutes

Information:

- This practice examination consists of two parts. You must answer all question sections.
- Pens, pencils, highlighters, erasers, rulers and a scientific calculator are permitted.
- You may use the VCAA Physics formula sheet for this task.

on Resources

🔗 **Weblink** VCAA Physics formula sheet

Additional practice examination information	
Type of radiation	Approximate quality factor
γ-rays and x-rays	1
β particles	1
Slow neutrons	3
Fast neutrons	10
α particles	10–20

All correct answers are worth 1 mark each; an incorrect answer is worth 0.

1. Isotopes have the same number of
   A. nucleons.
   B. electrons.
   C. protons.
   D. neutrons.

2. The nucleus is held together by
   A. gravity.
   B. the electromagnetic force.
   C. the weak nuclear force.
   D. the strong nuclear force.

3. A radioactive element has a half-life of 6.0 hours. A sample of this material initially contains 64 000 nuclei. How many radioactive nuclei are remaining after 48 hours?
   A. 200
   B. 250
   C. 500
   D. 125

4. The decay equation $^1_0n \rightarrow {}^1_1p + {}^0_{-1}e + \bar{\nu} + $ energy is an example of
   A. α decay.
   B. β^+ decay.
   C. β^- decay.
   D. γ decay.

5. The radioactive isotope $^{226}_{86}$Ra undergoes two consecutive α particle decays before becoming an isotope of the element X.
   The mass number of this isotope of element X is
   A. 224.
   B. 220.
   C. 218.
   D. 214.

6. An 80-kg person absorbs 0.096 J of energy due to ionising radiation.
   The absorbed dose is
   A. 1.2 J.
   B. 1.2 mGy.
   C. 7.68 J.
   D. 7.68 Sv.

7. A 50-kg person absorbs 0.125 J of energy due to ionising radiation.
   The equivalent dose, if the energy was delivered by fast neutrons, is
   A. 2.5 mJ.
   B. 2.5 mGy.
   C. 2.5 Sv.
   D. 25 mSv.

8. A sample of a radioisotope has a half-life of 30 minutes.
   The time it will take the radioisotope's activity to drop from 32 MBq to 8 MBq is
   A. 1 hour.
   B. 2 hours.
   C. 4 hours.
   D. 8 hours.

9. Which of the following radioisotopes is the most suitable for a three-hour diagnosis appointment?

    A. Nitrogen-16 (half-life: 7.13 seconds)

    B. Carbon-11 (half-life: 20.3 minutes)

    C. Technetium-99 m (half-life: 6 hours)

    D. Molybdenum-99 (half-life: 66 hours)

10. A radioisotope with a half-life of 6 hours is used in the diagnosis of a patient. A check 48 hours later reveals that 0.05 milligram of the radioisotope remains.

    The mass of the radioisotope that was used in the diagnosis is

    A. 4 mg.

    B. 0.128 mg.

    C. 128 mg.

    D. 12.8 mg.

11. A hospital radiographer calculates the equivalent dose of radiation absorbed by a patient during a scan of their brain to be 1.8 mGy. The mass of the brain is 1.2 kg.

    The energy absorbed by the brain during the scan is

    A. 2.16 mJ.

    B. 9 mJ.

    C. 2.16 mSv.

    D. 9 Gy.

12. A radiographer technician wears a dosimeter to monitor their exposure to ionising radiation. Over a year, they were exposed to 5 mSv.

    Assuming this radiation is all β particles with energy of 1 MeV, the number of β particles that pass through their body if their mass is 75 kg is

    A. 2.34×10^{12} each second.

    B. 2.34×10^{12} each year.

    C. 6.42×10^{18} each second

    D. 6.42×10^{18} each year.

Use the following information to answer questions 13 and 14.

When a matter particle and an antimatter particle collide, they annihilate each other and all matter, and antimatter is transformed into energy in the form of electromagnetic radiation.

Mass of electron: 9.1×10^{-31} kg

Mass of proton: 1.67×10^{-27} kg

Speed of light: 3.0×10^{8} m s^{-1}

13. Using the data provided, what is the best estimate of the energy released when an electron and a positron meet, at rest?

    A. 1.6×10^{-13} J

    B. 8.2×10^{-14} J

    C. 5.5×10^{-22} J

    D. 2.7×10^{-22} J

14. When a proton and an antiproton meet, at rest, the energy released is 3.0×10^{-10} J.

    What is the approximate equivalent amount of this energy in MeV?

    A. 1.9 MeV

    B. 19 MeV

    C. 190 MeV

    D. 1900 MeV

15. How many protons and neutrons are there in an alpha particle $^{4}_{2}\alpha$?

    A. 4 protons and 2 neutrons

    B. 2 protons and 4 neutrons

    C. 2 protons and 2 neutrons

    D. 2 protons and 6 neutrons

Use the following information to answer questions 16 and 17.

Strontium-90, $^{90}_{38}$Sr, is a beta emitter with a half-life of 29 years.

16. Element X is the daughter element of strontium-90.
    Which of the following best describes the nucleus of X following the beta decay of strontium-90?
    A. $^{90}_{39}$X
    B. $^{86}_{36}$X
    C. $^{90}_{38}$X
    D. $^{86}_{39}$X

17. A chemical mixture from the last century contains strontium-90. A sample of it has been analysed and it is found that there is only 12.5% of the original amount of strontium-90 left.
    How old is this chemical mixture?
    A. 29 years old
    B. 58 years old
    C. 87 years old
    D. 116 years old

18. The binding energy per nucleon for the deuteron $^2_1$H (or D) is 1.11 MeV.
    Which of the following best describes the binding energy of this nucleus?
    A. 0.556 MeV
    B. 1.11 MeV
    C. 2.22 MeV
    D. 3.33 MeV

19. Helium fusion occurs in stars where three nuclei of helium fuse to form a carbon nucleus. This process releases 1.16×10^{-12} J of energy per carbon nucleus formed.
    What is the best estimate of the amount of mass lost for each carbon nucleus formed?
    A. 3.9×10^{-21} kg
    B. 1.3×10^{-29} kg
    C. 3.9×10^{-29} kg
    D. 1.3×10^{-21} kg

20. A radioactive source has 240 000 active nuclei initially. After 32 hours, there are 15 000 active nuclei remaining.
    Which of the following best describes the half-life of this source?
    A. 2 hours
    B. 4 hours
    C. 6 hours
    D. 8 hours

SECTION B — Short answer questions

Question 21 (4 marks)

Molybdenum-99 has a half-life of 66 hours.

a. A hospital in Brisbane needs at least 60 mg of molybdenum-99, but the delivery will take 11 hours in total.
   What is the minimum amount of molybdenum-99 that the hospital should order? **(2 marks)**
b. Estimate how long would it take for 99% of the initial amount of molybdenum-99 to decay. **(2 marks)**

Question 22 (5 marks)

Americium-241, $^{241}_{95}$Am, is an alpha emitter that is commonly used in smoke detectors. The following graph shows the activity curve of a sample of americium-241. This sample had an initial strength of 6.0×10^5 Bq.

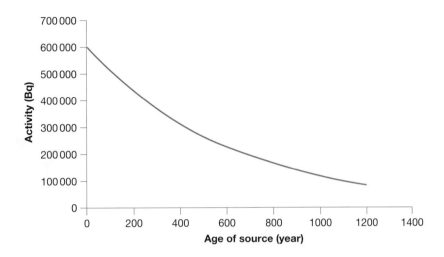

a. Use the graph to estimate the half-life of americium-241. Outline the two points on the graph that you used to arrive at your estimate. **(2 marks)**

b. Describe what is meant by the term 'alpha decay' (which occurs to an alpha emitter). **(1 mark)**

c. Americium-241 radioactively decays and its daughter element is neptunium, Np.
Write the full decay equation of americium. **(2 marks)**

Question 23 (6 marks)

Plutonium-239 is a fissile radioisotope that undergoes a fission reaction according to the following equation:

$$^{1}_{0}n + \, ^{239}_{94}Pu \rightarrow \, ^{134}_{54}Xe + \, ^{103}_{x}Zr + y\,^{1}_{0}n$$

a. What is the value of x in the equation? **(2 marks)**

b. How many neutrons, y, were produced in each reaction event? **(2 marks)**

c. Each reaction event results in a mass deficit of 3.1×10^{-28} kg.
What is the amount of energy produced per reaction event, expressed in MeV? **(2 marks)**

Question 24 (5 marks)

In the Sun, each fusion event liberates 17.56 MeV of energy with a corresponding loss of mass.

a. Calculate the mass lost in each fusion event. **(1 mark)**

b. If the Sun loses 4.3 million of tonnes per second due to fusion events, estimate how many fusion events occur per second. **(1 mark)**

c. If the combustion of 1 tonne of coal release 30 GJ, estimate how many tonnes of coal would need to be burnt to release as much energy as the Sun every second. **(3 marks)**

ASSESSMENT TASK — A MODELLING ACTIVITY

In this task, you will be required to explain the origins of atoms, the nature of subatomic particles and how energy can be produced by atoms.
- A scientific calculator is permitted for this task.
- Access to a spreadsheet program such as Excel will assist.

Total time: One lesson to complete the simulation; 50 minutes for the completion of the three listed tasks (5 minutes reading and 45 minutes writing)

Total marks: 35 marks

A MODELLING ACTIVITY FOR RADIOACTIVE SAMPLES

The radioactive decay of a single nucleus is a random event. However, for a large enough population of identical radioactive nuclei, the half-life can be found for the sample.

Even a small radioactive sample contains many billions of nuclei. Thus, the overall pattern of decay will appear quite ordered, with the source getting less intense over time.

Dice, when thrown, have a probability of landing with a 6 face up equal to $\frac{1}{6}$. Like radioactive material, the outcome of each die, or nucleus, is unpredictable, but a large sample will follow an ordered statistical behaviour.

In this experimental assessment task, you will produce a radioactive decay curve by simulating the radioactive decay process using dice. You will analyse this curve to determine the 'half-life' of your set of dice. The analogy is simple. Each trial corresponds to a set interval of time. If a die falls with a 6 face up it is said to have decayed. If not, it is deemed to be still radioactive in the sense that it has not decayed. By recording the number of dice remaining at the end of each successive trial you will trace the 'radioactive decay' of the sample of dice.

Materials

A box of dice, an appropriate table (either by paper or on a digital spreadsheet) and graph paper

Method

Firstly, count how many dice you have to start with; a good number would be 50 dice.

Create a table similar to the following. The first row has been done assuming there were 50 dice to begin with.

This table could be completed in a spreadsheet.

Number of trial, n	Number of dice remaining in sample
0	50
1	
2	

1. Toss all the dice randomly.
2. Count the number of 6s in the sample and remove them from the sample. Record the number of dice left after the first trial where $n = 1$.

3. Toss all the remaining dice and count the number of 6s. Again, remove them from the sample and record the number of dice left after the second trial, this time $n = 2$.
4. Repeat this process a further 13 times making a total of 15 trials.
5. If all the dice have been removed before the 15th trial, record the dice remaining in the last trials as 0.
6. Now perform the process another two times to obtain three independent simulations in total. You will need to make two more columns for your simulation data.

Results and discussion

Task 1

Create a spreadsheet or a table with column B recording the change in the sample size for trials 0 to 15 in your first simulation.

In columns C and D record the results of your second and third simulations.

In column E obtain the sum of each of the rows in each of the other three simulations to get the total results for your three simulations.

Task 2

Plot a graph of the total number of dice remaining (summed over three trials (column E)) versus the number of trials, 1 through 15. Draw a curve of best fit.

Task 3

Answer the following questions related to this investigation.
1. If there are 100 dice and they are tossed, what fraction on average would you expect to land with 6 face up?
2. Consequently, from 100 dice how many would you expect to be removed?
3. How many dice would be left if those dice were removed?
4. From those remaining after the first toss, repeat steps 1, 2 and 3, each time calculating the number of dice left after those with a face up value of 6 have been removed. Repeat this until there are less than 50 dice remaining.
5. Use this sequence of numbers (called a geometric sequence) to estimate the half-life.
6. Use your graph of your simulation to estimate the half-life for your population of dice. Ensure you show this on your graph and identify the result you obtained.
7. Describe how you used your graph to estimate this half-life.
8. What is meant by the half-life of a radioactive material?
9. What happens to the decay rate — that is, the number of atoms decaying — after each half-life? Explain.
10. One kilogram of Tc-99 (technetium-99) contains approximately 6.0×10^{24} atoms. The half-life of Tc-99 is 6 hours and it is a beta emitter. What does the term 'beta emitter' mean?
11. How long would it take for the Tc-99 sample to be reduced to a mass of:
    a. 500 grams (3.0×10^{24} atoms)
    b. 125 grams (1.5×10^{24} atoms)?
12. Tim has 2 kilograms of Tc-99. It takes 6 hours for the sample to be reduced to 1 kilogram. Tim states that every 6 hours, 1 kilogram will decay, so after 12 hours no sample will be left. Explain why the statement made by Tim is incorrect.
13. Compare the processes of nuclear fusion and nuclear fission. Explain why both fusion and fission are processes that can produce energy.
14. The process of nuclear fusion was fundamental during the development of the universe. Explain how this process lead to the formation of atoms, with specific reference to atoms and subatomic particles present during the formation of the universe and early nucleogenesis.
15. There are two types of forces holding the nucleus together: the strong nuclear force and the weak nuclear force. Describe these two forces and identify which is responsible for the decay simulated in this investigation.

 Resources

Digital document School-assessed coursework (doc-38059)

5 Concepts used to model electricity

KEY KNOWLEDGE

In this topic you will:

- apply concepts of charge (Q), electric current (I), potential difference (V), energy (E) and power (P), in electric circuits
- analyse different analogies used to describe electric current and potential difference
- investigate and analyse theoretically and practically electric circuits using the relationships:

$$I = \frac{Q}{t}, \ V = \frac{E}{Q}, \ P = \frac{E}{t} = VI$$

- model resistance in series and parallel circuits using
  - current versus potential difference (I–V) graphs
  - resistance as the potential difference to current ratio, including R = constant for ohmic devices.

Source: VCE Physics Study Design (2023–2027) extracts © VCAA; reproduced by permission.

PRACTICAL WORK AND INVESTIGATIONS

Practical work is a central component of VCE Physics. Experiments and investigations, supported by a **practical investigation eLogbook** and **teacher-led videos**, are included in this topic to provide opportunities to undertake investigations and communicate findings.

EXAM PREPARATION

▶ Access exam-style questions and their video solutions in every lesson, to ensure you are ready.

5.1 Overview

5.1.1 Introduction

What would your world be like without electricity? Would your mobile phone work or would you be able find your way around your house on a dark night? The transfer of electrical energy into other forms of energy helps us in many ways. The electricity that is transferred to heat in order to warm us on cold nights or the electricity that is transferred to move an electric bicycle is current electricity that has been created by the movement of electric charge.

Not all electric charge moves or flows. Static electricity can be created by the removal of electrons from a material. The discharge of static electricity, as occurs during a lightning storm, can be dramatic and sometimes dangerous.

FIGURE 5.1 Electricity is an integral part of modern life. Consider how the processes of transfer and transformation of energy occur in electric circuits.

LEARNING SEQUENCE

 Resources

Solutions	Solutions — Topic 5 (sol-0791)
Practical investigation eLogbook	Practical investigation eLogbook — Topic 5 (elog-1574)
Digital documents	Key science skills — VCE Physics Units 1–4 (doc-36950)
	Key terms glossary — Topic 5 (doc-36959)
	Key ideas summary —Topic 5 (doc-36960)
Exam question booklet	Exam question booklet — Topic 5 (eqb-0073)

5.2 Static and current electricity

5.2.1 Electric charge (in terms of the basic structure of matter)

Electric charge is a basic property of matter. Matter is all the substance that surrounds us — solid, liquid and gas.

You have probably experienced that small electric shock that happens when you touch a door handle after walking across carpet. This type of phenomenon has been observed for thousands of years. Objects such as glass, gemstones, tree resin and amber can become 'electrified' by friction when they are rubbed with materials such as animal fur and fabrics to produce a spark. Indeed, the words 'electric' and 'electricity' are derived from the Greek word for amber, *electron*.

Experiments in the early 1700s showed that:
- when rubbed together, both the object and the material became 'electrified' or 'charged'
- when charged objects were brought near each other, for some objects there was a force of attraction, while for others it was a force of repulsion.

electric charge the basic property of matter; it occurs in two states: positive (+) charge and negative (–) charge

FIGURE 5.2 a. Two positively charged objects repel each other. **b.** Oppositely charged objects attract each other.

a.

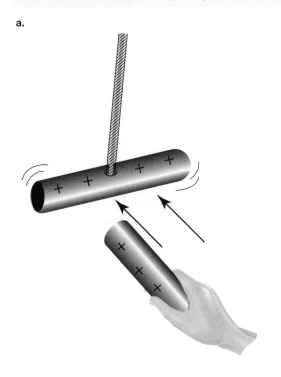

b.

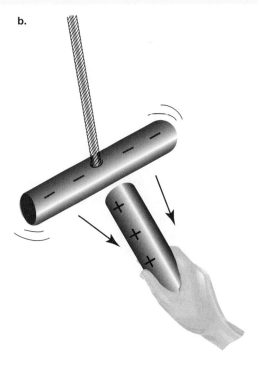

It was quickly observed that like-charged objects repelled each other while unlike-charged objects attracted each other. Charged objects exert a force on each other. The force between two stationary charged objects is called an **electrostatic force**.

All objects contain both positive and negative charges. If an object contains the same amount of both positive and negative charges, it is **neutral** or uncharged. If an object has more positive charges than negative charges, it has an overall positive charge. Similarly, an object with more negative charges than positive charges has an overall negative charge. The direction of the electrostatic force depends on the overall charge of an object, not the individual charges inside.

The direction of the electrostatic forces between electric charges act such that:
- two positive charges repel one another
- two negative charges repel one another
- a positive charge and a negative charge attract one another.

This is summarised as: *like charges repel; unlike charges attract.*

When a neutral glass rod is rubbed with a neutral silk cloth, electrons are transferred from the rod to the cloth. As the electric charges are static — or unmoving — after being transferred, this is called **static electricity**. As shown in figure 5.3, the rod becomes more positively charged (from losing electrons) and the cloth becomes more negatively charged. Therefore, they become attracted to one another as they have unlike charges. Neutral objects do not attract or repel other neutral objects. Experiments like these demonstrate that electric charge can be moved, while being neither created nor destroyed. Electric charge is conserved.

electrostatic force the force between two stationary charged objects

neutral an object that carries an equal amount of positive and negative charge

static electricity electricity produced as the result of an imbalance between negative and positive charges in objects containing charges that are usually static (unmoving)

FIGURE 5.3 Electric charge is conserved.

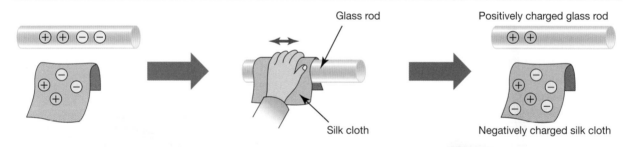

Glass rod

Positively charged glass rod

Silk cloth

Negatively charged silk cloth

Possible explanations for these observations abounded, but further experiments could not determine the exact nature of the electric charges. In the mid-1700s, Benjamin Franklin suggested that positively charged fluid was transferred from the silk to the glass, leaving the silk negative and the glass positive. Although a negatively charged fluid was equally plausible, Franklin's status as an eminent scientist ensured that the existence of a positive fluid was accepted. All developments in electrical engineering for the next 150 years were based on this convention. By 1897, when J.J. Thomson demonstrated that the negatively charged electron was responsible for electricity, it was too late to change the convention and all the associated labelling of meters.

5.2.2 Electric charge (in terms of the basic structure of atoms)

Our current ideas on electricity come from the structure of atoms. All matter is made up of atoms. Atoms in turn are made up of smaller particles called protons, neutrons and electrons. Protons and neutrons are found in the nucleus, while the electrons move in well-defined regions called orbits or shells.

FIGURE 5.5 **a.** The structure of an atom (electron shell) **b.** An atom showing orbits and shells

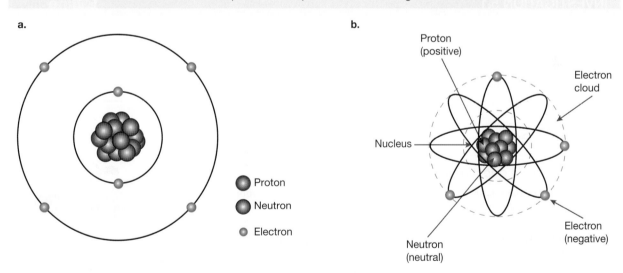

Protons and electrons possess a characteristic known as electric charge; because of their electric charge, these particles exert electric force on each other. Protons carry a positive charge and electrons carry a negative charge. The positive charge on a proton is equal in magnitude to the negative charge on an electron, meaning that the negative and positive charges are balanced and can **neutralise** each other. Neutrons have no electric charge; that is, they are uncharged or neutral.

neutralise when two opposite charges of the same magnitude cancel each other out

Conductors and insulators

Some materials are made up of atoms that are fixed together in such a way that all their electrons are bound tightly to the nucleus and are not free to travel through the material. Such substances are termed **electric insulators**. An insulator is a material that contains no charge carriers. If an insulator is given an electrostatic charge at a particular area on the insulator, the charge will remain at that area. Common examples of insulators include dry air, glass, plastics, rubber and ceramics. Insulators are necessary for creating static electricity.

An **electric conductor** is a material in which charged particles can move and travel freely through. These materials cannot be used to create static electricity, but instead are used for **current electricity**. Examples include:

- *metals*. These are materials whose outer electrons are so loosely bound to the nucleus that they are effectively free to move easily through the material. Usually this movement is random, but when the free electrons are forced to flow in one direction, a current can be created.
- *salt solutions*. These are liquids containing **ions** that are free to move through the solution. Pure water is not a conductor as it does not contain any ions. Tap water contains dissolved substances mixed with pure water and therefore contains ions, so tap water is a conductor.

A **semiconductor** is part of an important group of materials that allow electrons to move freely under certain conditions. Silicon, a common semiconductor, is mainly used in the construction of semiconducting photovoltaic cells. When the Sun shines on these cells in solar panels, they generate electricity. Semiconductors are also widely used in computers, mobile phones, touchscreens and other digital devices.

electric insulator a material in which the electrons are bound tightly to the nucleus and are not free to travel through the material

electric conductor a material that contains charge carriers; that is, charged particles can move and travel freely through the material

current electricity electricity in which electrons move though electrical conductors

ion a charged particle

semiconductor a material that allows electrons to pass through in certain conditions

elog-1622

INVESTIGATION 5.1

The Van de Graaff generator

Aim

To investigate electrostatic charge

5.2 Activities

learn on

5.2 Quick quiz on	5.2 Exercise	5.2 Exam questions

5.2 Exercise

1. After a plastic pen is rubbed with a piece of wool it can be used to attract small pieces of paper. Describe what has happened in terms of electric charge.
2. After rubbing a balloon on your clean dry hair, the balloon should try to stick to your hair when you attempt to remove it. Explain why this occurs.

3. If you separately rub two balloons on your hair and then hold them near each other, what will happen? Explain why this occurs.
4. After walking across a nylon carpet in woollen socks and then touching a metal doorknob it is possible to get an electric shock. Explain why this occurs.
5. What is an insulator?

5.2 Exam questions

▶ **Question 1 (1 mark)**

MC What term best describes the charge carriers when metals conduct electricity?
A. Protons   **B.** Cations   **C.** Anions   **D.** Electrons

▶ **Question 2 (1 mark)**

MC Two objects are placed next to each other and are attracted.

It is most correct to state that
A. one object contains only negative charges and the other contains only positive charges.
B. both objects must be neutral.
C. both objects contain the same number of electrons and protons.
D. each object will have an unequal amount of electrons and protons.

▶ **Question 3 (2 marks)**

Why are the wires that connect circuit components always made of metal?

▶ **Question 4 (2 marks)**

After rubbing a balloon in her hair, Chris brought it very close to an aluminium can lying on a flat table. When she slowly moved the balloon away from the can it started to roll and follow the balloon.

Describe why this happened.

▶ **Question 5 (3 marks)**

Many materials can be defined as conductors or insulators.
a. Define and provide an example of a *conductor*.   **(2 marks)**
b. Explain how these differ from semiconductors.   **(1 mark)**

More exam questions are available in your learnON title.

5.3 Electric charge and current

KEY KNOWLEDGE

- Apply concepts of charge (Q) and electric current (I) in electric circuits
- Explore different analogies used to describe electric current

Source: Adapted from VCE Physics Study Design (2023–2027) extracts © VCAA; reproduced by permission.

5.3.1 Measuring electric charge

Electricity is created by the movement of electrons, and the amount of charge fixed to an object or moving through an object is controlled by the charge on a single electron and the number of electrons being moved.

The unit for charge is the **coulomb** (C), named after the French physicist Charles-Augustin de Coulomb (1736–1806).

coulomb the unit of electric charge

One coulomb of charge is equal to the amount of charge carried by 6.24×10^{18} electrons. The charge carried by a single electron is equal to -1.6×10^{-19} C.

The -1.6×10^{-19} C charge possessed by an electron is the smallest free charge possible. All other charges are whole-number multiples of this value. This so-called elementary charge is equal in magnitude to the charge of a proton. The charge of an electron is negative, whereas the charge of a proton is positive.

The requirement for all charges to be whole-number multiples of the charge on a single electron is not an issue on a larger scale; it is not a problem for scientists to create a charge of 1.5 C, 10 C or even 1.3 μC. However, it is important when considering the charge on an individual atom or molecule. Atoms and molecules must gain or lose a whole number of electrons and therefore can only exist with a small range of charges.

tlvd-3751

SAMPLE PROBLEM 1 Calculating whole-number multiples of charges for electrons

Calculate how many electrons are needed to create a total charge of –6.4 C.

THINK	WRITE
1. All charges are composed of integer multiples of the charge on one electron.	Charge on one electron = -1.6×10^{-19} C
2. How many electrons are needed to create this charge?	$\dfrac{-6.4}{-1.6 \times 10^{-19}} = 4 \times 10^{19}$
3. State the solution.	4×10^{19} electrons

PRACTICE PROBLEM 1

Is it possible for a total charge of 1.2×10^{-18} C to exist? Justify your response with calculations.

EXTENSION: Small charges and quarks

Charges smaller than that carried by the electron are understood to exist, but they are not free to move as a current. Particles such as neutrons and protons are composed of quarks, with one-third of the charge of an electron, but these are never found alone.

5.3.2 Defining current

Electric current is the movement of charged particles from one place to another, and is measured as the rate of flow of charge. The charged particles may be electrons in a metal conductor or ions in a salt solution. Charged particles that move in a conductor can also be referred to as **charge carriers**.

electric current the movement of charged particles from one place to another

charge carrier a charged particle moving in a conductor

Electric current is a measure of the rate of flow of charge around a circuit. It can be expressed as:

$$I = \frac{Q}{t}$$

where:

I is the current, in amperes

Q is the quantity of charge flowing past a point in the circuit, in coulombs

t is the time interval, in seconds.

There are many examples of electric currents. Lightning strikes are examples of large currents. Nerve impulses that control muscle movement are examples of small currents. Charge flows in household and automotive electrical devices such as light globes and heaters. Both positive and negative charges flow in cells, in batteries and in the ionised gases of fluorescent lights. The solar wind is an enormous flow of protons, electrons and ions being blasted away from the Sun.

Not all moving charges constitute a current. There must be a net movement of charge in one direction for a current to exist. In a piece of metal conductor, electrons are constantly moving in random directions, but as there is no overall flow of charge in any one direction, there is no current. As water flows down a river there are millions of coulombs of charge moving with the water molecules, but there is no electrical current in this case because equal numbers of positive and negative charge are moving in the same direction. Therefore, until there is a net movement of charge in one direction, created by a battery or power source, there is no current.

For there to be a current in a circuit, there must be a complete conducting pathway around the circuit (so the charges have somewhere to move to when they flow) and a device to make the charged particles move, as shown in figure 5.6. When a **switch** in a circuit is open, the pathway, followed by the moving charged particles, is broken and the current stops almost immediately.

FIGURE 5.6 Current can flow when there is a complete conducting pathway and a net movement of electrons.

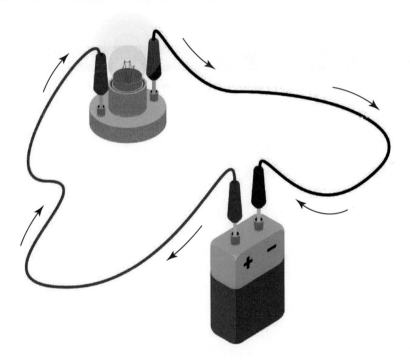

Electric current and its movement in an **electric circuit** will be further explored in topic 6.

The unit of current is the **ampere** (A). It is named in honour of the French physicist André-Marie Ampère (1775–1836). One ampere is the current in a conductor when 1 coulomb of charge passes a point in the conductor every second.

switch a device that stops or allows the flow of electricity through a circuit

electric circuit a closed loop of moving electric charge

ampere the unit of current

tlvd-0010

SAMPLE PROBLEM 2 Calculating the current of a conductor using charge and time

What is the current in a conductor if 10 coulombs of charge pass a point in 5.0 seconds?

THINK	WRITE
1. Current is the rate at which charge, Q, flows in the circuit. Give values for Q and t.	$Q = 10$ C, $t = 5.0$ s
2. Substitute values for Q and t into the formula $I = \dfrac{Q}{t}$.	$I = \dfrac{Q}{t}$ $= \dfrac{10 \text{ C}}{5.0 \text{ s}}$
3. Current is measured in amperes (A), where: $1 \text{ A} = 1 \text{ C s}^{-1}$.	$I = 2.0 \text{ C s}^{-1}$ $= 2.0$ A
4. State the solution.	The current in the conductor is 2.0 A.

PRACTICE PROBLEM 2

What is the current passing through a conductor if 15 coulombs of charge pass a point in 3.0 seconds?

tlvd-0011

SAMPLE PROBLEM 3 Calculating the charge through a load using current and time

How much charge passes through a load if a current of 3.0 A flows for 5 minutes and 20 seconds?

THINK	WRITE
1. To find the charge, Q, passing through the circuit, transpose the formula $I = \dfrac{Q}{t}$ making Q the subject.	$I = \dfrac{Q}{t}$ $\Rightarrow Q = It$
2. Give values for I and t. *Note:* Be sure to convert the time to seconds.	$I = 3.0$ A, $t = 5 \times 60 + 20 = 320$ s
3. Substitute values for I and t into the formula.	$Q = It$ $= 3.0 \text{ A} \times 320 \text{ s}$ $= 960$ C
4. Convert to scientific notation, to two significant figures.	$Q = 9.6 \times 10^2$ C
5. State the solution.	9.6×10^2 C charge passes through the load. (*Note:* As the solution is quite small in magnitude, leaving the solution as 960 C instead of in scientific notation is acceptable).

PRACTICE PROBLEM 3

For how long must a current of 2.5 amperes flow to make 7.5 coulombs of charge pass a point in a circuit?

tlvd-3752

SAMPLE PROBLEM 4 Determining the overall flow of charge

What is the overall flow of charge if 5 C flows to the right while 4 C flows to the left each second?

THINK	WRITE
1. The direction of movement of each charge is important — choose a positive direction and a negative direction.	Right is the positive direction. $+5$ C and -4 C
2. The overall flow of charge is the sum of the movement of each individual charge.	$+5 - 4 = +1$ C
3. State the solution.	The overall flow of charge is 1 C to the right.

PRACTICE PROBLEM 4

Determine the magnitude and direction of the overall charge flow if there is –7 C moving upwards and –3 C moving downwards.

In many household circuits, currents of the order of 10^{-3} A are common. To describe these currents, the milliampere (mA) is used. One milliampere is equal to 1×10^{-3} amperes. In many digital circuits, the current may be even smaller and may need to be measured in microamperes (μA) or even nanoamperes (nA).

The conversion of units is an important skill of Physics and will be further explored in topic 12. It is important to be able to convert between units, as many formulae require a specific unit. To convert between amperes and milliamperes:

$$\text{Amperes} \xrightleftharpoons[\div 1000]{\times 1000} \text{milliamperes}$$

tlvd-0012

SAMPLE PROBLEM 5 Converting millamperes to amperes

Convert 450 mA to amperes.

THINK	WRITE
1. To convert mA to A, divide by 1000 (or multiply by 10^{-3}).	$\dfrac{450 \text{ mA}}{1000} = 0.450 \text{ A}$
2. State the solution.	450 mA is equal to 0.450 A.

PRACTICE PROBLEM 5

Convert 280 mA to amperes.

5.3.3 Analogies and models for current

It is impossible for scientists to see the movement of charged particles in an electric circuit. One way to understand something we can't see is to use a **model**. A good scientific model uses objects and phenomena that we can see and understand or have experienced to explain things that we cannot see. Often a number of different models are used. Each model has its own strengths and weaknesses, with each better at explaining different aspects related to current. It is important to remember that models, including analogies, are often oversimplifications and as such have many limitations.

The electron drift model of current

Most circuits have metal conductors, which means that the charge carriers will be electrons.

Metal conductors can be considered to be a three-dimensional arrangement of atoms that have one or more of their electrons loosely bound. These electrons are so loosely bound that they tend to drift easily among the atoms. Metals are good conductors of both heat and electricity because of the ease with which these electrons are able to move, transferring energy as they go. Diagrammatically, the atoms are represented as positive ions (atoms that have lost an electron and have a net positive charge) in a 'sea' of free electrons.

When the ends of a conductor are connected to a battery, the free electrons drift towards the positive terminal. The electrons are attracted by the positive terminal and indeed accelerate, but constantly bump into atoms so on average they just drift along (see figure 5.7).

FIGURE 5.7 The electron drift model of current flowing through a metal. *Note:* Only two of the free electrons have been shown.

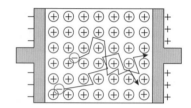

The hydraulic model of current

The flow of electrons through a metallic conductor can also be modelled by the flow of water through a pipe (see figure 5.8).

Electrons cannot be destroyed, nor, in a closed circuit, can they build up at a point. Therefore, if electrons are forced into one end of a conductor, an equal number will be forced out the other end. This is rather like pouring a cupful of water into one end of a full pipe. It forces a cupful of water to come out the other end.

Note that when water is put in one end of a pipe it is not the same water that comes out the other end, because the pipe was already full of water.

FIGURE 5.8 The hydraulic model for current flow. One cupful of water in one end of the pipe means one cupful out the other end.

The bicycle chain model of current

Other models are sometimes used; for example, the bicycle chain model. In this model the chain represents the circuit, and the links in the chain represent electrons. When the pedals are turned the chain moves and energy is transferred to the rear of the bicycle to move the rear wheel. The moving chain represents the movement of electrons around the circuit. Note that the transformation of energy from the pedals to the rear wheel is virtually instantaneous. The energy transfer from the pedals does not depend on particular chain links travelling from the pedals to the wheel. Similarly, the energy transfer in an electric circuit does not depend on particular electrons travelling to the **load**. Overall, the transfer of electrical energy is faster than the movement of electrons in the conducting wires.

model a representation of ideas, phenomena or scientific processes; can be a physical model, mathematical model or conceptual model

load a device in which electrical energy is converted into other forms to perform tasks, such as heating or lighting

5.3.4 How rapidly do electrons travel through a conductor?

As the different models of electron flow indicate, individual electrons do not move quickly through conductors or wires. A current is the result of an exceptionally large number of electrons creating an overall movement of charge in one direction.

The speed of electrons through the conductor depends on the cross-sectional area of the conductor, the number of electrons that are free to move, the electron charge and the size of the current.

For example, if a current of 10 A passes through a copper wire of cross-sectional area 1 mm^2, the electron speed is 0.16 mm s^{-1} or 1.6×10^{-4} m s^{-1}. This speed is known as the drift velocity (since the electrons are drifting through the wire) and is quite small.

tlvd-0013

SAMPLE PROBLEM 6 Determining the time taken for electrons to travel using distance and drift velocity

How long will it take an electron to travel from a car's battery to a rear light globe if it has a drift velocity of 1.0×10^{-4} m s^{-1} and there is 2.5 metres of metal to pass through? (Electrons travel from the negative terminal of the battery through the car body towards the circuit elements.)

THINK	WRITE
1. The drift velocity, v, equals the distance travelled, d, divided by the time taken to travel the distance, t.	$v = \dfrac{d}{t}$
2. Transpose $v = \dfrac{d}{t}$ to make t the subject.	$vt = d$ $\Rightarrow t = \dfrac{d}{v}$
3. Substitute the known values into the formula and solve for t.	$v = 1.0 \times 10^{-4}$ m s^{-1}, $d = 2.5$ m $t = \dfrac{d}{v}$ $= \dfrac{2.5 \text{ m}}{1.0 \times 10^{-4} \text{ m s}^{-1}}$ $= 2.5 \times 10^{4}$ s $= 25\,000$ s
4. State the solution.	It would take 25 000 seconds, which is more than 7 hours.

PRACTICE PROBLEM 6

How long will it take an electron to travel to a headset from a console if it has a drift velocity of 7.4×10^{-5} m s^{-1} and there is 1.2 metres of copper wire to pass through?

5.3 Activities

5.3 Quick quiz	5.3 Exercise	5.3 Exam questions

5.3 Exercise

1. State the difference between conventional current and electron current.
2. What is the difference between direct current and alternating current?
3. A steady direct current of 2.5 A flows in a wire connected to a battery for 15 seconds. How much charge enters or leaves the battery in this time?
4. Convert 45 mA to amperes.
5. Convert 2.3×10^{-4} A to milliamperes.
6. Is current used up in a light globe? Explain your answer.
7. A car light globe has a current of 3.5 A flowing through it. How much charge passes through it in 20 minutes?
8. The drift velocity is directly proportional to the current in the conductor. If electrons have a drift velocity of 1.6×10^{-4} m s^{-1} for a current of 10 A in a certain conductor, what would be their velocity if the current was 5.0 A?

5.3 Exam questions

Question 1 (1 mark)

MC What is the purpose of a switch in an electrical circuit?
A. To stop the current from flowing past a certain point in the circuit
B. To increase the current flowing through a load
C. To decrease the current flowing through a load
D. To allow current flow through the circuit to be controlled manually

Question 2 (1 mark)

MC A light bulb draws a current of 0.8 A.

How long must the light bulb be on for a total charge of 2 C to pass through it?
A. 2.5 seconds
B. 2.5 minutes
C. 1.6 seconds
D. 1.6 minutes

Question 3 (1 mark)

Convert 450 μA to amperes (1 μA = 1×10^{-6} A).

Question 4 (1 mark)

What is the current flowing through an extension cord if 15 C of charge passes through it in 50 seconds?

Question 5 (2 marks)

There is an electric current in a solution. In one second, positive ions carry a charge of +4.0 C from left to right. In the same time, negative ions carry a charge of −2.0 C from right to left.

What is the conventional current in the solution?

More exam questions are available in your learnON title.

5.4 Electric potential difference

5.4.1 Defining potential difference

Electric **potential difference** or **voltage drop** is a measurement of the energy carried by the charges creating a current. The energy is in the form of electrical potential energy and is provided by a power source such as a battery.

As the name suggests, electric potential difference is the difference in the energy of charges between two points. For a lightning bolt, this would be the difference in the energy of the charges in the sky and on the ground. For a battery, it is the difference between the energy of the charges at one terminal of the battery compared to the other terminal. This electric potential difference is measured in volts (V). A 9-V battery provides more energy to charges than a 1.5-V battery. A household power point, rated at 230 V, provides even more energy to charges.

Voltage measures the amount of energy carried by a specific charge. When 1 coulomb of charge has its electrical potential energy changed by 1 joule (J), it has a potential difference of 1 V. So a 9-V battery gives 9 J of energy to each coulomb of charge.

FIGURE 5.9 This 9-V battery provides 9 J of energy to each coulomb.

Voltage is the energy provided to each coloumb of charge:

$$V = \frac{E}{Q} \text{ or } E = VQ$$

where:

V is the potential difference (voltage), measured in volts

E is the energy, measured in joules

Q is the quantity of charge, measured in coulombs.

potential difference the amount of electrical potential energy, in joules, lost by each coulomb of charge in a given part of a circuit

voltage drop another term used to describe potential difference

tlvd-0014

a. How much energy does a 1.5-volt battery give to 0.50 coulombs of charge?
b. The charge on an electron is 1.6×10^{-19} coulombs. How much energy does each electron have as it leaves a 1.5-volt battery?

THINK	WRITE
a. 1. The energy is given by $E = VQ$.	a. $E = VQ$
2. Substitute the known values into the formula and solve for E.	$V = 1.5$ V, $Q = 0.50$ C $E = 1.5 \times 0.50$ $\quad = 0.75$ J
3. State the solution.	The battery would give 0.75 joules of energy to 0.50 coloumbs of charge.
b. 1. The energy is given by $E = VQ$.	b. $E = VQ$
2. Substitute the known values into the formula and solve for E.	$V = 1.5$ V, $Q = 1.6 \times 10^{-19}$ C $E = 1.5 \times 1.6 \times 10^{-19}$ $\quad = 2.4 \times 10^{-19}$ J
3. State the solution.	Each electron would have 2.4×10^{-19} joules of energy.

PRACTICE PROBLEM 7

A mobile phone battery has a voltage of 4.7 volts. During its lifetime, 4000 coulomb of charge leave the battery. How much energy did the battery originally hold?

Potential difference and methods used to measure this in circuits will be further explored in topic 6.

In many technologies, such as x-ray machines and particle accelerators, the energy of electrons needs to be determined. The number 1.6×10^{-19} joules is inconvenient, so another energy unit is used. It is called the **electron volt**, abbreviated eV, where
1 eV $= 1.6 \times 10^{-19}$ joules.

5.4.2 Electromotive force (emf)

The purpose of 'plug-in' power supplies, batteries or cells is to provide energy to the current. Such devices are said to provide an **electromotive force**, or emf. The term electromotive force is misleading since it does not refer to a 'force', measured in newtons. It is a measure of the energy supplied to each coulomb of charge passing through the power supply. The circuit symbol for emf is ε.

The unit of emf is the volt (V) because it is a measure of energy per coulomb. This is identical to potential difference. A power supply has an emf of X volts if it provides a current with X joules of energy for every coulomb of charge passing through the power supply.

The definitions of the emf and the potential difference are the same and can be used interchangeably. Emf is often used when referring to sources of energy such as power supplies, batteries, generators or power points. Potential difference is often used when referring to devices in which electrical potential energy is used up, such as light globes, computers, fridges and mobile phones.

electron volt a unit of energy, where 1 electron volt (eV) is equal to the energy of electrons

electromotive force (emf) a measure of the energy supplied to a circuit for each coulomb of charge passing through the power supply

5.4.3 Analogies and models for potential difference

As with current (explored in section 5.3.3), there are many models and analogies that can be used to describe potential difference. As with the analogies and models for current, these are simplifications to represent an idea and thus have limitations.

The hydraulic model of potential difference

As with current, a hydraulic model can be used to describe potential difference.

Current, the flow of electrons through a metallic conductor, can be modelled as the flow of water through a pipe. How can we use this analogy to explain potential difference?

Consider the water flowing through the pipe explored in section 5.3.3. The amount of water can be seen as the charge, and the flow of water from one end to the other is the current.

Voltage or potential difference can be compared to the difference of pressure at the start and the end of the pipe. At the start of the pipe there is high potential energy, as the water is entering here. However, at the end of the pipe, the potential energy is much lower. This difference in the potential energy between the start and end of the pipe is what allows for the water to flow in that direction. When this difference in pressure is higher, the water will flow faster, as each molecule of water will have more energy. In an electric circuit, if the potential difference is higher, each coulomb of charge has more energy to move through the metal conductors.

This can also be observed in the flow of water at a waterfall (such as Angel Falls, shown in figure 5.10). Think about a short and a tall waterfall. The water that flows down the longer waterfall and hits the ground will have more energy than the water in the shorter waterfall.

Greater heights can be seen as having greater gravitational potential energy. When the flowing water hits the ground, there is no longer gravitational potential energy. This difference in potential energy can be compared to potential difference. When the potential difference (or height of the water) is greater, each coulomb of charge (or drop of water) has more energy.

FIGURE 5.10 The movement of water in Angel Falls, in Venezuela, can be used as an analogy to model circuit electricity.

The bicycle chain model of potential difference

Another analogy that can be applied to potential difference is the bicycle chain model. In section 5.3.3, we explored that the chain represents the circuit and the links in the chain represent electrons. The movement of this chain represents the current. When the pedals are turned, the chain moves and energy is transferred to the rear of the bicycle to move the rear wheel. The amount of energy applied to the pedals affects the flow of the chain. As more energy is supplied by a cyclist pushing on the pedals, the chain moves faster as there is an increase in the supplied energy. This cyclist (or more specifically, the push on the pedals) can be used to represent the potential difference in a circuit (or the energy supplied to each coulomb of charge).

5.4 Activities

learnon

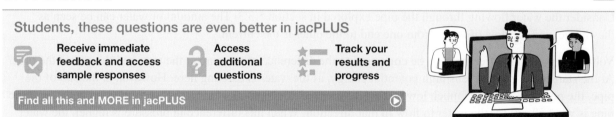

| 5.4 Quick quiz on | 5.4 Exercise | 5.4 Exam questions |

5.4 Exercise

1. What is the voltage supplied by a battery that gives 1.05 J of energy to 0.70 C of charge that passes through it?
2. Complete the following table by filling in the missing values from **a–f**.

Potential difference $\left(V = \dfrac{E}{Q}\right)$	Energy ($E = VQ$)	Charge $\left(Q = \dfrac{E}{V}\right)$
a.	32 J	9.6 C
b.	4.0 J	670 mC
9.0 V	c.	3.5 C
12 V	d.	85 mC
4.5 V	12 J	e.
240 V	7.5 kJ	f.

3. A 6.0-V source supplies 3.6×10^{-4} J of energy to a quantity of charge. Determine the quantity of charge in coulombs and microcoulombs.
4. Complete the following table by converting between the units of voltage provided.

a. 7.3 mV	V
b. 2300 V	kV
c. 0.00042 V	mV
d. 932 kV	V
e. 2.7 μV	V

5. Explain why a torch runs off a 3 V battery, while a car needs to be connected to a 9 V battery.
6. What is the emf of a battery that provides 9.0 J of energy to 6.0 C of charge?

5.4 Exam questions

Question 1 (2 marks)

As 4.0 C of charge passes through a light bulb, a quantity of electrical energy is transformed into 40 J of thermal energy and 8.0 J of light energy.

What is the voltage drop across the light bulb? Explain your reasoning.

Question 2 (1 mark)

How much electrical potential energy will 5.7 μC of charge transfer if it passes through a voltage drop of 6.0 V?

Question 3 (1 mark)

MC Which of the following supplies more energy to an appliance?
A. A 1.5-V battery with 2.7 C flowing through it
B. A 1.7-V battery with 2.5 C flowing through it
C. A 2.0-V battery with 2.2 C flowing through it
D. A 2.4-V battery with 1.9 C flowing through it

Question 4 (1 mark)

Describe the difference between a 1.5-V battery and a 9-V battery.

Question 5 (2 marks)

A battery is used to power a small device. The device transforms 540 J of electrical energy as 60 C of charge passes through it.

What is the emf of the battery? Explain your reasoning.

More exam questions are available in your learnON title.

5.5 Electrical energy and power

KEY KNOWLEDGE

- Apply concepts of energy (E) and power (P) in electric circuits
- Investigate and analyse theoretically and practically electric circuits using the relationships: $I = \dfrac{Q}{t}$, $V = \dfrac{E}{Q}$, $P = \dfrac{E}{t} = VI$

Source: Adapted from VCE Physics Study Design (2023–2027) extracts © VCAA; reproduced by permission.

5.5.1 Electrical energy transformations

Electrical devices transform electrical energy into other forms of energy. Since the potential difference is a measure of the loss in electrical potential energy by each coulomb of charge, the amount of energy (E) transformed by a charge (Q) passing through a load can be expressed as:

$$E = QV$$

since $V = \dfrac{E}{Q}$, where V is the potential difference across the load.

The amount of charge passing through a load in a time interval t can be expressed as:

$$Q = It$$

$$E = VIt$$

where:

E is the energy transferred, in joules

V is the potential difference, in volts

I is the current, in amperes

t is the time, in seconds.

tlvd-0015

SAMPLE PROBLEM 8 Calculating the potential difference using energy, current and time

What is the potential difference across a heater element if 3.6×10^4 J of heat energy is produced when a current of 5.0 A flows for 30 seconds?

THINK	WRITE
1. Recall the formula that calculates the amount of energy produced.	$E = VIt$
2. Transpose the formula to make the potential difference, V, the subject.	$V = \dfrac{E}{It}$
3. Substitute the known values into the formula and solve.	$E = 3.6 \times 10^4$ J, $I = 5.0$ A, $t = 30$ s $V = \dfrac{3.6 \times 10^4 \, \text{J}}{5.0 \, \text{A} \times 30 \, \text{s}}$ $= \dfrac{36\,000}{150}$ $= 240$ V
4. State the solution.	The potential difference is 240 V.

PRACTICE PROBLEM 8

What is the potential difference across a light globe if 1.44×10^3 J of heat is produced when a current of 2.0 A flows for 1 minute?

5.5.2 Power delivered by a circuit

In practice, it is the rate at which energy is transformed in an electrical load that determines its effect. The brightness of a light globe is determined by the rate at which electrical potential energy is transformed into light energy.

Power is the rate of doing work, or the rate at which energy is transformed from one form to another. Power is equal to the amount of energy transformed per second, or the amount of energy transformed divided by the time it took to do it.

power the rate of doing work, or the rate at which energy is transformed from one form to another

Power can be expressed as:

$$P = \frac{E}{t}$$

where:

P is the power delivered, in watts (W)

E is the amount of energy transformed, in joules

t is the time interval, in seconds.

1 watt = 1 joule per second = $1\ \mathrm{J\ s^{-1}}$

By substituting the formula for energy, $E = VIt$, into the formula for power, $P = \frac{E}{t}$, we find that:

$$P = \frac{VIt}{t}$$

$$\Rightarrow P = VI$$

This is a particularly useful formula because the potential difference, V, and electric current, I, can be easily measured in a circuit.

In household devices such as fridges, heaters and air conditioners, large power measurements are common. It is sometimes useful to use the unit kilowatt (kW), where 1 kilowatt = 1×10^3 watts. In many digital devices, small power measurements are normally needed. It can be useful to use milliwatts (mW) and sometimes even microwatts (μW).

tlvd-0016

SAMPLE PROBLEM 9 Calculating the power rating using voltage drop and current

What is the power rating of an electric heater if a current of 5.0 A flows through it when there is a voltage drop of 240 V across the heating element?

THINK	WRITE
1. Use the formula for power, $P = VI$.	$P = VI$
2. Substitute the known values into the formula and solve for P.	$V = 240\ \mathrm{V}, I = 5.0\ \mathrm{A}$ $P = VI$ $\quad = 240\ \mathrm{V} \times 5.0\ \mathrm{A}$ $\quad = 1200\ \mathrm{W}$
3. Remembering that 1 kW = 1000 watts, convert to kW by dividing the number of watts by 1000.	$P = 1.2\ \mathrm{kW}$
4. State the solution.	The power rating is 1.2 kW.

PRACTICE PROBLEM 9

What is the power rating of a CD player if it draws a current of 100 mA and is powered by four 1.5-V cells in series?

Transposing formulae

You can use the following triangles to assist transposing the formulae $P = VI$, $Q = It$ and $E = QV$.

Power formula triangle

Variants of the power formula triangle

For example, if you wish to transpose the formula $P = VI$ to make I the subject, cover the pronumeral you want to be the subject, in this case I. What is visible in the triangle shows what that pronumeral equals. In this example:

$$I = \frac{P}{V}$$

This method can also be used for any formula of the form $x = yz$.

tlvd-0017

SAMPLE PROBLEM 10 Calculating the energy supplied by a phone battery using voltage, time and current

How much energy is supplied by a mobile phone battery rated 3.7 V and 1200 mAh?
Note: **'mAh' stands for milliamp hours, which means that the battery would last for one hour supplying a current of 1200 mA, or two hours at 600 mA.**

THINK	WRITE
1. Recall the formula $E = VIt$, and state the known values.	$E = VIt$ $V = 3.7$ V, $I = 1200$ mA, $t = 1$ hour
2. Convert the current from mA to amperes by dividing by 1000. Convert the time to seconds.	$I = 1200$ mA $= \dfrac{1200}{1000}$ A $= 1.2$ A $t = 60 \times 60 = 3600$ s
3. Substitute the known values and solve for E.	$E = 3.7$ V $\times 1.2$ A $\times 3600$ s $= 16\,000$ J
4. Convert the number of joules to kJ by dividing by 1000.	$E = 16$ kJ
5. State the solution.	There are 16 kJ of energy supplied.

PRACTICE PROBLEM 10

A 3.7-V mobile phone battery has an energy capacity of 14 000 joules. In a talk mode test, the battery lasted for six hours. What was the average current?

---

elog-1623

tlvd-0807

tlvd-0018

SAMPLE PROBLEM 11 Determining the rate that energy is supplied from a battery

A 12-V car battery has a current of 2.5 A passing through it. At what rate is it supplying energy to the car's circuits?

THINK	WRITE
1. Recall the formula $P = VI$, and state the known values.	$P = VI$ $V = 12$ V, $I = 2.5$ A
2. As the energy supplied to the circuit equals the emf of the battery, substitute the known values for V and I into the formula $P = VI$ and solve.	$P = VI$ $\quad = 12\,\text{V} \times 2.5\,\text{A}$ $\quad = 30\,\text{W}$
3. State the solution.	The battery is supplying 30 W.

PRACTICE PROBLEM 11

At what rate is energy being supplied to a 3.0-V light when it is drawing a current of 4.0 A?

5.5 Activities

learn on

5.5 Quick quiz on	5.5 Exercise	5.5 Exam questions

5.5 Exercise

1. Calculate the current drawn by:
   a. a 60-W light globe connected to a 240 V source
   b. a 40-W globe with a voltage drop of 12 V across it
   c. a 6.0-V, 6.3-W globe when operating normally
   d. a 1200-W, 240-V toaster when operating normally.
2. How long will it take a 600-W microwave oven to transform 5.4×10^4 J of energy?
3. What is the power rating of an electric radiator if it draws a current of 10 A when connected to a 240-V AC household circuit?

4. An electric jug is connected to a 240-V supply and draws a current of 3.3 A. How long would it take to transfer 3.2×10^4 J of energy to its contents?
5. Mark wanted to calculate the energy used by a 4.5-kW air conditioner each week. The air conditioner was only turned on for Physics classes, which were one hour long and conducted four times a week. Mark believed the air conditioner used 18 kJ of energy.
   Was Mark correct? What mistake did he make when doing the calculation?
6. The element of a heater has a voltage drop of 240 V across it.
   a. In terms of energy, what does this mean?
   b. How much energy is transformed into thermal energy in the element if 25 C of charge flow through it?

5.5 Exam questions

▶ Question 1 (1 mark)

MC A device draws a current of 10 A for 20 s while operating with a potential difference of 6.0 V.

What is the best estimate of the electrical energy consumed?
A. 120 J
B. 200 J
C. 1000 J
D. 1200 J

▶ Question 2 (1 mark)

MC A device, operating with a voltage drop of 3.0 V across it, uses 150 J of electrical energy in 200 s.

What is the best estimate of the current?
A. 4.0 A
B. 2.25 A
C. 0.25 A
D. 0.20 A

▶ Question 3 (2 marks)

A battery is supplying 30 W of power to a circuit. The circuit is operating three devices: P, Q and R. Device P is operating at 15 W and device Q is operating at 7.0 W.

What is the power of device R? Explain your reasoning.

▶ Question 4 (3 marks)

A device is operating at 12 V and it uses 3600 J of electrical energy in a time of 2.5 minutes.

Calculate the current in the device.

▶ Question 5 (2 marks)

A rear window demister circuit draws 2.0 A of current from a 12-V battery for 30 minutes.
a. How much energy is transformed by the demister?   **(1 mark)**
b. What is the power rating of the demister?   **(1 mark)**

More exam questions are available in your learnON title.

5.6 Electrical resistance

5.6.1 Resistance

The **resistance** of a material or device is a measure of how difficult it is for a current to pass through it. The higher the value of resistance, the harder it is for the current to pass through the material or device, and therefore the more electrical potential energy is lost by the current and gained by the device.

> **resistance** the ratio of voltage drop, *V*, across a material or device to the current, *I*, flowing through it

The resistance, *R*, of a substance is defined as the ratio of the voltage drop, *V*, across it to the current, *I*, flowing through it.

$$R = \frac{V}{I}$$

The SI unit of resistance is the ohm (symbol Ω). One ohm is the resistance of a conductor in which a current of 1 ampere results from the application of a constant voltage drop of 1 volt across its ends.

$$1\,\Omega = 1\,\text{V A}^{-1}$$

Alternatively, 1 ohm is the resistance of a conductor in which a current of 1 ampere loses 1 volt of electrical potential energy as it passes through the resistor, with the resistor gaining 1 volt of electrical potential energy after a current of 1 ampere has passed through.

The ohm is named in honour of Georg Simon Ohm (1789–1854), a German physicist who investigated the effects of different materials in electric circuits.

FIGURE 5.11 Georg Simon Ohm

Resistance is a material property and it is temperature dependent. In general, the resistance of a metal conductor increases with temperature. Usually, the increases will not be significant over small temperature ranges, and most problems in this text ignore any temperature and resistance changes that might occur.

THE LIE DETECTOR

The lie detector, or polygraph, is a meter that measures the resistance of skin. The resistance of skin is greatly reduced by the presence of moisture. When people are under stress, as they may be when telling lies, they sweat more. The subsequent change in resistance is detected by the polygraph and is regarded as an indication that the person *may* be telling a lie.

---

elog-1624

tlvd-3753

INVESTIGATION 5.3   `online only`

The current-versus-voltage characteristics of a light globe

Aim

To design and construct a circuit that will enable you to find the current through the globe for a suitable range of voltages with length

INVESTIGATION 5.4

online only

Dependence of resistance on length of resistance wire

Aim

To investigate how the resistance of a resistance wire varies with length

5.6.2 Resistors

In many electrical devices, **resistors** are used to control the current flowing through, and the voltage drop across, parts of the circuits. Resistors have constant resistances ranging from less than 1 ohm to millions of ohms. There are three main types of resistors:

resistor a device used to control the current flowing through, and the voltage drop across, parts of a circuit

- 'Composition' resistors are usually made of the semiconductor carbon.
- Wire-wound resistors consist of a coil of fine wire made of a resistance alloy such as nichrome. (Nichrome is a heat-resistant alloy used in electrical heating elements. Its composition is variable, but is usually around 62% nickel, 15% chromium and 23% iron.)
- The metal film resistor consists of a glass or pottery tube coated with a thin film of metal.

FIGURE 5.12 a. Carbon or 'composition' resistors b. Wire-wound resistor c. Metal film resistor

a.

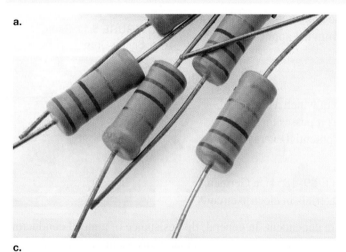

b.

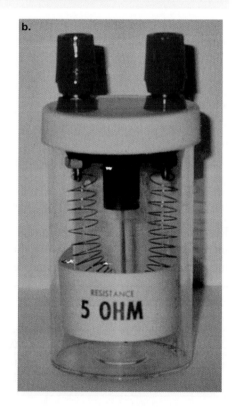

c.

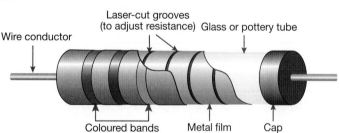

Some large resistors have their resistance printed on them. Others have a colour code to indicate their resistance, using the four coloured bands found on them. Further information on these bands can be accessed in the Extension box.

EXTENSION Determining the resistance using coloured bands

You will commonly use resistors in circuits. The resistance of many of these can be determined using the coloured bands found on them (figure 5.13). It is important to use an appropriate resistor in your circuits. Circuits will be examined further in topic 6.

The first two bands represent the first two digits in the value of resistance. The third band represents the power of ten by which the two digits are multiplied. The fourth band is the manufacturing tolerance. These values are shown in table 5.1.

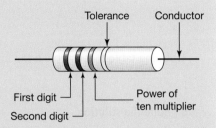

FIGURE 5.13 A resistor, showing the coloured bands

TABLE 5.1 The resistor colour code

Colour	Digit	Multiplier	Tolerance
Black	0	10^0 or 1	
Brown	1	10^1	
Red	2	10^2	±2%
Orange	3	10^3	
Yellow	4	10^4	
Green	5	10^5	
Blue	6	10^6	
Violet	7	10^7	
Grey	8	10^8	
White	9	10^9	
Gold		10^{-1}	±5%
Silver		10^{-2}	±10%
No colour			±20%

tlvd-0019

SAMPLE PROBLEM 12 Determining the resistance of different resistors (extension)

What is the resistance of the following resistors if their coloured bands are:
a. **red, violet, orange and gold**
b. **brown, black, red and silver?**

THINK

a. 1. Remember when holding a resistor to read its value, keep the gold or silver band on the right and read the colours from the left.

WRITE

a.

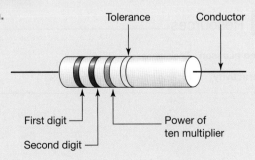

2. Using table 5.1, establish the first two digits.

Red = 2
Violet = 7
Hence, the first two digits are 27.

3. Using table 5.1, establish the multiplier.

The third band is orange, which means multiply the first two digits by 10^3.
The resistance is: $27 \times 10^3 = 27\,000$
$= 27 \text{ k}\Omega$

4. Using table 5.1, establish the tolerance.

The fourth band is gold, which means there is a tolerance of 5%.
$5\% \times 27\,000 \ \Omega = 1350 \ \Omega$

5. State the solution.

The true value is $27 \times 10^3 \text{ k}\Omega \pm 1350 \ \Omega$.

b. 1. Remember when holding a resistor to read its value, keep the gold or silver band on the right and read the colours from the left.

b.

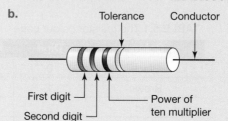

Tolerance — Conductor
First digit — Second digit — Power of ten multiplier

2. Using table 5.1, establish the first two digits.

Brown = 1
Black = 0
Hence, the first two digits are 10.

3. Using table 5.1, establish the multiplier.

The third band is red, which means multiply the first two digits by 10^2.
The resistance is: $10 \times 10^2 = 1000 \ \Omega$
$= 1\text{k} \ \Omega$

4. Using table 5.1, establish the tolerance.

The fourth band is silver, which means there is a tolerance of 10%.
$10\% \times 1000 \ \Omega = 100 \ \Omega$

5. State the solution.

The true value is $1.0 \times 10^3 \ \Omega \pm 100 \ \Omega$.

PRACTICE PROBLEM 12 (EXTENSION)

What are the resistances and tolerances of resistors with the colour codes:

a. orange, white, black, gold
b. green, blue, orange, silver
c. violet, green, yellow, gold?

 Resources

▶ **Video eLesson** Resistance (eles-2516)

✦ **Interactivity** Picking the right resistor (int-6391)

5.6.3 Ohm's Law

Georg Ohm established experimentally that the current, I, in a metal wire is proportional to the voltage drop, V, applied to its ends.

When he plotted his results on a graph of V versus I, he obtained a straight line.

$$I \propto V$$

FIGURE 5.14 Graphs of voltage versus current for two different metal wires

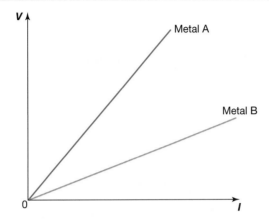

The equation of the line shown in figure 5.14 is known as Ohm's Law and can be written as:

$$V = IR$$

where R is numerically equal to the constant gradient of the line. This is known as the resistance of the metal conductor to the flow of current through it. Remember that the SI unit of resistance is the ohm (Ω).

You can use the triangle method for Ohm's Law.

Convert the quantity/pronumeral you want to be the subject; for example, R. What is visible in the triangle shows what the pronumeral equals.

The resistance, R, can also be expressed as:

$$R = \frac{V}{I}$$

tlvd-0020

SAMPLE PROBLEM 13 Calculating resistance using voltage and current

A transistor radio uses a 6.0-V battery and draws a current of 300 mA. What is the resistance of the radio?

THINK

1. From Ohm's Law, the resistance, R, can be found.

WRITE

$V = IR$

$\Rightarrow R = \dfrac{V}{I}$

2. State the known values and convert the current into amperes by dividing by 1000.	$V = 6.0$ V, $I = \dfrac{300\ \text{A}}{1000} = 0.300$ A
3. Substitute the values for V and I and solve to find R.	$R = \dfrac{6.0\ \text{V}}{0.300\ \text{A}}$ $= 20\ \Omega$
4. State the solution.	The resistance of the radio is 20 Ω.

PRACTICE PROBLEM 13

A 240-V kitchen appliance draws a current of 6.0 A. What is its resistance?

 Resources

 Weblink Ohm's Law app

5.6.4 Ohmic and non-ohmic devices

An **ohmic device** is one for which the resistance is constant for all currents under a range of changing physical conditions, including changing voltage, light level and temperature.

A **non-ohmic device** is one for which the resistance varies with changing physical conditions, including voltage, light level or temperature, but is constant for all currents passing through it under each set of physical conditions.

The graph in figure 5.14 has voltage on the y-axis and current on the x-axis. The graph is drawn this way so that the gradient of lines for the metals A and B gives the resistance of each. However, accepted convention graphs the independent variable on the x-axis and the dependent variable on the y-axis. So in the graph in figure 5.15a the gradient equals $\dfrac{1}{R}$.

> **ohmic device** a device for which the resistance is constant for all currents under a range of changing physical conditions, including changing voltage, light level and temperature
>
> **non-ohmic device** a device for which the resistance varies with changing physical conditions, including voltage, light level or temperature, but is constant for all currents passing through it under each set of physical conditions

FIGURE 5.15 The current-versus-voltage graphs for **a.** an ohmic resistor and **b.** a diode, which is a non-ohmic device

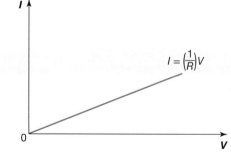

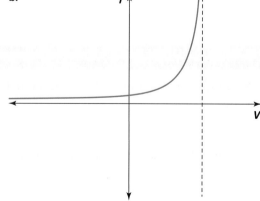

elog-1626

tlvd-0810

Ohmic and non-ohmic devices

Aim

To explore the resistance properties of a resistor and a light globe

Non-ohmic devices

Many non-ohmic devices are made from elements that are semiconductors. They are not insulators as they conduct electricity, though not as well as metals. Common semiconductor elements are silicon and germanium, which are in group 14 of the periodic table. Many new semiconductor devices are compounds of group 13 and group 15 elements such as gallium arsenide.

A **diode** is formed by joining two differently doped materials together. A diode allows current to flow through it in only one direction. This effect can be seen in the current–voltage graph for a diode in figure 5.15b, where a small positive voltage produces a current, while a large negative or reverse voltage produces negligible current.

Light-emitting diodes (LEDs) are diodes that give off light when they conduct. They are usually made from gallium arsenide. Gallium nitride is used in blue LEDs.

Thermistors are made from a mixture of semiconductors so they can conduct electricity in both directions. They differ from a metal conductor, whose resistance increases with temperature, as an increase in a thermistor's temperature increases the number of electrons available to move and the resistance can decrease. The resistance of a thermistor is also predictable and measurable, and changes by a significant amount compared to a metal conductor.

FIGURE 5.16 Circuit symbol for a diode

diode a device that allows current to pass through it in one direction only; its resistance changes with voltage

light-emitting diode (LED) a small semiconductor diode that emits light when a current passes through it

thermistor a device that has a resistance that changes with a change in temperature

light dependent resistor (LDR) a device that has a resistance that varies with the amount of light falling on it

FIGURE 5.17 a. Circuit symbol for a thermistor **b.** Resistance-versus-temperature graph for a thermistor

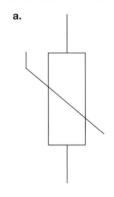

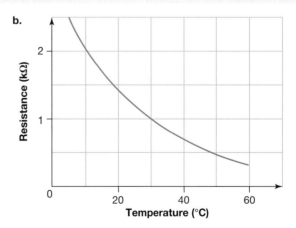

Light dependent resistors (LDRs) are like thermistors, except they respond to light. The resistance of an LDR decreases as the intensity of light shining on it increases. The axes in the graph for an LDR in figure 5.18 have different scales to the other graphs. As you move from the origin, each number is 10 times the previous one. This enables more data to fit in a small space.

FIGURE 5.18 a. Circuit symbols for an LDR **b.** Graph of resistance-versus-light intensity for an LDR, on a logarithmic scale

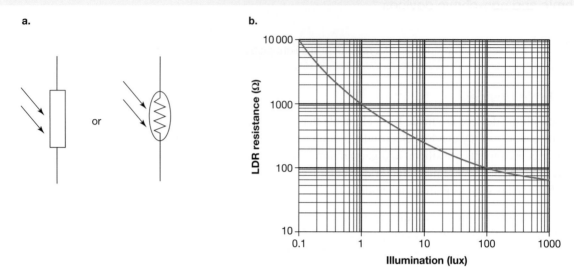

Note: A logarithmic scale is a non-linear scale and is commonly used as an effective way of displaying data that cover a large range of values on one graph. By plotting the log of resistance and the log of illumination, we are able to graphically plot two parameters.

5.6.5 Heating effects of currents

Whenever a current passes through a conductor, thermal energy is produced. This is due to the fact that the mobile charged particles — for example, electrons — make repeated collisions with the atoms of the conductor, causing them to vibrate more and produce an increase in the temperature of the material.

This temperature increase is not related to the direction of the current. A current in a conductor always generates thermal energy, regardless of which direction the current flows. Examples of devices that make use of this energy include radiators, electric kettles, toasters, stoves and fuses.

5.6.6 Power and resistance

Recall that the rate at which energy is dissipated by any part of an electric circuit can be expressed as:

$$P = VI$$

where:

P is the power, in watts

I is the current, in amperes

V is the voltage drop, in volts.

This relationship can be used, along with the definition of resistance, $R = \dfrac{V}{I}$, to deduce two different formulae describing the relationship between power and resistance:

$$R = \frac{V}{I}, \ V = IR, \ I\frac{V}{R}$$

Substituting $V = IR$ into the formula $P = VI$:

$$P = VI$$
$$= (IR)I$$

Thus:

$$P = I^2R \quad [1]$$

Substituting $I = \dfrac{V}{R}$ into the formula $P = VI$:

$$P = VI$$
$$P = V\left(\dfrac{V}{R}\right)$$

Thus:

$$P = \dfrac{V^2}{R} \quad [2]$$

You now have three different ways of determining the rate at which energy is transferred as charge flows through a voltage drop in an electric circuit:

$$P = VI \qquad P = I^2R \qquad P = \dfrac{V^2}{R}$$

In addition, the quantity of energy transferred, E, can be determined using:

$$E = VIt = I^2Rt = \dfrac{V^2t}{R}$$

These formulae indicate that in conducting wires with low resistance, very little energy is dissipated. If the resistance, R, is small, and the voltage drop, V, is small, the rate of energy transfer is also small.

tlvd-0021

SAMPLE PROBLEM 14 Determining the rate that electrical energy is transformed using resistance and voltage

A portable radio has a total resistance of 18 Ω and uses a 6.0-V battery consisting of four 1.5-V cells in series. At what rate does the radio transform electrical energy?

THINK	WRITE
1. Recall that power is the rate of energy use and use the formula containing the variables P, V and R.	$P = \dfrac{V^2}{R}$
2. Substitute the known values into the formula and solve for P.	$V = 6.0\text{ V}, R = 18\ \Omega$ $P = \dfrac{V^2}{R}$ $= \dfrac{(6.0\text{ V})^2}{18\ \Omega}$ $= 2.0\text{ W}$
3. State the solution.	The radio transforms energy at 2.0 W.

PRACTICE PROBLEM 14

What is the power rating of an electric jug if it has a resistance of 48 Ω when hot and is connected to a 240-V supply?

---

tlvd-0022

SAMPLE PROBLEM 15 Determining the normal operating current and resistance of a device

A pop-up toaster is labelled '240 V, 800 W'.
a. What is the normal operating current of the toaster?
b. What is the total resistance of the toaster while it is operating?

THINK	WRITE
a. 1. The three variables P, V and I are given in the formula for power.	**a.** $P = VI$
2. Transpose the formula to make I the subject.	$I = \dfrac{P}{V}$
3. Substitute the known values into the formula and solve for I.	$V = 240$ V, $P = 800$ W $I = \dfrac{800 \text{ W}}{240 \text{ V}}$ $= 3.33$ A
4. State the solution.	The normal operating current is 3.33 A.
b. 1. The three variables P, V and R are given in the formula for power.	**b.** $P = \dfrac{V^2}{R}$
2. Transpose the formula to make R the subject.	$R = \dfrac{V^2}{P}$
3. Substitute the known values into the formula and solve for R.	$V = 240$ V, $P = 800$ W $R = \dfrac{(240 \text{ V})^2}{800 \text{ W}}$ $= 72 \ \Omega$
4. State the solution.	The resistance is 72 Ω.

PRACTICE PROBLEM 15

A microwave oven is labelled '240 V, 600 W'.
a. What is the normal operating current of the microwave oven?
b. What is the total resistance of the microwave oven when it is operating?

5.6 Activities

| 5.6 Quick quiz | on | 5.6 Exercise | 5.6 Exam questions |

5.6 Exercise

1. How much energy is provided by a 6-V battery if a current of 3 A passes through it for 1 minute?
2. Complete the following table by filling in the missing values from **a–f**.

Potential difference	Current	Resistance
a.	8.0 A	4.0 Ω
b.	22 mA	2.2 kΩ
12 V	**c.**	6.0 Ω
240 V	**d.**	8.0×10^4 Ω
9.0 V	6.0 A	**e.**
1.5 V	45 mA	**f.**

3. The following graph shows the current-versus-voltage characteristic for an electronic device.

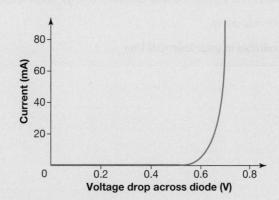

a. Is this device ohmic or non-ohmic? Justify your answer.
b. What is the current through the device when the voltage drop across it is 0.5 V?
c. What is the resistance of the device when the voltage drop across it is 0.5 V?
d. Estimate the voltage drop across the device and its resistance when it draws a current of 20 mA.
4. At what rate is thermal energy being transferred to a wire if it has a resistance of 5 Ω and carries a current of 0.30 A?
5. Calculate the resistance of the following globes if their ratings are:
   a. 240 V, 60 W
   b. 6.0 V, 6.3 W
   c. 12 V, 40 W.
6. What is the power rating of an electric jug if it has a resistance of 48 Ω when hot and is connected to a 240-V supply?

5.6 Exam questions

▶ **Question 1 (1 mark)**

MC An ohmic component is initially operating at voltage V_1 and power P_1. The voltage is now doubled ($V_2 = 2V_1$).

What is the new power (P_2) in the component?

A. $4P_1$
B. $2P_1$
C. $0.5P_1$
D. $0.25P_1$

▶ **Question 2 (3 marks)**

Bill and Ben are discussing Ohm's Law.

Bill describes it as 'voltage is directly proportional to current'. Ben replies, 'No, it should be expressed as $V = IR$.'

Which of these statements is correct? Or are they both valid? Justify your answer.

▶ **Question 3 (3 marks)**

A component draws a current of 0.50 A at a voltage drop of 12.5 V and a current of 1.2 A at a voltage drop of 30 V.

Is the component displaying ohmic behaviour? Justify your answer.

▶ **Question 4 (2 marks)**

A device with $R = 30\ \Omega$ is operated at $V = 9.0$ V for 100 s.

Calculate the total energy transformed. Show your reasoning.

▶ **Question 5 (2 marks)**

A device has a resistance of 80 Ω and it uses 960 J of electrical energy when operated for 5.0 minutes.

What is the current? Show your reasoning.

More exam questions are available in your learnON title.

5.7 Review

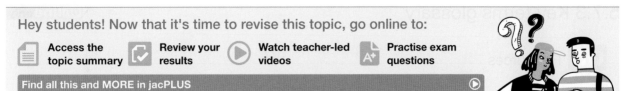
5.7.1 Topic summary

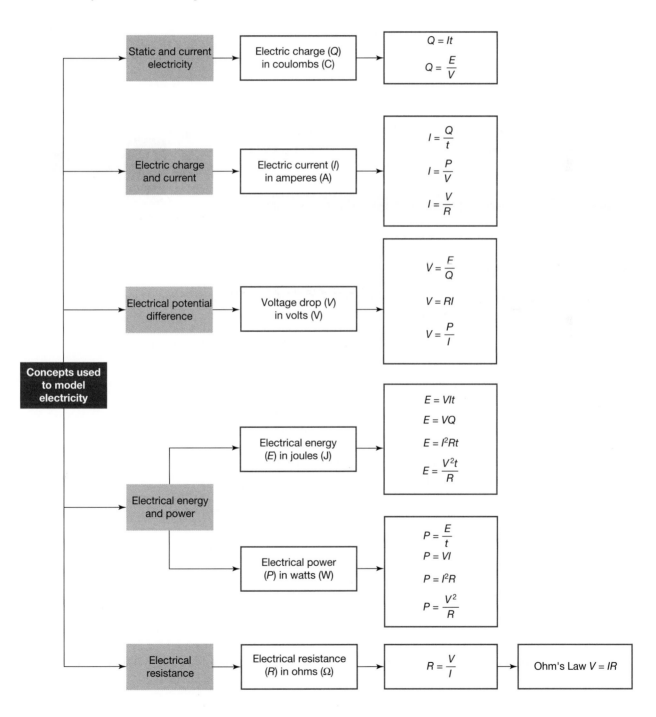

5.7.2 Key ideas summary

5.7.3 Key terms glossary

on Resources

Solutions	Solutions — Topic 5 (sol-0791)
Practical investigation eLogbook	Practical investigation eLogbook — Topic 5 (elog-1574)
Digital documents	Key science skills — VCE Physics Units 1–4 (doc-36950) Key terms glossary — Topic 5 (doc-36959) Key ideas summary — Topic 5 (doc-36960)
Exam question booklet	Exam question booklet — Topic 5 (eqb-0073)

5.7 Activities

learn on

Students, these questions are even better in jacPLUS

Receive immediate feedback and access sample responses

Access additional questions

Track your results and progress

Find all this and **MORE** in jacPLUS

5.7 Review questions

1. During a storm a lightning bolt discharges 3 million kJ of energy to Earth in 0.75 ms. The discharge involves the movement of 15 C of charge. What is the potential difference between the lightning source and Earth?

2. An electric kettle operating off a 240-V power supply uses 2.7 kW when boiling water.
   a. What is the current in the kettle?
   b. When the kettle is on for 2.5 minutes, how much energy does it use?

3. a. What is the voltage drop across a 44-Ω resistor carrying a current of 2.5 A?
   b. What would be the effect of connecting the same resistor to a larger power supply?

4. A children's toy comprises a 9-V battery, a switch and an electric motor in a simple circuit as shown in the following figure. The motor is labelled '9 V, 25 mA'.

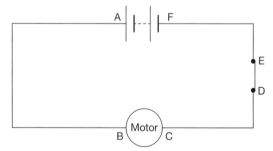

   a. When the switch is closed, what is the maximum rate at which the battery can supply energy to the motor?

b. By copying the axes shown, sketch a graph showing the energy held by a coulomb of charge as it moves around the circuit from A to F, then returning to A.

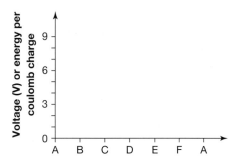

c. By copying the axes shown, complete the graph for current around the circuit.

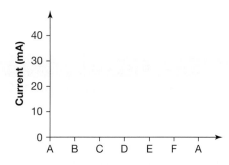

d. What is the resistance of the motor when operating at its maximum capacity?

5. A handheld fan is powered by two 1.5-V batteries. When tested at room temperature, 15 C of charge flowed through it every minute.

   a. What was the current in the device?
   b. What is the resistance of the device?
   c. At what rate was the device using energy?

6. **a.** What is the resistance of an 800-W toaster when a current of 3.3 A is flowing?
   b. If it takes 40 seconds to brown the toast, how much energy is used?

7. The voltage supplied to a tungsten globe is varied. The current and the potential difference across the lamp are measured and the results plotted in the following graph.

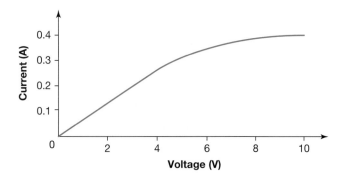

   a. Calculate the resistance of the globe when the current is 0.2 A.
   b. Would you describe the globe as ohmic or non-ohmic? Justify your answer.

8. The following graph shows the current-versus-voltage relationship for a non-ohmic device.

a. What is a non-ohmic device?

b. What is the current through the device when the voltage drop across it is 0.3 V?

c. What is the resistance of the device when the voltage drop across it is 0.3 V?

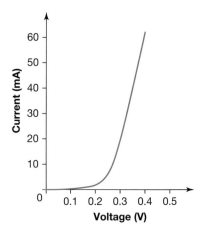

5.7 Exam questions

▶ Question 1

Which of the following statements best describes the transfer of electrical energy in a circuit?

A. Electric charge is moved in the direction of the current.

B. Free electrons pass all the way through the circuit to the load.

C. Free electrons dissipate their energy.

D. The battery pushes positive charges through the circuit.

▶ Question 2

Electron charge was removed from an object.

Which of the following could represent the amount of charge removed?

A. 1.5×10^{-19} C

B. 3.0×10^{-19} C

C. 4.8×10^{-19} C

D. 5.4×10^{-19} C

▶ Question 3

For 25 seconds a battery supplies a constant current of 5.0 A to a circuit.

Which of the following best represents the amount of charge leaving the battery?

A. 0 C

B. 0.20 C

C. 5.0 C

D. 125 C

Question 4

The current and voltage for an object in a circuit was collected and plotted on a graph.

What is the resistance of this object?

A. 0.60 Ω

B. 0.67 Ω

C. 1.50 Ω

D. 1.70 Ω

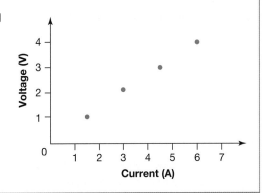

 Question 5

Each electron is given 1.44×10^{-18} joules of energy by a battery.

What is the electromotive force provided by the battery?

A. 1.5 V

B. 3.0 V

C. 9.0 V

D. 11 V

 Question 6

A 240-V rice cooker draws a current of 1.25 A.

How much energy is provided to the cooker in one minute?

A. 19.2 J

B. 300 J

C. 3.6 kJ

D. 18 kJ

 Question 7

At what rate is a 7.2-V battery supplying energy to a tablet device when it is drawing a current of 1200 mA?

A. 6.0 J

B. 6.0 W

C. 8.6 J

D. 8.6 W

 Question 8

When a 240-V microwave rated at 1200 W is operating, what is the current?

A. 5.0 mA

B. 6.0 mA

C. 5.0 A

D. 48.0 A

Question 9

When a 240-V microwave rated at 1200 W is operating, what is the total resistance?

A. 5 Ω
B. 6 Ω
C. 20 Ω
D. 48 Ω

Question 10

In the circuit shown, which of the following best describes the current passing through the ammeter?

A. The current is zero; it is used up in the 24-W load.
B. The current is 2 A.
C. The current is zero; the switch is open.
D. The current is non-constant.

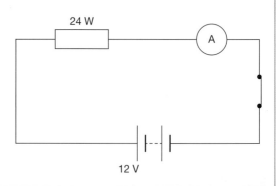

Section B — Short answer questions

Question 11 (5 marks)

Consider a 3.0-V battery.

a. How much energy does the battery supply to:
  i. one electron   **(1 mark)**
  ii. one coulomb of charge?   **(1 mark)**
b. The battery is used in a mobile telephone. A one-minute conversation uses the energy transferred by 0.04 C of charge.
  At what rate is energy being transferred to the telephone?   **(3 marks)**

Question 12 (2 marks)

To model the energy supply, current and load in an electrical circuit, a teacher and students stand in a circle loosely holding a loop of rope. The teacher models the energy supply by pulling the rope around while the students act as conductors by allowing the rope to readily move through their hands. One student, Luke, is asked to represent a globe in the circuit by making it more difficult for the rope to slide through his hands.

An observer suggests that in an electric circuit, the electrons must make their way to the globe before it lights up and that Luke must wait until the rope has travelled all the way from the teacher before he feels the pull on the rope.

In terms of both the model and how electricity behaves in a circuit, explain whether you agree or disagree with the observer.

Question 13 (2 marks)

An unknown electrical component is labelled '3.0 V, 2.4 W DC ONLY'.
a. What is the maximum current that the component can safely tolerate?   **(1 mark)**

b. A student places another component in a simple circuit to measure its voltage current characteristics, recording the results as shown in the provided graph.

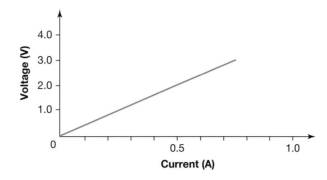

Calculate the resistance when the current is 600 mA. **(1 mark)**

Question 14 (2 marks)

An electric drill has a rechargeable 18-V battery that can store up to 6 MJ. When fully discharged it takes the battery 8 hours to fully recharge.

a. Calculate the total charge needed to fully charge the battery. **(1 mark)**
b. Calculate the average current drawn when charging the battery. **(1 mark)**

Question 15 (4 marks)

The following graph shows the current-versus-voltage characteristic for an electronic device.

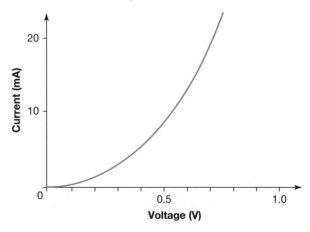

a. Is this device ohmic or non-ohmic? Justify your answer. **(1 mark)**
b. What is the current through the device when the voltage drop across it is 0.5 V? **(1 mark)**
c. What is the resistance of the device when the voltage drop across it is 0.5 V? **(1 mark)**
d. Estimate the rate at which energy is being used when the voltage drop across it is 0.5 V. **(1 mark)**

6 Circuit electricity

KEY KNOWLEDGE

In this topic, you will:
- justify the use of selected meters (ammeter, voltmeter, multimeter) in circuits
- model resistance in series and parallel circuits using:
  - current versus potential difference (I–V) graphs
  - resistance as the potential difference to current ratio, including R = constant for ohmic devices
  - equivalent resistance in arrangements in:
    - series: $R_{equivalent} = R_1 + R_2 + ... + R_n$ and
    - parallel: $\dfrac{1}{R_{equivalent}} = \dfrac{1}{R_1} + \dfrac{1}{R_2} + ... \dfrac{1}{R_n}$
- calculate and analyse the equivalent resistance of circuits comprising parallel and series resistance
- analyse circuits comprising voltage dividers
- investigate and apply theoretically and practically concepts of current, resistance, potential difference (voltage drop) and power to the operation of electronic circuits comprising resistors, light bulbs, diodes, thermistors, light dependent resistors (LDRs), light-emitting diodes (LEDs) and potentiometers (quantitative analysis restricted to use of $I = \dfrac{V}{R}$ and $P = VI$)
- investigate practically the operation of simple circuits containing resistors, variable resistors, diodes and other non-ohmic devices
- describe energy transfers and transformations with reference to resistors, light bulbs, diodes, thermistors, light dependent resistors (LDRs), light-emitting diodes (LEDs) and potentiometers in common devices
- compare power transfers in series and parallel circuits.

Source: VCE Physics Study Design (2023–2027) extracts © VCAA; reproduced by permission.

PRACTICAL WORK AND INVESTIGATIONS

Practical work is a central component of VCE Physics. Experiments and investigations, supported by a **practical investigation eLogbook** and **teacher-led videos**, are included in this topic to provide opportunities to undertake investigations and communicate findings.

EXAM PREPARATION

Access exam-style questions and their video solutions in every lesson, to ensure you are ready.

6.1 Overview

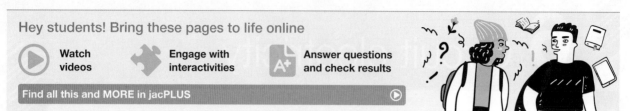
6.1.1 Introduction

In topic 5, the behaviour of current electricity was explored, and the concepts of current, potential difference, resistance, energy and power were introduced. These concepts are most relevant in electrical circuits. An electrical circuit is a number of electrical conductors connected to form a conducting path. A circuit can contain one or more sources of electromotive force (emf) to provide energy to the circuit.

FIGURE 6.1 Decorative lights that are connected in series, such as those on Luna Park, can contain shunts that allow the circuit to remain closed if the filament breaks.

If the conductors form a continuous closed path through which a current can circulate, the circuit is said to be a closed circuit, as shown in figure 6.1. If there is a break in the path so that charge cannot flow — for example, at a switch — the circuit is said to be an open circuit.

In many simple electric circuits, electrical energy is used for heating one or more loads. The temperature increase in a load occurs because the charge carriers, which are charged particles moving in a conductor, make repeated collisions with the atoms in the load. This increases the internal energy of the load and its temperature rises. The electrical energy transformed in the load originally came from some source of emf — for example, a battery, laboratory power pack or power point in the home. Examples of this type of circuit include a toaster plugged into a power point and a car demister circuit. The current flowing in these circuits depends on the resistance of the loads. The heating element in a toaster (often a thick nichrome wire) and other loads make it difficult for a current to flow. In this topic you will look at what happens in different types of electrical circuits and in the devices within these circuits.

LEARNING SEQUENCE

---

on Resources

📋 **Solutions**   Solutions — Topic 6 (sol-0792)

🔬 **Practical investigation eLogbook**   Practical investigation eLogbook — Topic 6 (elog-1575)

📄 **Digital documents**   Key science skills — VCE Physics Units 1–4 (doc-36950)
                                               Key terms glossary — Topic 6 (doc-36961)
                                               Key ideas summary — Topic 6 (doc-36962)

Exam question booklet   Exam question booklet — Topic 6 (eqb-0074)

6.2 BACKGROUND KNOWLEDGE
Electrical circuit rules

BACKGROUND KNOWLEDGE

- The sum of the currents flowing into a junction is equal in magnitude to the sum of the currents flowing out of that junction: $I_{in} = I_{out}$.
- In any closed loop of an electrical circuit, the sum of the voltage drops must equal the sum of the emfs in that loop.
- Ammeters and voltmeters are instruments for measuring current and voltage drop, respectively.

6.2.1 Circuit diagrams

A circuit diagram shows schematically the devices used in constructing an electrical circuit. Table 6.1 shows the symbols commonly used in drawing circuits. Figure 6.2 is a diagram of a circuit comprising a battery, **switch** and **resistor**.

switch a device that stops or allows the flow of electricity through a circuit

resistor a device used to control the current flowing through, and the voltage drop across, parts of a circuit

FIGURE 6.2 A circuit diagram

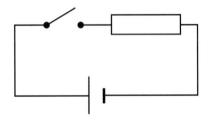

TABLE 6.1 Symbols used in circuit diagrams

Circuit component	Symbol	Circuit component	Symbol
Connection between conductors	■	Resistor with sliding contact to give a variable resistance	
Terminal	□	Semiconductor diode*	or
Conductors not connected*	or	Single pole switch (open)	
Conductors connected*	or	Button switch (open)	
Earth*	or	Voltmeter	(V)
Battery		Ammeter	(A)

(continued)

TABLE 6.1 Symbols used in circuit diagrams *(continued)*

Circuit component	Symbol	Circuit component	Symbol
Variable power supply*	or	Incandescent lamp*	or
Resistor	or	Light dependent resistor (LDR)	
Variable resistor*	or	Thermistor (heat dependent resistor)	

*The first of the two alternative symbols is used in this book.

6.2.2 Circuit rules

Accounting for electrons

Electric charge is conserved. At any point in a conductor, the amount of charge (usually electrons) flowing into that point must equal the amount of charge flowing out of that point. Electrons do not build up at a point in a conductor, nor will they magically disappear. You don't get traffic jams in **electrical circuits**.

In figure 6.3, $I_a + I_c = I_b + I_d + I_e$.

> **electric charge** the basic property of matter; it occurs in two states: positive (+) charge and negative (−) charge
>
> **electrical circuit** a closed loop of moving electric charge

The sum of the currents flowing into a junction is equal in magnitude to the sum of the currents flowing out of that junction:

$$I_{in} = I_{out}$$

FIGURE 6.3 Five wires soldered at a junction

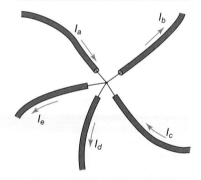

tlvd-0023

SAMPLE PROBLEM 1 Calculations involving conservation of electric charge

Calculate the magnitude and direction of the unknown current in the following figure, showing currents meeting at a junction.

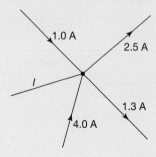

THINK	WRITE
1. From the diagram, calculate the total currents flowing into the junction.	$1.0\ \text{A} + 4.0\ \text{A} = 5.0\ \text{A}$
2. From the diagram, calculate the total currents flowing out of the junction.	$2.5\ \text{A} + 1.3\ \text{A} = 3.8\ \text{A}$
3. Recall that $I_{in} = I_{out}$. The unknown values must be a flow out of the junction. Solve for the unknown current out of the junction.	$I_{in} = I_{out}$ $5.0\ \text{A} = 3.8\ \text{A} + x$ $\Rightarrow x = 5.0\ \text{A} - 3.8\ \text{A}$ $\qquad = 1.2\ \text{A}$
4. State the solution.	The unknown current flowing out of the junction is 1.2 A.

PRACTICE PROBLEM 1

Determine the unknown current at the junction in the figure shown. State the direction of the current in each case.

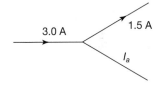

tlvd-0024

SAMPLE PROBLEM 2 Determining the values of various currents in a circuit

Determine the values of currents *a*, *b*, *c*, *d*, *e* and *f* as marked in the following figure.

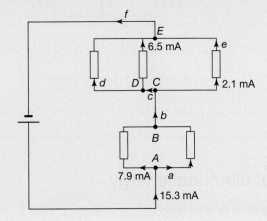

THINK	WRITE
1. Recall the formula $I_{in} = I_{out}$.	$I_{in} = I_{out}$
2. Determine the current at each junction.	$15.3\ \text{mA} = 7.9\ \text{mA} + a$ $\Rightarrow a = 15.3\ \text{mA} - 7.9\ \text{mA}$ $\qquad = 7.4\ \text{mA}$ $7.9\ \text{mA} + 7.4\ \text{mA} = b$ $\qquad \Rightarrow b = 15.3\ \text{mA}$

$$15.3 \text{ mA} = c + 2.1 \text{ mA}$$
$$\Rightarrow c = 15.3 \text{ mA} - 2.1 \text{ mA}$$
$$= 13.2 \text{ mA}$$
$$13.2 \text{ mA} = d + 6.5 \text{ mA}$$
$$\Rightarrow d = 13.2 \text{ mA} - 6.5 \text{ mA}$$
$$= 6.7 \text{ mA}$$
$$2.1 \text{ mA} = e$$
$$\Rightarrow e = 2.1 \text{ mA}$$
$$d + 6.5 \text{ mA} + e = f$$
$$\Rightarrow f = 6.7 \text{ mA} + 6.5 \text{ mA} + 2.1 \text{ mA}$$
$$= 15.3 \text{ mA}$$

3. State the solution.

The values of the currents are:
$a = 7.4$ mA, $b = 15.3$ mA, $c = 13.2$ mA, $d = 6.7$ mA, $e = 2.1$ mA, $f = 15.3$ mA.

PRACTICE PROBLEM 2

Determine the values of currents a and b as marked in the following figure.

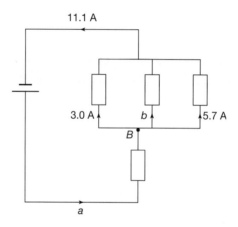

6.2.3 Conservation of electrical energy

Around a circuit electrical energy must be conserved.

In any closed loop of a circuit, the sum of the voltage drops must equal the sum of the **emfs** in that loop.

electromotive force (emf) a measure of the energy supplied to a circuit for each coulomb of charge passing through the power supply

SAMPLE PROBLEM 3 Calculating voltage drops in series circuits

Calculate the unknown voltage drop V_{bc} in the following figure.

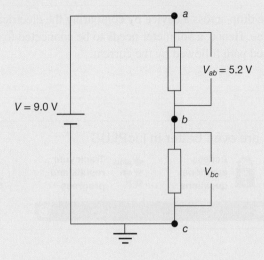

THINK

1. This circuit is a closed loop and the sum of the voltage drops within it must equal the voltage supplied by the battery.

2. State the solution.

WRITE

V = the sum of the voltage drops

= 9.0 V

$9.0\,V = V_{ab} + V_{bc}$

$= 5.2\,V + V_{bc}$

$\Rightarrow V_{bc} = 3.8\,V$

The unknown voltage drop is 3.8 V.

PRACTICE PROBLEM 3

Calculate the unknown voltage drop V_{ab} in the following figure.

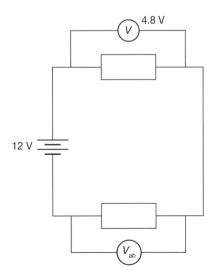

6.2.4 Use of meters in circuits

An ammeter measures the current flowing through a device and therefore needs to be connected into the circuit as part of the continuous closed path followed by the current.

A voltmeter measures the voltage drop across a device by comparing the electrical potential energy of the charges before and after the device. Hence, a voltmeter needs to be connected to two places in the circuit and is not part of the continuous closed path followed by the current.

6.2 Activities

learn on

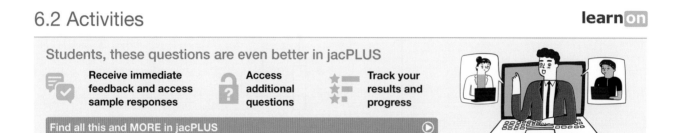

| 6.2 Quick quiz **on** | 6.2 Exercise | 6.2 Exam questions |

6.2 Exercise

1. Explain what is meant by the following terms, as they relate to electric circuits.
   a. Junction
   b. Current
   c. Voltage drop
   d. Conductor
2. Determine the missing currents in the following figures. State the direction of the current in each case.

   a.

   b.

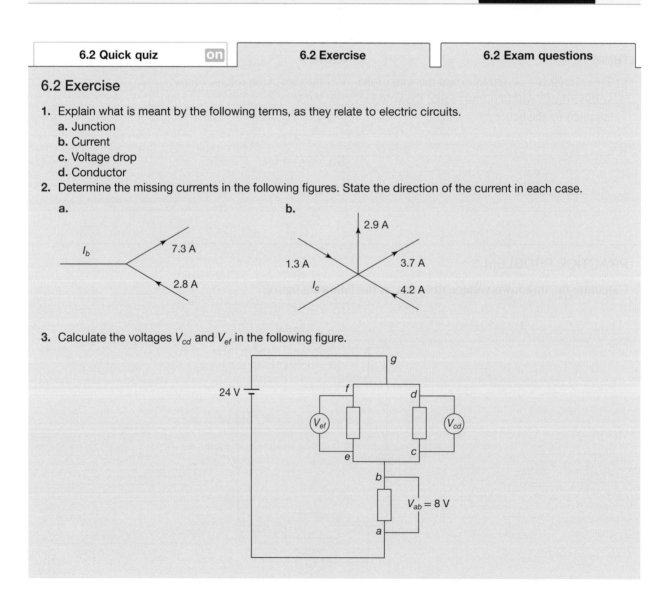

3. Calculate the voltages V_{cd} and V_{ef} in the following figure.

4. In the parallel circuit shown in the following figure, $I_1 = 0.20$ A, $R_1 = 60$ Ω and $I_T = 0.50$ A.

Determine values for:

a. V_{ab}

b. V_{cd}

c. ε

d. I_2

e. R_2.

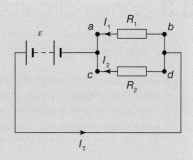

6.2 Exam questions

▶ Question 1 (6 marks)

Use the circuit diagram shown to answer the following questions.

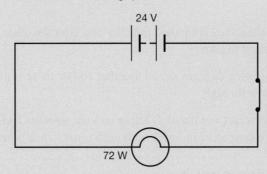

a. Draw an arrow to show the direction of the electron current. **(1 mark)**

b. Show where you would connect an ammeter to measure the current flowing through the globe. **(1 mark)**

c. Calculate the current flowing in the circuit. **(1 mark)**

d. When the switch is closed, what is the voltage drop across it? **(1 mark)**

e. Show where you would connect a voltmeter to confirm your prediction of the voltage drop across the switch. **(1 mark)**

f. Explain what happens when the switch is opened. **(1 mark)**

▶ Question 2 (2 marks)

A circuit comprises a 1.2-kW heating element, a switch and a voltage source.

a. Draw a circuit diagram demonstrating how to measure both the voltage across the heater and the current flowing through it. **(1 mark)**

b. If the voltage supplied is 240 V, what is the current passing through the heater? **(1 mark)**

▶ Question 3 (1 mark)

A doorbell connected to a battery comprises a button at the door, the bell and wires. The bell only sounds after the button is pushed.

Why doesn't the bell always sound?

▶ Question 4 (5 marks)

Consider two people connected in series with a 25-V source across them. The first person, R_1, has a resistance of 1000 Ω and the second person, R_2, has a resistance of 1500 Ω.

a. Determine the current through each person. **(1 mark)**

b. What is the voltage drop across each person? **(1 mark)**

The two people now connect themselves in parallel with the 25-V power source.

c. Determine the current through each person. **(2 marks)**

d. What is the voltage drop across each person? **(1 mark)**

▶ Question 5 (2 marks)

Prove mathematically that the resistance of a short-circuited resistor branch is 0.

More exam questions are available in your learnON title.

6.3 Series circuits

There are two ways in which circuit elements can be connected: in **series** and in **parallel**.

When devices are connected in series, they are joined together one after the other. There is only one path for the current to take.

When devices are connected in parallel, they are joined together so that there is more than one path for the current to flow through.

Many devices can be connected in series and parallel. These include resistors and cells. It is also possible to have circuits in which some devices are connected in series with each other, while others are connected in parallel.

series refers to devices joined together one after the other

parallel refers to devices connected so that one end of each device is joined at a common point, and the other end of each device is joined at another common point

6.3.1 Current in a series circuit

In a series circuit there is only one path for the current to take through the circuit. This means the current is the same in all parts of the circuit. Every resistor experiences the same current flowing through it.

Therefore, in figure 6.4, the current in R_1 equals the current in R_2 and in R_3. I_1 refers to the current in R_1, I_2 refers to the current in R_2, and so on. All these currents will be the same.

FIGURE 6.4 Resistors connected in series

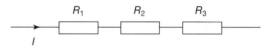

$$I = I_1 = I_2 = I_3$$

6.3.2 Voltage drop in a series circuit

In figure 6.4, V_1 is the voltage drop across R_1, V_2 is the voltage drop across R_2, and so on.

Since $V = IR$, and the current through each resistor is the same:

$$V_1 = IR_1$$

$$V_2 = IR_2$$

$$V_3 = IR_3$$

The total voltage drop, V_T, across resistors in series is equal to the sum of the voltage drops across each individual resistor.

$$V_T = V_1 + V_2 + V_3$$

The voltage in a series circuit is divided proportionally between the resistors. This means that a larger resistor will experience a larger voltage drop across it. Its larger **resistance** means the charges give it more electrical potential energy as they pass through. This happens because the same current is passing through each resistor.

resistance the ratio of voltage drop, V, across a material or device to the current, I, flowing through it

Dividing the voltage proportionally also means that if one resistor has twice the resistance of another, the first resistor will experience double the voltage drop. For example, a 40-Ω resistor will have a voltage drop twice as large as a 20-Ω resistor. If resistors have the same resistance, they will experience the same voltage drop.

6.3.3 Resistors in series

As the same current flows through all resistors in a series circuit and the total voltage drop, V_T, is the sum of the voltage drops across each individual resistor:

$$V_T = IR_1 + IR_2 + IR_3$$

$$V_T = I(R_1 + R_2 + R_3)$$

Since $V_T = R_{equivalent}I$ (where $R_{equivalent}$ is the equivalent resistance of all three resistors), the equivalent resistance offered by resistors in series is found by obtaining the sum of the individual resistances:

$$R_{equivalent} = R_1 + R_2 + R_3$$

Note: Equivalent resistance is sometimes called effective resistance, or the total resistance.

This means that the equivalent resistance of a circuit is increased by adding an extra resistor in series with the others. The resistance of a series circuit is greater than that for any individual resistor.

As the equivalent resistance of the circuit is increased by adding an extra resistor in series, the current through the circuit will decrease since the increased equivalent resistance of the circuit makes it harder for charges to travel through the circuit. The voltage drop across individual resistors will also decrease as the total voltage now needs to be divided proportionally among more resistors.

tlvd-0026

SAMPLE PROBLEM 4 Determining the equivalent resistance of resistors connected in series

Determine the equivalent resistance of a circuit comprising three resistors, having resistance values of 15 Ω, 25 Ω and 34 Ω, connected in series.

THINK	WRITE
1. When connected in series the equivalent resistance of the circuit is equal to the sum of the individual resistors.	$R_{equivalent} = \sum_{1}^{i} R_i$
2. Substitute the resistance values.	$R_{equivalent} = 15 \ \Omega + 25 \ \Omega + 34 \ \Omega$
3. Calculate the equivalent resistance.	$R_{equivalent} = 74 \ \Omega$
4. State the solution.	The equivalent resistance of the circuit is 74 Ω.

PRACTICE PROBLEM 4

Determine the equivalent resistance of a circuit comprising three resistors, having resistance values of 1.2 kΩ, 5.6 kΩ and 7.1 kΩ.

---

tlvd-0027

SAMPLE PROBLEM 5 Determining the current, voltage drop and resistance in a series circuit

In the series circuit shown, the emf of the power supply is 100 V; the current at point a, I_a, equals 1 A; and the value of R_2 is 60 Ω. Determine the:
a. current at point b
b. voltage drop across R_2
c. voltage drop across R_1
d. value of R_1.

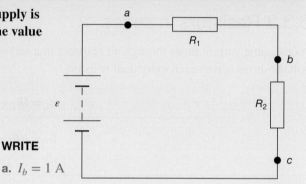

THINK

a. The current is the same at all points along the series circuit. Therefore, the current at point b, I_b, is 1 A.

b. Substitute known values in the equation for the voltage drop across the resistor, $V_2 = IR_2$.

c. This is a series circuit, so the voltage of the battery equals the sum of the voltage drops around the circuit. Therefore, $\varepsilon = V_1 + V_2$.

d. The resistance can be found using the relationship $V_1 = IR_1$ and making R the subject.

WRITE

a. $I_b = 1$ A

b. $V_2 = IR_2$
$\quad = 1\ \text{A} \times 60\ \Omega$
$\quad = 60\ \text{V}$

c. $\quad\varepsilon = V_1 + V_2$
$100\ \text{V} = V_1 + 60\ \text{V}$
$\Rightarrow V_1 = 40\ \text{V}$

d. $\quad V_1 = IR_1$
$40\ \text{V} = 1\ \text{A} \times R_1$
$\Rightarrow R_1 = 40\ \Omega$

PRACTICE PROBLEM 5

A 24-V battery supplies the energy to two objects connected in a series circuit as shown in the following figure.

The current at point a, I_a, is 0.6 A and the value of the known resistor is 15 Ω. Determine the:
a. current at point b
b. voltage drop across the known resistor
c. voltage drop across the unknown resistor, R_1
d. value of R_1.

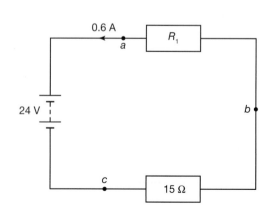

6.3.4 The voltage divider

The **voltage divider** is an example of resistors in series. It is used to divide or reduce a voltage to a value needed for a part of a circuit. A voltage divider is used in many control circuits; for example, turning on the heating in a house when the temperature drops.

The voltage divider has an input voltage, V_{in}, and an output voltage, V_{out}.

A general voltage divider is shown in figure 6.5.

The current, I, flowing through R_1 and R_2 is the same since R_1 and R_2 are in series.

$$V_{in} = I(R_1 + R_2)$$

$$\Rightarrow I = \frac{V_{in}}{R_1 + R_2}$$

$$V_{out} = IR_2$$

$$\Rightarrow V_{out} = \frac{V_{in}}{R_1 + R_2} \times R_2$$

This can be rewritten as:

$$V_{out} = \left[\frac{R_2}{R_1 + R_2}\right] V_{in}$$

FIGURE 6.5 A general voltage divider

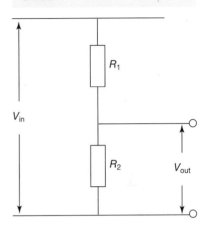

More generally:

$$V_{out} = \left[\frac{\text{resistance across which } V_{out} \text{ is taken}}{\text{sum of all resistances}}\right] V_{in} = \frac{R_{out}}{R_{total}} V_{in}$$

If R_1 and R_2 are equal in value, the voltage will be divided equally across both resistors. If R_1 is much greater than R_2, then most of the voltage drop will be across R_1.

---

elog-1798

tlvd-0813

INVESTIGATION 6.1

Determining emf and internal resistance

Aim

To determine the emf and internal resistance of circuits with old and new dry cells using resistors of varying resistances

---

 Resources

🔷 **Interactivity** Voltage dividers (int-6392)

SAMPLE PROBLEM 6 Calculating the values of an unknown resistor in a voltage divider

Calculate the value of the unknown resistor in the voltage divider shown in the following figure, if the output voltage is required to be **4.0 V**.

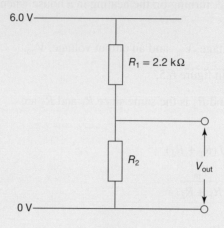

THINK

1. Substitute known values into the formula
$V_{out} = \dfrac{R_2 V_{in}}{R_1 + R_2}$ for a two-resistor voltage divider.

2. Solve by making R_2 the subject.

3. State the solution.

WRITE

$$V_{out} = \frac{R_2 V_{in}}{R_1 + R_2}$$

$$\Rightarrow 4.0\ V = \frac{6.0\ V \times R_2}{2.2\ k\Omega + R_2}$$

$$8.8\ k\Omega V + 4.0\ V R_2 = 6.0\ V R_2$$

$$2.0\ V R_2 = 8.8\ k\Omega V$$

$$\Rightarrow R_2 = 4.4\ k\Omega$$

The value of the unknown resistor in the voltage divider is 4.4 kΩ.

PRACTICE PROBLEM 6

Calculate the value of the unknown resistor in the voltage divider in sample problem 6 if the output voltage is to be 1.5 V.

6.3 Activities

learn on

Students, these questions are even better in jacPLUS

Receive immediate feedback and access sample responses

Access additional questions

Track your results and progress

Find all this and MORE in jacPLUS

| 6.3 Quick quiz on | 6.3 Exercise | 6.3 Exam questions |

6.3 Exercise

1. Determine the equivalent resistance of the following sets of resistors if they are connected in series.
   a. 12 Ω, 20 Ω, 30 Ω
   b. 1.2 kΩ, 3.2 kΩ, 11 kΩ

246 Jacaranda Physics 1 VCE Units 1 & 2 Fifth Edition

2. In the circuit shown, $R_1 = 1.0\ \Omega$, $R_2 = 2.0\ \Omega$ and the voltage drop across $R_1 = 3.0$ V. What is the value of V_b, the emf of the battery?

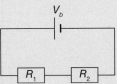

3. In the circuit shown, the emf of the battery is 6.0 V. The two resistors have resistances $R_1 = 5.0\ \Omega$ and $R_2 = X$. The current is 0.40 A. What is the resistance X?

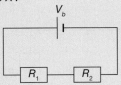

4. Calculate the unknown quantities in the series circuit shown in the following figure.

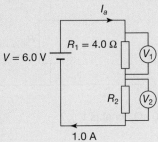

5. **MC** The following circuit diagram shows a voltage divider.

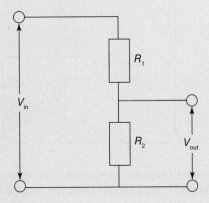

If this case, $V_{in} = 9.0$ V, $R_1 = 40\ \Omega$ and $R_2 = 60\ \Omega$.

Which of the following is the best estimate of V_{out}?
A. 3.6 V
B. 5.4 V
C. 6.0 V
D. 8.0 V

6. Calculate the voltage drop between *A* and *B* in each of the following voltage divider circuits.

a.

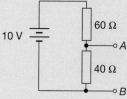

b.

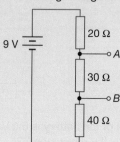

7. Consider the circuit shown.

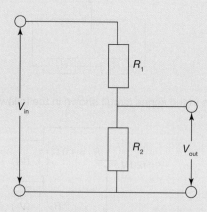

If $V_{in} = 12$ V, $R_1 = 5.0$ kΩ and $R_2 = 3.0$ kΩ, what is V_{out}?

8. Consider the voltage divider shown.

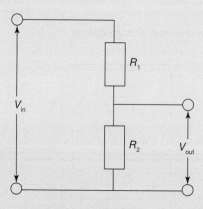

If the resistance of R_2 were to increase, what would happen to the value of V_{out}? Explain your answer fully.

9. a. Calculate the output voltage for the voltage divider shown in circuit a.
   b. What is the output voltage of circuit b if a load of resistance 4.4 kΩ is connected across the output terminals of the voltage divider?

a.

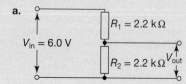

b.

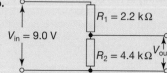

6.3 Exam questions

Question 1 (1 mark)

MC Consider the circuit shown. Three resistors, of resistance 1 Ω, 2 Ω and 3 Ω, are connected in series with a 12-V battery.

Which of the following gives the value of the current I_b ?
A. 12 A
B. 6.0 A
C. 2.0 A
D. 1.0 A

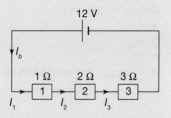

Question 2 (1 mark)

MC Three resistors of value 25 Ω, 15 Ω and 10 Ω are connected in series to a 10-V power supply.

What is the current in the circuit?
A. 0.20 A
B. 0.40 A
C. 0.70 A
D. 1.0 A

Question 3 (1 mark)

MC What is the value of the resistor shown in the following circuit?
A. 3 kΩ
B. 4 kΩ
C. 6 kΩ
D. 12 kΩ

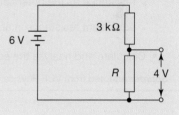

Question 4 (1 mark)

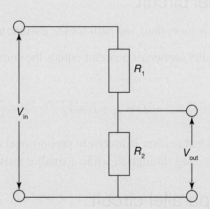

MC If V_{in} = 12 V, R_1 = X, R_2 = 50 Ω and V_{out} = 7.5 V, what is the resistance X?
A. 83 Ω
B. 50 Ω
C. 45 Ω
D. 30 Ω

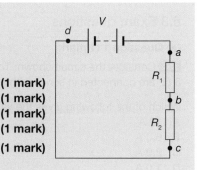

Question 5 (5 marks)

In the series circuit shown, V_{ab} = 20 V, R_2 = 30 Ω, and I_a = 2.0 A.

Determine values for:

a. I_b **(1 mark)**
b. V_{bc} **(1 mark)**
c. R_1 **(1 mark)**
d. $R_{equivalent}$, the equivalent resistance of the circuit   **(1 mark)**
e. V, the emf of the battery.   **(1 mark)**

More exam questions are available in your learnON title.

6.4 Parallel circuits

KEY KNOWLEDGE

- Justify the use of selected meters (ammeter, voltmeter, multimeter) in circuits
- Model resistance in series and parallel circuits using:
  - resistance as the potential difference to current ratio, including R = constant for ohmic devices
  - equivalent resistance in arrangements in parallel: $\dfrac{1}{R_{equivalent}} = \dfrac{1}{R_1} + \dfrac{1}{R_2} + ... \dfrac{1}{R_n}$
- Calculate and analyse the equivalent resistance of circuits comprising parallel and series resistance

Source: Adapted from VCE Physics Study Design (2023–2027) extracts © VCAA; reproduced by permission.

6.4.1 Current in a parallel circuit

In a parallel branch of a circuit, there is more than one path for the current to flow through.

The total current flowing into the parallel section of a circuit equals the sum of the individual currents flowing through each resistor.

$$I_T = I_1 + I_2 + I_3$$

The current flowing through an individual resistor is inversely proportional to the resistance of that resistor. A larger resistor will have less current flowing through it, while a smaller resistor will experience a larger current.

6.4.2 Voltage drop in a parallel circuit

As can be seen in figure 6.6, the left-hand sides of all the resistors are connected to point A and the right-hand sides of the resistors are connected to point B; therefore, each resistor in a parallel section of a circuit has the same voltage drop across it.

$$V_T = V_1 = V_2 = V_3$$

FIGURE 6.6 A parallel branch of a circuit

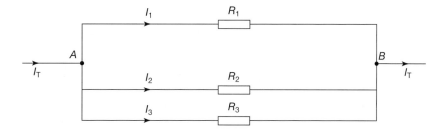

6.4.3 Resistors in parallel

In a parallel section of a circuit, the total current equals the sum of the individual currents and the voltage drops across each resistor are the same. It is possible to derive an expression for the equivalent resistance, $R_{\text{equivalent}}$, of a parallel section of a circuit.

$$I_T = I_1 + I_2 + I_3$$
$$\Rightarrow \frac{V}{R_T} = \frac{V}{R_1} + \frac{V}{R_2} + \frac{V}{R_3}$$

(since $I = \dfrac{V}{R}$ for each resistor and the whole section of the circuit)

Dividing both sides of the expression for the equivalent resistance by V gives:

$$\frac{1}{R_{\text{equivalent}}} = \frac{1}{R_1} + \frac{1}{R_2} + \frac{1}{R_3}$$

This means that the reciprocal of the equivalent resistance is equal to the sum of the reciprocals of the individual resistances.

The equivalent resistance is less than the smallest individual resistance. The more resistors there are added in parallel, the more paths there are for the current to flow through, and the easier it is for the current to flow through the parallel section.

As the equivalent resistance of the circuit is decreased by adding an extra resistor in parallel, the current through the circuit will increase, since the decreased equivalent resistance of the circuit makes it easier for charges to travel through the circuit. The voltage drop across individual resistors will remain the same as there has been no change to the amount of electrical potential energy supplied to the circuit.

Modelling resistors in parallel

One way to help understand this concept is to use the hydraulic model. Current is represented by water flowing in a pipe. Resistors are represented as thin pipes. The thinner the pipe, the greater the resistance; therefore, less water can flow in the circuit. A conductor is represented by a large pipe through which water flows easily. The source of emf is represented by a pump that supplies energy to the circuit.

If there is only one thin pipe, as in figure 6.7a, it limits the flow of water. Adding another thin pipe beside the first, as in figure 6.7b, allows more water to flow. The total resistance offered by the two thin pipes in parallel is less than that offered by an individual thin pipe.

a. **b.**

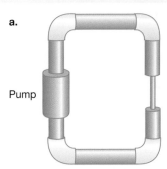 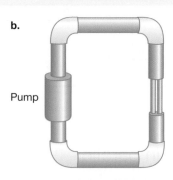

Pump   Pump

elog-1796

tlvd-3754

INVESTIGATION 6.2   online only

Parallel circuits

Aim

To observe the voltage and current in a circuit with resistors set up in parallel

 Resources

Digital document eModelling: Exploring resistors in parallel with a spreadsheet (doc-0047)

Weblinks   DC circuit water analogy

Simple electric circuits

tlvd-0028

SAMPLE PROBLEM 7 Determining the equivalent resistance of resistors connected in parallel

What is the equivalent resistance of three resistors connected in parallel if they have resistance values of 5.0 Ω, 10 Ω and 20 Ω?

THINK	WRITE
1. The three resistors are connected in parallel, so substitute the known values into the following: $\frac{1}{R_T} = \frac{1}{R_1} + \frac{1}{R_2} + \frac{1}{R_3}$.	$\frac{1}{R_{equivalent}} = \frac{1}{5.0\ \Omega} + \frac{1}{10\ \Omega} + \frac{1}{20\ \Omega}$
2. Solve for $R_{equivalent}$.	$\frac{1}{R_{equivalent}} = \frac{4}{20\ \Omega} + \frac{2}{20\ \Omega} + \frac{1}{20\ \Omega}$ $= \frac{7}{20\ \Omega}$ $\Rightarrow R_{equivalent} = \frac{20\ \Omega}{7}$ $= 2.9\ \Omega$

3. State the solution.

The equivalent resistance of the three resistors connected in parallel is 2.9 Ω.

Note: The equivalent resistance of a set of resistors connected in parallel is always less than the value of the smallest resistor used. Adding resistors in parallel increases the number of paths for current to flow through, so more current can flow and the resistance is reduced.

If there are n resistors of equal value, R, the equivalent resistance, $R_{\text{equivalent}}$, will be $R_{\text{equivalent}} = \dfrac{R}{n}$.

PRACTICE PROBLEM 7

Four resistors having values of 5 Ω, 5 Ω, 15 Ω and 20 Ω are connected in parallel. Calculate their equivalent resistance. Round your answer to the nearest whole number.

---

tlvd-0029

SAMPLE PROBLEM 8 Determining the current, voltage drop and resistance in a parallel circuit

Consider the parallel circuit shown in the following figure. The emf of the power supply is 9.0 V, R_2 has a resistance of 10 Ω and the current flowing through the power supply is 1.35 A. Determine:
a. the voltage drop across R_1 and R_2
b. I_2, the current flowing through R_2
c. I_1, the current flowing through R_1
d. the resistance of R_1
e. the equivalent resistance of the circuit.

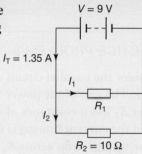

THINK	WRITE
a. 1. For a parallel circuit, $V_1 = V_2$.	**a.** $V_1 = V_2 = 9$ V
2. State the solution.	The voltage drop across R_1 and R_2 is 9 V.
b. 1. Determine the current in the resistor using the relationship $V = IR$ and making I the subject.	**b.** $I_2 = \dfrac{V}{R_2}$ $= \dfrac{9\text{ V}}{10\ \Omega}$ $= 0.90$ A
2. State the solution.	I_2, the current flowing through R_2, is 0.90 A.
c. 1. Recall the formula for the current leaving the battery.	**c.** $I_T = I_1 + I_2$
2. Make I_1 the subject and substitute the known values.	$I_1 = I_T - I_2$ $= 1.35\text{ A} - 0.90\text{ A}$ $= 0.45$ A
3. State the solution.	I_1, the current flowing through R_1, is 0.45 A.

d. 1. Determine the resistance by transposing the formula $V = IR$ to make R the subject.

d. $R_1 = \dfrac{V}{I_1}$

$= \dfrac{9 \text{ V}}{0.45 \text{ A}}$

$= 20 \text{ }\Omega$

2. State the solution.

The resistance of R_1 is 20 Ω.

e. 1. The resistors are in parallel.

e. $\dfrac{1}{R_{\text{equivalent}}} = \dfrac{1}{R_1} + \dfrac{1}{R_2}$

2. Substitute the known values.

$= \dfrac{1}{10 \text{ }\Omega} + \dfrac{1}{20 \text{ }\Omega}$

$= \dfrac{3}{20 \text{ }\Omega}$

3. $R_{\text{equivalent}}$ is the reciprocal.

$R_{\text{equivalent}} = \dfrac{20 \text{ }\Omega}{3}$

$= 6.7 \text{ }\Omega$

4. State the solution.

The equivalent resistance of the circuit is 6.7 Ω.

PRACTICE PROBLEM 8

Consider the parallel circuit shown in the following figure. The emf of the power supply is 24 V and the heater R_1 has a resistance of 40 Ω. When the switch is closed the current flowing is 0.8 A. Determine:

a. the voltage drop across R_1 and R_2
b. I_1, the current flowing through R_1
c. I_2, the current flowing through R_2
d. the resistance of R_2
e. the equivalent resistance of the circuit.

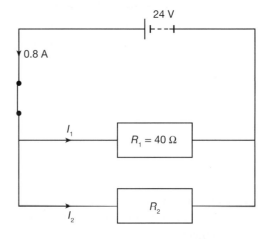

tlvd-0030

SAMPLE PROBLEM 9 Determining the equivalent resistance when resistors of different units are placed in parallel

Determine the equivalent resistance when a 10.0-Ω resistor is placed in parallel with a 10.0-kΩ resistor.

THINK

1. The two resistors are connected in parallel, so substitute the known values into the following:

$\dfrac{1}{R_{\text{equivalent}}} = \dfrac{1}{R_1} + \dfrac{1}{R_2}$.

WRITE

$\dfrac{1}{R_{\text{equivalent}}} = \dfrac{1}{R_1} + \dfrac{1}{R_2}$

$= \dfrac{1}{10.0 \text{ }\Omega} + \dfrac{1}{10\ 000 \text{ }\Omega}$

$$= \frac{1000}{10\ 000\ \Omega} + \frac{1}{10\ 000\ \Omega}$$

$$= \frac{1001}{10\ 000\ \Omega}$$

2. The reciprocal gives the value of the resistance.

$$R_{\text{equivalent}} = \frac{10\ 000\ \Omega}{1001}$$

$$= 9.99\ \Omega$$

3. State the solution.

The equivalent resistance is 9.99 Ω.

PRACTICE PROBLEM 9

Determine the equivalent resistance when a 1.2-kΩ resistor is placed in parallel with a 4.8-kΩ resistor.

Note: Adding a large resistance in parallel with a small resistance slightly reduces the equivalent resistance of that part of a circuit.

Parallel circuits are used extensively. Australian households are wired in parallel with an AC RMS (root mean square) voltage of 230 V. This is equivalent to a DC voltage of 230 V, and all the formulae that have been presented so far can be used for analysing AC circuits.

The advantage of having parallel circuits is that all appliances have the same voltage across them and the appliances can be switched on independently. If appliances were connected in series, they would all be on or off at the same time; and they would share the voltage between them, so no appliance would receive the full voltage. This would present problems when designing the devices, as it would not be known what voltage to allow for.

Car lights, front and rear, are wired in parallel for the same reason. If one lamp 'blows', the other lamps will continue functioning normally.

elog-1797

tlvd-3755

INVESTIGATION 6.3   online only

Series circuits

Aim

To observe the voltage and current in a circuit set up in series

6.4.4 Short circuits

A short circuit occurs in a circuit when a conductor of negligible resistance is placed in parallel with a circuit element. This element may be a resistor or a globe. The result of a short circuit is that virtually all the current flows through the conductor and practically none flows through the circuit element. Because there is effectively no voltage drop across the wire, there is also no voltage drop across the circuit element and no current flows through it. Think of what would happen in the hydraulic model if a conducting pipe were placed beside a thin pipe. This situation is represented in both ways in figure 6.8.

In this case, the current from the power supply passes through R_1, but then flows through the short circuit, effectively avoiding R_2 and R_3.

FIGURE 6.8 a. Circuit diagram showing a short circuit **b.** Hydraulic model of a short circuit

a.

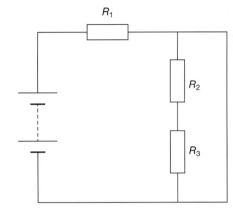

b.

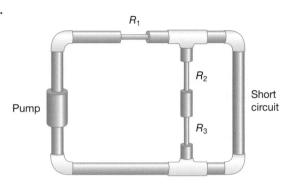

SAMPLE PROBLEM 10 Calculating the equivalent resistance with a short circuit

The following figure shows a 10-kΩ resistor that has been short-circuited with a conductor of 0 Ω resistance. Calculate the equivalent resistance of this arrangement.

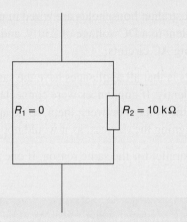

THINK

WRITE

1. The two resistors are connected in parallel, so substitute the known values into the following:

$$\frac{1}{R_{\text{equivalent}}} = \frac{1}{R_1} + \frac{1}{R_2}.$$

$$\frac{1}{R_{\text{equivalent}}} = \frac{1}{R_1} + \frac{1}{R_2}$$

$$= \frac{1}{0\ \Omega} + \frac{1}{10\ 000\ \Omega}$$

$$= \infty$$

2. The reciprocal gives the value of the resistance.

$$R_{\text{equivalent}} = 0\ \Omega$$

3. State the solution.

The equivalent resistance is 0 Ω.

PRACTICE PROBLEM 10

Determine the equivalent resistance when a 1.2-kΩ resistor is placed in parallel with a 4.8-kΩ resistor and a 500-Ω resistor. Round your answer to the nearest whole number.

6.4.5 Resistors in combinations of series and parallel

In many circuits, resistors may be in series with some resistors and parallel with others. An example is a set of decorative lights; the resistors in one section will be in series with each other and all turn on and off together; however, that section will be in parallel with other sections of lights that operate independently of each other.

When trying to calculate current, voltage or resistance in these more complex circuits it is important to carefully consider which resistors are in series with each other and which are in parallel. The circuit shown in figure 6.9 contains sets of three resistors in series with each other. Each group of four resistors is then in parallel with every other group. The rules of series and parallel circuits still apply — there are just multiple resistors in each set.

FIGURE 6.9 Example of a set of decorative lights

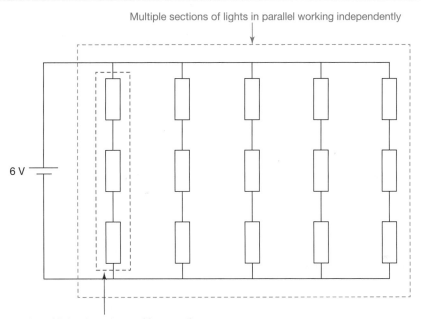

Multiple sections of lights in parallel working independently

6 V

One section of lights in series working together

tlvd-3376

SAMPLE PROBLEM 11 Calculating the equivalent resistance in a circuit with resistors in combination of series and parallel

a. **Calculate the equivalent resistance of the circuit shown.**

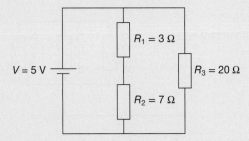

b. **Calculate the current, I_2, through R_2, the 7-Ω resistor.**
c. **Calculate the voltage, V_2, across R_2.**

THINK

WRITE

a. 1. The R_2 resistors are in series with each other and then in parallel with the R_3. Start by determining the resistance of the series section.

a. $R_{\text{series}} = R_1 + R_2$
$= 3 + 7$
$= 10 \ \Omega$

2. This group of resistors is in parallel with the 20-Ω resistor.

$\dfrac{1}{R_{\text{equivalent}}} = \dfrac{1}{R_{\text{series}}} + \dfrac{1}{R_3}$
$= \dfrac{1}{10} + \dfrac{1}{20}$
$= \dfrac{3}{20}$

3. Calculate the solution.

$R_{\text{equivalent}} = \dfrac{20}{3}$
$= 6.67 \ \Omega$

4. State the answer

The equivalent resistance of the circuit is 6.67 Ω.

b. 1. The 7-Ω resistor is in series with the 3-Ω resistor and both are in parallel with the rest of the circuit.

b. The voltage, V, across the series section containing R_1 and R_2 is 5 V.

2. The current I_1 through R_1 is the same as the current I_2 through R_2.
Using $V = IR_{\text{series}}$, calculate the current through R_2.

$I_1 = I_2 = I = \dfrac{V}{R_{\text{series}}} = \dfrac{5}{10} = 0.5 \ \text{A}$

3. State the answer.

The current through R_2 is 0.5 A.

c. 1. The current, I_2, though R_2 and the value of R_2 are known; using $V_2 = I_2R_2$ yields the value of V_2.

c. $V_2 = I_2R_2 = 0.5 \times 7 = 3.5 \ \text{V}$

2. State the answer.

The voltage across R_2 is 3.5 V.

PRACTICE PROBLEM 11

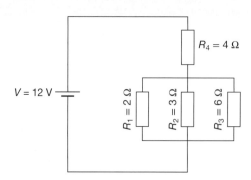

$R_4 = 4 \ \Omega$

$V = 12 \ \text{V}$

$R_1 = 2 \ \Omega$

$R_2 = 3 \ \Omega$

$R_3 = 6 \ \Omega$

a. For the circuit shown in the diagram above, calculate the equivalent resistance.
b. Calculate the current through R_2, the 3-Ω resistor.

6.4 Activities

| 6.4 Quick quiz on | 6.4 Exercise | 6.4 Exam questions |

6.4 Exercise

1. Determine the equivalent resistance when the following resistors are connected in parallel.
   a. 5.0 Ω, 10 Ω, 30 Ω
   b. 15 Ω, 60 Ω, 60 Ω
2. Two 10-Ω resistors are connected in parallel across the terminals of a 15-V battery.
   a. What is the equivalent resistance of the circuit?
   b. What current flows in the circuit?
   c. What is the current through each resistor?
3. Three resistors of 60 Ω, 30 Ω and 20 Ω are connected in parallel across a 90-V power source as shown in the following figure.
   a. Calculate the equivalent resistance of the circuit.
   b. Calculate the current flowing through the source.
   c. What is the current flowing through each resistor?

4. Calculate the unknown quantities in the parallel circuit shown in the following figure.

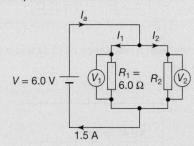

5. Three resistors, having resistance values of 6 Ω, 18 Ω and 9 Ω, are connected in parallel across a 36-V power supply.
   a. What current flows through each resistor?
   b. What total current flows in the circuit?
   c. What is the equivalent resistance of the circuit?
6. Four resistors are connected in the circuit below:

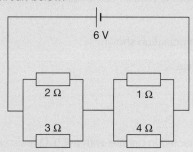

For this circuit, calculate:
a. the equivalent resistance
b. the current from the battery
c. the voltage across the 4-Ω resistor.

6.4 Exam questions

▶ **Question 1 (3 marks)**

The following figure shows an arrangement of switches and globes connected to a source of emf.

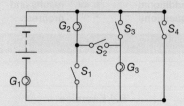

Which globes would light up if the following sets of switches are closed?

a. Switch S_2 only   **(1 mark)**
b. Switch S_3 only   **(1 mark)**
c. Switch S_4 only   **(1 mark)**

▶ **Question 2 (1 mark)**

When the following circuit is powered by an 18-V power supply, what is the current through the 6-Ω resistor?

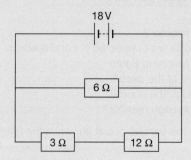

▶ **Question 3 (3 marks)**

The circuit shows a resistor, R_1, in parallel with a resistance of 15 Ω and a battery of emf 30 V. The current in R_1 is 3.0 A. The current in the 15-Ω resistor is I_2.

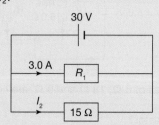

a. What is the resistance of R_1?   **(1 mark)**
b. What is the current I_2?   **(1 mark)**
c. What is the equivalent resistance of the parallel combination?   **(1 mark)**

▶ **Question 4 (2 marks)**

Consider the series–parallel circuit combination shown.

Calculate the total equivalent resistance.

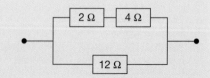

▶ **Question 5 (1 mark)**

MC An unknown resistor, R_X, is connected in parallel with a 2.0-Ω resistor. The equivalent resistance of the combination is 1.6 Ω.

What is the value of the unknown resistor?

A. 0.125 Ω **B.** 0.40 Ω **C.** 5.0 Ω **D.** 8.0 Ω

More exam questions are available in your learnON title.

6.5 Non-ohmic devices in series and parallel

6.5.1 Non-ohmic devices

Non-ohmic devices do not obey Ohm's Law. Their current-versus-voltage characteristics can be presented graphically.

The value of $\dfrac{V}{I}$ is not constant for non-ohmic devices.

The rules for series and parallel circuits still apply when analysing circuits containing non-ohmic devices, such as light bulbs, diodes, thermistors, light dependent resistors (LDRs) and light-emitting diodes (LEDs). Devices in series have the same current and share the voltage. Devices in parallel have the same voltage and the current is shared between them. The actual values of resistance, voltage or current are obtained from graphs for the devices. The specific type of graph and information provided depends on the type of non-ohmic device used; however, current-versus-voltage graphs are common.

> **non-ohmic device** a device for which the resistance varies with changing physical conditions, including voltage, light level or temperature, but is constant for all currents passing through it under each set of physical conditions

tlvd-0033

SAMPLE PROBLEM 12 Using a current-versus-voltage graph to determine voltage and current

The following figure shows the current-versus-voltage graph for two electrical devices.

If X and Y are in parallel and the current through X is 2 A, calculate:

a. the voltage across Y
b. the current through Y.

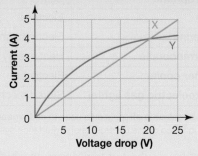

THINK

a. As X and Y are in parallel, the voltage across X equals the voltage across Y. Use the graph to determine the values.

b. Again, use the graph.

WRITE

a. When the current through X is 2 A, the voltage is 10 V, so the voltage across Y is also 10 V.

b. When the voltage across Y is 10 V, the current through Y is seen to be 3 A.

The following figures show two electrical devices, A and B, connected in a simple circuit, and the current-versus-voltage graph for A and B.

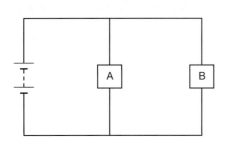

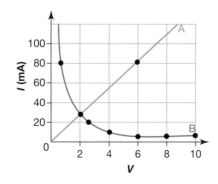

When the current through A is 80 mA, calculate:
a. the voltage across B
b. the current through B.

6.5.2 Transducers and sensors

Transducers are devices that convert energy from one form to another. They can be affected by, or can affect, the environment. The word *transducer* comes from the Latin for 'to lead across'. Table 6.2 lists a range of transducers and their energy conversions.

TABLE 6.2 Examples of transducers

| Transducer | Energy conversion | |
	From	To
Solar cell	Light	Electrical
Loudspeaker	Electrical	Sound
Microphone	Sound	Electrical
LED	Electrical	Light
Antenna	Electromagnetic	Electrical
Thermocouple	Heat energy (temp. difference)	Electrical
Peltier cooler	Electrical	Heat energy (temp. difference)
Piezoelectric gas lighter*	Stored mechanical energy due to pressure	Electrical

*The piezoelectric effect is also used in strain gauges to measure the stress in construction materials, and in the accelerometers found in video game controllers and guidance systems.

Sensors are a subset of transducers in which the energy conversion is to electrical; that is, to a variation in voltage. Some sensors generate the voltage directly; for example, piezoelectric devices. Other sensors whose resistance changes, such as **light dependent resistors (LDRs)** and **thermistors**, use a voltage divider circuit.

The resistance of an LDR at different light levels and the resistance of a thermistor at different temperatures can be represented on a graph. When analysing circuits containing these sensors it is important to first use the graph to find the correct resistance.

transducer a device that converts energy from one form to another form

light dependent resistor (LDR) a device that has a resistance that varies with the amount of light falling on it

thermistor a device that has a resistance that changes with a change in temperature

When sensors such as thermistors and LDRs are placed in a voltage divider circuit, the voltage supplied by the power source is shared between the two resistors depending on their relative resistances. This means that as the resistance of the LDR or thermistor changes, the voltage across both the LDR or thermistor and the resistor it is in series with will change. For example, as the light level changes and the resistance of an LDR increases, the voltage across it will increase, while the voltage across the other resistor in the voltage divider circuit will decrease. This can be used to turn lights on at night and off in the morning. As the temperature in a room changes and the resistance of a thermistor decreases, the voltage across it will also decrease, while the voltage across the other resistor in the voltage divider circuit will increase. This can be used to turn a heater or air conditioner on or off to maintain the temperature in a room. An increasing voltage can be used to turn devices on while a decreasing voltage will turn devices off.

tlvd-0034

SAMPLE PROBLEM 13 Calculations involving a voltage divider circuit with a thermistor 1

The resistance of a thermistor changes with temperature as shown in the following graph. In the voltage divider circuit shown, the thermistor is one of the two resistors; the other is a resistor with a fixed resistance value.

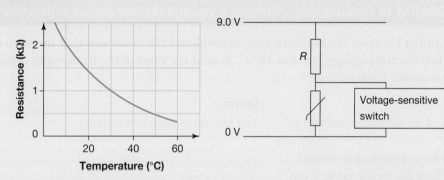

As the temperature drops, the resistance of the thermistor increases. As the thermistor's resistance increases, its share of the voltage from the power supply also increases, while that of the fixed value resistor will decrease.
A voltage-sensitive switch is placed across the thermistor. It is built to turn on a heater when the voltage across the thermistor is greater than 6 V. Your task as the circuit designer is to determine the resistance value required for the fixed-value resistor to turn on the heater at 19 °C.

THINK	WRITE
1. Determine the resistance of the thermistor at 19 °C using the graph.	From the graph, at 19 °C the thermistor has a resistance of 1.5 kΩ (1500 Ω).
2. Substitute the resistance into the voltage divider equation.	$V_{\text{out}} = \left[\dfrac{R_2}{R_1 + R_2} \right] V_{\text{in}}$ $6V = \left[\dfrac{1.5 \text{ k}\Omega}{R + 1.5 \text{ k}\Omega} \right] \times 9 \text{ V}$
3. Solve for R.	$6 \times (R + 1500) = 1500 \times 9$ $6R + 9000 = 13\,500$ $6R = 4500$ $\Rightarrow R = 750 \ \Omega$
4. State the solution.	The resistance required for the fixed-value resistor to turn on the heater at 19 °C is 750 Ω.

PRACTICE PROBLEM 13

The resistance-versus-temperature characteristics of a thermistor are shown in the following graph. The thermistor is connected in parallel with a fixed resistor similar to that shown in the circuit in sample problem 13.

The switch across the thermistor is designed to turn on a warning light when the voltage across the thermistor is 4 V or greater. Determine the resistance value required for the fixed-value resistor to turn on the warning light at 20 °C.

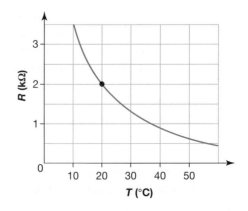

tlvd-0035

SAMPLE PROBLEM 14 Calculations involving a voltage divider circuit with a thermistor 2

To reduce the heating bill from the heater in sample problem 13, it is decided that the heater should be turned on when the temperature is below 18 °C. Should the value of the fixed-value resistor be increased or decreased? Explain.

THINK

The voltage to turn on the switch will still be 6 V, so the voltage across the two resistors will be unchanged. The ratio of their resistance values will therefore also be the same. From the graph in sample problem 13, it can be seen that at 18 °C the thermistor's resistance will be greater than it was at 19 °C. So to keep the ratio the same, R must increase.

This can also be explained using current. As the resistance of the thermistor is higher at the lower temperature, there will be less current through both resistors. As the voltage drop across R is to remain the same, its resistance will need to be greater ($V = IR$).

WRITE

The resistance must be increased.

PRACTICE PROBLEM 14

The thermistor and voltage-sensitive switch from sample problem 13 are to be used for a cooling system. The cooling system is to turn on when the temperature is greater than 24 °C. To which resistor — the thermistor or the fixed resistor — should the voltage-sensitive switch be connected so that the voltage is greater than 6 V for temperatures greater than 24 °C? Explain. What should be the value of the fixed resistor?

6.5.3 Diodes

A **diode** is a non-ohmic device whose resistance depends on the voltage drop it experiences. **Light-emitting diodes (LEDs)** are becoming very common; they are used in traffic lights, car headlights, television screens and to backlight keyboards. Other types of diodes are very common in electrical devices as they can be used to control the flow of current.

FIGURE 6.10 LEDs like these are commonly used for signage, and in the electronics and automotive industries.

A diode has a positive end and a negative end, meaning it needs to be connected the correct way around in a circuit. In figure 6.10, which shows some light-emitting diodes, you can see that one piece of wire connected to each diode is longer than the other. The longer wire needs to be connected to the positive side of the power source and the shorter wire needs to be connected to the negative side. When a diode is connected correctly it is **forward biased** and current can flow through the diode. When a diode is connected incorrectly it is **reverse biased** and current will not flow. This can be important for circuits connected to an AC power source.

The circuit symbol for a diode is shown in figure 6.11. The 'arrow' shows the direction of current flow when the diode is forward biased. In this image, the positive side of the diode is on the left and the negative side is on the right.

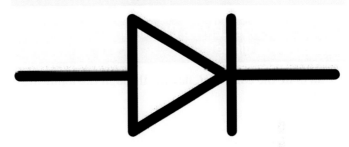

FIGURE 6.11 Circuit symbol for a diode

When the voltage drop across a diode is small, the resistance of a diode is very large and little current flows through the device. When the voltage drop across the diode reaches a value called the **switch-on voltage**, the resistance of the diode decreases to almost zero and current flows freely through the device. The voltage across a diode can never be larger than the switch-on voltage. The switch-on voltage for a diode depends on the materials used to construct the diode.

As the resistance of a diode is either very small or very large, graphs showing the resistance of diodes are rarely used. Instead, current-versus-voltage graphs of diodes are used. One of these graphs can be seen in figure 6.12. This graph shows zero current through the diode when the voltage across it is negative and the diode is reverse biased and at very small voltages. As the voltage increases, the current increases slightly until it becomes infinite, shown by the vertical line of the graph. The voltage at which the current becomes infinite is the switch-on voltage. For the device shown in figure 6.12, the switch-on voltage is 0.7 V.

As the resistance of a diode drops to almost zero when it experiences its switch-on voltage, the diode cannot control the amount of current flowing through it; it can only control whether current flows or not. As a result, a diode must always be connected in series with another resistor. The resistor prevents the current from becoming too large and damaging the diode.

diode a device that allows current to pass through it in one direction only; its resistance changes with voltage

light-emitting diode (LED) a small semiconductor diode that emits light when a current passes through it

forward biased refers to when a diode's positive side is connected to the positive output of a power source

reverse biased refers to when a diode's positive side is connected to the negative output of a power source

switch-on voltage the voltage needed to allow current to flow freely through a diode

FIGURE 6.12 Current-versus-voltage characteristic for a silicon diode

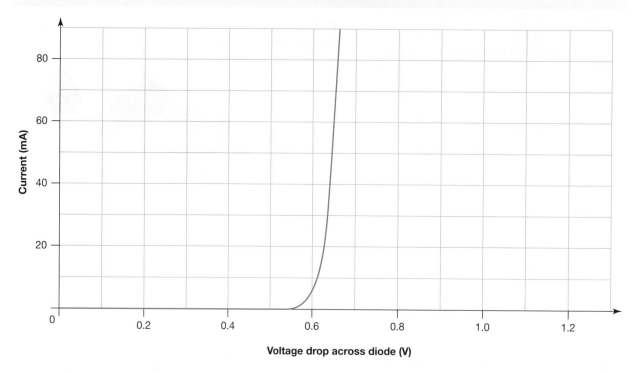

SAMPLE PROBLEM 15 Calculating current flowing through a circuit using the switch-on voltage of a diode

Calculate the current, I, flowing through the following circuit. The diode has a switch-on voltage, V_d, of 1.2 V.

THINK

1. Check whether the diode is forward biased or reverse biased.

2. Calculate the voltage across the 120-Ω resistor.

WRITE

1. The positive end of the diode is connected to the positive side of the battery, so the diode is forward biased.

2. $V_R = V - V_d$
   $= 4 - 1.2$
   $= 2.8 \text{ V}$

3. Use $V = IR$ for the resistor to calculate the current through the circuit. The circuit is a series circuit so the current is the same everywhere.

3. $I = I_R$

$$= \frac{V_R}{R}$$

$$= \frac{2.8}{120}$$

$$= 0.023 \text{ A}$$

4. State the answer.

4. The current through the circuit is 2.3×10^{-2} A.

PRACTICE PROBLEM 15

In the circuit shown, determine the switch-on voltage, V_d, of the diode if the current measured by the ammeter is 0.6 A.

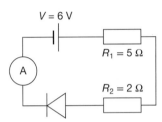

$V = 6 \text{ V}$

$R_1 = 5 \ \Omega$

A

$R_2 = 2 \ \Omega$

6.5.4 Potentiometers

A **potentiometer**, called a 'pot' for short, is a variable voltage divider. It consists of a fixed resistor, usually a length of wire, with a contact that can slide up and down, varying the amount of resistance in each arm of the voltage divider.

Potentiometers are commonly used as controls in radio equipment, either as a slide control or in a rotary form. They are also the basis of joysticks in game controls.

These circuits are very similar to circuits containing sensors. The difference is that only one resistor is drawn in the circuit, with the position of the arrow showing how the resistance is being divided between the two arms of the voltage divider circuit. Increasing the resistance in an arm will lead to an increase in the voltage across that arm, and therefore an increase in volume or light output. Decreasing the resistance in an arm will lead to a decrease in the voltage across that arm, which could send a message to a video game to move more slowly or in a different direction.

potentiometer a variable voltage divider, with a fixed resistor that can slide up and down

FIGURE 6.13 a. The symbol for a potentiometer **b.** A potentiometer

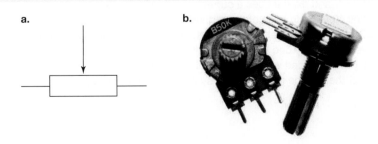

a.

b.

6.5 Activities

| 6.5 Quick quiz | 6.5 Exercise | 6.5 Exam questions |

6.5 Exercise

1. The voltage-versus-current characteristic graph for a non-ohmic device is shown.
   a. What is the device's current when the voltage drop across it is 100 V?
   b. What is the voltage drop across the device when the current through it is 16 mA?
   c. What is the resistance of the device when it carries a current of 16 mA?

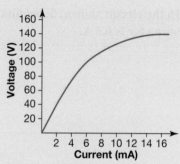

2. The device described in question 1 is placed in series with a 5.0-kΩ resistor and a voltage drop is applied across the combination. This arrangement is shown in the following figure. The current in the resistor is measured to be 6.0 mA.

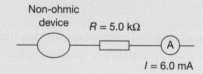

   a. What is the voltage drop across the resistor?
   b. What is the current in the device?
   c. What is the voltage drop across the device?
   d. What is the total voltage drop across the device and the resistor?
3. The device described in question 1 is now placed in parallel with a 5.0-kΩ resistor and a new voltage is applied across the combination. This arrangement is shown in the following figure. The current in the resistor is measured to be 20 mA.
   a. Calculate the voltage drop across the resistor.
   b. What is the voltage drop across the device?
   c. What is the current in the device?
   d. What is the total current in the circuit?

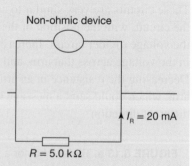

4. Explain how a potentiometer can be used to dim the lights in a bedroom.
5. One light-emitting diode (LED) used in a set of traffic lights has a switch-on voltage of 0.6 V and experiences a current of 2.4 mA. The set of traffic lights contains 40 LEDs in each of the red, amber and green sections. Before LEDs were used, a set of traffic lights used three 200-W light bulbs.
   a. Calculate the power emitted by one LED in the traffic light.
   b. Calculate the total power needed when the traffic lights are green.
   c. Suggest a reason for the change in the lighting system used in traffic lights.

6.5 Exam questions

▶ Question 1 (1 mark)

MC The diagram shows a voltage divider circuit in which the output voltage is across the thermistor.

The graph shows the resistance–temperature characteristic for this thermistor.

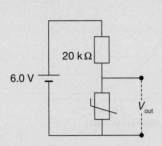

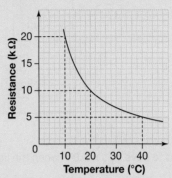

Which of the following is the temperature at which the output will be 3.0 V?
A. 10 °C
B. 15 °C
C. 20 °C
D. 40 °C

▶ Question 2 (1 mark)

MC The diagram shows a voltage divider circuit in which the output voltage is across the thermistor.

The graph shows the resistance–temperature characteristic for this thermistor.

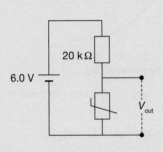

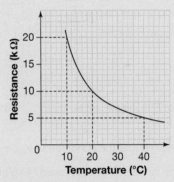

Initially the temperature is such that V_{out} = 4.0 V. The temperature now increases.

Which statement best describes what happens to V_{out}?
A. It depends on the initial temperature.
B. V_{out} now increases.
C. V_{out} stays the same.
D. V_{out} now decreases.

Question 3 (3 marks)

The voltage divider circuit shown has an unknown resistor, R, in series with a light dependent resistor (LDR). The LDR has the characteristic curve shown in the graph. The output voltage across the LDR is used to operate a switch.

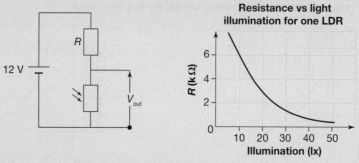

Resistance vs light illumination for one LDR

It is desired to have $V_{out} = 8.0$ V when the light intensity is 10 lx.

What is the required value of the resistance R? Reasoning must be shown.

Question 4 (2 marks)

The voltage divider circuit shown has an unknown resistor, R, in series with a light dependent resistor (LDR). The LDR has the characteristic curve shown in the graph. The output voltage across the LDR is used to operate a switch.

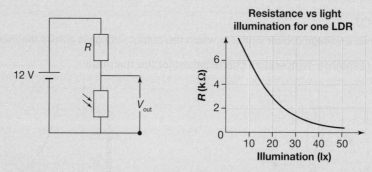

Resistance vs light illumination for one LDR

The value of R is initially set so that $V_{out} = 8.0$ V at a particular light intensity, L_1. (You do not need to know the value of R.)

It is then desired that the condition of $V_{out} = 8.0$ V should occur at a greater light intensity, L_2.

Should R be increased or decreased? Explain your reasoning.

Question 5 (7 marks)

A diode with a switch-on voltage of 0.7 V is connected to a battery and resistor as shown in the diagram.

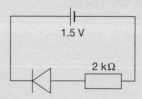

1.5 V

2 kΩ

a. Calculate the current flowing through the circuit. **(2 marks)**
b. What change could you make to increase the current flowing through the circuit? What would the current be after you had made this change? **(2 marks)**
c. Hira believes that removing the resistor would greatly increase the power output of the diode.

   Evaluate this statement. **(3 marks)**

More exam questions are available in your learnON title.

6.6 Power in circuits

KEY KNOWLEDGE

• Compare power transfers in series and parallel circuits

Source: VCE Physics Study Design (2023–2027) extracts © VCAA; reproduced by permission.

6.6.1 Total power in a circuit

Recall that the **power** being used in a circuit element is the product of the voltage drop across it and the current through it: $P = VI$. The total power being provided to a circuit is the sum of the power being used in, or 'dissipated by', the individual elements in that circuit. It does not matter if the elements are connected in series or in parallel.

power the rate of doing work, or the rate at which energy is transformed from one form to another

$$P_T = P_1 + P_2 + P_3 = \ldots$$

tlvd-0036

SAMPLE PROBLEM 16 Determining the total current flowing through a parallel household circuit

A household electrical circuit is wired in parallel. Determine the total current flowing in the circuit if the following appliances are being used: a 600-W microwave oven, a 450-W toaster and a 1000-W electric kettle. Household circuits provide a voltage drop of 230 V across each appliance.

THINK	WRITE
1. The total power being used in the circuit is the sum of the power used in each component.	$P_T = 600 + 450 + 1000$ $\quad = 2050 \text{ W}$
2. Transpose the formula $P = IV$ to make I the subject, then substitute the known values.	$I_T = \dfrac{P_T}{V}$ $\quad = \dfrac{2050 \text{ W}}{230 \text{ V}}$ $\quad = 8.91 \text{ A}$
3. State the solution.	The total current flowing in the circuit is 8.91 A.

PRACTICE PROBLEM 16

A household electrical circuit is wired in parallel. Determine the total current flowing through a household circuit when the following devices are being used: a 400-W computer, a 200-W blender, a 500-W television and a 60-W lamp.

6.6 Activities

learn

| 6.6 Quick quiz on | 6.6 Exercise | 6.6 Exam questions |

6.6 Exercise

1. Three resistors of values 25 Ω, 15 Ω and 10 Ω are connected in series to a 10-V power supply.
   a. Calculate the current in the circuit.
   b. What is the voltage drop across each resistor?
   c. At what rate is energy being transformed in each resistor?
   d. What is the total power rating of the circuit?
2. The three resistors in question 1 are connected in parallel.
   a. Calculate the current in the circuit.
   b. What is the voltage drop across each resistor?
   c. At what rate is energy being transformed in each resistor?
   d. What is the total power rating of the circuit?

6.6 Exam questions

Question 1 (1 mark)

The total power output of the following circuit is 500 mW.

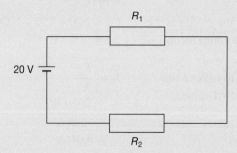

If $R_2 = 3R_1$, calculate the value of R_2.

Question 2 (2 marks)

A household circuit contains a 300-W toaster, a 1400-W fridge and a 1000-W microwave oven wired in parallel. The voltage supplied to the house is 230 V.

What is the total current flowing through the circuit?

Question 3 (4 marks)

A household circuit is designed to deliver a maximum of 20 A. The household has a 5-kW heater, a 3-kW fridge, a 1.8-kW oven and a 1.4-kW dishwasher.
a. What is the maximum power the household circuit can provide? **(2 marks)**
b. What is the maximum number of appliances that could be connected into the circuit? **(2 marks)**

⊙ Question 4 (4 marks)

Calculate the total power output in the circuit shown.

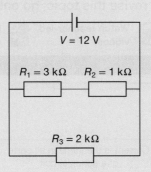

⊙ Question 5 (3 marks)

Calculate the unknown resistance, R_x, in the following circuit if the the total power output in the circuit is 200 W.

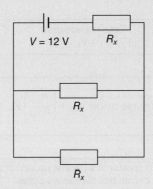

More exam questions are available in your learnON title.

6.7 Review

6.7.1 Topic summary

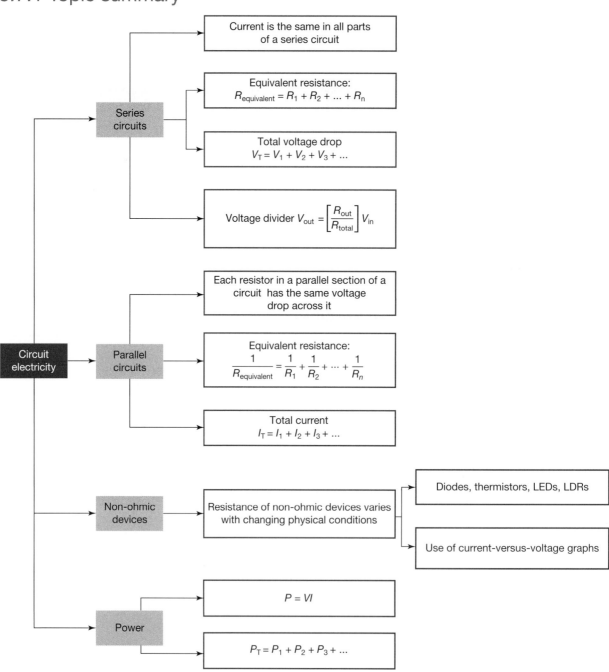

Current is the same in all parts of a series circuit

Equivalent resistance:
$R_{\text{equivalent}} = R_1 + R_2 + ... + R_n$

Total voltage drop
$V_T = V_1 + V_2 + V_3 + ...$

Series circuits

Voltage divider $V_{\text{out}} = \left[\dfrac{R_{\text{out}}}{R_{\text{total}}}\right] V_{\text{in}}$

Each resistor in a parallel section of a circuit has the same voltage drop across it

Equivalent resistance:
$\dfrac{1}{R_{\text{equivalent}}} = \dfrac{1}{R_1} + \dfrac{1}{R_2} + \cdots + \dfrac{1}{R_n}$

Parallel circuits

Total current
$I_T = I_1 + I_2 + I_3 + ...$

Circuit electricity

Non-ohmic devices

Resistance of non-ohmic devices varies with changing physical conditions

Diodes, thermistors, LEDs, LDRs

Use of current-versus-voltage graphs

Power

$P = VI$

$P_T = P_1 + P_2 + P_3 + ...$

6.7.2 Key ideas summary

6.7.3 Key terms glossary

6.7 Activities

learn on

6.7 Review questions

1. Some lighting circuits are controlled by two switches in different locations. Design a DC circuit to show how this could be achieved.

2. Determine the value of the unknown resistor in the voltage dividers shown.

   a.

   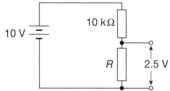

   b.

   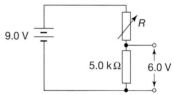

3. Consider the following circuit.
   a. What current passes through each resistor?
   b. How much power is dissipated in each resistor?

   The switch is now closed.

   c. What current passes through each resistor?
   d. How much power is dissipated in each resistor?

   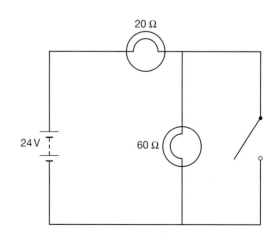

4. For the circuit shown, calculate:

   a. the equivalent resistance of the circuit
   b. the voltage, V_3, across R_3, the 10-Ω resistor
   c. the current, I_1, through R_1, the 8-Ω resistor.

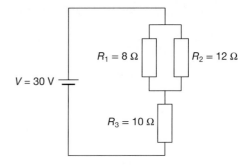

5. A student connects a variable resistor and a lamp in series in a circuit to a fixed power supply.

   a. What will happen in the lamp as the resistance of the variable resistor is increased?
   b. As the resistance of the variable resistor is increased, will the power consumed in the circuit change? Justify your answer.

   The lamp and the variable resistor are now connected in parallel.

   c. What will happen in the lamp as the resistance of the variable resistor is increased?
   d. What will happen to the power consumed in the circuit?

6. A thermistor has the temperature-versus-resistance characteristic shown by the bottom curve in the following graph. It is placed in the voltage divider shown in the circuit diagram.

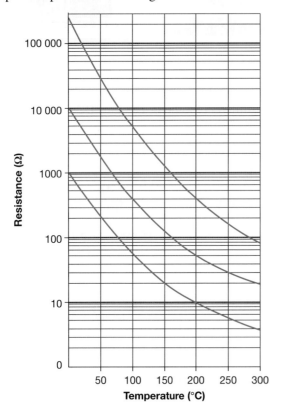

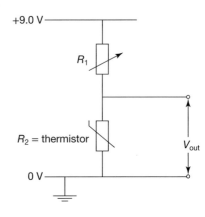

   a. What is the resistance of the thermistor when the temperature is 150 °C?
   b. What is the value of the variable resistor if the temperature is 200 °C and V_{out} is 6 V?

7. An electrical circuit contains only a diode, a 9-V battery and an unknown device. The diode has a switch-on voltage of 1.7 V and there is a current of 39 mA flowing through the circuit. Draw this circuit with all its components and label it.

8. Two lamps, labelled as 12 V, 9 W and 12 V, 24 W, are connected in parallel with a 12-V battery.

    a. What is the equivalent resistance of the circuit?
    b. Will the lamps be equally bright? Explain your answer.

9. Consider the following circuit.

    a. Determine the equivalent resistance of the circuit.
    b. Determine the current, I_3, through R_3, the 12-Ω resistor.
    c. Determine the voltage, V_2, across R_2, the 10-Ω resistor.

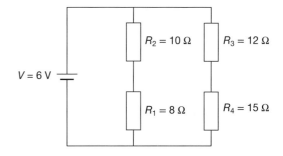

10. While recording a video for her students, Cathy is using a 100-W laptop, a 50-W lamp and a 28-W monitor, all plugged in parallel to the same power strip in her house. Calculate the total current flowing through this power strip.

6.7 Exam questions

All correct answers are worth 1 mark each; an incorrect answer is worth 0.

Use the following information to answer Questions 1–3.

Three resistors of 20 Ω, 30 Ω and 50 Ω are connected in series across a 9.0-V battery.

▶ Question 1

What is the equivalent resistance of the three resistors?

A. 0.01 Ω
B. 0.33 Ω
C. 100 Ω
D. 330 Ω

▶ Question 2

What current flows in the circuit?

A. 90 mA
B. 27 mA
C. 9 A
D. 27 A

▶ Question 3

What is the total voltage drop across the circuit?

A. 0 V
B. 0.09 V
C. 9 V
D. 900 V

Use the following information to answer Questions 4–6.

Three resistors of 6 Ω, 18 Ω and 9 Ω are connected in parallel across a 36-V power supply.

▶ Question 4

What current flows through the 9-Ω resistor?

A. 2 A
B. 4 A
C. 6 A
D. 12 A

▶ Question 5

What is the equivalent resistance of the circuit?

A. 3 Ω
B. 4.5 Ω
C. 12 Ω
D. 33 Ω

▶ Question 6

What total current flows in the circuit?
A. 2 A
B. 4 A
C. 6 A
D. 12 A

▶ Question 7

Three resistors of unknown resistance x Ω are connected in parallel to a 9-V power supply.

If the total current flowing through the circuit is 200 mA, the value of x is

A. 15 Ω.
B. 45 Ω.
C. 135 Ω.
D. 66.67 Ω.

▶ Question 8

In the circuit shown, what is the voltage drop between A and B?

A. 0 V
B. 4 V
C. 16 V
D. 20 V

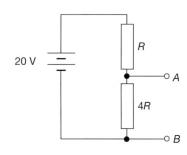

Question 9

What is the value of the resistor shown in the following circuit?

A. 3 kΩ
B. 4 kΩ
C. 6 kΩ
D. 12 kΩ

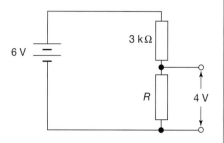

Question 10

Which behaviour is more likely when a thermistor is connected in an electrical circuit?

A. The resistance of the thermistor increases as the operating temperature increases.
B. The ratio of the voltage across the thermistor to the current flowing through it is constant.
C. When the potential difference across the thermistor is increased, the current flowing through it is also likely to increase.
D. The behaviour of the thermistor is not temperature dependent.

Section B — Short answer questions

Question 11 (4 marks)

Consider two people connected in series with a 12-V source across them. The first person, R_1, has a resistance of 1500 Ω, and the second person, R_2, has a resistance of 2500 Ω.

a. Determine the current through each person. **(1 mark)**
b. Calculate the voltage drop across each person. **(1 mark)**

The two people now connect themselves in parallel with the 12-V power source.

c. Determine the voltage drop across each person. **(1 mark)**
d. Calculate the current through each person. **(1 mark)**

Question 12 (3 marks)

Consider the following circuit.

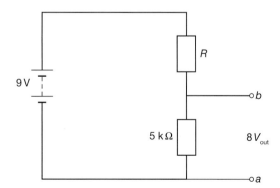

a. Determine the value of the unknown resistor. **(1 mark)**
b. Calculate the power of the circuit. **(1 mark)**

c. The resistor is now replaced by a variable resistor as shown.

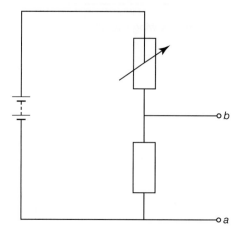

What happens to the voltage drop V_{ab}, when the resistance of the variable resistor decreases and the other resistance is unchanged? **(1 mark)**

Question 13 (3 marks)

A thermistor has the temperature-versus-resistance characteristics shown by the top curve in the following graph. It is placed in the voltage divider shown in the following circuit diagram.

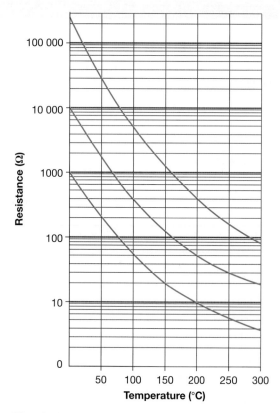

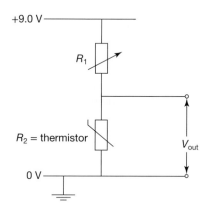

a. What is the resistance of the thermistor when the temperature is 200 °C? **(1 mark)**

b. Calculate the value of the variable resistor in the voltage divider if the temperature is 100 °C and the output voltage is 4.5 V. **(2 marks)**

a. Determine the value of R_2 in the voltage divider in the following figure that would give an output voltage of 2.0 V.　　　　　　　　　　　　　　　　　　　　　　　　　　　　**(2 marks)**

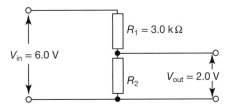

b. The resistor R_2 is now replaced by the thermistor with a temperature-versus-resistance curve, as shown by the middle curve in the following graph.

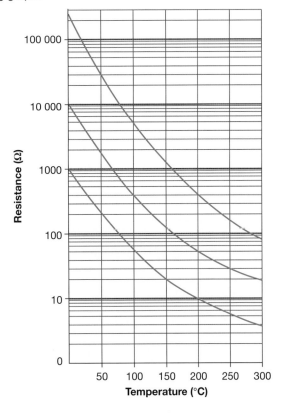

　　What is the value of V_{out} when the temperature of the thermistor is 100 °C?　　　　　　**(1 mark)**

⏵ **Question 15 (2 marks)**

Design a dimmer switch circuit for a light. Does the circuit consume more power when the light is bright or dull? Justify your answer.

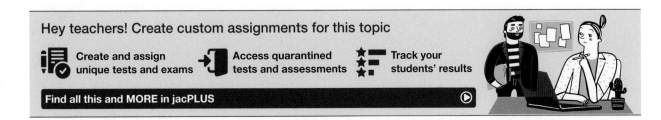

7 Using electricity and electrical safety

KEY KNOWLEDGE

In this topic, you will:
- model household (AC) electrical systems as simple direct current (DC) circuits
- explain why the circuits in homes are mostly parallel circuits
- model household electricity connections as a simple DC circuit comprising fuses, switches, circuit breakers, loads and earth
- apply the kilowatt-hour (kW h) as a unit of energy
- compare the operation of safety devices including fuses, circuit breakers and residual current devices (RCDs)
- describe the causes, effects and first aid treatment of electric shock and identify the approximate danger thresholds for current and duration.

Source: VCE Physics Study Design (2023–2027) extracts © VCAA; reproduced by permission.

PRACTICAL WORK AND INVESTIGATIONS

Practical work is a central component of VCE Physics. Experiments and investigations, supported by a **practical investigation eLogbook** and **teacher-led videos**, are included in this topic to provide opportunities to undertake investigations and communicate findings.

EXAM PREPARATION

▶ Access exam-style questions and their video solutions in every lesson, to ensure you are ready.

7.1 Overview

7.1.1 Introduction

Electricity is essential to our lives. It lights our homes, schools and workplaces, in addition to providing us with communication and information. It keeps indoor areas climate controlled, and facilitates our movement on the roads and through large areas and tall buildings. However, electricity can also be dangerous if used incorrectly. By understanding how electricity is used in our homes, schools and workplaces, we can protect ourselves from electric shock and appreciate the benefits of electrical safety systems. This topic looks specifically at the types of electric circuits in our homes and the ways they keep the lights on and the temperature appropriate, while also keeping us safe.

FIGURE 7.1 Light-duty all-electric and plug-in hybrid vehicles (combining electric drives with combustion engines) reduce reliance on petrol, increase efficiency and reduce greenhouse gas emissions.

LEARNING SEQUENCE

 Resources

Solutions	Solutions — Topic 7 (sol-0793)
Practical investigation eLogbook	Practical investigation eLogbook — Topic 7 (elog-1576)
Digital documents	Key science skills — VCE Physics Units 1–4 (doc-36950)
	Key terms glossary — Topic 7 (doc-36963)
	Key ideas summary — Topic 7 (doc-36964)
Exam question booklet	Exam question booklet — Topic 7 (eqb-0075)

7.2 Household electricity and usage

KEY KNOWLEDGE

- Model household (AC) electrical systems as simple direct current (DC) circuits
- Explain why the circuits in homes are mostly parallel circuits
- Model household electricity connections as a simple DC circuit comprising fuses, switches, circuit breakers, loads and earth
- Apply the kilowatt-hour (kW h) as a unit of energy

Source: VCE Physics Study Design (2023–2027) extracts © VCAA; reproduced by permission.

7.2.1 Household use of electricity

Houses connected to the main electrical grid are supplied with an AC voltage of 230 V_{RMS} at a **frequency** of 50 Hz. The term '230 V_{RMS}' means that the AC voltage produces the same effect when applied across a conductor as a DC voltage of 230 V applied across the same conductor. The actual value of the voltage oscillates between +325 V and −325 V. 'RMS' (root mean square) refers to the mathematical process by which the equivalent DC voltage is calculated. A 'frequency of 50 Hz' means that the full cycle is completed 50 times each second. The voltage supplied is sinusoidal in nature.

> **frequency** a measure of how many times per second an event happens, such as the number of times a wave repeats itself every second
>
> **period** the amount of time, measured in seconds, that one cycle or event takes, such as the time taken for an object moving in a circular path and at a constant speed to complete one revolution

---

Frequency (f) is a measure of how many times per second an event happens. The unit for frequency is the hertz (Hz). One hertz means one cycle or event per second.

Period (T) is the amount of time one cycle or event takes, measured in seconds.

Period is the reciprocal of frequency:

$$T = \frac{1}{f} \text{ or } f = \frac{1}{T}$$

A frequency of 50 Hz means that the period is 0.02 seconds, as shown in figure 7.2.

FIGURE 7.2 The variation of voltage with time for a supply of 230 V_{RMS}, 50 Hz

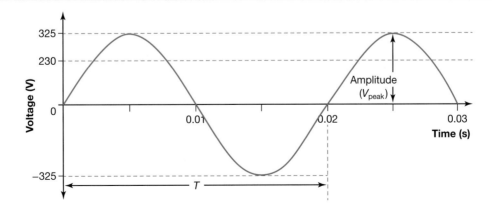

Electricity is fed into the home through underground cables or overhead lines. It enters the house through a switchboard. This contains a mains switch that can cut off the supply of electricity to the house. There is also a meter that measures the amount of electrical energy transformed in the house. From the meter the electricity enters a fuse box or circuit-breaker box, where it is divided among a number of parallel circuits. The role of fuses and circuit breakers is discussed later in this topic.

The structure of household circuits differs from the structure of DC circuits studied in previous topics, as there are normally three wires used when connecting appliances and devices instead of just two. Household circuits make use of the earth to complete the circuit. The **active wire** in a circuit is connected to the

FIGURE 7.3 Many houses have a meter and circuit breaker in their meter box. The circuit breaker is in the bottom left corner.

230 V_{RMS} supply at the switchboard. Its voltage oscillates periodically between +325 V and −325 V relative to a reference voltage called 'earth'. The earth is defined as having a voltage of 0 V; this means there is an electrical potential difference of 230 V between the active wire and the earth. The **neutral wire** is connected to the neutral link at the switchboard, which is connected to the earth through the supply wires and via a metal rod driven into the ground at the switchboard. The neutral wire is always at 0 V. The voltage drop between the active and the neutral wire oscillates between +325 V and −325 V. The earth wire is a direct connection to the ground, which absorbs any unwanted or unused electrical energy.

When an appliance is connected between the active wire and the neutral wire, current flows backwards and forwards between the active and neutral wires through the device, supplying it with energy. Unlike the DC circuits, the current doesn't make a complete loop from the power plant to the appliance and back.

Conventional current flows from a high voltage to a low voltage. When the active wire is positive, the current flows from the active to the neutral wire and so to the earth. When the active wire is negative, the neutral wire at 0 V will have the higher voltage and the current will flow from the neutral to the active wire.

active wire the wire connected to the 230 V_{RMS} supply at the switchboard

neutral wire the wire connected to the neutral link at the switchboard, which is connected to the earth

In lighting circuits, only the active and neutral wires are used if there are no metal fittings. The current oscillates through the filaments of the globes, transforming energy. If metal fittings are used, they must be earthed. If the active or neutral wire accidentally touch the metal fittings, the current will immediately be conducted to the ground. If a person touches the metal fittings there will be a reduced chance of electrocution. A typical lighting circuit is shown in figure 7.4.

FIGURE 7.4 A typical household lighting circuit

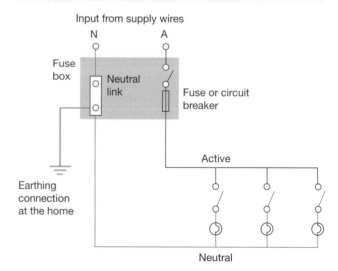

In power circuits, a third wire is used. This is called the **earth wire**. It connects the case of the appliance being used to the earth as a safety device. Its function is discussed in section 7.3.8. Otherwise, a power circuit operates in exactly the same way as a lighting circuit, with the current oscillating between the active and neutral wires through the appliance. Figure 7.5 shows how three-point power sockets are connected in a typical power circuit.

FIGURE 7.5 A typical power circuit

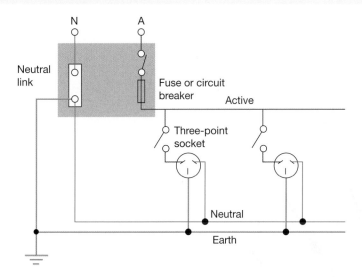

Figure 7.6 shows the connection of the earth wire to the case of an appliance.

FIGURE 7.6 The connection of the earth wire to the case of an appliance

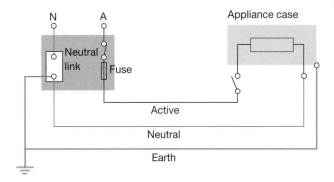

Note that in both the lighting and power circuits the switch is in the wire connecting the device to the active wire. A switch in the neutral wire would also turn off the device, but the functional parts of the device — such as the element of a toaster — would still be directly attached to the active wire and would be 'live'.

If you were to touch anything in contact with the active wire while you were in contact with the ground, there would be a voltage drop across you and a potentially lethal current could flow through you. For this reason, if ever it is necessary to tamper with the functional parts of an electrical device, it must be unplugged first. The three-point socket may have been wrongly wired, and it is not worth taking the risk.

Different-coloured insulating plastic is used to distinguish the three wires from each other. In the modern system, the active wire is brown, the neutral wire is blue and the earth wire is striped green and yellow. In the old system, red was used for the active, black for the neutral and green for the earth wires.

FIGURE 7.7 A three-point socket. The top-left point is the active connection (brown wire), the top-right point is neutral (blue wire) and the bottom point is the earth (green/yellow wire). The switch is connected in the active wire.

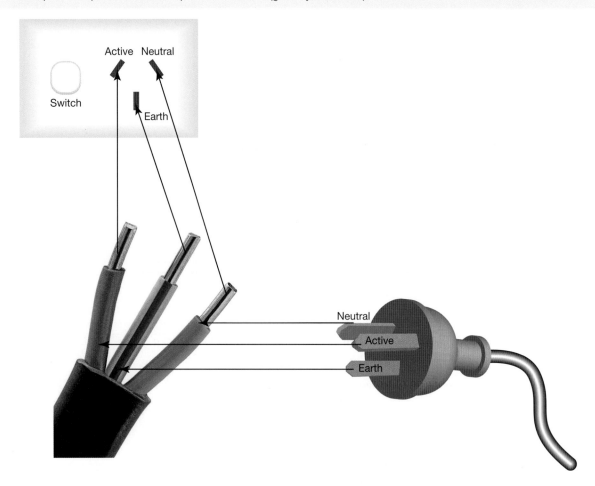

7.2.2 Power ratings

The total power used in an electric circuit, be it series or parallel, is the sum of the power used by each device in the circuit. Electrical appliances or devices are given a power rating, which is usually printed on them.

tlvd-0037

SAMPLE PROBLEM 1 Determining the current and resistance of a device

A toaster is rated at 1400 W, 230 V.
a. What current does the toaster draw when operating normally?
b. What is the resistance of the toaster element when hot?

THINK	WRITE
a. 1. Use the relationship between power, voltage and current to make I the subject.	a. $P = VI$ $\Rightarrow I = \dfrac{P}{V}$

2. Substitute the known values and solve for I.

$P = 1400 \text{ W}, V = 230 \text{ V}$

$$I = \frac{P}{V}$$

$$= \frac{1400 \text{ W}}{230 \text{ V}}$$

$$= 6.09 \text{ A}$$

3. State the solution.

When operating normally the toaster draws 6.09 A.

b. 1. Use the relationship between voltage, resistance and current to make R the subject.

b. $V = IR$

$$\Rightarrow R = \frac{V}{I}$$

2. Substitute the known values and solve for R.

$V = 230 \text{ V}, I = 6.09 \text{ A}$

$$R = \frac{V}{I}$$

$$= \frac{230 \text{ V}}{6.09 \text{ A}}$$

$$= 37.8 \, \Omega$$

3. State the solution.

The resistance of the toaster element when hot is 37.8 Ω.

Alternatively:

1. The resistance could also have been calculated using the relationship $P = \dfrac{V^2}{R}$.

$$P = \frac{V^2}{R}$$

2. Transpose to make R the subject.

$$R = \frac{V^2}{P}$$

3. Substitute in the values, $V = 230 \text{ V}$ and $P = 1400 \text{ W}$, and solve.

$$R = \frac{230^2}{1400}$$

$$= 37.8 \, \Omega$$

PRACTICE PROBLEM 1

A compact fluorescent light globe is rated at 15 W, 230 V.
a. Calculate the current through the globe when it is operating normally.
b. Calculate the resistance of the globe when it is operating normally.

7.2.3 Paying for electricity

The meter on a household switchboard is used to measure how much electrical energy has been consumed on the premises. The amount of electrical energy used in a household can be determined by multiplying the rate of power transformation by the time. Since power is equal to the voltage drop multiplied by current, and the voltage drop across a household is 230 V, the meter on the switchboard records the total current that has passed through the premises over a certain period of time. This amount is converted into the amount of energy 'consumed' or transformed.

For domestic and commercial electrical consumption the joule is too small to appropriately measure the amount of energy being used. The unit used for measuring energy in these cases is the **kilowatt-hour (kW h)**. This is the amount of energy transformed by a 1000 W appliance when used for one hour.

kilowatt-hour (kW h) the amount of energy transformed by a 1000 W appliance when used for one hour

tlvd-0038

SAMPLE PROBLEM 2 Calculating joules from kilowatt-hours

How many joules does 1 kilowatt-hour represent?

THINK	WRITE
1. Energy (kW h) = power (kW) × time (h)	$1 \text{ kW h} = 1 \text{ kW} \times 1 \text{ h}$ $= 1000 \text{ W} \times (60 \times 60) \text{ s}$ $= 3.6 \times 10^6 \text{ J}$ $= 3.6 \text{ MJ}$
2. State the solution.	1 kW h represents 3.6 MJ.

PRACTICE PROBLEM 2

An oven used 8 kW h of energy. In joules, how much energy was used?

The cost to consumers of 1 kW h of electrical energy can be found on electricity accounts and can vary depending on the time of day, time of year and the amount of power consumed. To calculate the cost of running a particular appliance, calculate the energy in kW h by multiplying the power rating of the appliance, in kilowatts, by the number of hours that it was used for. Then multiply this by the cost per kW h.

FIGURE 7.8 Many appliances have a sticker outlining their energy rating and the average kW h of energy consumed per year.

If the power rating of an appliance is unknown the energy consumed can be found using the formula:

$$E = VIt$$

where:

E = the energy consumed, in joules

V = the voltage drop across the device, in volts

I = the current flowing through the device, in amperes

t = the time, in seconds.

The amount of energy is then converted into kW h by dividing by 3.6×10^6.

To calculate the cost, use the formula:

$$\text{cost} = \text{energy} \times \text{rate}$$

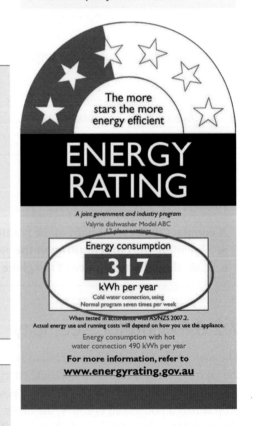

The more stars the more energy efficient

ENERGY RATING

A joint government and industry program
Valyrie dishwasher Model ABC
12 place settings

Energy consumption

317

kWh per year
Cold water connection, using
Normal program seven times per week

When tested in accordance with AS/NZS 2007.2.
Actual energy use and running costs will depend on how you use the appliance.

Energy consumption with hot
water connection 490 kWh per year

For more information, refer to
www.energyrating.gov.au

SAMPLE PROBLEM 3 Determining the power rating and energy consumption of a device

tlvd-0039

A television draws 0.37 A of current when connected to a 230 V supply.
a. What is the power rating of the television?
b. How much energy does the television consume if it is operated for 5 hours a day for 4 weeks?
c. What is the cost of running the television for this period of time if the consumer is charged at a rate of 16.381 cents per kW h?

THINK	WRITE
a. 1. Substitute in $P = VI$.	a. $P = 230\text{ V} \times 0.37\text{ A}$ $= 85\text{ W}$
2. State the solution.	The power rating of the television is 85 W.
b. 1. Energy = power × time	b. Power = 85 W Time = 5 hours per day for 28 days $= 140\text{ h}$ Energy $= 85\text{ W} \times 140\text{ h}$ $= 12\,000\text{ W h}$ $= 12\text{ kW h}$
2. State the solution.	The television consumes 12 kW h if it is operated for 5 hours a day for 4 weeks.
c. 1. Cost = energy × rate	c. Cost $= 12\text{ kW h} \times 16.381\text{ cents}$ $= 197\text{ cents}$ $= \$1.97$
2. State the solution.	The cost of running the television is $1.97.

PRACTICE PROBLEM 3

A home sound system consumes 2.4 W of electric power when it is on standby and connected to a 230-V supply.
a. Calculate the current that flows through the system when it is on standby.
b. Calculate the energy consumed by the system if it is left on standby for one week.
c. Calculate the cost of leaving the system on standby for one week if electricity is priced at 12 cents per kW h.

 Resources

 Weblink Operating costs of electrical appliances

7.2 Activities

Students, these questions are even better in jacPLUS

Receive immediate feedback and access sample responses

Access additional questions

Track your results and progress

Find all this and MORE in jacPLUS ▶

7.2 Quick quiz on	7.2 Exercise	7.2 Exam questions

7.2 Exercise

1. What is meant when it is said that a house is supplied with electricity at 230 V_{RMS}, 50 Hz?
2. Why is an overload in a household circuit potentially dangerous?
3. What coloured insulation is used for the active, neutral and earth wires in modern houses?
4. Sketch a power point and plug. Label the active, neutral and earth in each case.
5. Why do many appliances need to be connected to both the neutral and earth wires?
6. When is the earth wire used in household lighting circuits?
7. What is the cost of running a 300-W refrigerator for a year (365 days) if the refrigerator operates on average for 12 hours a day, and electricity costs 31.28 cents per kW h?
8. The following table gives the power consumption of various products when they are on standby.

Product	Power (W)
Laptop computer	14.5
Modem	3.4
Cordless phone equipment	3.7
DVD player	2.4
Television	6.2

   a. Calculate the energy used by each product if it is left on standby for one year.
   b. Calculate the mass of greenhouse gases produced by these products if they are left on standby for one year, assuming that 1 kW h of energy produces greenhouse gases that are equivalent to 1.444 kilograms of CO_2.

7.2 Exam questions

▶ Question 1 (1 mark)

MC An appliance is rated at 3600 W at 230 V_{RMS}.

Which of the following best gives the value of its resistance?
A. 0.060 Ω
B. 15 Ω
C. 16 Ω
D. 32 Ω

▶ Question 2 (2 marks)

The mains power supply is rated at 240 V_{RMS} at 50 Hz.

What is the period (time for one cycle) of the mains voltage? Show your reasoning.

▶ Question 3 (2 marks)

In a correctly wired house, it should be safe to touch the neutral wire. (Do not try this at home!)

Explain how the household wiring brings this about.

7.3 Electrical safety

> **KEY KNOWLEDGE**
>
> - Compare the operation of safety devices including fuses, circuit breakers and residual current devices (RCDs)
> - Describe the causes, effects and first aid treatment of electric shock and identify the approximate danger thresholds for current and duration
>
> *Source:* VCE Physics Study Design (2023–2027) extracts © VCAA; reproduced by permission.

7.3.1 A shocking experience

An **electric shock** is a violent disturbance of the nervous system caused by an electrical discharge or current through the body.

There are various factors that contribute to the severity of an electric shock. The first of these is the size of the voltage involved. Also, the human body is far more sensitive to alternating current than direct current. Voltages as low as 32 V AC and 115 V DC can be fatal.

It is not the voltage alone that causes damage to the human body. When you slide across a car seat, you can generate a voltage of several thousand volts. When you get out of the car and touch the ground, you are discharged and experience a shock, but with no serious consequences. The voltage drop across a person is one factor in determining the seriousness of an electric shock, but clearly other factors are involved.

The following information refers to shocks involving alternating currents with a frequency of 50 Hz.

FIGURE 7.9 A victim of electric shock in a home accident, possibly caused by overloading the circuit

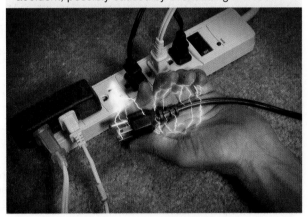

electric shock a violent disturbance of the nervous system caused by an electrical discharge or current through the body

7.3.2 Resistance of the human body

One contributing factor to the severity of an electric shock is the resistance of the human body. The interior of the body is a good conductor of electricity. The tissues and fluids beneath the skin conduct electricity due to the presence of ions in the fluids.

The skin provides the main resistance to the flow of electricity in the body. The resistance of the skin ranges in value from 10^6 Ω for dry skin to 1500 Ω for a person with wet hands, and about 500 Ω for a person sitting in a bath. This is a good reason for keeping electrical appliances away from water in the bathroom; tap water is a reasonably good conductor as it contains ions.

Resistance to current flow is offered by the skin up to about 600 V, at which point the skin is punctured and offers little resistance to current flow. Breaks or cuts in the skin also reduce the resistance of the skin.

One of the main reasons for the high resistance of skin is the poor contact that is made between the skin and the electrical source. Water improves the contact. In hospitals, a conducting gel is used when a good electrical contact is required — for example, when using an electrocardiogram.

7.3.3 The effect of current

The second and most important factor to be considered in respect to the severity of an electric shock is the amount of current flowing through the body. This is important because impulses within the nervous system are themselves electrical in nature, and even very small currents passing along nerves make muscles contract. Skeletal muscles (muscles attached to the bones) work in pairs. To raise your forearm, for example, the biceps muscle contracts and the triceps muscle relaxes. To lower your forearm, the biceps muscle relaxes and the triceps muscle contracts. This arrangement of muscles is shown in figure 7.10.

One effect of passing a small current through the body is to make muscles contract. Another effect stimulates the nerves that send pain signals to the brain, causing the painful sensations associated with shocks.

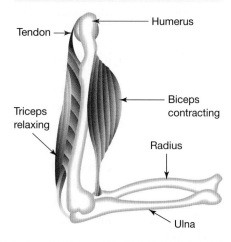

FIGURE 7.10 Diagram showing the biceps and triceps muscles in the human arm

Tendon — Humerus — Biceps contracting — Triceps relaxing — Radius — Ulna

---

EXTENSION: As a matter of fact

You may have heard of someone who received an electric shock being 'thrown across the room'. This is not due to any explosion, but to the violent contraction of the person's muscles.

A current of 9 mA AC across the chest causes shock. A current of twice that amount causes difficulty in breathing. A current of 20 mA causes muscles to become paralysed: they contract and stay contracted. A person unfortunate enough to touch a live conductor with the palm of their hand will grip onto the conductor and not be able to let go. Some electricians, if unsure whether a wire is live, may bring the back of the hand towards the wire. Any shock they receive will contract the muscles so that the hand is pulled away from the wire. This procedure is definitely not recommended.

A current as low as 25 mA through the trunk of the body can cause **fibrillation**. This is the disorganised, rapid contraction of separate parts of the heart so that it pumps no blood, and death may soon follow. Sometimes fibrillation subsides when the external voltage is removed.

fibrillation the disorganised, rapid contraction of separate parts of the heart that prevent it from pumping blood; death may follow

Defibrillation is a medical intervention technique carried out on victims of heart attack. If the cardiac monitor shows that fibrillation is occurring, a current of 20 A at 3000 V is passed through the heart for about 5 milliseconds. This produces a major contraction of all the muscles in the heart, which usually jolts them back into the proper rhythm. The shock is applied above and below the heart via two large electrodes called paddles. Conducting gel is used to make good contact with the body. It is important that the operator and other staff are well insulated from the patient.

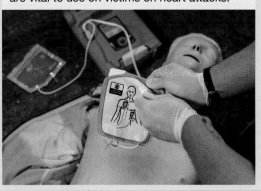

FIGURE 7.11 Automatic external defibrillators are vital to use on victims on heart attacks.

7.3.4 The effect of current path

The third factor affecting the severity of an electric shock is the path of the current through the body. Respiratory arrest generally requires the current to pass through the back of the head. Ten milliamps of current through the forearm muscles make them contract sufficiently to hold the victim to any live conductor he or she is gripping. The most dangerous pathway for current is through the trunk of the body. The least dangerous pathway for current is through an arm or a leg, where there are no organs essential for survival; however, all electrical shocks are serious no matter which part of the body is involved.

7.3.5 Time of exposure

The final factor contributing to the severity of a shock is the time of exposure to the current. The longer the current flows through the body, the greater the damage to tissue will be. Table 7.1 shows what effects current size and time duration have on the heart.

TABLE 7.1 Electric shock current-versus-time parameters

Current (mA)	Time (ms)	Effect
50	10–200	Usually no dangerous effect
50	> 4000	Fibrillation possible
100	10–100	Usually no dangerous effect
100	> 600	Fibrillation possible
500	> 40	Fibrillation possible

7.3.6 In the event of a shock

The first priority when helping a victim of electric shock is to make sure that the victim is not still connected to the electrical source. If the person is still connected and you grab them, you could be electrocuted too. (**Electrocution** is death brought about by an electric shock.) Your muscles may contract and you will not be able to let go of the victim, becoming a victim yourself. Turn off the electric circuit or knock the victim away from the live conductor with an insulating material; for example, a wooden chair. Make sure you don't step in any water or touch a wet area, as tap water is a good conductor of electricity.

Artificial respiration should be given to the victim if breathing has ceased. Respiratory failure is a common cause of death among shock victims. This is true of people who have been struck by lightning also. An ambulance should be called immediately.

electrocution death brought about by an electrical shock

7.3.7 Safety in household circuits

Every year many lives are lost and much property is damaged or destroyed because of electrical 'accidents', or through electrical faults, in both industrial and domestic situations. Accidents occur because basic safety precautions are not followed in dealing with electricity. The effects of electricity on the human body have already been discussed. This section looks at some common electrical faults and the safety devices employed to reduce the danger to people.

One common pitfall is overloading a circuit (see figure 7.12). This happens when multiple adapters or powerboards are connected to each other or to the same power outlet. In parallel circuits, the total current flowing in the circuit is the sum of the individual currents flowing through the devices in the circuit. Too many appliances operating on a single power circuit will produce a large current in the conducting wires. The wires will get hot and melt their insulation, potentially causing a fire in the walls or ceilings of the building.

A **short circuit** can occur when frayed electrical cords (see figure 7.13) or faulty appliances allow the current to flow from one conductor to another with little or no resistance. This allows the current to increase rapidly, with the same results as an overload. Appliances with damaged power cords and damaged charging cables should not be used for this reason.

Cheap extension cords are another source of overheating. They are not designed to carry more than 7 A safely, and exceeding this amount may result in the insulation melting and allow arcing to occur.

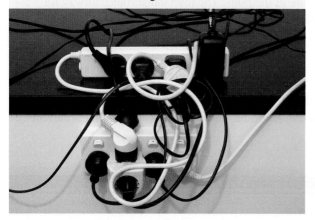

FIGURE 7.12 Overloading of an electric circuit

FIGURE 7.13 Frayed electrical cords

Fuses and circuit breakers

In domestic applications, each circuit is protected by either a **fuse** or a **circuit breaker**. A fuse is a short length of conducting wire or strip of metal that melts when the current through it reaches a certain value. The most common type of fuse is the plug-in type illustrated in figure 7.14. This has a ceramic body with metal prongs projecting from each end. A short piece of special fuse wire connects the metal prongs. When the current through the fuse exceeds a predetermined value the wire melts, or 'blows', breaking the circuit. (Figure 7.14 shows a plug that includes a cartridge fuse.)

A circuit breaker carries out the same function as a fuse. It breaks the circuit when the current through the circuit exceeds a particular value. The advantage circuit breakers have over fuses is that they can be reset easily. There are two types of circuit breaker available: thermal and electromagnetic.

short circuit a malfunction or fail that can occur when frayed electrical cords or faulty appliances allow the current to flow from one conductor to another with little or no resistance, resulting in a rapid increase of the current and potentially causing the wires to get hot and start a fire

fuse a short length of conducting wire or strip of metal that melts when the current through it reaches a certain value, breaking the circuit

circuit breaker a device that breaks a circuit when the current through it exceeds a certain value, carrying out the same function as a fuse

FIGURE 7.14 A plug-in fuse

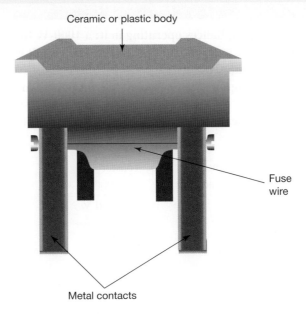

Ceramic or plastic body

Fuse wire

Metal contacts

When a current flows through a thermal circuit breaker, like the one illustrated in figure 7.15, the heating coil heats the bimetallic strip. The two metals in the bimetallic strip expand at different rates when heated, causing the strip to bend. When the current exceeds the predetermined amount, the bimetallic strip bends so much that it opens the catch and the circuit is broken. Because of the time it takes to heat up the strip, these circuit breakers will not trip if the current surge is of a short duration. This type of circuit breaker is not satisfactory if a short circuit occurs, and offers little assistance in preventing electrocutions.

FIGURE 7.15 A thermal circuit breaker

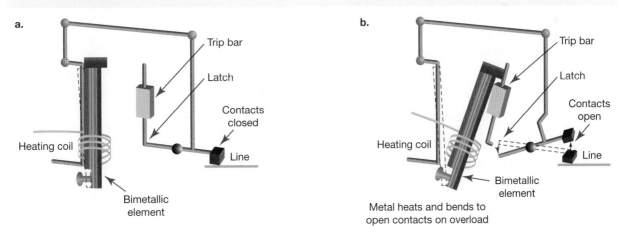

a.

Trip bar

Latch

Contacts closed

Line

Heating coil

Bimetallic element

b.

Trip bar

Latch

Contacts open

Line

Heating coil

Bimetallic element

Metal heats and bends to open contacts on overload

The electromagnetic type of circuit breaker uses the magnetic effects of electric currents: it uses an electromagnet to lift the catch and break the circuit. The bigger the current is in the coil, the stronger the electromagnetic force will be on the lever system. Again, these circuit breakers are designed to break the circuit at predetermined values of the current. To prevent this type of circuit breaker tripping when a short-duration current surge occurs, the switching mechanism is usually restrained in some way. The magnetic circuit breaker will trip almost instantly when a heavy overload occurs. It provides good protection against short circuits.

Both fuses and circuit breakers are placed in the active wire at the meter box. Light circuits are generally designed to take a maximum safe current of 5 A, whereas power circuits have a maximum safe current of 15 A.

SAMPLE PROBLEM 4 Determining the current flowing through a fuse in a circuit with multiple appliances

A kitchen circuit has the following appliances operating in it: a 1000-W toaster, a 312-W refrigerator, a 1200-W kettle, a 600-W microwave oven and a 60-W juicer. The circuit is protected by a 15-A fuse and is connected to a 230-V, 50 Hz supply.

a. What is the current flowing through the fuse when all the appliances are operating at the same time? Will the fuse 'blow'?

b. Will the fuse 'blow' if a 2400-W heater is used at the same time as the other appliances?

THINK	WRITE
a. 1. First, calculate the current through each appliance using their power ratings and the formula $P = VI$ or $I = \dfrac{P}{V}$.	a. Toaster: $$I = \frac{1000 \text{ W}}{230 \text{ V}} = 4.35 \text{ A}$$ Refrigerator: $$I = \frac{312 \text{ W}}{230 \text{ V}} = 1.36 \text{ A}$$ Kettle: $$I = \frac{1200 \text{ W}}{230 \text{ V}} = 5.22 \text{ A}$$ Microwave oven: $$I = \frac{600 \text{ W}}{230 \text{ V}} = 2.61 \text{ A}$$ Juicer: $$I = \frac{60 \text{ W}}{230 \text{ V}} = 0.26 \text{ A}$$
2. The total current in the circuit is the sum of the individual currents of the appliances. Add the currents to determine the current through the fuse. If the current is less than the capacity of the fuse, the fuse will not melt.	$4.35 + 1.36 + 5.22 + 2.61 + 0.26 = 13.8$ A
3. State the solution.	The circuit is protected by a 15-A fuse, so the fuse will not 'blow'.
b. 1. Calculate what the additional current is using $I = \dfrac{P}{V}$.	b. $$I = \frac{2400 \text{ W}}{230 \text{ V}} = 10.4 \text{ A}$$
2. If the current is larger than 15 A it is at risk of 'blowing'.	The 2400 W heater will draw an additional 10.4 A. Total current in the circuit: 10.4 A + 13.8 A = 24.2 A
3. State the solution.	24.2 A is much greater than 15 A, so the fuse will 'blow'.

PRACTICE PROBLEM 4

A bathroom fan, light and heater system consists of one 75-W light globe, four 150-W heat lamps and one 100-W fan. It is connected to a 230-V supply.

a. Calculate the total current through the system when the fan and light globe are in use.

b. Calculate the total current through the system when the fan and two heat lamps are in use.

c. Calculate the total current through the system when all the devices are in use.

7.3.8 Earthing

The earth wire is another safety measure used for power circuits (see figure 7.16). It connects the metal chassis of an appliance to the earth, which is at 0 V. This connection is made via a metal rod driven into the ground at the switchboard.

An electrical fault could occur if the active wire were to come into contact with the metal case of an appliance. The case would then carry an AC voltage, and anyone touching the case would receive a shock. The earth wire provides a lower-resistance conducting path to the earth than the appliance and the person. The low resistance involved produces a large current in the circuit and the fuse 'blows' or the circuit breaker trips.

The earth wire does not provide the most reliable protection. As can be seen in figure 7.17, the amount of time it takes to 'blow' a fuse depends on the size of the current. A quicker method of breaking the circuit is needed if lives are to be saved.

FIGURE 7.16 A three-pin plug with its back cover removed. The earth wire is covered in a green and yellow coating. Also note the cartridge fuse inside the plug.

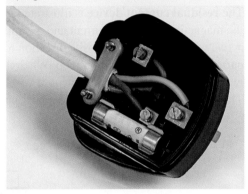

FIGURE 7.17 Time before a typical 10 A fuse 'blows', as a function of (RMS) current

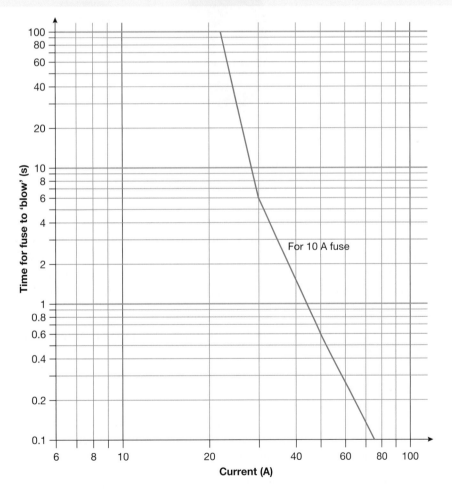

If the current is flowing through a person to the earth, the following principles should be followed: the current should be as small as possible and the time of exposure to the current should be as short as possible. Fuses are not designed to meet these requirements. Their main function is to prevent fires in buildings due to the overheating of wires when they carry too great a current.

7.3.9 Residual current device

The **residual current device** is illustrated in figure 7.18. It operates by making use of the magnetic effects of a current and is similar to a transformer. The current in the active wire flows in the opposite direction to the current in the neutral wire. Both currents pass through the iron loop. When the current in the active wire is equal in magnitude to the current in the neutral wire, each wire produces a magnetic field. These fields are equal in magnitude, but opposite in direction, and have no overall effect.

> **residual current device** a current-activated circuit breaker; it operates by making use of the magnetic effects of a current to break a circuit in the event of an electrical fault

FIGURE 7.18 A residual current device

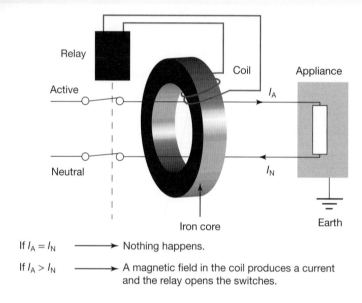

If $I_A = I_N$ ⟶ Nothing happens.

If $I_A > I_N$ ⟶ A magnetic field in the coil produces a current and the relay opens the switches.

However, if there is an electrical fault and a residual current flows to the earth via the earth wire or a person, the current in the active wire will be greater than in the neutral wire. The residual current is the difference between the active and neutral currents. The magnetic effects of the two currents will no longer cancel. A current is then produced in the relay circuit, and both the active and neutral wires of the circuit are broken by a switch.

A residual current device operates in about 40 milliseconds, limiting the current to 30 mA. At such values the shock will be perceptible, but not likely to have any harmful effects. The residual current device is useful only when the current flows to earth, not if the current flows through the person between the active and neutral wires.

7.3.10 Double insulation

Handheld electrical tools and appliances, such as electric drills and hair dryers, are protected by double insulation. These appliances have only a two-pin plug, using only the active and neutral wires. They should not be earthed. The symbol ▣ on the casing of an appliance (see figure 7.19) means that it is double insulated.

As the name implies, double insulation means that the accessible metal parts cannot become live unless two independent layers of insulation fail. The inner layer is called the functional insulation. This layer has both electrical insulation and heat-resisting properties. The outer layer is called protective insulation and often forms part of the casing.

FIGURE 7.19 An electric hair dryer with the double insulation symbol seen on the casing

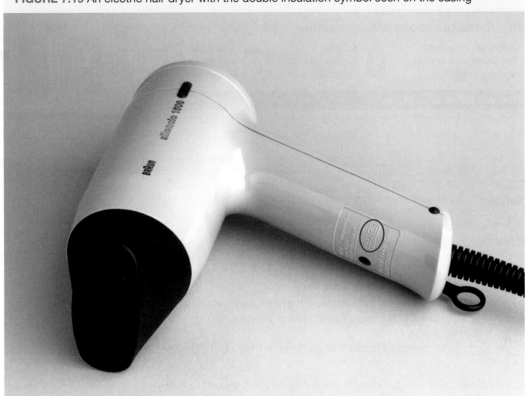

elog-1818

tlvd-0814

INVESTIGATION 7.1 online only

Examination of an electrical device

Aim

To examine an electrical device and report on how it functions and what safety features it has

elog-1819

INVESTIGATION 7.2 online only

Model circuits

Aim

To apply the principles of series and/or parallel circuits to either household or car electrical systems by designing a circuit

 Resources

 Video eLesson Electrical safety (eles-2517)

Weblink Electrical safety

7.3 Activities

| 7.3 Quick quiz **on** | 7.3 Exercise | 7.3 Exam questions |

7.3 Exercise

1. What is the difference between an electric shock and electrocution?
2. Describe factors that reduce the resistance of human skin.
3. Why is the amount of current flowing through the body important in determining the severity of an electric shock?
4. What is fibrillation?
5. What would happen if you touched a shock victim who was still conducting an electrical current?
6. How is the severity of a shock related to the time of exposure?
7. What is 'double insulation'?
8. A worker touched an overhead power line and was electrocuted. A newspaper reported the incident as follows:

   'He touched the cable and 50 000 V of electricity surged through his body.'

   Evaluate this statement.

7.3 Exam questions

Question 1 (1 mark)

MC An appliance is said to have 'double insulation'.

What is meant by this statement?
A. There is twice the usual thickness of insulation around the casing.
B. There are two layers of insulation between the live wire and external metal parts.
C. There are two earth wires connected to any external metal parts.
D. There is double thickness of insulation of the power leads.

Question 2 (3 marks)

Explain why the bathroom is a more dangerous place than the lounge room for electric shocks.

Question 3 (3 marks)

State which of the two quantities, voltage or current, is more important in causing electric shock. Explain your answer. Discuss both voltage and current.

Question 4 (3 marks)

Explain when and how the earth wire functions to protect humans from electric shock.

Question 5 (3 marks)

Explain when and how a residual current device functions to protect humans from electric shock.

More exam questions are available in your learnON title.

7.4 Review

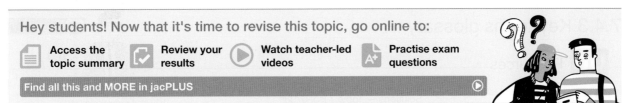

Hey students! Now that it's time to revise this topic, go online to:

Access the topic summary

Review your results

Watch teacher-led videos

Practise exam questions

Find all this and MORE in jacPLUS

7.4.1 Topic summary

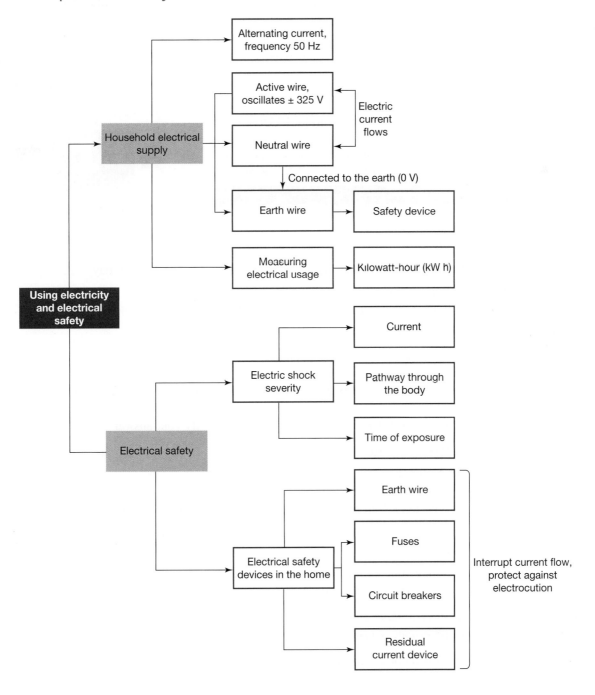

7.4.2 Key ideas summary

7.4.3 Key terms glossary

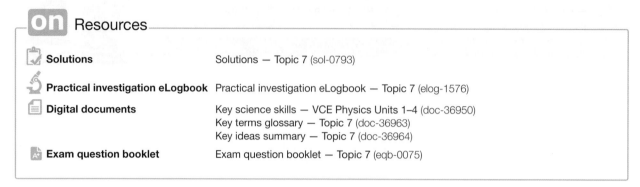

on Resources

☑ **Solutions**	Solutions — Topic 7 (sol-0793)
🔬 **Practical investigation eLogbook**	Practical investigation eLogbook — Topic 7 (elog-1576)
📄 **Digital documents**	Key science skills — VCE Physics Units 1–4 (doc-36950) Key terms glossary — Topic 7 (doc-36963) Key ideas summary — Topic 7 (doc-36964)
🅰 **Exam question booklet**	Exam question booklet — Topic 7 (eqb-0075)

7.4 Activities

learn on

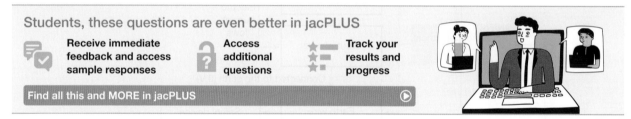

Students, these questions are even better in jacPLUS

💬 **Receive immediate feedback and access sample responses**

🔓 **Access additional questions**

⭐ **Track your results and progress**

Find all this and MORE in jacPLUS ▶

7.4 Review questions

1. A 3.6-kW domestic electric hot water system typically heats water for three hours per day.
   a. How much energy will this appliance consume in one year?
   b. If the cost of electricity is 28 cents per kW h, how much will it cost to provide hot water for one year?

2. Two 10-A speakers are connected to a 230-V sound system.
   a. Should these speakers be connected in series or parallel? Justify your answer.
   b. How much power will these speakers use?
   c. If the cost of electricity is 32 cents per kW h, how much will it cost to operate the speakers at full volume for six hours?
   d. There is a fuse in each speaker. What size fuse should be used? Justify your answer.

3. An oil heater is rated at 1000 W and runs off a 230-V supply.
   a. What current does the heater draw?
   b. What is the effective resistance of the heater?
   c. If electricity costs 17 cents per kW h, what does it cost to run the heater for five hours?

4. Some household appliances such as electric kettles are encased with metal. When a person touches the metal casing they do not get a shock.
   a. What stops someone from getting a shock when they touch the metal casing?
   b. If someone did get a shock, why would this occur?

5. a. Define both the terms *electric shock* and *electrocution*.
   b. Explain under what circumstances an electric shock is likely to be fatal.

6. Describe the difference in operation of, and advantages of, a fuse in a circuit and a circuit breaker.

7. A bedroom circuit has the following appliances operating in it: a 150-W gaming console, a 140-W television, a 60-W lamp, a 55-W DVD player and a 200-W mini-fridge. The circuit is protected by a 15-A fuse and it is connected to a 230-V, 50 Hz supply.

   a. What is the current flowing through the fuse when all the appliances are operating at the same time?
   b. Will the fuse 'blow' if a 2400-W air conditioner is used at the same time as the other appliances?

8. Why is the length of time a current is flowing through the body important in determining the severity of an electric shock?

9. An individual uses a hair dryer immediately after getting out of the shower to dry their hair, with their hands still wet. Explain why having wet hands leads to a greater risk of electric shock than having dry hands.

10. Compare the operation of residual current devices and circuit breakers, and how they improve safety of electric circuits.

7.4 Exam questions

Section A — Multiple choice questions

All correct answers are worth 1 mark each; an incorrect answer is worth 0.

▶ Question 1

Which colour wire carries the larger voltage in an electrical cable connected to a 230-V AC supply?

A. All carry the same voltage.
B. Green and yellow stripes
C. Brown
D. Blue

▶ Question 2

Which of the following best describes the reason for building fuses into electrical devices?

A. Fuses are not used in electrical devices.
B. To protect the device from damage
C. To protect the circuit from current overload
D. To protect against electric shock

▶ Question 3

Under what condition is a fuse in a circuit most likely to 'blow'?

A. When the current is too large
B. When the voltage is too large
C. When the current is too small
D. When the voltage is too small

▶ Question 4

Which of the following statements best describes the function of a circuit breaker?

A. A circuit breaker diverts excess current to other parts of the circuit.
B. A circuit breaker is a resettable switch.
C. A circuit breaker comprises many fuses that are resettable.
D. A circuit breaker must be replaced after a circuit has blown.

Question 5

What current is drawn by a 1.84-kW sandwich maker when it is connected to a 230-V supply?

A. 0.3 A
B. 2.8 A
C. 8.0 A
D. 14.7 A

Question 6

When charged at 30 cents per kW h, what will be the cost of leaving a 250-W computer switched on for two days?

A. $1.80
B. $3.60
C. $5.40
D. $7.20

Question 7

An electric hand tool with a power rating of 2 kW has a built-in fuse.

Which of the following fuses would be most suitable?

A. 3 A
B. 5 A
C. 7 A
D. 9 A

Question 8

Which of the following correctly describes the appearance and function of the earth wire?

A. It is coated in blue plastic and connects the appliance to the earth.
B. It is coated in brown plastic and provides a lower-resistance conducting path to the earth than the appliance and the person.
C. It is coated in green and yellow plastic and connects the case of the appliance being used to the earth as a safety device.
D. It is coated in green plastic and is a safety measure in electric circuits.

Question 9

Many hair dryers have a double insulation symbol (▣) found on them.

What does it mean for a hair dryer to have double insulation?

A. It has a two-pin plug with no neutral wire.
B. There are two layers that separate the live wire and the external metal casings.
C. There is a single insulating layer coating the wire, but it is double thickness.
D. It has double the number of wires than usual (two neutral, two earth and two live).

Question 10

What is the first priority when someone is a victim of electric shock?

A. Calling an ambulance
B. Making sure the victim is not still connected to the electric source
C. Giving immediate CPR
D. Making sure that a defibrillator is available

Section B — Short answer questions

Question 11 (3 marks)

A person is holding an electric appliance that is connected to a 230-V AC supply. If the insulation between the case and the rest of the appliance is short circuited, the person could receive an electric shock.

Describe the factors that will determine the severity of the shock they could receive.

Question 12 (5 marks)

An appliance with a metal case is unsafely wired directly into a simple AC circuit, as shown in the following figure.

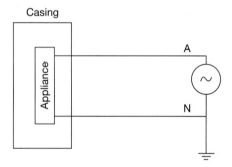

a. On the diagram, show what safety features should be provided to help avoid electric shocks and possible damage to the appliance. **(3 marks)**
b. For the safety features you identify, explain how they improve the safety of the system. **(2 marks)**

Question 13 (3 marks)

The metal-cased appliance in **Question 12** is to be re-wired to a three-pin plug so that it can be plugged into a domestic 230-V AC socket.

On the following diagram, show where each wire should be connected to the appliance.

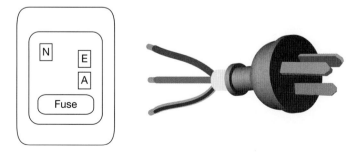

Question 14 (6 marks)

The following table gives the power consumption of various products when they are on standby.

Product	Power (W)
Laptop computer	14.5
Microwave	4.2
Laser printer	8.5
Set-top box	11.2
Television	6.2

a. Calculate the energy used by each product if it is left on standby for one year. **(3 marks)**
b. Calculate the mass of greenhouse gases produced by these products if they are left on standby for one year, assuming that 1 kW h of energy produces greenhouse gases that are equivalent to 1.444 kg of CO_2. **(2 marks)**
c. If you were charged 50 cents per kW h, how much would you be required to pay in one year if the devices were left on standby? **(1 mark)**

 Question 15 (5 marks)

A person stands on a rubber mat that connects an electric appliance to a 230-V mains power outlet. However, the appliance, which has a metal casing, has short circuited.

a. What is likely to happen if the person has a resistance of 600 kΩ? Justify your answer. **(2 marks)**
b. What is the smallest voltage on the case of the appliance that could cause fibrillation of the heart? **(1 mark)**
c. What would most likely happen if the person stands on a wet floor in bare feet, where their resistance is effectively 3 kΩ? Justify your answer. **(2 marks)**

AREA OF STUDY 3 How can electricity be used to transfer energy?

OUTCOME 3

Investigate and apply a basic DC circuit model to simple battery-operated devices and household electrical systems, apply mathematical models to analyse circuits, and describe the safe and effective use of electricity by individuals and the community.

PRACTICE EXAMINATION

STRUCTURE OF PRACTICE EXAMINATION		
Section	Number of questions	Number of marks
A	20	20
B	3	20
Total		40

Duration: 50 minutes

Information:
- This practice examination consists of two parts. You must answer all question sections.
- Pens, pencils, highlighters, erasers, rulers and a scientific calculator are permitted.
- You may use the VCAA Physics formula sheet for this task.

 Resources

🔗 **Weblink** VCAA Physics formula sheet

SECTION A — Multiple choice questions

All correct answers are worth 1 mark each; an incorrect answer is worth 0.

1. Students in a class are each provided with a battery. Each battery has an emf of 15 V. In an experiment, a student measured that a total 375 C of charge has passed through her battery.
   How much energy has been supplied by the battery to these charges?
   A. 1625 J
   B. 2650 J
   C. 5625 J
   D. 6250 J

Use the following information to answer questions 2 and 3.
A model electric motor is connected with a 12-V battery. When the motor is running normally, the current through the motor is measured to be 2.5 A.

2. What is the resistance of the motor?
   A. 2.0 Ω
   B. 4.8 Ω
   C. 7.2 Ω
   D. 9.6 Ω

3. How much energy was consumed by the electric motor after 30 seconds?

    A. 75 J

    B. 360 J

    C. 720 J

    D. 900 J

4. An electrical pump operating normally has 9 A of current flowing through it. The electrical resistance of the pump is 8 Ω.
Which of the following is the best estimate of its power?

    A. 72 W

    B. 576 W

    C. 648 W

    D. 764 W

5. Anja and Cris are provided with a thermistor for a Physics investigation. The characteristic curve of the thermistor is shown as follows.

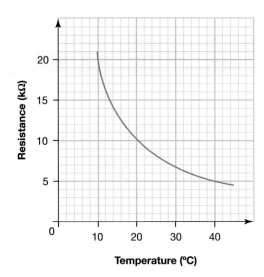

They measured the resistance of the thermistor when the temperature is 25 °C.
What is the expected value of the resistance?

    A. 5 kΩ

    B. 8 kΩ

    C. 12 kΩ

    D. 15 kΩ

6. An LED is connected in series with a 290-Ω resistor and a 9.0-V battery. The diode has a switch-on voltage of 3.20 V and is forward biased in this circuit.
Which of the following is the best estimate of the current in the circuit?

    A. 2.0 mA

    B. 11 mA

    C. 20 mA

    D. 31 mA

7. An ohmic device with a resistance of 7.5 Ω is connected to a circuit, and the current through the device and the voltage across it are measured.
Which of the following combinations of voltage and current could be approximate readings of this set-up?

    A. $V = 20.0$ V and $I = 3.2$ A

    B. $V = 24.0$ V and $I = 4.2$ A

    C. $V = 28.8$ V and $I = 3.8$ A

    D. $V = 32.4$ V and $I = 4.8$ A

8. The following diagram shows part of an electrical circuit with the values of the current in different parts of the circuit.

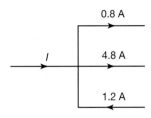

Which of the following is closest to the value of the current, *I*?

A. 3.6 A

B. 4.4 A

C. 5.6 A

D. 6.8 A

9. The following diagram is a battery circuit. When the switch is closed, the voltages across the different resistors are shown on the diagram.

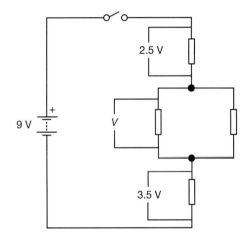

Which of the following is closest to the value of the voltage, *V*?

A. 2.5 V

B. 3.0 V

C. 3.5 V

D. 4.0 V

Use the following information to answer questions 10 and 11.

Charlene and Deen found a box full of electrical resistors with resistances of either 7 Ω or 12 Ω. There are no other types of resistors. They decide to construct an electrical circuit with a laboratory power supply and connecting wires.

10. Charlene and Deen connected a 7.00-Ω resistor and a 12.0-Ω resistor in parallel and measured their equivalent resistance.
Which of the following is closest to the value of the reading on the resistance meter?

A. 4.42 Ω

B. 5.42 Ω

C. 6.42 Ω

D. 7.42 Ω

11. One electrical circuit requires a resistance of 20.0 Ω.
Which of the following combinations of 7.00-Ω and 12.0-Ω resistors could give them the required 20.0 Ω?

A. Series connection of a 7.00-Ω and a 12.0-Ω resistor

B. Parallel connection of two 7.00-Ω resistors, then in series with a 12.0-Ω resistor

C. Parallel connection of two 12.0-Ω resistors, then in series with a 7.00-Ω resistor

D. Parallel connection of two 12.0-Ω resistors, then in series with two 7.00-Ω resistors

12. The following circuit shows a 12-V battery connected with two parallel resistors, each with a resistance of 16 Ω.

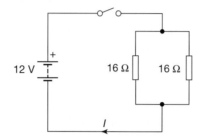

Which of the following is closest to the value of the current, I?

A. 0.75 A

B. 1.5 A

C. 3.0 A

D. 4.5 A

13. The following circuit shows a 12-V battery connected in series with a 3.0-Ω resistor and another resistor with an unknown resistance, R. The current, I, is 1.5 A.

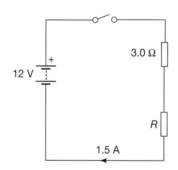

Which of the following is closest to the value of R?

A. 4.0 Ω

B. 5.0 Ω

C. 6.0 Ω

D. 8.0 Ω

14. The following circuit shows a 12-V battery connected with two parallel resistors, each of which has an unknown resistance, R. The current, I, is 2.4 A.

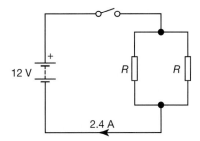

12 V
2.4 A

Which of the following is closest to the value of R?

A. 7.2 Ω

B. 8.4 Ω

C. 10 Ω

D. 14 Ω

15. The following circuit is a voltage divider circuit with a 9-V battery connected to a thermistor in series with a 4-kΩ resistor. The output voltage, V_{out}, is 6 V.

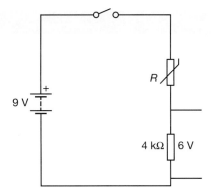

9 V
R
4 kΩ   6 V

Which of the following is closest to the value of the resistance of the thermistor, R?

A. 1 kΩ

B. 2 kΩ

C. 3 kΩ

D. 4 kΩ

16. The Australian Standards specify that the AC electrical supply to households has a root mean square voltage of 230 V.
What does this mean in relation to the power supply?

A. It has a peak voltage of 230 V.

B. It has a peak-to-peak voltage of 460 V.

C. It will provide the same heating effect as a DC supply of 325 V.

D. It will provide the same heating effect as a DC supply of 230 V.

17. What colour is the active wire in household electrical wiring?

A. Blue

B. Brown

C. Green and yellow

D. Red

18. A dishwasher is rated at 1800 W and is operated for 2.5 hours. The cost of energy is 30 cents per kW h. How much does it cost to operate the dishwasher?

    A. 75 cents

    B. 135 cents

    C. 75 000 cents

    D. 135 000 cents

19. What is the operating principle of a residual current device (RCD)?

    A. It measures the difference between the current of the active wire and the current of the neutral wire.

    B. It melts due to the dangerously high current.

    C. It provides an alternative path to current that flows through the casing of an electrical appliance.

    D. It activates an electromagnet to switch off the circuit.

20. Which of the following values is in the range of the electrical resistance of a dry hand?

    A. $500\ \Omega$

    B. $2000\ \Omega$

    C. $500\ k\Omega$

    D. $2\ M\Omega$

SECTION B — Short answer questions

Question 21 (8 marks)

Consider the following circuit.

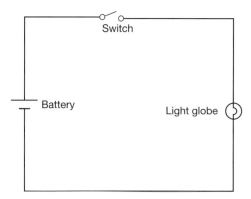

A battery is connected to a light globe and a switch. The emf of the battery is 4.2 V. The switch is currently open.

a. What is the voltage across the switch? **(1 mark)**

The switch is now closed. During each second, 1.4 C of charge pass through the battery when the switch is closed. $1\ C = 6.24 \times 10^{18}$ electrons.

b. What is the current flowing out of the battery? **(1 mark)**

c. How much energy does the battery supply to each coulomb of charge? **(1 mark)**

d. State the potential difference across the switch when it is closed. **(1 mark)**

e. Calculate the number of electrons that flow through the battery each second. **(1 mark)**

f. What is the power that the battery is transferring to the circuit? **(1 mark)**

g. How much electrical potential energy is transformed into light and heat by the globe in 3 minutes? **(1 mark)**

h. Calculate the resistance of the light globe. **(1 mark)**

Question 22 (8 marks)

Consider the following circuit where a 15-V battery is connected to a switch, S, and there are three light globes with the labelled resistances.

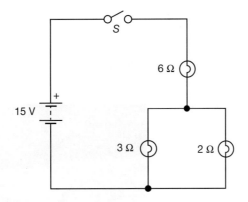

a. Determine the equivalent resistance of the circuit. **(2 marks)**
b. Calculate the current flowing through the 6-Ω light globe. **(2 marks)**
c. What is the potential difference across the 2-Ω light globe? **(2 marks)**
d. Find the current flowing through the 3-Ω light globe. **(1 mark)**
e. Calculate the power supplied by the battery. **(1 mark)**

Question 23 (4 marks)

An electrician decided to break open an old toaster and examine the various safety features.

a. Within the plug, he noted the presence of an earth wire.
Describe the appearance of the earth wire, and explain how this works as a safety feature. **(2 marks)**
b. The electrician decides the toaster is still fine to use and plugs it back into his kitchen circuit. The circuit already has a 500-W fridge, a 800-W coffee machine and a 1000-W fan plugged into it. This is connected to a 230-V supply and protected by a 10.5-A fuse.
If the toaster is 700 W, will the fuse blow when the toaster is added to the circuit? Justify your response using calculations. **(2 marks)**

PRACTICE SCHOOL-ASSESSED COURSEWORK

ASSESSMENT TASK — PROPOSED SOLUTION TO A SCIENTIFIC OR TECHNOLOGICAL PROBLEM

In this task, you will be required to investigate and apply a basic DC circuit model to household electrical systems, apply mathematical models to analyse circuits and describe the safe and effective use of electricity by individuals.

- A scientific calculator is permitted.

Total time: 50 minutes (5 minutes reading and 45 minutes writing)

Total marks: 40 marks

SUPPLYING ELECTRICAL POWER TO A SHED

A farmer, Alisa, is considering how to supply electrical power to a shed containing farming equipment. The farming equipment is designed to work safely from a power supply with a voltage of 30 V to 50 V DC.

The main problem is that the shed is 1 km from the power lines that service her farmhouse.

Alisa has two choices:
1. To purchase a generator and position it next to the shed
2. To construct additional power lines that connect the shed to the existing power lines that service her farmhouse.

Part A: Using a generator

A review of generators shows that they come with different attributes, which need to be analysed to assist Alisa in solving her problem.
1. One generator states that it is an AC generator while a second model states that it is a DC generator. In what way are the two generators different in terms of the electrical current they produce when connected to a load such as a piece of farming equipment?

Alisa reviews two DC generators, X and Y, as a possible solution to her problem.
2. Alisa finds that DC generator X outputs a voltage of 32 V, while generator Y outputs a voltage of 48 V. Explain how the generators differ by clearly explaining in terms of energy transfer what the different voltages mean.
3. Each generator has a cut-off safety switch if the current drawn exceeds the maximum specified current. The maximum current of the 32-V generator is 36 A and is slightly lower in the 48-V generator, at 32 A. Which generator, X or Y, is capable of producing the greatest power output at maximum current? Use an appropriate calculation to support your choice.
4. What would be the electrical resistance of the farming equipment if 36 A were drawn from the 32-V generator?
5. What would be the electrical resistance of the farming equipment if 32 A were drawn from the 48-V generator?

Alisa measures the electrical resistance of the farming equipment and finds it to be 1.20 Ω.

6. If Alisa connected the farming equipment to each of the generators in turn, what current would each generator supply to the equipment?

7. a. How much power would the farming equipment consume when using the 32-V generator?
   b. How much power would the farming equipment consume when using the 48-V generator?
   c. The farming equipment is required to run for 8 hours per day. What is the total energy delivered to the farming equipment if generator X is used? Express your answer in joules and kW h.

8. a. Is there a problem using either of the generators with the existing farming equipment?
   b. Of the two generators Alisa has reviewed, which one would be more suitable? Explain your choice.

Part B: Using power lines

The potential difference of the power lines to Alisa's farmhouse is 240 V DC. It would normally be an AC voltage, but for this assessment we will simplify the task and make the voltage a 240 DC voltage.

9. If Alisa is to use power lines, how much cable (in km) will she need to supply electricity to the shed from her farmhouse?

10. Alisa finds that she can purchase the cable, but it has a resistance of 1.60 Ω per km. What would be the equivalent resistance of the power lines?

11. Draw a circuit of a 240-V DC power supply and power lines that connects to the farming equipment. In your diagram include a switch S that can isolate the farming equipment from the 240-V power supply.

Alisa has previously measured the resistance of the farming equipment to be 1.20 Ω.

12. What would be the equivalent resistance of the circuit containing the power lines and the farming equipment?

13. Determine the current that would be in the circuit if the switch S were closed and the farming equipment turned on.

14. a. By determining the voltage across the farming equipment, if the switch were closed, assess whether the system would be safe. Remember that the farming equipment is designed to operate safely with a voltage of between 30 V and 50 V.
   b. If you find the system would not be safe, give one possible modification that would change the situation and explain how it would enable this. If you find the system would be safe to operate, explain why.

Part C: Proposing a solution

15. Sum up your findings from Part A and B, identifying which of the two choices you would recommend to solve Alisa's problem. Justify your response, ensuring you compare both choices.

 Resources

 Digital document School-assessed coursework (doc-38060)

Source: VCE Physics Study Design (2023–2027) extracts © VCAA; reproduced by permission.

8 Analysing motion

KEY KNOWLEDGE

In this topic, you will:
- identify parameters of motion as vectors or scalars
- analyse graphically, numerically and algebraically, straight-line motion under constant acceleration: $v = u + at$, $v^2 = u^2 + 2as$, $s = \frac{1}{2}(u + v)t$, $s = ut + \frac{1}{2}at^2$, $s = vt - \frac{1}{2}at^2$

- analyse, graphically, non-uniform motion in a straight line

Source: VCE Physics Study Design (2023–2027) extracts © VCAA; reproduced by permission.

Note: Vector quantities are bolded.

PRACTICAL WORK AND INVESTIGATIONS

Practical work is a central component of VCE Physics. Experiments and investigations, supported by a **practical investigation eLogbook** and **teacher-led videos**, are included in this topic to provide opportunities to undertake investigations and communicate findings.

EXAM PREPARATION

Access exam-style questions and their video solutions in every lesson, to ensure you are ready.

8.1 Overview

8.1.1 Introduction

The study of the motion of objects relies upon an accepted and reliable means for describing and analysing motion. In this topic you will explore the key quantities used to describe motion, their characteristics and interrelationships. By the end of this topic you should be able to distinguish between scalar and vector quantities. You will be able to describe motion in terms of position, displacement, speed, velocity and acceleration. You will analyse motion using the most suitable method, including numerical calculations, estimations, algebraic analysis and graphical analysis. This area of study is often broadly referred to as kinematics. The skills developed through this topic provide the foundation for studies of the role of forces and energy in the understanding of motion in the topics that follow.

FIGURE 8.1 When driving a vehicle, the driver needs to be aware of their movements in terms of position, speed, direction and acceleration.

LEARNING SEQUENCE

---

on Resources

 Solutions   Solutions — Topic 8 (sol-0794)

 Practical investigation eLogbook   Practical investigation eLogbook — Topic 8 (elog-1577)

Digital documents   Key science skills — VCE Physics Units 1–4 (doc-36950)
                             Key terms glossary — Topic 8 (doc-36965)
                             Key ideas summary — Topic 8 (doc-36966)

Exam question booklet   Exam question booklet — Topic 8 (eqb-0076)

8.2 Describing movement

KEY KNOWLEDGE

- Identify parameters of motion as vectors or scalars
- Analyse numerically and algebraically, straight-line motion under constant acceleration

Source: Adapted from VCE Physics Study Design (2023–2027) extracts © VCAA; reproduced by permission.

8.2.1 Vectors and scalars

The study of the way an object moves is the starting point for developing an understanding of the nature of forces and their relationship with motion.

There are many parameters we explore when we look at motion. These can often be divided into two main groups:

- **Scalars**, or scalar quantities, are those that have a **magnitude** but no direction. Examples of scalar quantities include:
  - distance
  - speed
  - time
  - work
  - energy.

  When writing variables for scalar quantities, we usually represent this using an italicised symbol. For example, the scalar quantity 'time' is represented as t.
- **Vectors**, or vector quantities, are those that have both a magnitude and direction. Examples of vectors include:
  - displacement
  - velocity
  - acceleration
  - force
  - momentum
  - impulse.

  When writing variables for vector quantities, we usually represent this using an italicised and bolded symbol. For example, the vector quantity 'velocity' is represented as $\boldsymbol{v}$. (*Note:* As an alternate to bolded typeface, an arrow may be placed above or below the variable.)

Many of these quantities will be explored throughout this topic, and in topics 9 and 10.

8.2.2 Distance and displacement

Distance is a measure of the length of the path taken during the change in position of an object. Distance is a scalar quantity. It does not specify a direction.

Displacement is a measure of the change in position of an object from the starting position. Displacement is a vector quantity. In order to describe a displacement fully, a direction must be specified as well as a magnitude. The path taken by the fly in figure 8.2 as it escapes the swatter illustrates the difference between distance and displacement. The displacement of the fly is 60 centimetres to the right, while the distance travelled is well over 1 metre.

In a 100-metre sprint, the runners travel in a straight line; therefore the magnitude of the displacement is the same as the distance. However, it is the displacement that fully describes the change in position of the runner because it specifies the direction.

scalar a quantity that specifies magnitude (size) but not direction

magnitude the size or quantity of an object or variable

vector a quantity that specifies direction as well as magnitude (size)

distance a measure of the length of the path taken by an object; it is a scalar quantity

displacement a measure of the change in position of an object; it is a vector quantity

FIGURE 8.2 Distance and displacement are different quantities.

FIGURE 8.3 Runners in the 100-metre sprint travel in a straight line.

The displacement of an object that has moved from position x_1 to position x_2 is expressed as:

$$\text{Displacement} = \text{change in position}$$
$$= \text{final position} - \text{initial position}$$
$$\Delta x = x_2 - x_1$$

Displacement can also be represented by the symbols x or s. A displacement vector is pointing in the same direction as the movement of the object.

Note: Vectors can be represented using either bold text or a right arrow.

tlvd-0069

SAMPLE PROBLEM 1 Calculating distance and displacement

A hare and a tortoise decide to have a race along a straight 100-metre stretch of highway. They both head due north. However, at the 80-metre mark, the hare realises that he dropped his phone at the 20-metre mark. He dashes back, grabs his phone and resumes the race, arriving at the finishing line at the same time as the tortoise. (It was a very fast tortoise!)

a. **What was the displacement of the hare during the entire race?**
b. **What was the distance travelled by the hare during the race?**
c. **What was the distance travelled by the tortoise during the race?**
d. **What was the displacement of the hare during his return to collect his phone?**

THINK	WRITE
a. 1. Recall the relationship $\Delta x = x_2 - x_1$.	a. $\Delta x = x_2 - x_1$
2. Identify the variables. Using the start as the reference point, the displacement was 100 m north. In symbols, this calculation can be done by denoting north as positive and south as negative.	$x_1 = 0$, $x_2 = 100$ m north
3. Substitute into the relationship to find Δx.	$\Delta x = 100 - 0$ $= 100$ m north
4. State the solution.	The displacement of the hare is 100 metres north.

b. 1. The distance is the length of the path taken.

b. The hare travels a total distance of 80 m (before noticing his phone) + 60 m (running back to the 20-m mark) + 80 m (from the 20-m mark to the finishing line).

Distance $= 80 + 60 + 80$

$\qquad\quad = 220 \text{ m}$

2. State the solution.

The total distance that the hare travels is 220 metres.

c. 1. The distance travelled by the tortoise is the length of the path taken.

c. Length of the path taken $= 100$ m

2. State the solution.

The tortoise travels a total distance of 100 metres.

d. 1. Recall the relationship $\Delta x = x_2 - x_1$.

d. $\Delta x = x_2 - x_1$

2. The hare returns from a position 80 m north of the reference point (or start) back to a position 20 m north of the reference point.

$\Delta x = 20 \text{ m} - 80 \text{ m}$

$\qquad = -60 \text{ m}$

3. State the solution.

The displacement is 60 metres south.

PRACTICE PROBLEM 1

A cyclist rides 4 kilometres due west from home, then turns right to ride a further 4 kilometres due north. She stops, turns back and rides home along the same route.

a. What distance did she travel during the entire ride?

b. What was her displacement at the instant that she turned back?

c. What was her displacement from the instant she commenced her return journey until she arrived home?

8.2.3 Speed

Speed is a measure of the rate at which an object moves over a distance. When you calculate the speed of a moving object, you need to measure the distance travelled over a time interval.

The average speed of an object can be calculated by dividing the distance travelled by the time taken:

$$\text{Average speed} = \frac{\text{distance travelled}}{\text{time interval}}$$

The speed obtained using this formula is the average speed during the time interval. Speed is a scalar quantity as it does not include direction. The unit of speed is m s^{-1} (or m/s) if SI units are used for distance and time. However, it is often more convenient to use other units, such as cm s^{-1} or km h^{-1} (which can also be written as cm/s and km/h respectively).

speed the rate at which distance travelled changes over time; it is a scalar quantity

CASE STUDY: Snail's pace

A snail would lose a race with a giant tortoise! A giant tortoise can reach a top speed of 0.37 km h⁻¹. However, its 'cruising' speed is about 0.27 km h⁻¹. The world's fastest snail covers ground at the breathtaking speed of about 0.05 km h⁻¹. However, the common garden snail is more likely to move at a speed of about 0.02 km h⁻¹. Both of these creatures are slow compared with light, which travels through the air at 1080 million km h⁻¹, and sound, which travels through the air (at sea level) at about 1200 km h⁻¹.

How long would it take the snail, giant tortoise, light and sound respectively to travel once around the equator, a distance of 40 074 km?

FIGURE 8.4 A common garden snail travels at 0.02 km h⁻¹.

8.2.4 Converting units of speed

It is often necessary to convert units that are not derived from SI units (such as km h⁻¹) to and from units that are derived SI units, such as m s⁻¹.

TABLE 8.1 Converting units of speed

To convert 60 km h⁻¹ to m s⁻¹	To convert 30 m s⁻¹ to km h⁻¹
$$= \dfrac{60\,\text{km}}{1\,\text{h}}$$ $$= \dfrac{60\,000\,\text{m}}{3600\,\text{s}}$$ $$= 16.7\,\text{m s}^{-1}$$	$$\dfrac{30\,\text{m}}{1\,\text{s}}$$ $$= \dfrac{0.030\,\text{km}}{\frac{1}{3600}\,\text{h}}$$ $$= \dfrac{3600 \times 0.030\,\text{km}}{1\,\text{h}}$$ $$= 108\,\text{km h}^{-1}$$
The speed in km h⁻¹ has been multiplied by $\dfrac{1000}{3600}$, or divided by 3.6.	The speed in m s⁻¹ has been multiplied by $\dfrac{3600}{1000}$; that is, by 3.6.

To quickly convert between speeds in m s⁻¹ and km h⁻¹ you can simply multiply or divide by 3.6, depending on which way you are converting.

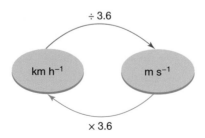

tlvd-3780

A plane carrying passengers from Melbourne to Perth flies at an average speed of 250 m s⁻¹. The flight takes 3 hours. Use this information to determine the approximate distance by air between Melbourne and Perth.

THINK	WRITE
1. Recall the relationship and rearrange to find distance.	$\text{Average speed} = \dfrac{\text{distance travelled}}{\text{time interval}}$ $\Rightarrow \text{Distance travelled} = \text{average speed} \times \text{time interval}$
2. Identify the values, and convert into the similar units. *Note*: Alternatively, the distance could be calculated in metres and converted to kilometres. Multiply the time interval by 3600 to convert h to s.	$\text{Time interval} = 3\,\text{h}$ $\text{Average speed} = 250\,\text{m s}^{-1}\ (\times 3.6 \text{ to convert to km h}^{-1})$ $\qquad\qquad\quad = 900\,\text{km h}^{-1}$
3. Substitute into the relationship to find distance.	$\text{Distance travelled} = 900\,\text{km h}^{-1} \times 3\,\text{h}$ $\qquad\qquad\qquad\quad = 2700\,\text{km}$
4. State the solution.	The approximate distance by air between Melbourne and Perth is 2700 kilometres.

PRACTICE PROBLEM 2

A car takes 8 hours to travel from Canberra to Ballarat at an average speed of 25 m s⁻¹. What is the road distance from Canberra to Ballarat?

8.2.5 Velocity

In everyday language, the word *velocity* is often used to mean the same thing as speed. In fact, velocity is not the same quantity as speed; **velocity** is a measure of the rate of displacement, or rate of change in position, of an object. Because displacement is a vector quantity, velocity is also a vector quantity. The velocity has the same direction as the displacement. The symbol v is used to denote velocity. (Unfortunately, the symbol v is often used to represent speed as well, which can be confusing.) To make a distinction between vectors and scalars, this text mostly displays vector symbols in bold. When bold can't be used to distinguish a vector (for example, when writing by hand), it is common practice to place a vector symbol above $(\vec{v})$ or below $(\underset{\sim}{v})$ vector quantities.

> **velocity** the rate at which displacement changes over time, or the rate of change in position; it is a vector quantity

---

The average velocity of an object, v_{av}, during a time interval t can be expressed as:

$$v_{av} = \frac{\Delta x}{\Delta t}$$

where Δx represents the displacement (change in position).

Note: A velocity vector points in the same direction as the change of position. The change of position, or displacement, gives the direction of velocity.

For motion in a straight line in one direction, the magnitude of the velocity is the same as the speed. The motion of the fly in figure 8.5 illustrates the difference between velocity and speed. If the fly takes two seconds to complete its flight, its average velocity is:

$$v_{av} = \frac{\Delta x}{\Delta t}$$

$$= \frac{60 \text{ cm to the right}}{2 \text{ s}}$$

$$= 30 \text{ cm s}^{-1} \text{ to the right}$$

The path taken by the fly is about 180 cm. Its average speed is:

$$\text{Average speed} = \frac{\text{distance travelled}}{\text{time travelled}}$$

$$= \frac{180 \text{ cm}}{2 \text{ s}}$$

$$= 90 \text{ cm s}^{-1}$$

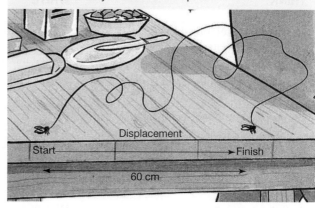

FIGURE 8.5 When motion is not in a straight line, in one direction, velocity is different to speed.

---

tlvd-3781

SAMPLE PROBLEM 3 Calculating the average speed and average velocity

Refer to sample problem 1. Assume the hare takes 20 seconds to complete the race. Calculate the
a. **average speed of the hare**
b. **average velocity of the hare.**

THINK	WRITE
a. 1. Recall the relationship.	a. $\text{Average speed} = \dfrac{\text{distance travelled}}{\text{time interval}}$
2. Identify the values.	Distance travelled $= 220 \text{ m}$ Time interval $= 20 \text{ s}$
3. Substitute into the relationship to find the average speed.	$\text{Average speed} = \dfrac{220 \text{ m}}{20 \text{ s}}$ $= 11 \text{ m s}^{-1}$
4. State the solution.	The average speed of the hare is 11 m s^{-1}.
b. 1. Recall the relationship.	b. Average velocity: $v_{av} = \dfrac{\Delta x}{\Delta t}$
2. Identify the values.	Displacement $= 100 \text{ m north}$ Time interval $= 20 \text{ s}$
3. Substitute into the relationship to find v_{av}.	$v_{av} = \dfrac{100 \text{ m north}}{20 \text{ s}}$ $= 5.0 \text{ m s}^{-1} \text{ north}$
4. State the solution.	The average velocity of the hare is 5 m s^{-1} north.

PRACTICE PROBLEM 3

During the final 4.2-kilometre run stage of a triathlon, a participant runs 2.8 kilometres east, then changes direction to run a further 1.4 kilometres in the opposite direction, completing the stage in 20 minutes. Calculate the participant's:

a. average speed

b. average velocity.

Express both answers in m s^{-1} to two significant figures.

elog-1663

INVESTIGATION 8.1   **online** only

Going home

Aim

To analyse your journey from home to school

8.2.6 Instantaneous speed and velocity

Neither the average speed nor the average velocity provide information about movement at any particular instant of time. For example, when Jamaican athlete Usain Bolt broke the 100-metre world record in 2009 with a time of 9.58 seconds, his average speed was 10.4 m s^{-1}; however, he was not travelling at that speed throughout his run. He would have taken a short time to reach his maximum speed and would not have been able to maintain it throughout the run. His maximum speed would have been much more than 10.4 m s^{-1}.

The speed at any particular instant of time is called the **instantaneous speed**. Note that as speed is calculated as the ratio of distance to time, a tiny distance and time interval must be measured in order to approximate instantaneous speed. The velocity at any particular instant of time is, not surprisingly, called the **instantaneous velocity**. If an object moves with a constant velocity during a time interval, its instantaneous velocity throughout the interval is the same as its average velocity.

> **instantaneous speed** the speed at a particular instant of time
>
> **instantaneous velocity** the velocity at a particular instant of time

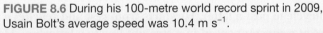

FIGURE 8.6 During his 100-metre world record sprint in 2009, Usain Bolt's average speed was 10.4 m s^{-1}.

8.2.7 Acceleration

When the velocity of an object changes, it is helpful to describe it in terms of the rate at which the velocity is changing. In everyday language, the word *accelerate* is used to mean 'speed up'. The word *decelerate* is used to mean 'slow down'. However, if you wish to describe motion precisely, these words are not adequate. The rate at which the velocity of an object changes is called **acceleration**. Acceleration is a vector quantity. An acceleration vector points in the direction of change in velocity. That is, if an object slows down, the final velocity is smaller than the initial velocity; therefore the acceleration points in the opposite direction to the way that the object is heading. It also points in the same direction as the way in which a force is acting.

> **acceleration** the rate at which the velocity of an object changes; it is a vector quantity

A car starting from rest and reaching a velocity of 60 km h^{-1} north in 5 seconds has an average acceleration of 12 km h^{-1} per second or 12 (km h^{-1})s^{-1} north. This is expressed in words as 12 km per hour per second. In simple terms, it means that the car increases its speed in a northerly direction by an average of 12 km h^{-1} each second.

> The average acceleration of an object, a_{av}, can be expressed as:
> $$a_{av} = \frac{\Delta v}{\Delta t}$$
> The change in velocity during the time interval Δt can be expressed as:
> $$\Delta v = \text{final velocity} - \text{initial velocity}$$
> This is commonly also written as:
> $$\Delta v = v - u$$

The direction of the average acceleration is the same as the direction of the change in velocity. The unit of acceleration is m s^{-2} if SI units are used for distance and time. The superscript of −2 represents that this unit is metres per second per second, showing the change in the speed or velocity of an object (in m s^{-1}) every second. However, it may sometimes be more convenient to use other units, such as km h^{-2}. In some cases, two different time units may be used, such as km h^{-1}s^{-1}.

tlvd-0072

SAMPLE PROBLEM 4 Determining average acceleration using time and velocity

Spiro leaves home on his bicycle to post a letter. He starts from rest and reaches a speed of 10 m s^{-1} in 4 seconds. He then cycles at a constant speed in a straight line to a letterbox. He brakes at the letterbox, coming to a stop in 2 seconds, posts the letter and returns home at a constant speed of 8 m s^{-1}. On reaching home he brakes, coming to rest in 2 seconds. The direction away from home towards the letterbox is assigned as positive.

a. What is Spiro's average acceleration before he reaches his 'cruising speed' of 10 m s^{-1} on the way to the letterbox?
b. What is Spiro's average acceleration as he brakes at the letterbox?
c. What is Spiro's average acceleration as he brakes when arriving home?
d. During which two parts of the trip is Spiro's acceleration negative?
e. Does a positive acceleration always mean that the speed is increasing? Explain.

THINK	WRITE
a. 1. Recall the relationship.	a. Average acceleration: $a_{av} = \dfrac{\Delta v}{\Delta t}$
2. Identify the values.	$\Delta v = 10 \text{ m s}^{-1}$ $\Delta t = 4 \text{ s}$

3. Substitute into the relationship to find the average acceleration.

$$\text{Average acceleration} = \frac{10\,\text{m s}^{-1}}{4\,\text{s}}$$
$$= 2.5\,\text{m s}^{-2}$$

4. State the solution.

Spiro's average acceleration on his journey to cruising speed is 2.5 m s^{-2} towards the letterbox.

b. 1. Recall the relationship.

b. Average acceleration: $a_{av} = \dfrac{\Delta v}{\Delta t}$

2. Identify the values.

$\Delta v = -10\,\text{m s}^{-1}$
$\Delta t = 2\,\text{s}$

3. Substitute into the relationship to find the average acceleration.

$$\text{Average acceleration} = \frac{-10\,\text{m s}^{-1}}{2\,\text{s}}$$
$$= -5\,\text{m s}^{-2}$$

4. State the solution.

Spiro's average acceleration when braking at the letterbox is 5 m s^{-2} away from the letterbox.

c. 1. Recall the relationship.

c. Average acceleration: $a_{av} = \dfrac{\Delta v}{\Delta t}$

2. Identify the values.

$\Delta v = 8\,\text{m s}^{-1}$
$\Delta t = 2\,\text{s}$

3. Substitute into the relationship to find the average acceleration.

$$\text{Average acceleration} = \frac{8\,\text{m s}^{-1}}{2\,\text{s}}$$
$$= 4\,\text{m s}^{-2}$$

4. State the solution.

Spiro's average acceleration when braking at the end of his journey is 4 m s^{-2} towards the letterbox.

d. 1. Recall the positive and negative directions.

d. Positive direction is away from the house, negative direction is towards the house.

2. Acceleration is negative when the velocity in the direction towards the house is increasing.

The two sections of the trip where acceleration is negative are braking at the letterbox and accelerating back towards home from the letterbox.

e. 1. Recall the definition of acceleration.

e. Acceleration is equal to change in velocity.

2. As speed does not take into account direction, an increasing speed does not necessarily equate to a positive acceleration.

A decreasing speed in the negative direction is a positive change in velocity, and hence a positive acceleration.

PRACTICE PROBLEM 4

a. A cheetah (the fastest land animal) takes 2 seconds to reach its maximum speed of 30 m s^{-1}. What is the magnitude of its average acceleration?

b. A drag-racing car reaches a speed of 420 km h^{-1} from a standing start in 6 seconds. What is its average acceleration in:

  i. km h^{-1} s^{-1}

  ii. m s^{-2}?

EXTENSION: Changing direction

A non-zero acceleration does not always result from a change in speed. Consider a car travelling at 60 km h^{-1} in a northerly direction turning right and continuing in an easterly direction at the same speed. Assume that the complete turn takes 10 seconds. The average acceleration during the time interval of 10 seconds is given by:

$$a_{av} = \frac{\Delta v}{\Delta t}$$

The change in velocity must be determined first. Thus:

$$\Delta v = v - u$$
$$= v + -u$$

The vectors v and $-u$ are added together to give the resulting change in velocity.

The magnitude of the change in velocity is calculated using Pythagoras' theorem or trigonometric ratios to be 85 km h^{-1}. Alternatively, the vectors can be added using a scale drawing and then measuring the magnitude and direction of the sum. The direction of the change in velocity can be seen in figure 8.7 to be south-east.

Note that to calculate the direction of the change of velocity, you can use the ratio 'opposite side to adjacent side' to determine the tangent of the angle. When stating the direction, always start with north or south and then use the angle to determine how far to the east or west the object has moved.

$$a_{av} = \frac{\Delta v}{\Delta t}$$
$$= \frac{85 \text{ km h}^{-1} \text{ south-east}}{10 \text{ s}}$$
$$= 8.5 \text{ km h}^{-1} \text{ s}^{-1} \text{ south-east}$$

In fact, the direction of the average acceleration is the same as the direction of the average net force on the car during the 10-second interval. The steering wheel is used to turn the wheels to cause the net force to be in this direction.

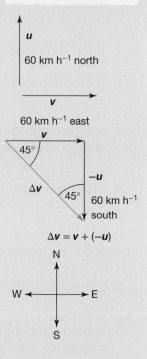

FIGURE 8.7 A change in acceleration can occur even if there is no change in speed.

8.2 Activities

learnon

Students, these questions are even better in jacPLUS

 Receive immediate feedback and access sample responses

 Access additional questions

 Track your results and progress

Find all this and MORE in jacPLUS ▶

| **8.2 Quick quiz** on | **8.2 Exercise** | **8.2 Exam questions** |

8.2 Exercise

1. State whether each of the following is a vector quantity or a scalar quantity.
   a. Distance
   b. Displacement
   c. Speed
   d. Velocity
   e. Acceleration

2. The speed limit on Melbourne's suburban freeways is 100 km h^{-1}. Express this speed in m s^{-1}.

3. If Emma McKeon's average speed while swimming a 100-metre freestyle race is about 1.5 m s^{-1}, calculate what her average speed would be in km h^{-1}.

4. The speed limit on US freeways is 55 miles per hour. One mile is approximately 1.6 km. Express this speed in each of the following measures:
   a. km h^{-1}
   b. m s^{-1}.

5. A jogger heads due north from his home and runs 400 metres along a straight footpath before realising he has forgotten his sunscreen and runs straight back to get it.
   a. What distance has the jogger travelled by the time he gets back home?
   b. What was the displacement of the jogger when he started to run back home?
   c. What was the jogger's displacement when he arrived back home to pick up the sunscreen?

6. The world records for some women's track events (as at early 2022) are listed in the following table.

World records for women's track events		
Athlete	**Event**	**Time**
Florence Griffith Joyner (USA)	100 m	10.49 s
Florence Griffith Joyner (USA)	200 m	21.34 s
Marita Koch (German Democratic Republic)	400 m	47.60 s
Jarmilia Kratochvilova (Czechoslovakia)	800 m	1 min 53.28 s
Genzebe Dibaba (Ethiopia)	1500 m	3 min 50.07 s
Wang Junxia (China)	3000 m	8 min 06.11 s
Letesenbet Gidey (Ethiopia)	5000 m	14 min 06.62 s
Letesenbet Gidey (Ethiopia)	10 000 m	29 min 01.03 s

   a. Calculate the average speed (to three significant figures) of each of the athletes listed in the table.
   b. Why is there so little difference between the average speeds of the world-record holders of the 100-metre and 200-metre events despite the doubling of the distance?
   c. How long would it take Genzebe Dibaba to complete the marathon if she could maintain her average speed during the 1500-metre event for the entire 42.2 km course? (The world record for the women's marathon, set on 13 October 2019, is 2 h 15 min 4 s.)
   d. Which of the athletes in the table has an average speed that is the same as the magnitude of her average velocity? Explain.

7. a. A jogger takes 30 minutes to cover a distance of 5 km. What is the jogger's average speed in:
      i. km h^{-1}
      ii. m s^{-1}?
   b. How long does it take for a car travelling at 60 km h^{-1} to cover a distance of 200 metres?

8. In 2020, cyclist Chloé Dygert, of the USA, set a world record of 3 min, 16.037 s for the 3000-metre pursuit.
   a. What was her average speed?
   b. How long would it take her to cycle from Melbourne to Bendigo, a distance of 151 km, if she could maintain her 3000-metre pursuit average speed for the whole distance?
   c. How long does it take a car to travel from Melbourne to Bendigo if its average speed is 80 km h^{-1}?
   d. A car travels from Melbourne to Bendigo and back to Melbourne in 4 hours.
      i. What is the car's average speed?
      ii. What is the car's average velocity?

9. Once upon a time, a giant tortoise had a bet with a hare that she could beat him in a foot race over a distance of 1 kilometre. The giant tortoise can reach a speed of about 7.5 cm s^{-1}. The hare can run as fast as 20 m s^{-1}. Both animals ran at their maximum speeds during the race. However, the hare was a rather arrogant creature and decided to have a little nap along the way. How long did the hare sleep if the result was a tie?

10. A Year 11 student arrives at school late and runs from the front gate of the school to the Physics laboratory. He runs the first 120 metres at an average speed of 6 m s^{-1}, the next 120 metres at an average speed of 4 m s^{-1} and the final 120 metres at an average speed of 2 m s^{-1}. What was the student's average speed during his attempt to arrive at his class on time?

11. While on holiday, a Physics teacher drives her old Volkswagen from Melbourne to Wodonga, a distance of 300 kilometres. Her average speed was 80 km h^{-1}. She trades in her old Volkswagen and purchases a brand-new Toyota Prius. She proudly drives her new car back home to Melbourne at an average speed of 100 km h^{-1}. Calculate the average speed for the entire journey.

12. For each of the following scenarios, calculate:
    i. the change in speed
    ii. the change in velocity.
    a. The driver of a car heading north along a freeway at 100 km h^{-1} slows down to 60 km h^{-1} as the traffic gets heavier.
    b. A fielder catches a cricket ball travelling towards him at 20 m s^{-1}.
    c. A tennis ball travelling at 25 m s^{-1} is returned directly back to the server at a speed of 30 m s^{-1}.

13. A car travelling east at a speed of 10 m s^{-1} turns right to head south at the same speed. Has the car undergone an acceleration? Explain your answer with the aid of a diagram.

14. Estimate the acceleration of a car, in m s^{-2}, as it resumes its journey through the suburbs after stopping at traffic lights.

15. Use the data in the table in question **6** to help you estimate the average acceleration of a world-class 100-metre sprinter at the beginning of a race.

8.2 Exam questions

Question 1 (1 mark)

MC Consider the following four scenarios for the motion of an object:

P. Velocity is positive and acceleration is positive.

Q. Velocity is positive and acceleration is negative.

R. Velocity is negative and acceleration is positive.

S. Velocity is negative and acceleration is negative.

Which of the following best describes the two scenarios in which the speed of the object is decreasing?
A. P and Q
B. Q and R
C. P and R
D. P and S

Question 2 (2 marks)

An object moving in a straight line has an initial displacement of +12 m followed by one of −20 m. The time interval is 4.0 s.

Calculate the average velocity.

Question 3 (2 marks)

An object travels at a constant velocity of 20 m s^{-1} north-west for 2.0 minutes.

Calculate the displacement.

Question 4 (2 marks)

A car, moving in a straight line at 10 m s^{-1}, slows down and comes to rest in a time of 5.0 s.

Calculate the average acceleration of the car.

Question 5 (2 marks)

A car is moving around a circular track at a constant speed.

Is the car accelerating or not? Explain your answer.

More exam questions are available in your learnON title.

8.3 Analysing motion graphically

The motion of an object often varies with time. To help analyse the motion of objects it can be very useful to construct graphs of the characteristics of that motion against time.

8.3.1 Position-versus-time graphs

Bolter Beryl and Steady Sam decide to race each other on foot over a distance of 100 metres. They run due west. Timekeepers are instructed to record the position of each runner after each 3-second interval.

TABLE 8.2 The progress of Bolter Beryl and Steady Sam

Time (seconds)	Position (distance from starting line) in metres	
	Bolter Beryl	**Steady Sam**
0.0	0	0
3.0	43	20
6.0	64	40
9.0	78	60
12.0	90	80
15.0	100	100

FIGURE 8.8 The position-versus-time graph of the race provides valuable information about the motion of the two runners.

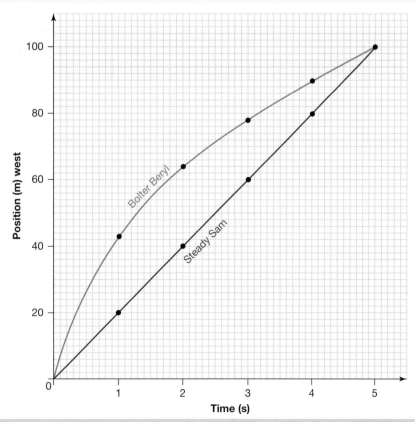

The points indicating Bolter Beryl's position after each 3-second interval are joined with a smooth curve. It is reasonable to assume that her velocity changes gradually throughout the race.

A number of observations can be made from the graph of position versus time.

- Both runners reach the finish at the same time. The result is a dead heat. Bolter Beryl and Steady Sam each have the same average speed and the same average velocity.
- Steady Sam, who has an exceptional talent for steady movement, maintains a constant velocity throughout the race. In fact, his instantaneous velocity at every instant throughout the race is the same as his average velocity. Steady Sam's average velocity and instantaneous velocity are both equal to the gradient of the position-versus-time graph since:

$$v_{av} = \frac{\Delta x}{\Delta t}$$

$$= \frac{100 \, m \, west}{15 \, s}$$

$$= 6.7 \, m \, s^{-1} \, west$$

$$= \frac{rise}{run}$$

$$= gradient$$

Steady Sam's velocity throughout the race is $6.7 \, m \, s^{-1}$ west. As Steady Sam's velocity is the same throughout, we refer to this as **uniform motion**.

- Bolter Beryl makes a flying start; however, after her initial 'burst', her instantaneous velocity decreases throughout the race as she tires. Her average velocity is also $6.7 \, m \, s^{-1}$ west.

A more detailed description of Bolter Beryl's motion can be given by calculating her average velocity during each 3-second interval of the race (see table 8.3). Bolter Beryl's velocity is changing throughout the race. Therefore, we refer to this as **non-uniform motion**.

TABLE 8.3 Bolter Beryl's changing velocity

Time interval (s)	Displacement Δx (m west)	Average velocity during interval $v_{av} = \frac{\Delta x}{\Delta t}$ ($m \, s^{-1}$ west)
0.0–3.0	43 − 0 = 43	14.0
3.0–6.0	64 − 43 = 21	7.0
6.0–9.0	78 − 64 = 14	4.7
9.0–12.0	90 − 78 = 12	4.0
12.0–15.0	100 − 90 = 10	3.3

The average velocity during each interval is the same as the gradient of the straight line joining the data points representing the beginning and end of the interval. An even more detailed description of Bolter Beryl's run could be obtained if the race was divided into, say, 100 time intervals. The average velocity during each time interval (and the gradient of the line joining the data points defining it) would be a very good estimate of the instantaneous velocity in the middle of the interval. In fact, if the race is progressively divided into smaller and smaller time intervals, the average velocity during each interval would become closer and closer to the instantaneous velocity in the middle of the interval.

uniform motion motion in which the velocity remains constant over time

non-uniform motion motion in which the velocity changes over time

Figure 8.9 shows how this process of using smaller time intervals can be used to find Bolter Beryl's instantaneous velocity at an instant 4 seconds from the start of the race.

FIGURE 8.9 The first 9 seconds of Bolter Beryl's run. Note that measurements were further divided into 1-second intervals compared to figure 8.8.

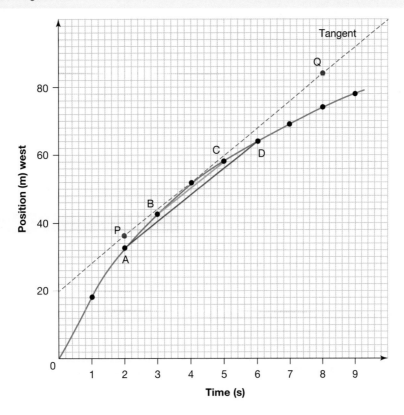

Bolter Beryl's instantaneous velocity is not the same as the average velocity during the 3-second to 6-second time interval shown in table 8.3. However, it can be estimated by drawing the line AD (shown in orange) and finding its gradient. The gradient of the line BC (shown in green) would provide an even better estimate of the instantaneous velocity. If you continue this process of decreasing the time interval used to estimate the instantaneous velocity, you will eventually obtain a line that is a tangent to the curve. The gradient of the tangent to the curve is equal to the instantaneous velocity at the instant represented by the point at which it meets the curve.

The gradient of the tangent to the curve at 4 seconds in figure 8.9 can be determined by using the points P and Q (shown in pink).

$$\text{Gradient} = \frac{\text{rise}}{\text{run}}$$

$$= \frac{(84 - 36)\,\text{m}}{(8 - 2)\,\text{s}}$$

$$= \frac{48\,\text{m}}{6\,\text{s}}$$

$$= 8\,\text{m}\,\text{s}^{-1}$$

Bolter Beryl's instantaneous velocity at 4 seconds from the start of the race is therefore 8 m s^{-1} west.

Just as the gradient of a position-versus-time graph can be used to determine the velocity of an object, a graph of distance versus time can be used to determine its speed. Because Bolter Beryl and Steady Sam were running in a straight line and in one direction only, their distance from the starting point is the magnitude of their change in position. Their speed is equal to the magnitude of their velocity.

The *gradient* of the tangent of a *distance-versus-time* graph gives the *instantaneous speed* of the object.

This gradient can be calculated as:

$$\frac{\text{rise (the change on the } y\text{-axis)}}{\text{run (the change on the } x\text{-axis)}}$$

on Resources

📄 **Digital document** eModelling: Numerical model of motion 1: finding speed from position–time data (doc-0048)

8.3.2 Velocity-versus-time graphs

The race between Bolter Beryl and Steady Sam described by the position-versus-time graph in figure 8.8 can also be described by a graph of velocity versus time. Steady Sam's velocity is 6.7 m s⁻¹ due west throughout the race. The curve describing Bolter Beryl's motion can be plotted by determining the instantaneous velocity at various times during the race. This can be done by drawing tangents at a number of points on the position-versus-time graph in figure 8.9. Table 8.4 shows the data obtained using this method. The velocity-versus-time graph in figure 8.10 describes the motion of Bolter Beryl and Steady Sam.

TABLE 8.4 Bolter Beryl's velocity during the race

Time (s)	Velocity (m s⁻¹ west)
0.0	18.0
2.0	12.0
4.0	8.0
6.0	5.4
8.0	4.7
10.0	4.2
12.0	3.5
14.0	3.1

FIGURE 8.10 Velocity-versus-time graph for the race

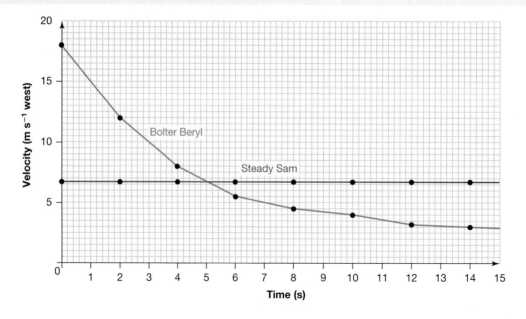

The velocity-versus-time graph confirms what you already knew by looking at the position-versus-time graph, namely that:

- Steady Sam's velocity is constant and equal to his average velocity
- the magnitude of Bolter Beryl's velocity is decreasing throughout the race.

The velocity-versus-time graph allows you to estimate the velocity of each runner at any time. It provides a much clearer picture of the way that Bolter Beryl's velocity changes during the race, namely that:

- the magnitude of her velocity decreases rapidly at first, but less rapidly towards the end of the race
- for most of the duration of the race, she is running more slowly than Steady Sam. In fact, Bolter Beryl's speed (the magnitude of her velocity) drops below that of Steady Sam's after only 4.7 seconds.

Displacement from a velocity-versus-time graph

In the absence of a position-versus-time graph, a velocity-versus-time graph provides useful information about the change in position, or displacement, of an object. Steady Sam's constant velocity, the same as his average velocity, makes it very easy to determine his displacement during the race.

$$\Delta x = v_{av}\Delta t \ \left(\text{since} \ v_{av} = \frac{\Delta x}{\Delta t} \right)$$
$$= 6.7 \text{ m s}^{-1} \text{ west} \times 15 \text{ s}$$
$$= 100 \text{ m west}$$

This displacement is equal to the area of the rectangle under the graph depicting Steady Sam's motion.

$$\text{Area} = \text{length} \times \text{width}$$
$$= 15 \text{ s} \times 6.7 \text{ m s}^{-1} \text{ west}$$
$$= 100 \text{ m west}$$

In fact, the area under any part of the velocity-versus-time graph is equal to the displacement during the interval represented by that part.

FIGURE 8.11 The displacement of Steady Sam between 0 and 6 seconds is equal to the area under the velocity-versus-time graph. This is 40.2 m in this case.

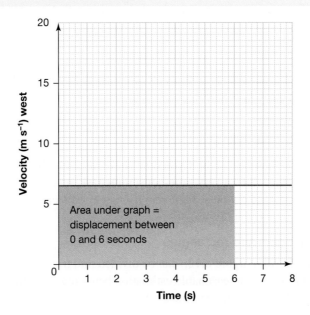

The *area* under a *velocity-versus-time* graph gives the *displacement* (or change in position) of the object during that time interval. To determine the position of the object you must know its starting position.

Note: If the velocity-versus-time graph is in the negative velocity range, you can get a negative value for the area. Although this may seem counterintuitive, it indicates a negative displacement. With a velocity-versus-time graph and an object moving forward, the displacement is added cumulatively. Thus, to calculate the displacement for the fifth second, for instance, you need to subtract the area from 0–4 seconds to the area for 0–5 seconds.

The *area* under a *speed-versus-time* graph gives the *distance* travelled by the object during that time interval.

Because the race was a dead heat, Bolter Beryl's average velocity was also 6.7 m s^{-1}. Her displacement during the race can be calculated in the same way as Steady Sam's:

$$\Delta x = v_{av}\Delta t$$
$$= 6.7\,\mathrm{m\,s^{-1}} \times 15\,\mathrm{s}$$
$$= 100\,\mathrm{m\ west}$$

However, Bolter Beryl's displacement can also be found by calculating the area under the velocity-versus-time graph depicting her motion.

There are two methods by which this can be done.

Counting squares to determine area

The number of squares under the velocity-versus-time graph can help determine the displacement. In this case, each individual square under the velocity-versus-time graph can be counted. Two half squares can be combined to make one full square.

The displacement can then be determined by the following formulae:

Displacement = number of squares × height of one square × width of one square

as the area of one square can be determined using:

area of one square = its height × its width

So for example, in figure 8.10, each square is 0.5 m s^{-1} in height and 0.2 s in width. If you count the number of squares underneath the curve between 0 and 15 seconds, approximately 1000 squares can be counted.

Therefore, the displacement between 0 and 15 seconds is equal to:

$$\Delta x = \text{number of squares} \times \text{height of one square} \times \text{width of one square}$$
$$= 1000 \times 0.5\ \mathrm{m\,s^{-1}} \times 0.2\ \mathrm{s}$$
$$= 100\ \mathrm{m}$$

The area under Beryl's velocity-versus-time graph is, not surprisingly, 100 metres.

The counting squares method can be very useful if you don't have large numbers of squares to count. In the above case, counting 1000 squares will be a very time-consuming task, and often involves some level of estimation. In this case, we can instead divide a shape into triangles and rectangles and calculate the area of each of these.

Dividing the shape into triangles and rectangles

Many curves, such as the one shown in figure 8.12, can be divided into triangles and rectangles (as shown between 2 and 4 seconds). This leads to a reasonable estimation being made for the displacement during that time by calculating the area of these triangles and rectangles.

FIGURE 8.12 The area under a velocity-versus-time graph where velocity is not constant can be more easily calculated by dividing the shape into rectangles and triangles.

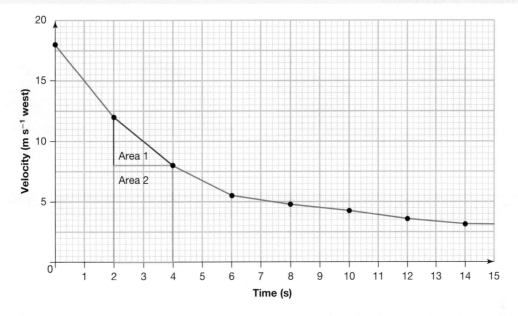

In figure 8.12, for example, we can determine the displacement of the individual between 2 and 4 seconds using the triangle and rectangle shown.

$$\text{Area 1 (triangle)} = \frac{bh}{2}$$
$$= \frac{2\,\text{s} \times 4\,\text{m s}^{-1}}{2}$$
$$= 4\,\text{m}$$

$$\text{Area 2 (rectangle)} = lw$$
$$= 2\,\text{s} \times 8\,\text{m s}^{-1}$$
$$= 16\,\text{m}$$

$$\text{Total displacement} = 4 + 16$$
$$= 20\,\text{m}$$

If, for example, this was done using numerous triangles and rectangles under the curve between 0 and 15 seconds, a value close to 100 m would be calculated.

EXTENSION: Calculating the area under a graph with time intervals

When an object travels with a constant velocity, it is obvious that the displacement of the object is equal to the area under a velocity-versus-time graph of its motion. However, it is not so obvious when the motion is not constant. The graphs in figure 8.13 describe the motion of an object that has an increasing velocity. The motion of the object can be approximated by dividing it into time intervals of Δt and assuming that the velocity during each time interval is constant. The approximate displacement during each time interval is equal to $\Delta x = v_{av}\Delta t$, which is the same as the area under each rectangle. The approximate total displacement is therefore equal to the total area of the rectangles.

To better approximate the displacement, the graph can be divided into smaller time intervals. The total area of the rectangles is approximately equal to the displacement. By dividing the graph into even smaller time intervals, even better estimates of the displacement can be made. In fact, by continuing the process of dividing the graph into smaller and smaller time intervals, it can be seen that the displacement is, in fact, equal to the area under the graph.

FIGURE 8.13 By dividing the velocity-versus-time graph into rectangles representing small time intervals, the displacement can be estimated.

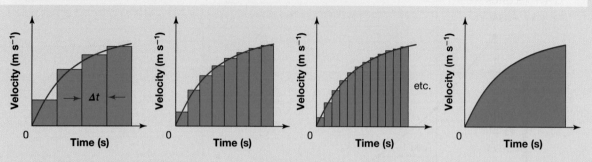

tlvd-3782

SAMPLE PROBLEM 5 Comparing displacement using velocity-versus-time graphs

At what time in the race between Bolter Beryl and Steady Sam did Bolter Beryl's speed drop below Steady Sam's speed? What is Steady Sam's displacement at this point in time?

THINK	WRITE
1. See the velocity-versus-time graph (figure 8.10) to determine when Bolter Beryl's speed dropped below Steady Sam's.	The time at which Bolter Beryl's speed dropped below Steady Sam's is approximately 4.7 seconds.
2. Determine Steady Sam's displacement by calculating the area under the graph.	Area under the graph $= 4.7 \times 6.7$ $\qquad\qquad\qquad\quad = 31\,\text{m}$ Steady Sam's displacement was 31 metres at 4.7 seconds.

PRACTICE PROBLEM 5

The following graph is of a section of figure 8.10 showing the first 4.7 seconds of Bolter Beryl's motion.

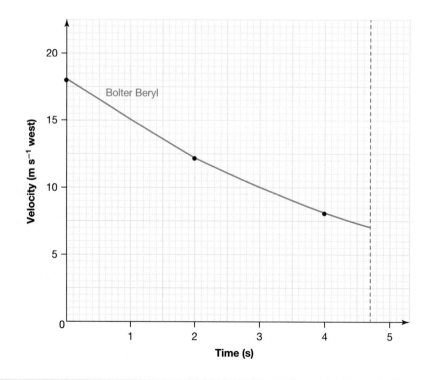

a. **Estimate Bolter Beryl's displacement after 2 seconds.**
b. **At 4.7 seconds into the race, determine how far ahead Bolter Beryl was of Steady Sam (whose velocity was a constant 6.7 m s⁻¹).**

Acceleration from a velocity-versus-time graph

The graph in figure 8.14 describes the motion of an elevator as it moves from the ground floor of a building to the top floor and back down again. The elevator stops briefly at the top floor to pick up a passenger. In this graph, upward displacement from the ground floor is defined as positive. The graph has been divided into seven sections labelled A to G.

FIGURE 8.14 The motion of an elevator

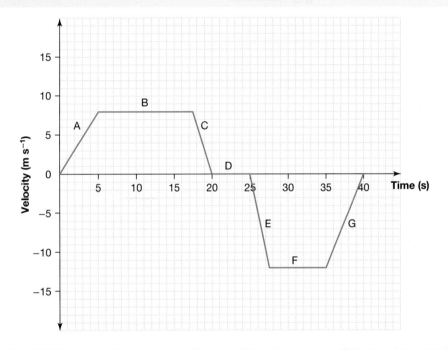

Finding the acceleration

The acceleration at any instant during the motion can be determined by calculating the gradient of the graph. This is a consequence of the definition of acceleration. The gradient of a velocity-versus-time graph is a measure of the rate of change of velocity just as the gradient of a position-versus-time graph is a measure of the rate of change of position.

> The *gradient* of a velocity-versus-time graph gives the *acceleration* of the object.

Throughout interval A (see figure 8.14) the acceleration, a, of the elevator is:

$$a = \frac{\text{rise}}{\text{run}}$$
$$= \frac{+8.0\,\text{m s}^{-1}}{5.0\,\text{s}}$$
$$= +1.6\,\text{m s}^{-2} \text{ or } 1.6\,\text{m s}^{-2} \text{ up}$$

During intervals B, D and F, the velocity is constant and the gradient of the graph is zero. The acceleration during each of these intervals is, therefore, zero.

Throughout interval C, the acceleration is:

$$a = \frac{-8.0 \text{ m s}^{-1}}{2.5 \text{ s}}$$
$$= -3.2 \text{ m s}^{-2} \text{ or } 3.2 \text{ m s}^{-2} \text{ down}$$

Throughout interval E, the acceleration is:

$$a = \frac{-12 \text{ m s}^{-1}}{2.5 \text{ s}}$$
$$= -4.8 \text{ m s}^{-2} \text{ or } 4.8 \text{ m s}^{-2} \text{ down}$$

Throughout interval G, the acceleration is:

$$a = \frac{+12 \text{ m s}^{-1}}{5.0 \text{ s}}$$
$$= +2.4 \text{ m s}^{-2} \text{ or } 2.4 \text{ m s}^{-2} \text{ up}$$

Notice that during interval G the acceleration is positive (up) while the velocity of the elevator is negative (down). The direction of the acceleration is the same as the direction of the *change* in velocity.

Finding the displacement

As explained previously, the area under the graph is equal to the displacement of the elevator. Dividing the area into triangles and rectangles and working from left to right yields an area of:

$$\text{Area} = \left(\frac{1}{2} \times 5.0 \text{ s} \times 8.0 \text{ m s}^{-1} \right) + \left(12.5 \text{ s} \times 8.0 \text{ m s}^{-1} \right) + \left(\frac{1}{2} \times 2.5 \text{ s} \times 8.0 \text{ m s}^{-1} \right) +$$
$$\left(\frac{1}{2} \times 2.5 \text{ s} \times -12 \text{ m s}^{-1} \right) + \left(7.5 \text{ s} \times 12 \text{ m s}^{-1} \right) + \left(\frac{1}{2} \times 5.0 \text{ s} \times -12 \text{ m s}^{-1} \right)$$
$$= 20 \text{ m} + 100 \text{ m} + 10 \text{ m} - 15 \text{ m} - 90 \text{ m} - 30 \text{ m}$$
$$= -5 \text{ m}$$

This represents a downwards displacement of 5 metres, which is consistent with the elevator finally stopping two floors below the ground floor.

elog-1664

tlvd-0820

INVESTIGATION 8.2   on line only

Let's play around with some graphs

Aim

To demonstrate the motion represented by position-versus-time graphs and velocity-versus-time graphs

---

on Resources

Digital documents eModelling: Numerical model of motion 2: Finding position from speed–time data (doc-0049)
eModelling: Numerical model for acceleration (doc-0050)

Video eLessons Motion with constant acceleration (eles-0030)
Ball toss (eles-0031)

8.3.3 Acceleration-versus-time graphs

The graph in figure 8.15 represents the acceleration of the elevator described in the previous section.

FIGURE 8.15 An acceleration-versus-time graph for the elevator

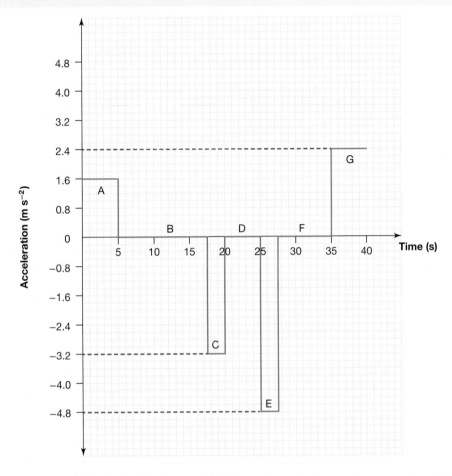

The *area* under an *acceleration-versus-time* graph gives the *change in velocity* of the object during that time interval. To determine the velocity of the object you must know its starting velocity.

Just as the area under a velocity-versus-time graph is equal to the change in position of an object, the area under an acceleration-versus-time graph is equal to the change in velocity of an object. The area under the part of the graph representing the entire upwards part of the journey (between 0 and 20 m) can be calculated to determine the velocity over this time.

This area can be divided into three further areas:
- Area A — from 0 to 5 seconds, where the acceleration is 1.6 m s^{-2}
- Area B — from 5 to 17.5 seconds, where the acceleration is 0 m s^{-2}
- Area C — from 17.5 to 20 seconds, where the acceleration is −3.2 m s^{-2}.

Therefore, the velocity for the upwards journey is given by:

$$\text{Area A} + \text{Area B} + \text{Area C} = \left(5.0\,\text{s} \times 1.6\,\text{m s}^{-2}\right) + 0 + \left(2.5\,\text{s} \times -3.2\,\text{m s}^{-2}\right)$$
$$= +8.0\,\text{m s}^{-1} + -8.0\,\text{m s}^{-1}$$
$$= 0$$

This indicates that change in velocity during the upward journey is zero. This is consistent with the fact that the elevator starts from rest and is at rest when it reaches the top floor. Similarly, the area under the whole graph is zero.

The change in velocity during intervals C, D and E is given by the sum of areas C, D and E. Thus:

$$\text{Area C} + \text{Area D} + \text{Area E} = \left(2.5\,\text{s} \times -3.2\,\text{m s}^{-2}\right) + 0 + \left(2.5\,\text{s} \times -4.8\,\text{m s}^{-2}\right)$$
$$= -8.0\,\text{m s}^{-1} + -12\,\text{m s}^{-1}$$
$$= -20\,\text{m s}^{-1}$$

The change in velocity is -20 m s^{-1} or 20 m s^{-1} down.

At the beginning of time interval C the velocity was 8 m s^{-1} upwards. A change of velocity of -20 m s^{-1} would result in a velocity at the end of interval E of 12 m s^{-1} downwards. This is consistent with the description of the motion in the velocity-versus-time graph in figure 8.10.

The change in velocity during interval G is 5 s × 2.4 m s^{-1}, or 12 m s^{-1} upwards, which is consistent with the elevator coming to rest at the end of its journey.

8.3.4 Working with motion graphs

Position-versus-time graphs
- The instantaneous velocity of an object can be obtained from a graph of the object's position versus time by determining the gradient of the curve at the point representing that instant. This is a direct consequence of the fact that velocity is a measure of the rate of change of position.
- Similarly, the instantaneous speed of an object can be obtained by determining the gradient of a graph of the object's distance travelled from a reference point versus time.

Velocity-versus-time graphs
- The displacement of an object during a time interval can be obtained by determining the area under the velocity-versus-time graph representing that time interval. The actual position of an object at any instant during the time interval can be found only if the starting position is known.
- Similarly, the distance travelled by an object during a time interval can be obtained by determining the corresponding area under the speed-versus-time graph for the object.
- The instantaneous acceleration of an object can be obtained from a graph of the object's velocity versus time by determining the gradient of the curve at the point representing that instant. This is a direct consequence of the fact that acceleration is defined as the rate of change of velocity.

Acceleration-versus-time graphs
- The change in velocity of an object during a time interval can be obtained by determining the area under the acceleration-versus-time graph representing that time interval. The actual velocity of the object can be found at any instant during the time interval only if the initial velocity is known.

TABLE 8.5 Summary of motion graphs

		Position-versus-time graphs	Velocity-versus-time graphs	Acceleration-versus-time graphs
Quantities that you can read directly from the graph	Horizontal axis	Time	Time	Time
	Vertical axis	Position	Velocity	Acceleration
Quantities that you can calculate from the graph	Gradient of tangent	Instantaneous velocity	Instantaneous acceleration	
	Area under the graph		Change in position (displacement)	Change in velocity

elog-1665

On your bike or on your own two feet

Aim

To record and analyse the motion of a bicycle or runner over a distance of 100 metres on a straight track

8.3 Activities   **learn on**

8.3 Quick quiz on	**8.3 Exercise**	**8.3 Exam questions**

8.3 Exercise

1. The position-versus-time graph shown describes the motion of five different objects, labelled A to E.

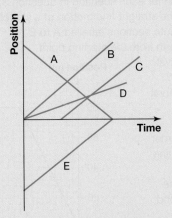

   a. Which two objects start from the same position, but at different times?
   b. Which two objects start at the same position at the same time?
   c. Which two objects are travelling at the same speed as each other, but with different velocities?
   d. Which two objects are moving towards each other for the whole period shown on the graph?
   e. Which of the five objects has the lowest speed?

2. **a.** Describe in words the motion shown for each of scenarios A, B and C shown in the following figure.

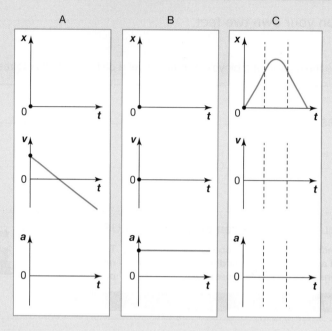

   b. Copy the incomplete graphs for each scenario into your workbooks and then complete each graph.
3. Sketch a velocity-versus-time graph to illustrate the motion described in each of the following situations.
   a. A bicycle is pedalled steadily along a road. The cyclist stops pedalling and allows the bicycle to come to a stop.
   b. A parachutist jumps out of a plane and opens his parachute midway through the fall to the ground.
   c. A ball is thrown straight up into the air and is caught at the same height from which it was thrown.
4. Sketch a position-versus-time graph for each scenario in question **3**.
5. The following graph is a record of the straight-line motion of a skateboarder during an 80-second time interval. The time interval has been divided into sections labelled A to E.
   The skateboarder initially moves north from the starting point.

   a. During which section of the interval was the skateboarder stationary?
   b. During which sections of the interval was the skateboarder travelling north?
   c. At what instant did the skateboarder first move back towards the starting line?
   d. What was the displacement of the skateboarder during the 80-second interval?
   e. What distance did the skateboarder travel during the 80-second interval?
   f. During which section of the interval was the skateboarder speeding up?
   g. During which section of the interval was the skateboarder slowing down?

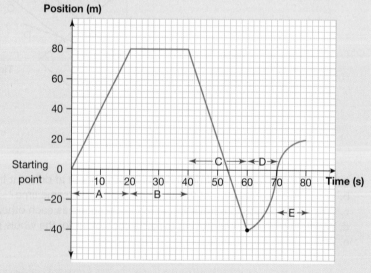

   h. What was the skateboarder's average speed during the entire 80-second interval?
   i. What was the velocity of the skateboarder throughout section C?
   j. Estimate the velocity of the skateboarder 65 seconds into the interval.

6. The following graph is a record of the motion of a battery-operated toy robot during an 80-second time interval. The interval has been divided into sections labelled A to G.

a. During which sections is the acceleration of the toy robot zero?

b. What is the displacement of the toy robot during the 80-second interval?

c. What is the average velocity of the toy robot during the entire interval?

d. At what instant did the toy robot first reverse direction?

e. At what instant did the toy robot first return to its starting point?

f. During which intervals did the toy robot have a negative acceleration?

g. During which intervals did the toy robot decrease its speed?

h. Explain why your answers to parts f and g are different from each other.

i. What is the acceleration of the toy robot throughout section E?

j. What is the average acceleration during the first 20 seconds?

k. Describe the motion of the toy robot in words.

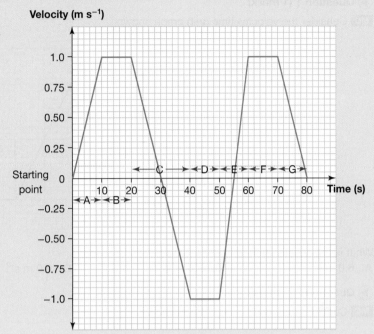

7. The following graph compares the straight-line motion of a jet ski and a car as they each accelerate from an initial speed of 5 m s⁻¹.

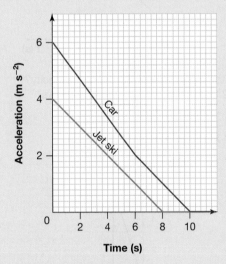

a. Which is first to reach a constant speed — the jet ski or the car — and when does this occur?

b. What is the final speed of:
   i. the jet ski
   ii. the car?

c. Draw a speed-versus-time graph describing the motion of either the jet ski or the car.

8.3 Exam questions

▶ Question 1 (1 mark)

MC Consider the velocity–time (*v*–*t*) graph shown.

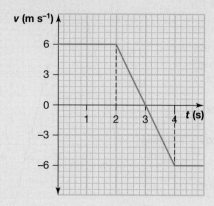

What is the acceleration at time $t = 1.0$ s?

A. 6.0 m s^{-2} **B.** 3.0 m s^{-2} **C.** 2.0 m s^{-2} **D.** 0

▶ Question 2 (1 mark)

MC Consider the position–time graph shown.

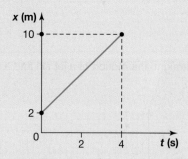

Which of the following best gives the magnitude of the average velocity?

A. 2.0 m s^{-1} **B.** 2.25 m s^{-1} **C.** 2.5 m s^{-1} **D.** 3.0 m s^{-1}

▶ Question 3 (3 marks)

The velocity–time graph shows an idealised motion for an object.

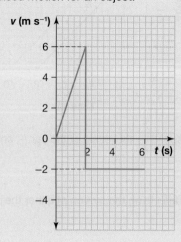

Calculate the total displacement for the first 4.0 s.

Question 4 (4 marks)

Consider the position–time (x–t) graph shown.

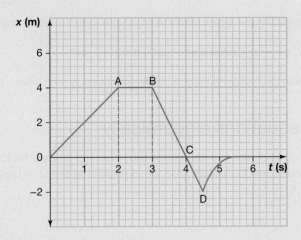

a. What is the instantaneous velocity at the time $t = 4$ s? Explain your answer. **(2 marks)**

b. During which one-second time interval (e.g. 0–1 s, 1–2 s, etc.) does the maximum positive instantaneous velocity occur? Explain your answer. **(2 marks)**

Question 5 (6 marks)

Consider the velocity–time (v–t) graph shown.

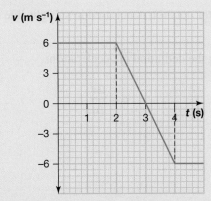

a. Calculate the displacement for the interval $t = 2$ s to $t = 4$ s. Show your working. **(3 marks)**

b. Find the average velocity for the first 3.0 s of the motion ($t = 0$ to $t = 3$ s). Show your working. **(3 marks)**

More exam questions are available in your learnON title.

8.4 Equations for constant acceleration

8.4.1 Deriving the equations algebraically

In the absence of a graphical representation, a number of formulae can be used to describe straight-line motion as long as the acceleration is constant. These formulae are expressed in terms of the quantities used to describe such motion. The terms are:

- initial velocity, u
- final velocity, v
- acceleration, a
- time interval, t
- displacement, s.

Displacement, velocity and acceleration are all vector quantities. The direction of their vectors can be expressed simply as negative or positive quantities, or by a specific direction such as south, up or left.

Note that, in some resources, this vector notation is not used in straight-line motion, but it has been included in this section for consistency.

The first formula is found by restating the definition of acceleration:

$$a = \frac{\Delta v}{\Delta t}$$

where:
Δv = the change in velocity
Δt = the time interval.

Thus:

$$a = \frac{v - u}{t}$$
$$\Rightarrow \ v - u = at$$
$$\Rightarrow \ v = u + at \qquad\qquad [1]$$

The second formula is found by restating the definition of average velocity:

$$v_{av} = \frac{\Delta s}{\Delta t}$$

where:
Δs = the change in position.

But:

$$v_{av} = \frac{u + v}{2}$$

Thus:

$$\frac{u+v}{2} = \frac{s}{t}$$

$$\Rightarrow s = \frac{1}{2}(u+v)t \qquad [2]$$

Three more formulae are obtained by combining formulae [1] and [2]:

$$s = \frac{1}{2}(u+u+at)t \qquad \text{(substituting } v = u + at \text{ from formula [1] into formula [2])}$$

$$= \frac{1}{2}(2u+at)t$$

$$\qquad\qquad\qquad\qquad\qquad\qquad\qquad\qquad [3]$$

$$= \left(u + \frac{1}{2}at\right)t$$

$$\Rightarrow s = ut + \frac{1}{2}at^2$$

$$s = \frac{1}{2}(v-at+v)t \qquad \text{(substituting } u = v - at \text{ from formula [1] into formula [2])}$$

$$= \frac{1}{2}(2v-at)t$$

$$= \left(v - \frac{1}{2}at\right)t$$

$$\Rightarrow s = vt - \frac{1}{2}at^2 \qquad [4]$$

A final formula can be found by eliminating t from formula [2]:

$$s = \frac{1}{2}(u+v)t \qquad \text{(formula [2])}$$

But:

$$t = \frac{v-u}{a} \qquad \text{(rearranging formula [1])}$$

$$\Rightarrow s = \frac{1}{2}(u+v)\left(\frac{v-u}{a}\right)$$

$$= \frac{1}{2}\left(\frac{v^2-u^2}{a}\right) \qquad \text{(expanding the difference of two squares)}$$

$$\Rightarrow 2as = v^2 - u^2$$

$$\Rightarrow v^2 = u^2 + 2as \qquad [5]$$

In summary, the formulae for straight-line motion under constant acceleration are:

$$v = u + at \qquad [1]$$

$$s = \frac{1}{2}(u + v)t \qquad [2]$$

$$s = ut + \frac{1}{2}at^2 \qquad [3]$$

$$s = vt - \frac{1}{2}at^2 \qquad [4]$$

$$v^2 = u^2 + 2as \qquad [5]$$

8.4.2 Deriving the equations graphically

Each of the five formulae derived here allow you to determine an unknown characteristic of straight-line motion with a constant acceleration as long as you know three other characteristics. Although the formulae have not been derived from graphs, they are entirely consistent with a graphical approach.

Acceleration = gradient of velocity-versus-time graph

$$= \frac{\text{rise}}{\text{run}}$$

$$= \frac{v - u}{t}$$

$$\Rightarrow a = \frac{v - u}{t}$$

$$\Rightarrow v = u + at \qquad [1]$$

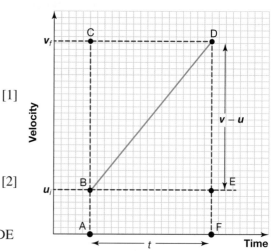

FIGURE 8.16 A velocity-versus-time graph for an object travelling in a straight line with constant acceleration

Displacement = area under graph
= area of trapezium ABDF

$$s = \frac{1}{2}(u + v)t \qquad [2]$$

Displacement = area under graph
= area of rectangle ABEF + area of triangle BDE

$$= ut + \frac{1}{2}t \times at \qquad (v - u = at \text{ from } [1]) \qquad [3]$$

$$s = ut + \frac{1}{2}at^2$$

Displacement = area under graph
= area of rectangle ACDF − area of triangle BCD

$$= vt - \frac{1}{2} \times t \times at \qquad (v - u = at \text{ from } [1])$$

$$s = vt - \frac{1}{2}at^2$$

Formula [5] can be derived by combining formula [1] with any of formulae [2], [3] or [4].

8.4.3 Problem-solving steps

The following steps may help you when solving problems using the constant acceleration equations.

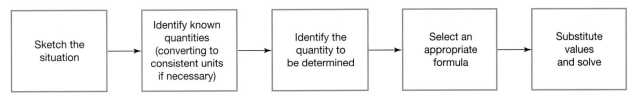

It is possible to rearrange each of the equations to make different variables the subject. Table 8.6 summarises all possible versions of the equations and may be useful when solving problems.

TABLE 8.6 Equations for solving problems

		Variables that are involved in the problem				
		u v a t	*u v a s*	*u v t s*	*u a t s*	*v a t s*
Variable that is to be calculated	**u**	$u = v - at$	$u^2 = v^2 - 2as$	$u = \dfrac{2s}{t} - v$	$u = \dfrac{s}{t} - \dfrac{at}{2}$	–
	v	$v = u + at$	$v^2 = u^2 + 2as$	$v = \dfrac{2s}{t} - u$	–	$v = \dfrac{s}{t} + \dfrac{at}{2}$
	a	$a = \dfrac{v - u}{t}$	$a = \dfrac{v^2 - u^2}{2s}$	–	$a = \dfrac{2(s - ut)}{t^2}$	$a = \dfrac{2(vt - s)}{t^2}$
	t	$t = \dfrac{v - u}{a}$	–	$t = \dfrac{2s}{(u + v)}$	Find **v** then solve	Find **u** then solve
	s	–	$s = \dfrac{v^2 - u^2}{2a}$	$s = \dfrac{1}{2}(u + v)t$	$s = ut + \dfrac{1}{2}at^2$	$s = vt - \dfrac{1}{2}at^2$

tlvd-0074

SAMPLE PROBLEM 6 Calculating velocity using speed and time

Ying drops a coin into a wishing well and takes 3.0 seconds to make a wish. The coin splashes into the water just as she finishes making her wish. The coin accelerates towards the water at a constant 9.8 m s^{-2}.

a. What is the coin's velocity as it strikes the water?
b. How far does the coin fall before hitting the water?

THINK	WRITE
a. 1. Recall the appropriate constant acceleration formula.	a. $v = u + at$
2. Identify the values.	$u = 0$, $a = 9.8 \,\mathrm{m\,s^{-2}}$, $t = 3.0 \,\mathrm{s}$
3. Substitute into the formula to find the final velocity.	$v = 0 + 9.8 \times 3.0 = 29 \,\mathrm{m\,s^{-1}}$
4. State the solution.	The coin had a velocity of 29 m s^{-1} when it hit the water.
b. 1. Recall the appropriate constant acceleration formula.	b. $s = ut + \dfrac{1}{2}at^2$
2. Identify the values.	$u = 0$, $a = 9.8 \,\mathrm{m\,s^{-2}}$, $t = 3.0 \,\mathrm{s}$
3. Substitute into the formula to find the distance.	$s = 0 \times 3.0 + \dfrac{1}{2} \times 9.8 \times 3.0^2 = 44 \,\mathrm{m}$
4. State the solution.	The coin fell 44 metres before hitting the water.

PRACTICE PROBLEM 6

A parked car with the handbrake off rolls down a hill in a straight line with a constant acceleration of 2 m s^{-2}. It stops after colliding with a brick wall at a speed of 12 m s^{-1}.
a. For how long was the car rolling?
b. How far did the car roll before colliding with the wall?

 Resources

📄 **Digital document**   eModelling: Solving problems with a graphics calculator (doc-0051)

tlvd-0075

SAMPLE PROBLEM 7 Calculating speed and acceleration using distance and time

The driver of a car was forced to brake in order to prevent serious injury to a neighbour's cat. The car skidded in a straight line, stopping just 2 centimetres short of the startled but lucky cat. The driver (who happened to be a Physics teacher) measured the length of the skid mark to be 12 metres. His passenger (also a Physics teacher with an exceptional skill for estimating small time intervals) estimated that the car skidded for 2 seconds.
a. At what speed was the car travelling as it began to skid?
b. What was the acceleration of the car during the skid?

THINK	WRITE
a. 1. Recall the appropriate constant acceleration formula.	a. $s = \dfrac{1}{2}(u+v)t$
2. Identify the values.	$v = 0,\ t = 2\,\text{s},\ s = 12\,\text{m}$
3. Substitute into the relationship to find the initial velocity.	$12 = \dfrac{1}{2}(u+0) \times 2$ $\Rightarrow u = 12\,\text{m s}^{-1}$
4. State the solution.	The initial speed of the car was 12 m s^{-1}.
b. 1. Recall the appropriate constant acceleration formula.	b. $s = vt - \dfrac{1}{2}at^2$
2. Identify the values.	$v = 0,\ t = 2\,\text{s},\ s = 12\,\text{m}$
3. Substitute into the relationship to find the acceleration.	$12 = 0 \times 2 - \dfrac{1}{2}a \times 2^2$ $12 = -2a$ $\Rightarrow a = -6\,\text{m s}^{-2}$
4. State the solution.	The acceleration during the skid was −6 m s^{-2}; that is, the car was decelerating at 6 m s^{-2} during the skid.

PRACTICE PROBLEM 7

A car travelling at 24 m s^{-1} brakes to come to a stop in 1.5 seconds. If its acceleration was constant, what was the car's:
a. **stopping distance**
b. **acceleration?**

It is worth noting that sample problems 6 and 7 could both have been solved without the use of the constant acceleration formulae. Both examples could have been completed with a graphical approach and a clear understanding of the definitions of velocity and acceleration. Go ahead and try to answer both problems without the formulae.

8.4 Activities

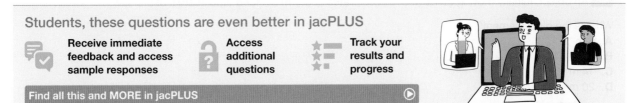
8.4 Quick quiz on	8.4 Exercise	8.4 Exam questions

8.4 Exercise

1. How long does it take for:
   a. a car to accelerate on a straight road at a constant 6 m s^{-2} from an initial speed of 60 km h^{-1} (17 m s^{-1}) to a final speed of 100 km h^{-1} (28 m s^{-1})
   b. a downhill skier to accelerate from rest at a constant 2 m s^{-2} to a speed of 10 m s^{-1}?

2. In Acapulco, on the coast of Mexico, professional high divers plunge from a height of 36 metres above the water. (The highest diving boards used in Olympic diving events are 10 metres above the water.) Assuming that throughout their dive, the divers are falling vertically from rest with an acceleration of 10 m s^{-2}, estimate:
   a. the length of the time interval during which the divers fall through the air
   b. the speed with which the divers enter the water.

3. A skateboarder travelling down a hill notices the busy road ahead and comes to a stop in 2.0 seconds over a distance of 12 metres. Assume a constant negative acceleration.
   a. What was the initial speed of the skateboarder?
   b. What was the acceleration of the skateboarder as they came to a stop?

4. A car is travelling at a speed of 100 km h^{-1} when the driver sees a large fallen tree branch in front of her. At the instant that she sees the branch, it is 50 metres from the front of her car. The car travels a distance of 48 metres *after* the brakes are applied before coming to a stop.
   a. What is the average acceleration of the car while the car is braking?
   b. How long does the car take to stop once the brakes are applied?
   c. What other information do you need in order to determine whether the car stops before it hits the branch? Make an estimate of the missing item of information to predict whether or not the car is able to stop in time.

5. A dancer in a school musical is asked to leap 80 centimetres into the air, taking off vertically on one beat of the music and landing with the next beat. If the music beats every 0.5 seconds, is the leap possible? The acceleration of the dancer during the leap can be assumed to be 10 m s^{-2} downwards.

6. A brand-new Rolls Royce rolls off the back of a truck as it is being delivered to its owner. The truck is travelling along a straight road at a constant speed of 60 km h^{-1}. The Rolls Royce slows down at a constant rate, coming to a stop over a distance of 240 metres. It is a full minute before the truck driver realises that the precious load is missing. The driver brakes immediately, leaving a 25-metre skid mark on the road. The driver's reaction time (the time interval between noticing the problem and depressing the brake) is 0.5 seconds. How far behind is the Rolls Royce when the truck stops?

7. A girl at the bottom of a 100-metre high cliff throws a tennis ball vertically upwards. At the same instant, a boy at the very top of the cliff drops a golf ball so that it hits the tennis ball while both balls are still in the air. The acceleration of both balls can be assumed to be 10 m s^{-2} downwards.
   a. With what speed is the tennis ball thrown so that the golf ball strikes it at the top of its path?
   b. What is the position of the tennis ball when the golf ball strikes it?

8.4 Exam questions

▶ Question 1 (1 mark)

MC An object, initially moving at 4.0 m s^{-1}, accelerates at 3.0 m s^{-2} for a time interval of 2.0 s.

What is the displacement?
A. 6 m
B. 12 m
C. 14 m
D. 20 m

▶ Question 2 (1 mark)

MC An object, initially moving at 3.0 m s^{-1}, accelerates at 4.0 m s^{-2} for a time interval of 5.0 s.

What is the final speed?
A. 23 m s^{-1}
B. 20 m s^{-1}
C. 17 m s^{-1}
D. 13 m s^{-1}

▶ Question 3 (2 marks)

A car is moving at 20 m s^{-1} in a straight line. It then brakes smoothly to a stop in a displacement of 40 m.

Calculate the acceleration of the car.

▶ Question 4 (3 marks)

A small ball is projected vertically upward with an unspecified initial velocity. The acceleration of the ball is 9.8 m s^{-2} vertically downward. The ball reaches its highest point after 2.0 s.

Calculate the maximum height reached. Show your working. (*Hint*: This can be found using just one equation.)

▶ Question 5 (9 marks)

An object has an initial velocity of 6.0 m s^{-1}. It then accelerates at a constant rate for 4.0 s, reaching a final velocity of 14 m s^{-1}.
   a. What is the magnitude of the acceleration? **(1 mark)**
   b. What is the displacement? **(2 marks)**
   c. What is the average velocity? **(1 mark)**
   d. While travelling at 14 m s^{-1} the object is then subjected to a constant braking effect that brings it to rest in a distance of 24.5 m.

   What is the acceleration while braking? **(3 marks)**
   e. While travelling at 14 m s^{-1} the object is then subjected to a constant braking effect that brings it to rest in a distance of 24.5 m.

   What is the time taken for the object to come to rest? **(2 marks)**

More exam questions are available in your learnON title.

8.5 Review

8.5.1 Topic summary

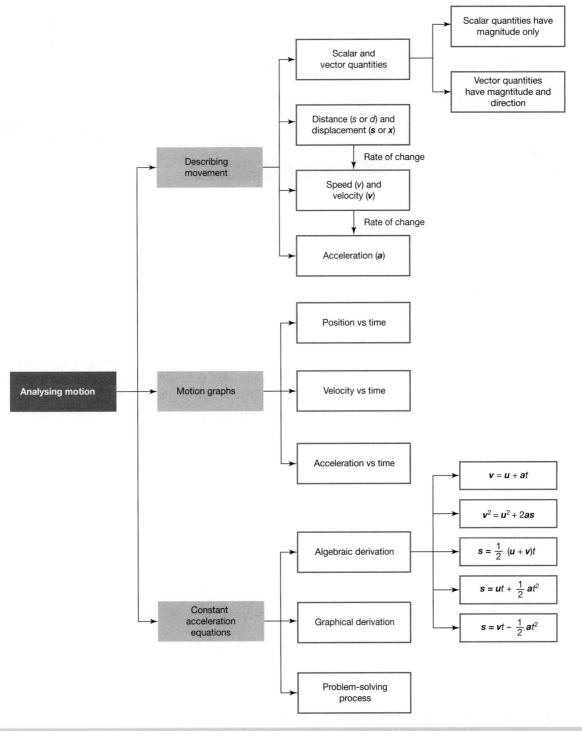

8.5.2 Key ideas summary

8.5.3 Key terms glossary

8.5 Activities

learn on

8.5 Review questions

1. A student is doing a 'shuttle run' activity where they run backwards and forwards along a straight line that runs north–south. They run 10 metres north, then 5 metres south, then 7 metres north, then 9 metres south. What distance have they travelled? What is their displacement?

2. An engineer takes a direct flight from Munich to Detroit on an Airbus A350. The flight takes 10 hours and 30 minutes and covers a distance of 6980 kilometres. What is the average speed of the aircraft during this flight?

3. A shuttlecock in a game of badminton is travelling at 330 km h^{-1} when it strikes the racquet of an opposing player and rebounds at 264 km h^{-1}. Assume that the collision takes 0.25 seconds. Determine the average acceleration that the shuttlecock experiences during the collision with the racquet.

4. The following graph records the straight-line motion of a skateboarder.

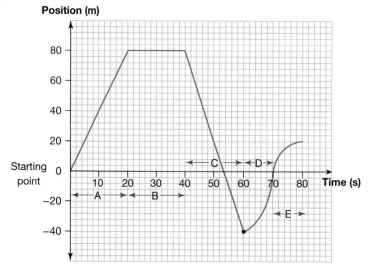

Calculate the velocity of the skateboarder during section A.

5. The following velocity-versus-time graph was sketched by a Physics student from data collected during a practical investigation.

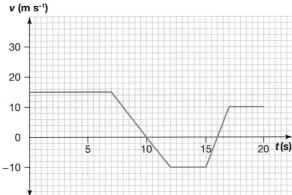

Determine the change in position of the object during the 20-second motion.

6. The student is analysing their data presented in question **5**. Use the graph to estimate the largest magnitude acceleration that the object experiences during this motion.

7. A Physics student records the acceleration versus time for a dynamics trolley colliding with a bumper. Use the following graph to estimate the change in velocity of the trolley during the collision.

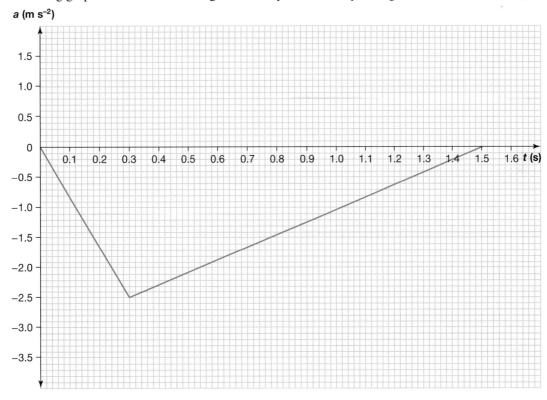

8. A Physics student has their very first experience of cross-country skiing on a school camp. They tentatively ski in a straight line down a gentle slope, with constant acceleration, starting from rest. They travel 50 metres in 40 seconds. What was their final speed?

9. A model rocket is launched directly upwards from a school oval. The engine switches off when the rocket is 45 metres above the ground and travelling at 90 m s^{-1} directly upwards. Assume that it continues to travel directly upwards with a constant downwards acceleration due to gravity of 9.8 m s^{-2}. What is the total height above the oval that the rocket reaches?

10. A car travelling in a 40 km h^{-1} speed limit zone brakes hard to avoid hitting a kangaroo that hops onto the road. The car comes to a complete stop 4.5 seconds after the brakes are applied and leaves a skid mark that is 27 metres long. Was the driver travelling faster than the speed limit when they braked? Use a calculation to support your answer.

8.5 Exam questions

▶ Question 1

Which of the following does a vector quantity have?

A. Size (magnitude) only
B. Direction only
C. Both size (magnitude) and direction
D. None of the above

▶ Question 2

A track cyclist warming up for a race completes three laps of a 250-metre velodrome track.

What is their displacement?

A. 0 m
B. 250 m
C. 500 m
D. 750 m

▶ Question 3

The Airbus A380 aircraft has a cruising speed of 1060 km h^{-1}.

How many metres does it travel in 1 second when cruising at this speed?

A. 3816 m
B. 1060 m
C. 294 m
D. 106 m

▶ Question 4

A hairy-nosed wombat that is feeling threatened runs as fast as its legs can take it. At its top speed it travels 21.0 metres in 1.90 seconds.

What is its average velocity during this interval?

A. 0.09 m s^{-1}
B. 39.9 m s^{-1}
C. 1.11 m s^{-1}
D. 11.1 m s^{-1}

A car is travelling at 12 m s⁻¹.

How fast is this in km h⁻¹?

A. 3.3 km h⁻¹
B. 43.2 km h⁻¹
C. 33.3 km h⁻¹
D. 4.3 km h⁻¹

A Tesla Model S accelerates from a standstill to a speed of 97 km h⁻¹ in 2.28 seconds.

What is its average acceleration during this time interval?

A. 12 m s⁻²
B. 43 m s⁻²
C. 153 m s⁻²
D. 27 m s⁻²

The following graph describes the motion of an elevator. The acceleration of the elevator is zero at which intervals?

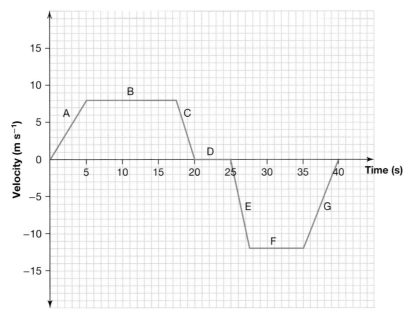

A. Intervals E, F and G
B. Interval D
C. Intervals B, D and F
D. Intervals B and F

The following graph describes the motion of an elevator. The elevator reaches the maximum speed for the time period shown during which interval?

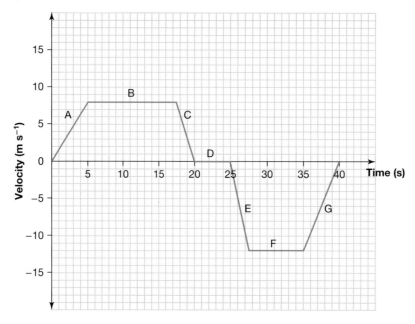

A. Interval B

B. Interval C

C. Interval E

D. Interval F

A student drops a stone into a wishing well (starting at rest). It falls at a rate of 9.80 m s^{-2}. The student hears the stone hit the water 2.00 seconds after they dropped it.

How far down the well did the stone travel from the student's hand until it hit the water?

A. 9.80 m

B. 19.6 m

C. 39.2 m

D. 78.4 m

A shopping trolley rolls down a hill in a straight line with a constant acceleration of 3.00 m s^{-2}. It starts at rest beside a car and travels 100 metres before it collides with a picket fence.

How fast is the shopping trolley travelling the instant before it collides with the fence?

A. 17.3 m s^{-1}

B. 600 m s^{-1}

C. 24.5 m s^{-1}

D. 300 m s^{-1}

▶ Question 11 (1 mark)

Jo is jogging around a track at a leisurely constant speed. He is joined by a friend who matches his speed and challenges him to a race to the finish. Jo accelerates to run as fast as he can and then continues at this speed until the finish line.

Sketch a position-versus-time graph of Jo's motion.

▶ Question 12 (2 marks)

A cyclist makes a left turn around a corner, keeping their speed constant.

Do they experience an acceleration during this motion? Justify your response.

▶ Question 13 (5 marks)

Two runners, Alex and Bo, are exchanging a baton in a 4 × 100-metre track relay. Their motion is shown in the following velocity-versus-time graph.

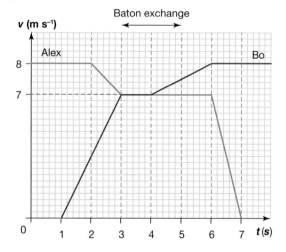

a. Calculate how far ahead Bo is of Alex at the instant Bo starts to run. **(3 marks)**

b. The rules of the race require that the baton exchange takes place over a maximum of 20 metres distance and that the second runner can start running a maximum of 10 metres before the exchange.

Use the graph to determine if Alex and Bo are likely to have complied with these rules. **(2 marks)**

▶ Question 14 (2 marks)

A diver is standing at the top of a 75-metre tall cliff. They leap off the cliff with an initial vertical velocity of 3 m s^{-1}. Assume that their acceleration throughout the dive is 9.8 m s^{-2} directly downwards.

How long will it be from the beginning of their leap until the instant they hit the water below?

During the filming of a new movie, a stuntman has to chase a moving bus and jump into it. The stuntman is required to stand still until the bus passes him and then start chasing it. The following velocity-versus-time graph describes the motion of the stuntman and the bus from the instant that the bus door passes the stationary stuntman.

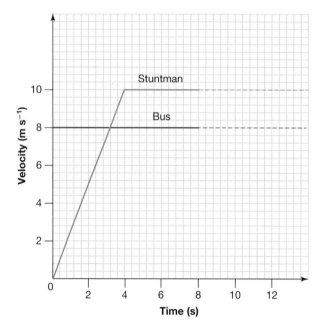

a. At what instant did the stuntman reach the same speed as the bus? **(1 mark)**
b. What is the magnitude of the acceleration of the stuntman during the first 4 seconds? **(1 mark)**
c. At what instant did the stuntman catch up with the bus door? **(2 marks)**
d. How far did the stuntman run before he reached the door of the bus? **(1 mark)**

 Resources

 Teacher-led videos Teacher-led videos for every exam question

9 Forces in action

KEY KNOWLEDGE

In this topic, you will:
- apply the vector model of forces, including vector addition and components of forces, to readily observable forces including the force due to gravity, friction and normal forces
- model the force due to gravity, F_g, as the force of gravity acting at the centre of mass of a body, $F_{\text{on body by Earth}} = mg$, where g is the gravitational field strength (9.8 N kg^{-1} near the surface of Earth)
- apply Newton's three laws of motion to a body on which forces act:

$$a = \frac{F_{\text{net}}}{m}, F_{\text{on A by B}} = -F_{\text{on B by A}}$$

- model forces as vectors acting at the point of application (with magnitude and direction), labelling these forces using the convention 'force on A by B' or $F_{\text{on A by B}} = -F_{\text{on B by A}}$
- apply concepts of momentum to linear motion: $p = mv$
- explain changes in momentum as being caused by a net force: $\Delta p = F_{\text{net}}\Delta t$
- calculate torque, $\tau = r_\perp F$
- analyse translational and rotational forces (torques) in simple structures in translational and rotational equilibrium.

Source: VCE Physics Study Design (2023–2027) extracts © VCAA; reproduced by permission.

Note: Vector quantities are bolded.

PRACTICAL WORK AND INVESTIGATIONS

Practical work is a central component of VCE Physics. Experiments and investigations, supported by a **practical investigation eLogbook** and **teacher-led videos**, are included in this topic to provide opportunities to undertake investigations and communicate findings.

EXAM PREPARATION

Access exam-style questions and their video solutions in every lesson, to ensure you are ready.

9.1 Overview

Hey students! Bring these pages to life online

Watch videos

Engage with interactivities

Answer questions and check results

Find all this and MORE in jacPLUS

9.1.1 Introduction

Have you ever seen somebody parachuting down from a skydive or BASE jump? They can land without injury as the force of the air resisting the motion of the parachute is large enough to slow them to a safe speed.

This explanation stems from the concept of forces, which are central to our understanding and analysis of motion. In his groundbreaking book *Philosophia Naturalis Principia*, published in 1687, Sir Isaac Newton proposed three laws of motion. These laws accurately explain the motion of objects on Earth and throughout the universe. In this topic we will use Newton's laws to explore the nature of forces and their relationship to motion.

FIGURE 9.1 BASE jumpers use a high point such as a cliff to launch themselves. The forces involved in the jump must be carefully calculated to allow the parachute to open in time.

LEARNING SEQUENCE

on Resources

Solutions	Solutions — Topic 9 (sol-0795)
Practical investigation eLogbook	Practical investigation eLogbook — Topic 9 (elog-1578)
Digital documents	Key science skills — VCE Physics Units 1–4 (doc-36950)
	Key terms glossary — Topic 9 (doc-36967)
	Key ideas summary — Topic 9 (doc-36968)
Exam question booklet	Exam question booklet — Topic 9 (eqb-0077)

9.2 Forces as vectors

9.2.1 Describing a force

A **force** is an interaction between two objects that can cause a change, commonly a push, pull or twist applied to one object by another, measured in newtons (N). Some types of force require contact. For example, the force applied by your hand on a netball or basketball when shooting a goal requires contact between your hand and the ball. Some forces do not require contact. For example, the force of gravity applied on your body by Earth is present even when you are not in contact with Earth.

In diagrams showing forces, such as figure 9.3, a labelled arrow should be drawn from the **centre of mass** of the object upon which the force acts (this is not included in figure 9.2 for the sake of simplicity). The length of the arrow should indicate the relative size of the force.

To fully describe a force, you need to specify its direction as well as its magnitude or size. A quantity that can be fully described only by specifying a magnitude and direction is called a vector quantity. Force is a vector quantity.

> **force** an interaction between two objects that can cause a change; commonly a push, pull or twist applied to one object by another, measured in newtons (N); it is a vector quantity
>
> **centre of mass** the point at which all of the mass of an object can be considered to be situated

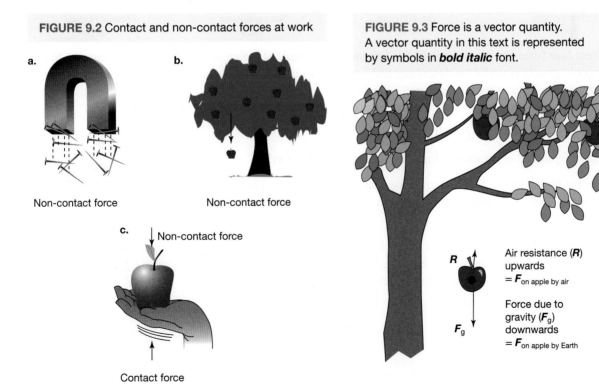

FIGURE 9.2 Contact and non-contact forces at work

a.

Non-contact force

b.

Non-contact force

c.   Non-contact force

Contact force

FIGURE 9.3 Force is a vector quantity. A vector quantity in this text is represented by symbols in **_bold italic_** font.

R Air resistance (R) upwards
= $F_{\text{on apple by air}}$

Force due to gravity (F_g) downwards
= $F_{\text{on apple by Earth}}$

F_g

When labelling forces, it helps to describe the force as $\boldsymbol{F}_{\text{on A by B}}$; for example, the arrow representing the force due to gravity on the apple in figure 9.3 is labelled $\boldsymbol{F}_{\text{on apple by Earth}}$.

The SI unit of force is the newton (N) and $1 \text{ N} = 1 \text{ kg m s}^{-2}$. The force of gravity on a 100-gram apple is about 1 N downwards. A medium-sized car starting from rest is subjected to a forward force of about 4000 N.

Quantities that can be described without specifying a direction are called scalar quantities. Mass, energy, time and temperature are all examples of scalar quantities.

The present understanding of our universe posits that there are four fundamental forces or interactions (see figure 9.4).

FIGURE 9.4 The four fundamental forces

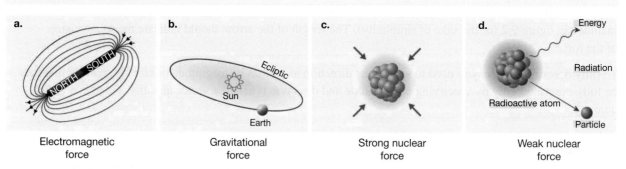

| Electromagnetic force | Gravitational force | Strong nuclear force | Weak nuclear force |

While these four forces or interactions are fundamental to our understanding of the physics of interactions in the universe, only gravity is useful for describing and analysing everyday motion at a human scale. The following sections introduce some everyday categories of forces.

9.2.2 Force due to gravity ($\boldsymbol{F}_g$)

The apple in figure 9.3 is attracted to Earth by the **force due to gravity**.

The force due to gravity is a force of attraction that exists between any pair of objects that have mass. These two objects do not need to be in contact with one another.

force due to gravity the force applied to an object due to gravitational attraction

The force on an object due to the pull of gravity is usually given the symbol $\boldsymbol{F}_g$. In simple scenarios where objects of relatively small mass (e.g. humans, buildings, whales) are in close proximity to a very massive object (e.g. Earth), the force due to gravity can be represented by:

$$\boldsymbol{F}_g = m\boldsymbol{g}$$

where:

m = the mass of the smaller object in kg
$\boldsymbol{g}$ = the gravitational field strength in N kg^{-1} due to the larger object.

The **gravitational field strength** is defined as the force of gravity on a unit of mass. Gravitational field strength is a vector quantity. The direction of the force due to gravity on an object is towards the centre of the source of attraction (e.g. the centre of Earth).

The magnitude of gravitational field strength at Earth's surface is, on average, 9.8 N kg^{-1}. The magnitude of **g** decreases as altitude (height above sea level) increases. It also decreases as one moves from the poles towards the equator. Table 9.1 shows the magnitude of **g** at several different locations.

TABLE 9.1 Variation in gravitational field strength

Location	Altitude (m)	Latitude	Magnitude of *g* (N kg^{-1})
Equator	0	0°	9.780
Sydney	18	34°S	9.797
Melbourne	12	37°S	9.800
Denver	1609	40°N	9.796
New York	38	41°N	9.803
North Pole	0	90°N	9.832

The magnitude of **g** at Earth's surface will be taken as 9.8 N kg^{-1} throughout this text. At the surface of the Moon, the magnitude of **g** is 1.60 N kg^{-1}.

tlvd 3783

SAMPLE PROBLEM 1 Calculating the force due to gravity

What is the force due to gravity acting on a 50-kilogram student:
a. **on Earth**
b. **on the Moon?**

THINK	WRITE
a. 1. Recall the formula for force due to gravity.	a. $F_g = mg$
2. Substitute the values to determine the force due to gravity.	$F_g = 50 \times 9.8$ $= 490 \text{ N downwards}$
3. State the solution.	The force due to gravity acting on a 50-kilogram student on Earth is 490 N downwards.
b. 1. Recall the formula for force due to gravity.	b. $F_g = mg$
2. Substitute the values to determine the force due to gravity.	$F_g = 50 \times 1.60$ $= 80 \text{ N downwards}$
3. State the solution.	The force due to gravity acting on a 50-kilogram student on the Moon is 80 N downwards.

PRACTICE PROBLEM 1

a. What is the difference between the force due to gravity by Earth acting on a 70-kilogram person at the North Pole and at the equator?
b. A hospital patient is very accurately measured to have a mass of 64.32 kilograms and the force due to gravity by Earth acting on them is 630.08 N. In which of the locations in table 9.1 could the patient be?

EXTENSION: Weighing in

Bathroom scales are designed for use only on Earth. Fortunately (at this point in time), that's where most of us live.

If a 60-kilogram student stood on bathroom scales on the Moon, the reading would be only about 10 kilograms. Yet the mass of the student remains 60 kilograms. Bathroom scales measure force, not mass.

However, scales are designed so that you can read your mass in kilograms. Otherwise, you would have to divide the measured force by 9.8 to determine your mass. The manufacturer of the bathroom scales saves you the trouble of having to do this.

The 60-kilogram student experiences a force due to gravity from Earth of about 588 N. However, on the Moon the force due to gravity from the Moon is only about 100 N. The reading on the scales will be 100 N divided by 9.8 N kg^{-1}, giving the result of 10.2 kilograms.

The term *weight* is often used in high-school Physics texts in order to draw a distinction between the force due to gravity on an object and its mass. However, the accepted definition of the term *weight* in Physics is more complex than this, and there are different conventions for how it is defined. Confusion that this may cause is further compounded by inconsistent use of the term *weight* in everyday life. As such, the term *weight* is not used in this text or the VCE Physics Study Design.

INVESTIGATION 9.1

online only

The relationship between mass and the force due to gravity

Aim

To examine the relationship between mass and the force due to gravity

9.2.3 Friction (F_{fr})

Friction is a force that surfaces exert on each other when they 'rub' together. The magnitude of the friction force depends greatly on the nature of each of the two surfaces. Smooth surfaces experience smaller friction forces than rough surfaces. However, even very smooth surfaces are rough on a microscopic scale.

It is this roughness that is mostly responsible for the resistance to motion that we call friction. As two surfaces move across each other, they intermesh, resisting the motion.

Friction can seem to be a real nuisance at times. For instance, it causes wear and tear in car engines and can make them overheat.

Friction is also a necessary force in many situations. If you have ever walked on ice, you know the importance of friction to walking.

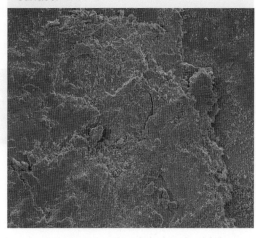

FIGURE 9.5 This scanning electron micrograph shows a magnified metal surface.

Friction is also needed for safe driving in a car. A large friction force is needed to start moving, change direction and stop. The rubber tyres of a car have a deep tread to ensure that the friction force is still present on a wet road when water forms a lubricating film between the road and the tyres. Smooth tyres would slide across a wet road, making it difficult to stop, turn or accelerate. The deep grooves in tyres pick up the water from the road and throw it backwards, providing a drier road surface and greater friction.

elog-1767

tlvd-0822

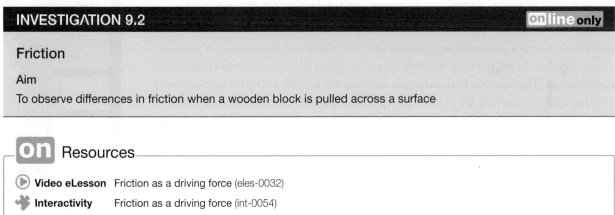

INVESTIGATION 9.2

online only

Friction

Aim

To observe differences in friction when a wooden block is pulled across a surface

on Resources

▶ **Video eLesson** Friction as a driving force (eles-0032)

🧩 **Interactivity** Friction as a driving force (int-0054)

9.2.4 Forces from fluid motion

The motion of an object through fluids such as air or water results in forces acting on the object. This resistance to the motion of objects through air and water is given a variety of names including fluid friction, fluid resistance, drag or **air resistance**

Because fluid friction increases with the speed of the object, streamlining is particularly important in cars, planes, watercraft and bicycles (see figure 9.6). Streamlining involves creating a shape that reduces the slowing effect of collisions with particles of the fluid.

While air resistance can be a seen as a problem, it is an absolute necessity for parachutists and paragliders, who rely on it to slow their descent and land safely.

friction the force applied to the surface of an object when it is pushed or pulled against the surface of another object

air resistance the force applied to an object, opposite to its direction of motion, by the air through which it is moving

FIGURE 9.6 Racing cyclists reduce the effects of fluid friction (air resistance) by wearing streamlined costumes and using aerodynamically designed bicycles.

9.2.5 The normal force (F_N)

At this moment you are probably sitting on a chair with your feet on the floor. The material in the chair has been compressed and is pushing back up (see figure 9.7). This force is called the **normal force**, F_N, as it acts at right angles to the surface.

9.2.6 Compression and tension in materials

Forces acting on materials, such as a concrete pillar or a steel cable, can cause the material to experience compression or tension. These are types of loadings on a material.

In a material that is experiencing compression, the atoms and molecules are pushed closer together. In the photo of the Pole House in figure 9.8, the concrete pillar is in compression. The house is pushing down on the pillar and the rocky hillside below is pushing up on it, causing the pillar to be squashed in compression. An object that is in compression will be pushing against the objects that are compressing it. Some materials are best suited for withstanding compression, such as concrete, stone, brick and bone.

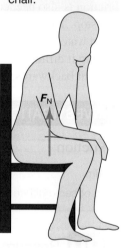

FIGURE 9.7 The normal force acts perpendicular to the surface of the chair.

normal force the force that acts perpendicularly to a surface as a result of an object applying a force to the surface

FIGURE 9.8 The Pole House located on the Great Ocean Road is supported by a concrete column that goes several metres into the ground.

Similarly, when atoms and molecules in a material are pulled further apart by a loading, it experiences **tension**. In the photo of the crane in figure 9.9, the steel cable is in tension. The crane is pulling up on the cable and the load that it is supporting is pulling down on the cable, causing it to be stretched in tension. An object that is in tension will be pulling against the objects that are stretching it.

tension a pulling force along the length of an object (such as a rope or cable) that is being stretched

FIGURE 9.9 Cranes use metal cables in tension to lift large loads.

$F_{\text{by crane on cable}}$

$F_{\text{by cable on crane}}$

$F_{\text{by cable on mass}}$

$F_{\text{by Earth on mass}}$

9.2.7 Free-body diagrams

A free-body diagram or force diagram is used to depict all external forces acting upon an object (see figures 9.8 and 9.9). The key to developing this skill is practice.

Steps to consider to create a free-body diagram

- Represent the object with a suitable simplified shape or, if the problem is simple enough, a single point (e.g. a car could be represented very simply as a rectangular block).
- Identify each interaction that the object is experiencing and the force that represents that interaction (e.g. gravitational interaction with Earth represented by the force due to gravity, F_g).
- Represent each force with an arrow drawn at the point where it acts on the object (e.g. a friction force on a car by the road will act where the car tyres contact the road; the force due to gravity will act at the centre of mass of an object).
- The direction of the arrow should be in the expected direction of the force. If the direction is not known, make an assumption (e.g. the force in a connection between a car and trailer could be either forwards or backwards).
- The length of the arrow should represent the expected magnitude (size) of the force relative to the other forces in the diagram. If unknown, make an assumption.

SAMPLE PROBLEM 2 Drawing force diagrams

Draw a force diagram for each of the following.
a. **An ice skater moving at a constant velocity**
b. **A ball hanging straight down on a string**
c. **A trolley slowing down**
d. **A block stationary on slope**

THINK	WRITE
a. The only forces acting on the ice skater are the normal force and the force due to gravity.	a. 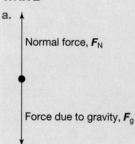
b. The forces acting on a ball hanging on a string are the force due to gravity and a tension force.	b. 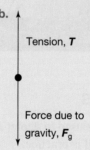
c. The forces acting on a trolley that is slowing down are the force due to gravity, the normal force and a friction force.	c. 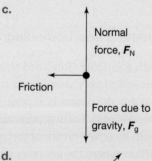
d. The forces acting on a stationary block on a slope are the force due to gravity, the normal force and a friction force in the direction up the slope.	d. 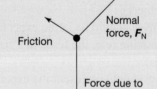

PRACTICE PROBLEM 2

Draw a force diagram for each of the following.
a. A falling stone
b. A mass being pulled at a steady speed
c. A piece of iron on a string, which is hanging near a strong magnet

9.2.8 The net force (F_{net}) or sum of forces

The **net force** is the result of combining all the forces into one force. You may also see this referred to as the *sum of forces* or the *resultant force*.

> **net force** the vector sum of the forces acting on an object

To understand whether a net force is present, consider whether an object accelerates.

If the object is moving at constant velocity or if it is stationary then the net force is equal to zero. This does not mean that forces are not present; it just means that if they are, then they are balanced and in opposing directions.

When more than one force acts on an object, the net force is found by the vector addition of the forces. Figures 9.10b, c and d are examples of two forces acting in different directions, while figure 9.10a shows them acting in the same direction. The net force, usually denoted by F_{net}, is indicated in each of the examples.

It is critical to distinguish that the net force is not a particular type or category of force that acts on an object. It is simply the combination of all forces.

To distinguish the net force vector from the vectors for the actual forces, the arrow for F_{net} is often drawn in another colour or as a dashed line. A geometric method of adding vectors uses a parallelogram, where the vector sum is the diagonal. This is shown in figure 9.10d.

FIGURE 9.10 Adding forces together

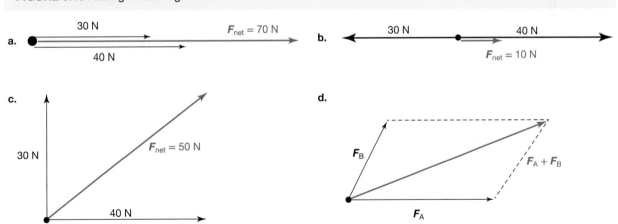

Note that in more complex cases, it is far easier to list the x and y components of each vector, sum the x components and sum the y components, and then use Pythagoras' theorem to calculate the resultant force.

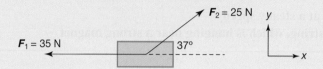

SAMPLE PROBLEM 3 Calculating forces and net forces acting on an object

The following free-body diagram shows the forces acting on an object.

$F_2 = 25$ N y

$F_1 = 35$ N   37°

x

Calculate the:
a. component of F_2 that acts in the x direction
b. component of F_2 that acts in the y direction
c. magnitude of the net force on the object.

THINK

a. 1. The force F_2 acts at an angle of 37° from the x-axis. Its x component can be found using trigonometry.

2. State the solution.

b. 1. Similarly the y component of F_2 can be found using sin.

2. State the solution.

c. 1. To find the net force, consider the force components in the x and y directions separately.

2. Combining these two forces, Pythagoras' theorem can be used to find the resultant net force.

WRITE

a. $\cos\theta = \dfrac{\text{adjacent}}{\text{hypotenuse}}$

$\cos(37°) = \dfrac{F_{2x}}{25}$

$\Rightarrow F_{2x} = 25\cos(37°)$

≈ 20 N

The component of F_2 that acts in the x direction is 20 N.

b. $\sin\theta = \dfrac{\text{opposite}}{\text{hypotenuse}}$

$\sin(37°) = \dfrac{F_{2y}}{25}$

$\Rightarrow F_{2y} = 25\sin(37°)$

≈ 15 N

The component of F_2 that acts in the y direction is 15 N.

c. In the x direction:
$F_{\text{net}, x} = F_{2x} - F_1$
$= 20 - 35$
$= -15$ N
In the y direction:
$F_{\text{net}, y} = F_{2y}$
$= 15$
$= 15$ N

F_{net} 15 N

15 N

$c^2 = a^2 + b^2$

$F_{\text{net}}^2 = 15^2 + 15^2$
$= 450$

$$\Rightarrow F_{net} = \sqrt{450}$$
$$\approx 21.2 \, N$$

3. State the solution.

The magnitude of the net force on the object is 21.2 N.

Note: This problem only requires the magnitude of the net force. If the direction was required it would be necessary to calculate the angle that the force is acting at.

PRACTICE PROBLEM 3

The following free-body diagram shows the forces acting on an object.

Calculate the magnitude of the net force acting on the object.

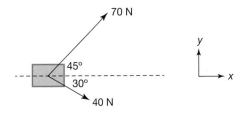

elog-1768

tlvd-3785

9.2 Activities

learnon

Students, these questions are even better in jacPLUS

Receive immediate feedback and access sample responses

Access additional questions

Track your results and progress

Find all this and MORE in jacPLUS

| 9.2 Quick quiz   on | 9.2 Exercise | 9.2 Exam questions |

9.2 Exercise

1. Describe the difference between a vector quantity and a scalar quantity.
2. Which of the following are vector quantities?
   I. Mass
   II. Force due to gravity
   III. Gravitational field strength
   IV. Time
   V. Energy
   VI. Temperature

3. A car has a mass of 1400 kilograms with a full petrol tank.
   a. What is the magnitude of the force due to gravity acting on it at the surface of Earth?
   b. What would be the magnitude of the force due to gravity acting on it on the surface of Mars, where the magnitude of the gravitational field strength is 3.6 N kg^{-1}?
   c. What is the mass of the car on the surface of Mars?
4. Estimate the magnitude of the force due to gravity acting at the surface of Earth on:
   a. an apple
   b. a textbook
   c. your Physics teacher.
5. Estimate your own mass in kilograms and determine:
   a. the magnitude of the force due to gravity acting on you at the surface of Earth
   b. the magnitude of the force due to gravity acting on you at the surface of Mars, where the magnitude of the gravitational field strength is 3.6 N kg^{-1}
   c. your mass on the planet Mars.
6. Draw force diagrams for:
   a. an open parachute falling slowly to the ground.
   b. a thrown basketball approaching its maximum height before coming down into the basket
   c. a car approaching a red light rolling slowly to a stop
   d. a rocket during liftoff.
7. Draw force diagrams for each of the following figures. For the figures shown in parts a–c, show the forces acting on the rock. For the figure shown in part d, show the force acting on the block.

a.

b.

c.

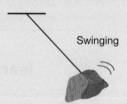

Swinging

d.

Stretched spring

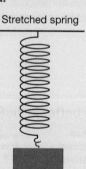

8. A person is standing on a horizontal floor. Draw and label in the form $F_{\text{on A by B}}$ all of the forces acting on the person, the floor and Earth.
9. Determine the net force in the following situations.

a.

3 N 2 N
      4 N

N
W ← → E
S

b.

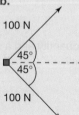

100 N
      45°
      45°
100 N

10. In the following diagrams, the net force is shown along with all but one of the contributing forces. Determine the magnitude and direction of the missing force.

a.

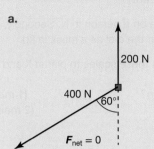

b.

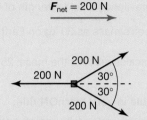

11. A car is moving north on a horizontal road at a constant speed of 60 km h^{-1}. Draw a diagram showing all of the significant forces acting on the car. Show all of the forces as if they were acting through the centre of mass.

12. Determine the magnitude of the horizontal components of each of the forces shown in the following diagrams.

a.

b.

c.

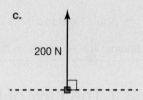

9.2 Exam questions

▶ Question 1 (1 mark)

MC A force of 3 N east is added to a force of 4 N south.

What is the magnitude of the net force?
- A. 7 N
- B. 5 N
- C. 1 N
- D. −7 N

▶ Question 2 (2 marks)

F_A = 5.0 N north and F_B = 5.0 N east.

Calculate the net force when these two forces are added together. Give both magnitude and direction.

▶ Question 3 (3 marks)

Three forces are added together: A is 8.0 N east, B is 2.0 N west and C is x N south. The net force, D, has magnitude 7.5 N.

Calculate the magnitude of x.

▶ Question 4 (2 marks)

For gravitational field strengths on Earth and the Moon, take g_E = 9.80 N kg^{-1} and g_M = 1.60 N kg^{-1}.

An object experiences a force due to gravity of 128 N on the Moon.

What would be the force due to gravity acting on it if it was relocated to the surface of Earth?

9.3 Newton's First Law of Motion

KEY KNOWLEDGE

- Apply Newton's three laws of motion to a body on which forces act: $a = \dfrac{F_{net}}{m}$, $F_{\text{on A by B}} = -F_{\text{on B by A}}$

Source: VCE Physics Study Design (2023–2027) extracts © VCAA; reproduced by permission.

9.3.1 The Law of Inertia

It is difficult to explain the motion of objects without an understanding of force and **inertia**.

inertia the resistance of any physical object to a change in its speed or direction of motion

CASE STUDY: Understanding inertia

The ancient Greek philosopher Aristotle (384–322 BCE) concluded from his observations that a moving object would come to rest if no force was pushing it. Aristotle thought that steady motion required a constant force and that 'being at rest' was the natural state of matter. This view held sway for almost 2000 years, although contrary views were expressed by philosophers such as Epicurus and Lucretius.

It was not until Galileo (1564–1642) that this explanation of motion was seriously challenged. Galileo argued that if a ball rolled down an inclined plane gained speed and a ball rolled up an inclined plane lost speed, a ball rolled along a horizontal plane should neither gain nor lose speed. Galileo knew that this did not really happen. He claimed that if there was a lot of friction, the ball slowed down quickly; if there was little friction, the ball slowed down more gradually. However, he predicted that if there were no friction at all, the ball would continue to move with a constant speed forever unless something else caused it to slow down or stop.

Galileo introduced the concept of friction as a force and concluded that objects retain their velocity unless a force, often friction, acts upon them. Galileo stated in his *Discorsi* (1638):

> *A body moving on a level surface will continue in the same direction at constant speed unless disturbed.*

Sir Isaac Newton (1643–1727) was able to refine Galileo's ideas about motion. In 1687, he published his *Philosophia Naturalis Principia*, which included three laws of motion.

Newton's First Law of Motion

Every object continues in its state of rest or uniform motion unless made to change by a non-zero net force.

A bowling ball rolling down a bowling alley (see figure 9.11) changes its motion because the net force on it is not zero. In fact, it slows down because the direction of the net force is opposite to the direction of motion. The vertical forces, gravity and the normal force balance each other. The only 'unbalanced' force is that of friction.

FIGURE 9.11 The changing net force on this bowling ball determines its state of motion.

EXTENSION: Sir Isaac Newton

Sir Isaac Newton was one of many famous scientists who were not outstanding students at school or university. Newton left school at 14 years of age to help his widowed mother on the family's farm. He turned out to be unsuited to farming and spent much of his time reading. At the age of 18, he went to Cambridge University, where he showed no outstanding ability.

FIGURE 9.12 Sir Isaac Newton

When Cambridge University was closed down in 1665 due to an outbreak of the plague, Newton went home and spent the next two years studying and writing. During this time, he developed the laws of gravity that explain the motion of the planets, and his three famous laws of motion. Over the same period, he put forward the view that white light consisted of many colours and he invented calculus. Newton's laws of gravity and motion were not published until about 20 years later.

Newton later became a member of Parliament, a warden of the Mint and President of the Royal Society. After his death in 1727, he was buried in Westminster Abbey, London, alongside many English kings, queens, political leaders and poets.

9.3 Activities

 learn on

Students, these questions are even better in jacPLUS

Receive immediate feedback and access sample responses

Access additional questions

Track your results and progress

Find all this and MORE in jacPLUS ▶

| 9.3 Quick quiz **on** | 9.3 Exercise | 9.3 Exam questions |

9.3 Exercise

1. With reference to Newton's First Law of Motion, explain why a smartphone sliding across a table does not move at a constant speed.
2. When you are standing on a bus, train or tram that stops suddenly, you lurch forwards. Explain why this happens in terms of Newton's First Law.
3. If the bicycle that you are riding runs into an obstacle such as a large rock, you may be flung forwards over the handlebars. Explain in terms of Newton's First Law why this happens.
4. A removalist is pushing some heavy boxes across the floor of their van at a constant speed. They exert a force of 125 N on the boxes. Calculate the magnitude of the friction force acting on the boxes during this motion.
5. A Physics student who is also in the circus is practicing juggling balls whilst travelling in a train. They misjudge a throw and one of the balls goes out an open window. It quickly disappears beside the track behind the train while the other balls remain inside. Use your knowledge of Newton's First Law to explain why the horizontal motion of the ball changed once it exited the train.

9.3 Exam questions

▶ Question 1 (2 marks)

A ball that is rolling along a perfectly horizontal and very smooth surface on Earth will eventually slow down to a stop.

Identify the forces causing this and use Newton's First Law to explain why this is the case.

▶ Question 2 (2 marks)

A high-altitude skydiver has reached terminal velocity and is falling at a constant velocity towards Earth. The force due to gravity acting on them at this point in their fall is 600 N.

State the magnitude and direction of the total air resistance force acting on them.

▶ Question 3 (2 marks)

A space probe travelling in deep space moves at a constant velocity without using any rockets or other propulsion. Yet an aircraft on Earth cruising at constant speed must keep its engines producing thrust so that it does not slow down.

Explain how both situations are consistent with Newton's First Law.

▶ Question 4 (2 marks)

A boat is pulling a raft along a river at constant speed. The retarding forces on the boat and raft are 600 N and 300 N respectively.

Calculate the driving force on the boat. Explain your reasoning.

Question 5 (3 marks)
A car is travelling quickly along the highway when it runs into an obstacle and stops abruptly. One of the passengers is not wearing a seat belt. Describing the incident to a friend later, he says that he was thrown forward by the collision. The friend questions how the collision could produce a forward force on the passenger.

Using Newton's First Law, explain what happened to the passenger in the collision.

More exam questions are available in your learnON title.

9.4 Newton's Second Law of Motion

KEY KNOWLEDGE

- Apply Newton's three laws of motion to a body on which forces act: $a = \dfrac{F_{net}}{m}$, $F_{\text{on A by B}} = -F_{\text{on B by A}}$

Source: VCE Physics Study Design (2023–2027) extracts © VCAA; reproduced by permission.

9.4.1 The Law of Mass and Acceleration

Casual observations indicate that the acceleration of a given object increases as the net force on the object increases. It is also clear that lighter objects change their velocity at a greater rate than heavier objects when the same force is applied.

It can be shown experimentally that the acceleration, a, of an object is:
- proportional to the net force, F_{net}, acting on it

$$a \propto F_{net}$$

- inversely proportional to its mass, m.

$$a \propto \frac{1}{m}$$

Thus, the relationship between force, mass and acceleration can be seen as:

$$a \propto \frac{F_{net}}{m}$$

The SI unit of force, the newton (N), is defined such that a net force of 1 N causes a mass of 1 kilogram to accelerate at 1 m s^{-2}. The value of the constant, k, is 1. It has no units.

Newton's Second Law of Motion

The acceleration of an object as produced by a net force is directly proportional to the magnitude of the net force, in the same direction as the net force, and inversely proportional to the mass of the object.

$$a = \frac{F_{net}}{m}$$
$$F_{net} = ma$$

This statement of Newton's Second Law allows you to:
- determine the net force acting on an object without knowing any of the individual forces acting on it
- deduce the net force as long as you can measure or calculate (using formulae or graphs) the acceleration of a known mass
- determine the mass of an object. You can do this by measuring the acceleration of an object on which a known net force is exerted.
- predict the effect of a net force on the motion of an object of known mass.

tlvd-3786

SAMPLE PROBLEM 4 Application of Newton's Second Law of Motion

A 65-kilogram Physics teacher, starting from rest, glides down a slide in the local playground. The net force on her during the slide is a constant 350 N. How fast will she be travelling at the bottom of the 8-metre slide?

THINK	WRITE
1. Recall Newton's Second Law of Motion.	$F_{net} = ma$
2. Substitute the mass and force to find the acceleration.	$F_{net} = ma$ $350 = 65a$ $\Rightarrow a = \dfrac{350}{65}$ $\approx 5.38 \, \text{m s}^{-2}$
3. As the acceleration is constant, a constant acceleration formula can be used.	$v^2 = u^2 + 2as$ $= 0 + 2 \times 5.38 \times 8$ $= 86.08$ $\Rightarrow v = \sqrt{86.08}$ $\approx 9.3 \, \text{m s}^{-1}$
4. State the solution.	She will be travelling at approximately 9.3 m s^{-1} at the bottom of the slide.

PRACTICE PROBLEM 4

a. What is the magnitude of the average force applied by a tennis racquet to a 58-gram tennis ball during service if the average acceleration of the ball during contact with the racquet is 1.2×10^4 m s^{-2}?

b. A toy car is pulled across a smooth, polished horizontal table with a spring balance. The reading on the spring balance is 2 N and the acceleration of the toy car is measured to be 2.5 m s^{-2}. What is the mass of the toy car? (*Note*: Because the table is described as smooth and polished, friction can be ignored.)

 Resources

 Video eLesson Newton's Second Law (eles-0033)

9.4.2 Applying Newton's Second Law in real life

The practice problems presented in the previous section include assumptions and simplifications of what really happens. For instance, the table surface in part b of practice problem 4 is smooth. This description was deliberately included so that you would know the force of friction on the toy car was **negligible**, which means that it can be ignored and so simplifies the problem.

The assumptions made to answer each of the questions asked in the sample problem and practice problem are called **idealisations**. Making reasonable idealisations makes it possible to simplify a problem whilst also still obtaining results that are a close approximation of what will occur in reality. However, caution is needed when making idealisations. For example, it would be unreasonable to ignore the air resistance on a tennis ball while it was soaring through the air at 150 km h^{-1} (42 m s^{-1}) after the serve was completed.

Most applications of Newton's Second Law are not as simple as those given in the practice problems in the previous section. Some more typical examples are presented in the following sample and practice problems.

> **negligible** a quantity so small that it can be ignored when modelling a phenomenon or an event
>
> **idealisation** assuming ideal conditions that don't exactly match the real situation to make modelling a phenomenon or event easier

tlvd-0080

SAMPLE PROBLEM 5 Applications of Newton's Second Law of Motion and the constant acceleration formula

When the head of an 80-kilogram bungee jumper is 24 metres from the surface of the water below, her velocity is 16 m s^{-1} downwards and the tension in the bungee cord is 1200 N. Air resistance can be assumed to be negligible.
a. **What is her acceleration at that instant?**
b. **If her acceleration remained constant during the rest of her fall, would she stop before hitting the water?**

THINK

a. 1. Calculate the force due to gravity, then draw a diagram to show the forces acting on the bungee jumper.

2. Calculate the net force.

WRITE

a. $F_g = mg$
   $= 80 \times 9.8$
   $= 784\,N$

$F_{net} = 1200 - 784$
   $= 416\,N$ upwards

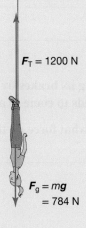

$F_T = 1200\,N$

$F_g = mg$
   $= 784\,N$

3. Use Newton's Second Law to calculate the acceleration.

$$F_{net} = ma$$
$$416 = 80a$$
$$\Rightarrow a = \frac{416}{80}$$
$$= 5.2 \,\mathrm{m\,s^{-2}} \text{ upwards}$$

4. State the solution.

Her acceleration at that instant is $5.2 \,\mathrm{m\,s^{-2}}$.

b. 1. If the jumper's acceleration were constant, one of the constant acceleration formulae could be used to answer this question. Assign down as positive for this part of the question as the bungee jumper has a downwards initial velocity and displacement during the time period being considered.

b. $v^2 = u^2 + 2as$
$u = 16 \,\mathrm{m\,s^{-1}}$, $v = 0$, $a = -5.2 \,\mathrm{m\,s^{-2}}$, $s = ?$

2. Substitute the values into the formula and solve for s.

$$v^2 = u^2 + 2as$$
$$0 = 16^2 + 2 \times -5.2 \times s$$
$$0 = 256 - 10.4\,s$$
$$10.4\,s = 256$$
$$\Rightarrow s = \frac{256}{10.4}$$
$$\approx 24 \,\mathrm{m}$$

3. State the solution.

The bungee jumper will not stop in time. However, don't be upset. In practice, the acceleration of the bungee jumper would not be constant. The tension in the cord would increase as she fell. Therefore, the net force on her would increase and her upwards acceleration would be greater in magnitude than the calculated value. She will therefore almost certainly come to a stop in a distance considerably less than that calculated.

PRACTICE PROBLEM 5

A 1200-kilogram sports car is testing its brakes by driving at a constant speed of 100 km h^{-1} and then braking hard. To pass the test it needs to come to a complete stop in a distance of 50 metres.

If the friction is a constant 1000 N, what force do the brakes need to apply for the sports car to pass the test?

SAMPLE PROBLEM 6 Determining net force and resistance forces

A waterskier of mass 80 kilograms, starting from rest, is pulled in a northerly direction by a horizontal rope with a constant tension of 240 N. After 6 seconds, he has reached a speed of 12 m s⁻¹.

a. What is the net force on the skier?

b. If the tension in the rope were the only horizontal force acting on the skier, what would his acceleration be?

c. What is the sum of the resistance forces on the skier?

THINK	WRITE
a. 1. A good first step is to draw a labelled force diagram. The force due to gravity can be calculated by the formula $F_g = mg$.	**a.** $F_g = mg$ $\quad = 80 \times 9.8$ $\quad = 784 \, \text{N}$

N ➤

Normal force = 784 N

Tension = 240 N

Resistance forces ←

Force due to gravity = *mg*
= 784 N

2. The net force cannot be determined by adding the individual force vectors because the resistance forces are not given, nor is there any information in the question to suggest that they can be ignored. To calculate the net force we will first need to find the acceleration.	$v = u + at$ $12 = 0 + a \times 6$ $\Rightarrow a = \dfrac{12}{6}$ $\quad = 2 \, \text{m s}^{-2} \, \text{north}$
3. Use Newton's Second Law to calculate the net force.	$F_{net} = ma$ $\quad = 80 \times 2$ $\quad = 160 \, \text{N north}$
4. State the solution.	The net force on the skier is 160 N north.
b. 1. As the vertical forces cancel out, the net force is determined by the horizontal forces. If the tension were the only horizontal force acting on the waterskier it would be the net force.	**b.** $F_{net} = 240 \, \text{N north}$
2. Use Newton's Second Law to calculate the acceleration.	$F_{net} = ma$ $240 = 80a$ $\Rightarrow a = \dfrac{240}{80}$ $\quad = 3 \, \text{m s}^{-2} \, \text{north}$

3. State the solution.

If the tension in the rope were the only horizontal force acting on the waterskier, his acceleration would be 3 m s^{-2} north.

c. 1. The sum of the resistance forces (friction caused by the water surface and air resistance) on the waterskier is the difference between the net force and the tension.

c. Sum of resistance forces $= F_{net} -$ tension
$= 160\,\text{N north} - 240\,\text{N north}$
$= 80\,\text{N south}$

2. State the solution.

The sum of the resistance forces on the waterskier is 80 N south.

PRACTICE PROBLEM 6

A loaded sled with a mass of 60 kilograms is being pulled across a level snow-covered field with a horizontal rope. It accelerates from rest over a distance of 9 metres, reaching a speed of 6 m s^{-1}. The tension in the rope is a constant. The frictional force on the sled is 200 N. Air resistance is negligible.
a. What is the acceleration of the sled?
b. What is the magnitude of the tension in the rope?

---

tlvd-3787

SAMPLE PROBLEM 7 Determining the acceleration using a velocity-versus-time graph

The following velocity-versus-time graph describes the motion of a 45-kilogram girl on rollerblades as she rolls on a horizontal concrete path for 6 seconds before crossing onto a rough horizontal gravel path for the remaining 4 seconds.
a. What was the magnitude of the net force on the girl on the concrete surface?
b. If the only horizontal force acting on the blades is the friction force applied by the path, what is the value of the following ratio?

$$\frac{\text{Friction force of gravel path on rollerblades}}{\text{Friction force of concrete path on rollerblades}}$$

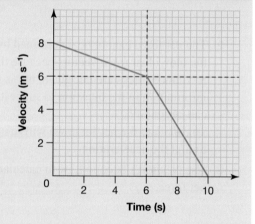

THINK

a. 1. The acceleration of the girl while she was on the concrete surface is given by the gradient of the corresponding section of the velocity-versus-time graph.

WRITE

a. $a = \dfrac{\text{rise}}{\text{run}}$

$= \dfrac{-2}{6}$

$= \dfrac{-1}{3}\,\text{m s}^{-2}$

---

2. Use Newton's Second Law to calculate the magnitude of the net force on the girl while on the concrete surface.

$$F_{net} = ma$$
$$= 45 \times \frac{-1}{3}$$
$$= -15\,\text{N}$$

3. State the solution.

The magnitude of the net force on the girl on the concrete surface is −15 N.

b. 1. If the only horizontal force acting on the rollerblades is friction, the net force on the girl is the same as the friction force on the blades.

b. $$\frac{\text{Friction force of gravel path on rollerblades}}{\text{Friction force of concrete path on rollerblades}}$$
$$= \frac{F_{net} \text{ on girl while on gravel}}{F_{net} \text{ on girl while on concrete}}$$

2. Apply Newton's Second Law and cancel out common terms.

$$\frac{F_{net} \text{ on girl while on gravel}}{F_{net} \text{ on girl while on concrete}} = \frac{ma \text{ on gravel}}{ma \text{ on concrete}}$$
$$= \frac{a \text{ (during last 4 s)}}{a \text{ (during first 6 s)}}$$

3. Use the gradient of the graph to evaluate the ratio of accelerations.

$$\frac{a \text{ (during last 4 s)}}{a \text{ (during first 6 s)}} = \frac{\text{gradient (for last 4 s)}}{\text{gradient (for first 6 s)}}$$
$$= \frac{\left(\frac{-6}{4}\right)}{\left(\frac{-1}{3}\right)}$$
$$= \frac{36}{8}$$
$$= 4.5$$

4. State the solution.

The value of the ratio is 4.5.

PRACTICE PROBLEM 7

If the velocity-versus-time graph in sample problem 7 was applied to a car of mass 1200 kg on two road surfaces, what net force (in magnitude) acts on the car during:
a. the first 6 seconds
b. the final 4 seconds?

9.4.3 Falling down

Objects that are falling (or rising) through the air in the atmosphere near the surface of Earth are subjected to two forces — the force due to gravity and air resistance. The force due to gravity of the object is effectively constant. The magnitude of the air resistance, however, is not constant. It depends on many factors, including the object's speed, surface area and density. It also depends on the density of the body of air through which the object is falling. The air resistance is always opposite to the direction of motion. The net force on a falling object of mass m and force due to gravity F_g can therefore be expressed as:

$$F_{net} = ma \quad \text{(where } a \text{ is the acceleration of the object)}$$
$$F_g - \text{air resistance} = ma$$

When dense objects fall through small distances near the surface of Earth it is usually quite reasonable to assume that the air resistance is negligible. Thus:

$$\boldsymbol{F_g} = m\boldsymbol{a}$$
$$\Rightarrow m\boldsymbol{g} = m\boldsymbol{a} \quad \text{(where } \boldsymbol{g} \text{ is the gravitational field strength)}$$
$$\Rightarrow \boldsymbol{g} = \boldsymbol{a}$$

The acceleration of a body in free fall in a vacuum or where air resistance is negligible is equal to the gravitational field strength. At Earth's surface, where $\boldsymbol{g}$ = 9.8 N kg^{-1}, this acceleration is 9.8 m s^{-2}. The units N kg^{-1} and m s^{-2} are equivalent.

---

on Resources

📄 **Digital documents** Simulation of basketball throw (doc-0052)
eModelling: Skydiver spreadsheet (doc-0054)

▶ **Video eLesson** Air resistance (eles-0035)

EXTENSION: Terminal velocity

Note that this content is useful if you are choosing to study option 2.3 (How do heavy things fly?) in Unit 2, Area of Study 2.

Terminal velocity is reached when the forces acting on a falling object are balanced and it stops accelerating. When considering objects falling vertically downwards in the atmosphere near the surface of Earth, the two forces are air resistance or drag, $\boldsymbol{F_D}$, acting to oppose the motion of the object, and the force due to gravity, $\boldsymbol{F_g}$, acting vertically downwards (see figure 9.13).

The force due to gravity and air resistance are in balance when an object is at terminal velocity.

The force due to gravity, $\boldsymbol{F_g}$, depends only on the mass of the object, m, and the gravitational field strength, $\boldsymbol{g}$, which can both be assumed to be roughly constant during the fall of an object near the surface of Earth.

FIGURE 9.13 Skydivers accelerate until the drag force equals the force due to gravity, at terminal velocity.

The air resistance or drag, $\boldsymbol{F_D}$, acting on an object depends on a number of factors:
- C_D is the drag coefficient, which measures the ease with which air can move over an object. In simple terms, this indicates how streamlined an object is. This is a constant value for a particular object shape and orientation, and is often determined experimentally or via computer analysis.
- ρ is the air density measured in kg m^{-3}, which can be assumed to be constant; however, in reality it will vary with altitude, temperature, humidity and pressure.
- v is the velocity that the object is moving through the air measured in m s^{-1}.
- A is the cross-sectional or reference area that is perpendicular to the direction of motion of the object measured in m^2. For an object falling vertically this will be the horizontal cross-section.

When terminal velocity occurs, these two forces must be in balance (net force is zero, acceleration is zero).

$$\boldsymbol{F_D} = \boldsymbol{F_g}$$
$$\frac{1}{2}C_D\rho v^2 A = m\boldsymbol{g}$$

FIGURE 9.14 Forces acting on a falling object at terminal velocity

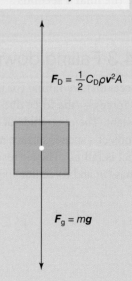

$$\boldsymbol{F_D} = \frac{1}{2}C_D\rho v^2 A$$

$$\boldsymbol{F_g} = m\boldsymbol{g}$$

This can be rearranged to determine a relationship for the terminal velocity.

$$v = \sqrt{\frac{2mg}{C_D \rho A}}$$

Provided that reasonable estimates can be made of the quantities involved, this relationship can be used to calculate the theoretical terminal velocity of falling objects. This can provide an interesting focus for an extended practical investigation.

If a bowling ball, a golf ball and a table tennis ball were dropped at the same instant from a height of 2 metres in a vacuum, they would all reach the ground at the same time. Each ball would have an initial velocity of zero, an acceleration of 9.8 m s^{-2} and a downward displacement of 2 metres.

If, however, the balls are dropped either in a classroom or outside, the table tennis ball will reach the ground a moment later than the other two balls.

The acceleration of each of the balls is:

$$a = \frac{F_{net}}{m}$$

$$= \frac{mg - F_D}{m} \quad \text{(where } F_D \text{ is air resistance)}$$

$$= \frac{mg}{m} - \frac{F_D}{m}$$

$$= g - \frac{F_D}{m}$$

$\frac{F_D}{m}$ is very small for the bowling ball and the golf ball. Even though the air resistance on the table tennis ball is small, its mass is also small and $\frac{F_D}{m}$ is not as small as it is for the other two balls.

WARNING: Do not drop a bowling ball from a height of 2 metres indoors. If you wish to try this experiment, replace the bowling ball with a medicine ball and keep your feet out of the way!

FIGURE 9.15 A bowling ball, a golf ball and a table tennis ball dropped from a height of 2 metres. Which one would you expect to reach the ground first?

9.4 Activities

learnon

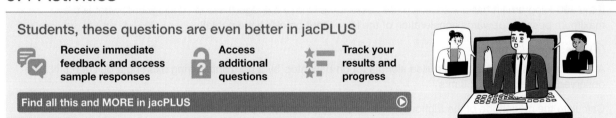

Students, these questions are even better in jacPLUS

Receive immediate feedback and access sample responses

Access additional questions

Track your results and progress

Find all this and MORE in jacPLUS

| 9.4 Quick quiz on | 9.4 Exercise | 9.4 Exam questions |

9.4 Exercise

1. What is an idealisation? Provide an example of an idealism that could be used to simplify a physics problem.
2. When a space shuttle takes off, its initial acceleration is 3.0 m s^{-2}. It has an initial mass of about 2.2×10^6 kg.
   a. Determine the magnitude of the net force on the space shuttle as it takes off.
   b. What is the magnitude of the upward thrust as it takes off?

3. A 6-kilogram bowling ball and a 60-kilogram gold bar are dropped at the same instant from the third floor of the Leaning Tower of Pisa. Use Newton's Second Law to explain why:
   a. they both reach the ground at the same time
   b. a 6-kilogram doormat dropped from the same location at the same time takes significantly longer to reach the ground.
4. A bungee jumper with a mass of 70 kilograms leaps from a bridge.
   a. What is the force due to gravity acting on the bungee jumper?
   b. During which part of the jump is:
      i. the upwards force on the jumper due to the tension in the bungee cord greater than the force due to gravity on the jumper
      ii. the force due to gravity on the jumper greater than the upward pull of the bungee cord?
   c. What tension in the bungee cord is needed for the jumper to travel at a constant speed? Does this occur at any time during the jump? Explain.
5. A car of mass 1200 kilograms starts from rest on a horizontal road with a forward driving force of 10 000 N. The resistance to motion due to road friction and air resistance totals 2500 N.
   a. What is the magnitude of the net force on the car?
   b. What is the magnitude of the acceleration of the car?
   c. What is the speed of the car after 5 seconds?
   d. How far has the car travelled after 5 seconds?
6. A train of mass 8.0×10^6 kilograms travelling at a speed of 25 m s^{-1} is required to stop over a maximum distance of 360 metres. What frictional force must act on the train when the brakes are applied if the train is to do this?
7. A short-sighted skier of mass 70.0 kilograms suddenly realises while travelling at a speed of 12.0 m s^{-1} that there is a steep cliff 50.0 metres straight ahead. What frictional force is required on the skier if he is to stop just before he skis off the edge of the cliff?
8. A Physics teacher decides, just for fun, to stand on some bathroom scales (calibrated in newtons) in a lift. The scales provide a measure of the force with which they push up on the teacher. When the lift is stationary, the reading on the bathroom scales is 700 N. What will be the reading on the scales when the lift is:
   a. moving upwards at a constant speed of 2.0 m s^{-1}
   b. accelerating downwards at 2.0 m s^{-2}
   c. accelerating upwards at 2.0 m s^{-2}?
9. A ball of mass 0.50 kilograms is thrown vertically upwards.
   a. What is the velocity of the ball at the top of its flight?
   b. What is the magnitude of the ball's acceleration at the top of its flight?
   c. What is the net force on the ball at the top of its flight?

9.4 Exam questions

▶ Question 1 (3 marks)

The cable holding a lift would break if the tension in it were to exceed 25 000 N.

If the 480-kilogram lift has a load limit of 24 passengers whose average mass is 70 kilograms, what is the maximum possible upwards acceleration of the lift without breaking the cable?

▶ Question 2 (3 marks)

A block of mass 2.5 kg is being pulled along a rough benchtop by a horizontal string exerting a pull of 6.5 N. The observed acceleration is 2.0 m s^{-2}.

Calculate the magnitude of the friction force exerted on the benchtop by the block.

▶ Question 3 (3 marks)

In a large warehouse, a golf cart is used to pull a small trailer by means of a horizontal connecting rod.

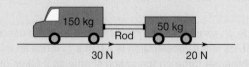

The masses of the cart and trailer are 150 kg and 50 kg respectively. The retarding friction forces on the cart and trailer are 30 N and 20 N respectively. The driving force is provided by the cart. At one time, the acceleration of the cart and trailer is 1.5 m s^{-2}.

Determine the tension in the connecting rod at this time.

▶ **Question 4 (3 marks)**

A student applies a pulling force of 6.0 N to a block of mass 2.0 kg on a wooden benchtop. She measures the acceleration of the block as 2.5 m s^{-2}. The student comments that she had expected a larger value of acceleration.

What is the force that the student has overlooked?

Calculate the magnitude of this force.

▶ **Question 5 (3 marks)**

A car, of mass 1200 kg, is pulling a trailer, of mass 600 kg, on a straight, horizontal road using a towrope. The total retarding force on the car is 800 N and on the trailer it is 400 N.

Calculate the tension in the towrope when the car and the trailer are both accelerating forward at 1.5 m s^{-2}.

More exam questions are available in your learnON title.

9.5 Newton's Third Law of Motion

KEY KNOWLEDGE

- Apply Newton's three laws of motion to a body on which forces act: $a = \dfrac{F_{net}}{m}$, $F_{on\,A\,by\,B} = -F_{on\,B\,by\,A}$
- Model forces as vectors acting at the point of application (with magnitude and direction), labelling these forces using the convention 'force on A by B' or $F_{on\,A\,by\,B} = -F_{on\,B\,by\,A}$

Source: VCE Physics Study Design (2023–2027) extracts © VCAA; reproduced by permission.

9.5.1 Forces in pairs

Newton said that forces always act in pairs. Forces represent or model the interaction between two objects. Each object acts on the other, so where there is one force by one object, there is always another force by the other object, as illustrated in figure 9.16.

The pairing of these forces is apparent in the symmetry of their labels. If the label for one force is $F_{on\,B\,by\,A}$, then the label for the other is $F_{on\,A\,by\,B}$. It is important to note that in these pairs of forces, the forces act on different objects.

Newton not only identified forces acting in pairs, he also said that these two forces act in opposite directions and are equal in magnitude or size. This statement became Newton's Third Law of Motion.

FIGURE 9.16 Forces act in pairs.

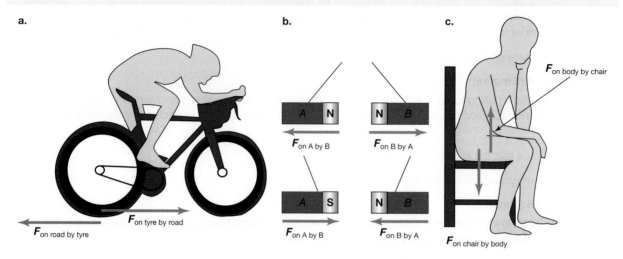

a.

$F_{\text{on road by tyre}}$ $F_{\text{on tyre by road}}$

b.

$F_{\text{on A by B}}$ $F_{\text{on B by A}}$

$F_{\text{on A by B}}$ $F_{\text{on B by A}}$

c.

$F_{\text{on body by chair}}$

$F_{\text{on chair by body}}$

Newton's Third Law of Motion:

If object B applies a force to object A, then object A applies an equal and opposite force to object B.

$$F_{\text{on A by B}} = -F_{\text{on B by A}}$$

This symmetry between the pair of forces can be used to identify the other of the pair if only one is given.

Action and reaction

Some texts summarise Newton's Third Law as 'For every action, there is an equal and opposite reaction'. This version is not preferred and can cause confusion. The word 'reaction' here has a different meaning to its use in 'normal force'. The statement also implies one force in the pair is a response to the other, which is incorrect.

There are four forces acting as two force pairs when you sit on a comfortable chair. Earth pulls down on you, and the compressed springs and foam push up on you. So, one force pair is the upward push by the springs on you and the downward push by the bones in your pelvis on the chair. The second force pair is Earth pulling down on you and you pulling Earth upwards.

The net force on a person sitting in a chair is the vector sum of all the forces acting on the person. The net force is zero because the upward push by the chair, $F_{\text{on student by chair}}$, is balanced by the downward, and of equal size, pull of the force of gravity, $F_{\text{on student by Earth}}$.

FIGURE 9.17 Forces while sitting

$F_{\text{on student by chair}}$
Chair pushes student

$F_{\text{on chair by student}}$
Student pushes chair

SAMPLE PROBLEM 8 Action and reaction: Newton's Third Law of Motion

Two ice skaters, Jack and Jill, push each other away. The figure shows the force on Jill by Jack with a black arrow. Draw an arrow and label it to show the force on Jack by Jill.

$F_{\text{on Jill by Jack}}$

THINK

1. Recall Newton's Third Law of Motion.

2. Draw the force on Jack by Jill as an arrow of equal length in the opposite direction.

WRITE

$$F_{\text{on A by B}} = -F_{\text{on B by A}}$$

$F_{\text{on Jack by Jill}}$

PRACTICE PROBLEM 8

Draw and label arrows for the other forces in the following force pairs.

a.

$F_{\text{on boy by girl}}$

b.

$F_{\text{on gas by rocket}}$

c.

$\mathbf{F}_{\text{on apple by Earth}}$

d.

$\mathbf{F}_{\text{on ball by club}}$

9.5.2 Moving forward

The rowing boat in figure 9.18 is propelled forward by the push of water on the oars. As the face of each oar pushes backwards on the water, the water pushes forward with an equal and opposite force on each oar. The force on the oars, which are connected to the boat, propels the rowers and their boat forward. A greater push on the water results in a greater push on the oar.

FIGURE 9.18 This rowing team relies on a reaction force to propel itself forward.

In fact, none of your forward motion — whether you are on land, water or in the air — could occur without a Newton's Third Law force pair.

- When you swim, you push the water backwards with your hands, arms and legs. The water pushes in the opposite direction, propelling you forwards.
- In order to walk or run, you push your feet backwards and down on the ground. The ground pushes in the opposite direction, pushing forwards and up on your feet.
- The forward driving force on the wheels of a car is the result of a backwards push on the road by the wheels.
- A jet or a propeller-driven plane is thrust forwards by air. The jet engines or propellers are designed to push air backwards with a very large force. The air pushes forwards on the plane with an equally large force.

 **Resources**

9.5 Activities

| 9.5 Quick quiz on | 9.5 Exercise | 9.5 Exam questions |

9.5 Exercise

1. Copy the following table into your workbook. Describe fully the missing half of the following force pairs.

Force 1	Pair of force 1
You push on a wall with the palm of your hand.	
Your foot pushes down on a bicycle pedal.	
The ground pushes up on your feet while you are standing.	
Earth pulls down on your body.	
You push on a broken-down car to try to get it moving.	
A hammer pushes down on a nail.	

2. Label all of the forces in the following figures in the form $F_{\text{on A by B}}$.

a.

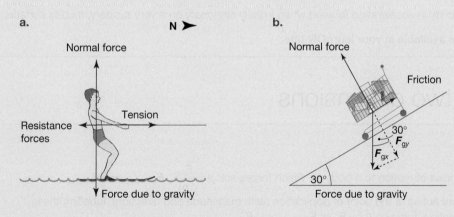

b.

3. Identify the force pairs in the following figure.

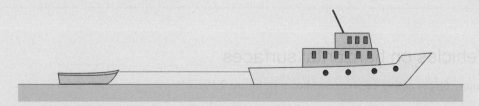

4. Explain, in terms of Newton's First and Third Laws of Motion, why a freestyle swimmer moves faster through the water than a breaststroke swimmer.

5. A student says that the friction forces on the front and back tyres of a car are an example of Newton's Third Law of Motion. Is the student correct? Explain.

9.5 Exam questions

▶ Question 1 (1 mark)
A car is travelling along a highway.

Identify two Newton's Third Law pairs of forces acting on the car using the convention $F_{\text{on A by B}}$.

▶ Question 2 (1 mark)
MC Which of the following statements about Newton's Third Law of Motion is *false*?
A. The two forces are opposite in direction.
B. The two forces act on the same object.
C. The two forces are equal in magnitude.
D. The two forces act on different objects.

▶ Question 3 (1 mark)
MC A small car has a head-on collision with a very heavy truck.

How do the magnitudes of the forces from the collision on each vehicle compare?
A. There is a larger magnitude force on the small car.
B. There is a larger magnitude force on the heavy truck.
C. The forces are equal in magnitude.
D. There is insufficient information to make a comparison.

▶ Question 4 (3 marks)
Two skaters, P and Q, are standing facing each other on smooth ice. P has twice the mass of Q. They push one another apart and Q moves away with an initial acceleration of magnitude 3.0 m s^{-2}.

Calculate the magnitude of the initial acceleration of P. Show your reasoning.

▶ Question 5 (3 marks)
Explain why a car has difficulty in accelerating forward when initially stationary on a very slippery, muddy surface.

More exam questions are available in your learnON title.

9.6 Forces in two dimensions

KEY KNOWLEDGE

- Apply Newton's three laws of motion to a body on which forces act: $a = \dfrac{F_{\text{net}}}{m}$, $F_{\text{on A by B}} = -F_{\text{on B by A}}$
- Model forces as vectors acting at the point of application (with magnitude and direction), labelling these forces using the convention 'force on A by B' or $F_{\text{on A by B}} = -F_{\text{on B by A}}$

Source: VCE Physics Study Design (2023–2027) extracts © VCAA; reproduced by permission.

9.6.1 Vehicles on horizontal surfaces

There are many forces acting on a car, as shown in figure 9.19.

The forces acting on a car being driven along a straight, horizontal road are:
- *the force due to gravity.* The force applied by Earth on the car acts through the centre of mass, or balancing point, of the car.
- *the normal force.* The force applied on the car by the road is a reaction to the force applied on the road by the car. A normal force pushes up on all four wheels.

- *the driving force*. This is provided by the road and is applied to the driving wheels. As a tyre on a driven wheel pushes back on the road, the road pushes forward on the tyre, propelling the car forward. The push of the road on the tyre is a type of friction commonly referred to as traction, or grip.
- *road friction*. This is the retarding force applied by the road on the tyres of the non-driving wheels.
- *air resistance*. The drag, or air resistance acting on the car, increases as the car moves faster. As a fluid friction force, air resistance can be reduced by streamlining the vehicle or reducing its size.

FIGURE 9.19 Forces acting on a car. The state of motion of a front-wheel-drive car on a horizontal road depends on the net force acting on it.

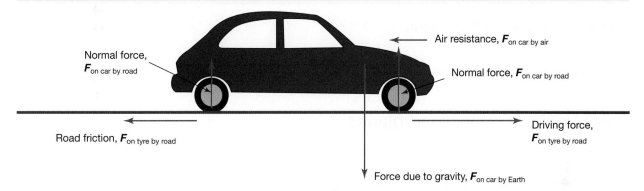

Normal force, $F_{\text{on car by road}}$

Air resistance, $F_{\text{on car by air}}$

Normal force, $F_{\text{on car by road}}$

Road friction, $F_{\text{on tyre by road}}$

Driving force, $F_{\text{on tyre by road}}$

Force due to gravity, $F_{\text{on car by Earth}}$

The net force acting on the car in figure 9.19 is zero. It is therefore moving along the road at constant speed. We know that it is moving to the right because both the air resistance and road friction act in a direction opposite to that of motion. If the car were stationary, neither of these forces would be acting at all.
- If the driving force were to increase, the car would speed up until the sum of the air resistance and road friction grew large enough to balance the driving force. Then, once again, the car would be moving at a constant, although higher, speed.
- If the driver stopped pushing down on the accelerator, the motor would stop turning the driving wheels and the driving force would become zero. The net force would be to the left. As the car slowed down, the air resistance and road friction would gradually decrease until the car came to a stop. The net force on the car would then be zero until such time as the driving force was restored.

CASE STUDY: Computer-controlled brake systems

An anti-lock brake system (ABS) allows the wheels to keep rolling no matter how hard the brakes are applied. A small computer attached to the braking system monitors the rotation of the wheels. If the wheels lock and rolling stops, the pressure on the brake pads (or shoes) that stops the rotation is reduced briefly. This action is repeated up to 15 times each second. Anti-lock brake systems are most effective on wet roads. However, even on a dry surface, braking distances can be reduced by up to 20 per cent.

FIGURE 9.20 ABS on a motorcycle

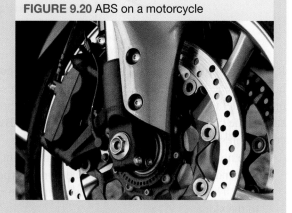

When brakes in an older car without ABS are applied too hard, as they often are when a driver panics, the wheels lock. The resulting sliding friction is less than the friction acting when the wheels are still rolling. The car skids, steering control is lost and the car takes longer to stop than if the wheels were still rolling. Before ABS became commonplace, drivers were often advised to 'pump' the brakes in wet conditions to prevent locking. This involves pushing and releasing the brake pedal in quick succession until the car stops. This, however, is very difficult to do in an emergency situation.

Since the advent of ABS, a myriad additional computer-controlled systems have been developed to improve safety when braking. One such system, often referred to as Brake Assist, monitors the application of force to the brake pedal to detect an emergency and apply additional brake pressure. More advanced systems use a range of technology to monitor conditions on the road ahead and can warn the driver to apply the brakes, and in some systems even apply the brakes automatically if there is no driver response. The development of systems such as these over recent decades has been a vital precursor to the rapid development of autonomous cars that is underway at present.

elog-1769

INVESTIGATION 9.4

Static, sliding and rolling friction

Aim

To compare the relative sizes of different forms of friction

9.6.2 Vehicles on inclined planes

A car left parked on a hill will begin to roll down the hill with increasing speed if it is left out of gear and the handbrake is off. Figure 9.21a shows the forces acting on such a car. In order to simplify the diagram, all of the forces are drawn as if they were acting through the centre of mass of the car. The forces on the car can then be modelled as acting on a single point. The direction of net force acting on the car is down the hill. It is clear that the force of gravity is a major contributor to the downhill motion of the car.

FIGURE 9.21 a. A simplified diagram of the forces acting on a car rolling down a slope **b.** Vectors can be resolved into components. In this case, the force due to gravity has been resolved into two components. The net force is parallel to the slope.

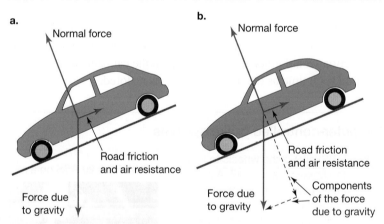

It is often useful to divide vectors into parts called **components**. Figure 9.21b shows how the force due to gravity can be broken up, or resolved, into two components — one parallel to the slope and one perpendicular to the slope. Notice that the vector sum of the components is the force due to gravity. By resolving it into these two components, two useful observations can be made:

components parts of a vector; a vector can be resolved into a number of components, and when all the components are added together, the result is the original vector

1. The normal force is balanced by the component of force due to gravity that is perpendicular to the surface. The net force perpendicular to the road surface is zero. This must be the case because there is no change in motion perpendicular to the slope.
2. The magnitude of the net force is simply the difference between the magnitude of the component of the force due to gravity that is parallel to the surface and the sum of the road friction and air resistance.

SAMPLE PROBLEM 9 Determining the magnitude of road friction on an incline

A car of mass 1600 kilograms left parked on a steep but rough road begins to roll down the hill. After a short while it reaches a constant speed. The road is inclined at 15° to the horizontal. The car's speed is sufficiently slow that the air resistance is insignificant and can be ignored. Determine the magnitude of the road friction on the car while it is rolling at constant speed.

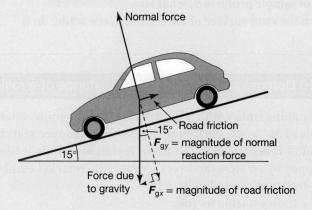

THINK	WRITE
1. Because the car is rolling at constant speed, the net force acting on it is zero.	$F_{net} = 0$
2. As the net force is zero, the magnitude of the friction must be equal to the magnitude of the force due to gravity that is in the direction of the slope.	$\text{Friction} = F_{gx}$
3. The force due to gravity in the direction of the slope can be calculated using the trigonometric ratio sine.	$\sin(15°) = \dfrac{F_{gx}}{F_g}$ $\Rightarrow F_{gx} = F_g \sin(15°)$
4. The force due to gravity can be found using the formula $F_g = mg$.	$F_g = mg$ $= 1600 \times 9.8$ $= 15\,680 \, \text{N}$
5. Substitute the force due to gravity into the formula for F_{gx}.	$F_{gx} = F_g \sin(15°)$ $= 15\,680 \sin(15°)$ $\approx 4058 \, \text{N}$

The road friction on the car is 4058 N.

Note: It is useful to consider the effect on the net force of the angle of the incline to the horizontal. If the angle is greater than 15°, the component of the force due to gravity parallel to the slope increases and the net force will no longer be zero. The speed of the car will therefore increase. The component of the force due to gravity perpendicular to the slope decreases and the normal force decreases by the same amount.

PRACTICE PROBLEM 9

a. A 5000-kilogram truck is parked on a road surface inclined at an angle of 20° to the horizontal. Calculate the component of the force due to gravity on the truck that is:
  i. down the slope of the road
  ii. perpendicular to the slope of the road.
b. In the case of the car in sample problem 9, what is:
  i. the component down the road surface of the normal force acting on it
  ii. the normal force?

SAMPLE PROBLEM 10 Determining the speed and distance of a rolling shopping trolley

tlvd-0085

A loaded supermarket shopping trolley with a total mass of 60 kilograms is left standing on a footpath that is inclined at an angle of 30° to the horizontal. As the tired shopper searches for his car keys, he fails to notice that the trolley is beginning to roll away. It rolls in a straight line down the footpath for 9 seconds before it is stopped by an alert (and very strong) supermarket employee. Find the:
a. speed of the shopping trolley at the end of its roll
b. distance covered by the trolley during its roll.
Assume that the footpath exerts a constant friction force of 270 N on the runaway trolley.

THINK

a. 1. Draw a diagram to show the three forces acting on the shopping trolley. Air resistance is not included as it is negligible. The forces should be shown as acting through the centre of mass of the loaded trolley, as in the diagram shown. The components of the force due to gravity, which are parallel and perpendicular to the footpath surface, should also be shown on the diagram.

2. Calculate the net force.

3. Use Newton's Second Law to calculate the acceleration.

4. Use a constant acceleration formula to calculate the final speed.

5. State the solution.

WRITE

a.

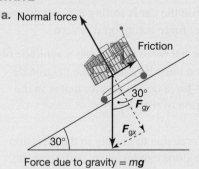

Force due to gravity $= mg$
$= 588$ N

$F_{net} = F_{gx} - \text{friction}$

$= mg\sin(30°) - 270\,\text{N}$

$= 588\,\text{N} \times \sin(30°) - 270\,\text{N}$

$= 294\,\text{N} - 270\,\text{N}$

$= 24\,\text{N}$

$F_{net} = ma$
$24 = 60a$
$\Rightarrow a = \dfrac{24}{60}$
$= 0.4\,\text{m s}^{-2}$ down the slope

$v = u + at$
$= 0 + 0.4 \times 9$
$= 3.6\,\text{m s}^{-1}$

The speed of the trolley at the end of its roll is 3.6 m s^{-1}.

b. 1. Use a constant acceleration formula to calculate the distance covered by the trolley.	**b.** $s = ut + \dfrac{1}{2}at^2$	
	$= 0 + \dfrac{1}{2} \times 0.4 \times 9^2$	
	$= 16.2 \text{ m}$	
2. State the solution.	The trolley travels 16.2 metres before it is stopped.	

PRACTICE PROBLEM 10

A cyclist rolls freely from rest down a slope inclined at 20° to the horizontal. The total mass of the bicycle and cyclist is 100 kilograms. The bicycle rolls for 12 seconds before reaching a horizontal surface. The surface exerts a constant friction force of 300 N on the bicycle.

a. What is the net force on the bicycle (including the cyclist)?
b. What is the acceleration of the bicycle?
c. What is the speed of the bicycle when it reaches the horizontal surface?

Resources

▶ **Video eLesson** Motion down an inclined plane (eles-0034)

🔗 **Weblink** Inclined plane

9.6.3 Connected objects

In many situations, Newton's laws need to be applied to more than one body.

Figure 9.22 shows a small dinghy being pulled by a larger boat. The forces acting on the larger boat are labelled in red, while the forces acting on the small dinghy are labelled in green. Newton's Second Law can be applied to each of the two boats separately.

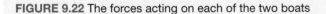

FIGURE 9.22 The forces acting on each of the two boats

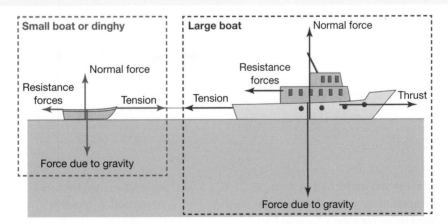

Figure 9.23 shows only the forces acting on the whole system of the two boats and the rope joining them. When Newton's Second Law is applied to the whole system, the system is considered to be a single object.

The thrust that acts on the larger boat and the system is provided by the water. The propeller of the larger boat pushes backwards on the water and the water pushes forwards on the propeller blades. The only force that can cause the small dinghy to accelerate forwards is the tension in the rope.

If the tension in the rope is greater than the resistance forces on the dinghy, the dinghy will accelerate. If the tension in the rope is equal to the resistance forces on the dinghy, it will move with a constant velocity. If the tension in the rope is less than the resistance forces on the dinghy, it will slow down; that is, its acceleration will be negative.

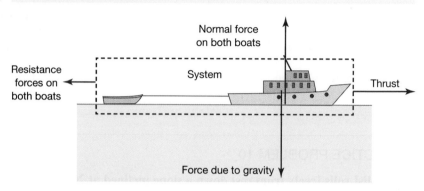

FIGURE 9.23 The forces acting only on the whole system. The system consists of the two boats and the rope joining them.

The rope pulls back on the larger boat with the same force that it applies in a forward direction on the small dinghy. This is consistent with Newton's Third Law. Through the rope, the larger boat pulls forwards on the small dinghy with a force that is equal and opposite to the force with which the small dinghy pulls backwards on the large boat.

tlvd-0086

SAMPLE PROBLEM 11 Calculating the acceleration and force of a car pulling a trailer

A car of mass 1400 kilograms towing a trailer of mass 700 kilograms accelerates at a constant rate on a horizontal road. A thrust of 5400 N is provided by the forward push of the road on the driving wheels of the car. The road friction on the car is 800 N, while that on the trailer is 400 N. The air resistance on both the car and the trailer is negligible. Determine the:

a. acceleration of both the car and trailer
b. force with which the trailer is pulled by the car (labelled P in the figure in the solution).

THINK

a. 1. Draw a diagram to show the forces acting on the car and trailer.

2. Consider the car and trailer as a system. The acceleration of the car and trailer can be calculated using Newton's Second Law if the net force on the system is known.

WRITE

a.

$$F_{net} = \text{driving force} - \text{road friction (car)}$$
$$- \text{road friction (trailer)}$$
$$= 5400\,\text{N} - 800\,\text{N} - 400\,\text{N}$$
$$= 4200\,\text{N}$$

3.	Use Newton's Second Law to calculate the acceleration.	$F_{net} = ma$ $4200 = 2100a$ $\Rightarrow a = \dfrac{4200}{2100}$ $= 2 \text{ m s}^{-2}$ to the right
4.	State the solution.	The acceleration of both the car and trailer is 2 m s^{-2} to the right.
b. 1.	Calculate the force with which the trailer is pulled by the car, P, by considering the net force on the trailer.	b. $\quad F_{net} = ma$ $P - 400 = 700 \times 2$ $\Rightarrow P = 700 \times 2 + 400$ $= 1800 \text{ N}$
2.	State the solution.	The force with which the trailer is pulled by the car is 1800 N.

PRACTICE PROBLEM 11

A boat of mass 2000 kilograms tows a small dinghy of mass 100 kilograms with a thick rope. The boat's propellers provide a forward thrust of 4700 N. The total resistance forces of air and water on the boat and dinghy system amount to 400 N and 100 N respectively.
a. **What is the acceleration of the boat and dinghy?**
b. **What is the net force on the dinghy?**
c. **What is the magnitude of the tension in the rope?**

9.6 Activities

9.6 Quick quiz	9.6 Exercise	9.6 Exam questions

9.6 Exercise

1. A ball rolls down a hill with an increasing speed.
   a. Draw a diagram to show all of the forces acting on the ball.
   b. What is the direction of the net force on the ball?
   c. What is the largest single force acting on the ball?
   d. When the ball reaches a horizontal surface, it slows, eventually coming to a stop. Explain, with the aid of a diagram, why this happens.
2. When you try to push a broken-down car with its handbrake still on, it does not move. What other forces are acting on the car to produce a net force of zero?

3. Redraw figure 9.19 for a car with rear-wheel drive.
4. A cyclist of mass 60 kilograms is riding up a hill inclined at 30° to the horizontal at a constant speed. The mass of the bicycle is 20 kilograms. The provided figure shows the forces acting on the bicycle–cyclist system.

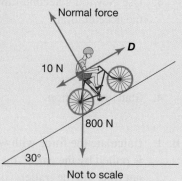

a. What is the net force on the bicycle–cyclist system?
b. What is the magnitude of the component of the force due to gravity on the system that is parallel to the road surface?
c. The sum of the magnitudes of the road friction and air resistance on the system is 10 N. What is the magnitude of the driving force, **D**?
d. What is the magnitude of the normal force on the bicycle–cyclist system?

5. An experienced downhill skier with a mass of 60 kilograms (including skis) is moving down a slope inclined at 30° with increasing speed. She is moving in a straight line down the slope.
a. What is the direction of the net force on the skier?
b. Draw a diagram showing the forces acting on the skier. Show all of the forces as if they were acting through her centre of mass.
c. What is the magnitude of the component of the skier's force due to gravity that is parallel to the slope?
d. If the sum of the forces resisting the movement of the skier down the slope is 8 N, what is the magnitude of the net force on her?

6. A skateboarder of mass 60 kilograms accelerates down a slope inclined at an angle of 30° to the horizontal. Her acceleration is a constant 2.0 m s^{-2}. What is the magnitude of the friction force resisting her motion?

7. A roller-coaster carriage (and occupants) with a total mass of 400 kilograms rolls freely down a straight track inclined at 40° to the horizontal with a constant acceleration. The frictional force on the carriage is a constant 180 N. What is the magnitude of the acceleration of the carriage?

8. A skateboarder of mass 56 kilograms is rolling freely down a straight incline. The motion of the skateboarder is described in the graph shown.
a. What is the magnitude of the net force on the skateboarder?
b. If the friction force resisting the motion of the skateboarder is a constant 140 N, at what angle is the slope inclined to the horizontal?

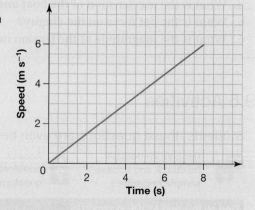

9. What force provides the forward force that gets you moving when you are:
a. ice skating
b. downhill skiing
c. waterskiing
d. skateboarding
e. swimming
f. rowing?

10. Front-wheel-drive cars have a number of advantages over rear-wheel-drive cars. Compare and comment on the forces acting on the tyres in the two different types of car while being driven at a constant speed on a horizontal road.

11. The magnitude of the force due to air resistance, **R**, on a car of mass 1200 kilograms can be approximated by the formula $R = 0.6v^2$, where **R** is measured in newtons and **v** is the speed of the car in m s^{-1}.
a. Design a spreadsheet to calculate the magnitude of the force of air resistance and the net force on a car for a range of speeds as it accelerates from 20 km h^{-1} to 50 km h^{-1} on a horizontal road. Assume that, while accelerating, the driving force is a constant 1800 N and the road friction on the non-driving wheels is a constant 300 N.
b. Use your spreadsheet to plot a graph of the net force versus speed for the car.
c. Modify your spreadsheet to show how the net force on the car changes when the same acceleration from 20 km h^{-1} to 60 km h^{-1} is undertaken while driving down a road at an angle of 10° to the horizontal.

12. A well-coordinated rollerblader is playing with a yoyo while accelerating on a horizontal surface. When the yoyo is at its lowest point for several seconds, it makes an angle of 5° with the vertical, as shown in the provided figure. Determine the acceleration of the rollerblader.

13. A student argues that since there are friction forces on the front and back wheels of a bicycle that act in opposite directions, the bicycle cannot move. Explain how the bicycle moves.

14. Two loaded trolleys of masses 3.0 kilograms and 4.0 kilograms, which are joined by a light string, are pulled by a spring balance along a smooth, horizontal laboratory bench as shown in the following figure. The reading on the spring balance is 14 N.

    a. What is the acceleration of the trolleys?
    b. What is the magnitude of the tension in the light string joining the two trolleys?
    c. What is the net force on the 4.0-kilogram-trolley?
    d. What would be the acceleration of the 4.0-kilogram trolley if the string was cut?

15. A warehouse worker applies a force of 420 N to push two crates across the floor as shown in the following figure. The friction force opposing the motion of the crates is a constant 2.0 N for each kilogram.
    a. What is the acceleration of the crates?
    b. What is the net force on the 40-kilogram crate?
    c. What is the force exerted by the 40-kilogram crate on the 30-kilogram crate?
    d. What is the force exerted by the 30-kilogram crate on the 40-kilogram crate?
    e. Would the worker find it any easier to give the crates the same acceleration if the positions of the two blocks were reversed? Support your answer with calculations.

9.6 Exam questions

▶ Question 1 (3 marks)

A cyclist is riding his bike along a straight, flat road. A trailer is connected to the bike by means of a horizontal tow bar.

The cyclist and bike have a total mass of 80 kg; the trailer has a mass of 20 kg.

The driving force on the bike is 220 N.

The total retarding force on the cyclist and bike is 150 N; the retarding force on the trailer is 50 N.

Calculate the acceleration of the trailer. Show your reasoning.

▶ Question 2 (1 mark)

MC A cyclist is moving forward at constant speed. The friction force on the rear tyre is 40 N forward. The retarding force of friction on the tyres is 10 N.

Which of the following is the best estimate of the drag force on the bike and rider?
A. 50 N
B. 40 N
C. 30 N
D. 10 N

Question 3 (1 mark)

MC The box on the ramp in the provided figure shown has a weight of 98 N. It is sliding at a constant speed down the slope.

Which of the following best gives the magnitude of the normal reaction force exerted on the box by the ramp?
A. 49 N
B. 63 N
C. 75 N
D. 98 N

Question 4 (1 mark)

MC A bike is moving forwards at constant speed.

What is the friction force of the road on the rear tyre?
A. Backwards
B. Forwards
C. Upwards
D. Downwards

Question 5 (1 mark)

MC A box sits at rest, on a slope inclined at 30° to the horizontal. The mass of the box is 4.0 kg.

Which of the following is the best estimate of the normal reaction force on the box?
A. 19.6 N
B. 22.6 N
C. 33.9 N
D. 39.2 N

More exam questions are available in your learnON title.

9.7 Momentum and impulse

KEY KNOWLEDGE

- Apply concepts of momentum to linear motion: $p = mv$
- Explain changes in momentum as being caused by a net force: $\Delta p = F_{net}\Delta t$

Source: VCE Physics Study Design (2023–2027) extracts © VCAA; reproduced by permission.

9.7.1 Momentum

How difficult is it to stop a moving object? How difficult is it to make a stationary object move? The answer to both of these questions depends on two physical characteristics of the object:

- The object's mass
- How fast the object is moving, or how fast you want it to move.

> **momentum** the product of the mass of an object and its velocity; it is a vector quantity

The product of these two physical characteristics is called **momentum**.

The momentum, p, of an object of mass m with a velocity v is defined as:

$$p = mv$$

Momentum is a vector quantity and has SI units of kg m s^{-1}.

SAMPLE PROBLEM 12 Calculating the momentum of a train

What is the momentum of a train of mass 8×10^6 kilograms that is travelling at a speed of 15 m s^{-1} in a northerly direction?

THINK	WRITE
1. Recall the formula for momentum.	$p = mv$
2. Substitute the mass and velocity in to find the momentum.	$p = mv$ $= 8 \times 10^6 \times 15$ $= 1.2 \times 10^8 \text{ kg m s}^{-1}$
3. State the solution.	The momentum of the train is 1.2×10^8 kg m s^{-1} north.

PRACTICE PROBLEM 12

A car of mass 1200 kilograms travels east with a constant speed of 15 m s^{-1}. It then undergoes a constant acceleration of 3 m s^{-2} for 2 seconds. What is the momentum of the car:
a. before it accelerates
b. at the end of the 2-seconds acceleration?

9.7.2 Impulse

Making an object stop, or causing it to start moving, requires a non-zero net force. The relationship between the net force acting on an object and its momentum can be explored by applying Newton's Second Law to the object.

$$F_{net} = ma$$
$$\Rightarrow F_{net} = m\left(\frac{\Delta v}{\Delta t}\right)$$
$$\Rightarrow F_{net}\Delta t = m\Delta v$$

The product $F_{net}\Delta t$ is called the **impulse** of the net force. The impulse of any force is defined as the product of the force and the time interval over which it acts. Impulse is a vector quantity with SI units of N s.

impulse the product of the force and the time interval over which it acts; it is a vector quantity

$$m\Delta v = m(v - u)$$
$$= mv - mu$$
$$= p_f - p_i$$

where:

p_f = the final momentum of the object

p_i = the initial momentum of the object.

Thus, the effect of a net force on the motion of an object can be summarised by the statement:

$$\text{Impulse = change in momentum}$$

In fact, when translated from the original Latin, Newton's Second Law reads:

The rate of change of momentum is directly proportional to the magnitude of the net force and is in the direction of the net force.

This is expressed algebraically as:

$$F_{net} = \frac{\Delta p}{\Delta t}$$

The effect of a net force on the motion of an object can be summarised by the statement:

Impulse = change in momentum

$$F_{net}\Delta t = m\Delta v$$

tlvd-0088

SAMPLE PROBLEM 13 Calculating the change in momentum and impulse of a ball

A 30-gram squash ball hits a wall horizontally at a speed of 15 m s^{-1} and bounces back in the opposite direction at a speed of 12 m s^{-1}. It is in contact with the wall for an interval of 1.5×10^{-3} seconds.
a. What is the change in momentum of the squash ball?
b. What is the impulse on the squash ball?
c. What is the magnitude of the force exerted by the wall on the squash ball?

THINK

a. 1. Change in momentum is calculated from the initial and final momentums. Consider towards the wall as the positive direction. (*Note:* This decision is arbitrary — you could choose the positive direction to be away from the wall and your answer will have the same magnitude but opposite sign.)

   2. State the solution.

b. 1. Impulse of the net force on the squash ball = change in momentum of the squash ball

   2. State the solution.

WRITE

a. $p_i = mu$ $p_f = mv$
    $= 0.03 \times 15$ $= 0.03 \times -12$
    $= 0.45\,\text{kg m s}^{-1}$ $= -0.36\,\text{kg m s}^{-1}$

$\Delta p = p_f - p_i$
   $= -0.36 - 0.45$
   $= -0.81\,\text{kg m s}^{-1}$

The change in momentum of the squash ball is 0.81 kg m s^{-1} away from the wall.

b. $I = \Delta p$
   $= -0.81\,\text{N s}$

The impulse on the squash ball is 0.81 N s away from the wall.

c. 1. Recall the formula for the net force.

c. $F_{\text{net}} = \dfrac{\Delta p}{\Delta t}$

2. Substitute the change in momentum and change in time into the equation to find the net force.

$F_{\text{net}} = \dfrac{\Delta p}{\Delta t}$

$= \dfrac{-0.81}{1.5 \times 10^{-3}}$

$= 540 \, \text{N}$

3. State the solution.

The magnitude of the force exerted by the wall on the squash ball is 540 N.

PRACTICE PROBLEM 13

During a crash test a 1400-kilogram car travelling at 16 m s^{-1} collides with a steel barrier and rebounds with an initial speed of 4.0 m s^{-1} before coming to rest. The car is in contact with the barrier for 1.4 seconds. What is the magnitude of:
a. the change in momentum of the car during contact with the barrier
b. the impulse applied to the car by the barrier
c. the force exerted by the barrier on the car?

INVESTIGATION 9.5

Impulse, momentum and Newton's Second Law of Motion

Aim

To measure and record the velocity of the trolley (or glider) at two separate instants as the load is falling

---

9.7.3 Determining impulse through graphical analysis

The force that was determined in sample problem 13 was actually the average force on the squash ball. In fact, the force acting on the squash ball changes, reaching its maximum magnitude when the centre of the squash ball is at its smallest distance from the wall. The impulse (I) delivered by a changing force is given by:

$$I = F_{\text{av}} \, \Delta t$$

If a graph of force versus time is available, the impulse can be determined from the area under the graph.

SAMPLE PROBLEM 14 Determining the magnitude of an impulse using velocity-versus-time graphs

The following graph describes the changing horizontal force on a 40-kilogram ice skater as she begins to move from rest. Estimate her velocity after 2 seconds.

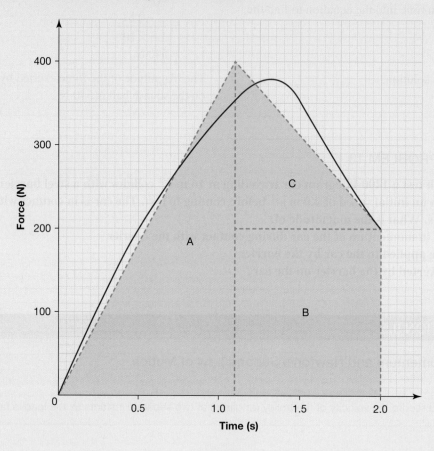

THINK

1. The magnitude of the impulse on the skater can be determined by calculating the area under the graph.

2. Use the formula for impulse to calculate the change in velocity.

3. State the solution.

WRITE

Magnitude of impulse = area A + area B + area C

$$= \frac{1}{2} \times 1.1 \times 400 + 0.9 \times 200$$

$$+ \frac{1}{2} \times 0.9 \times 200$$

$$= 220 + 180 + 90$$

$$= 490 \, \text{N}$$

$$I = \Delta p$$
$$= m\Delta v$$
$$490 = 40\Delta v$$
$$\Rightarrow \Delta v = \frac{490}{40}$$
$$= 12.25 \, \text{m s}^{-1}$$

As the skater started at rest, her velocity after 2 seconds will be equal to the change in velocity. The skater's velocity after 2 seconds is 12.25 m s⁻¹.

PRACTICE PROBLEM 14

Consider the motion described in sample problem 14.
a. Estimate the velocity of the skater after 1.1 seconds.
b. What is the acceleration of the skater during the first 1.1 seconds?
c. What constant force would produce the same change in velocity after 2.0 seconds?

EXTENSION: Linking back to velocity-versus-time graphs

You might recall that the displacement of an object can be determined by calculating the area under its velocity-versus-time graph, and displacement = $v_{av}\Delta t$. Similarly, the change in velocity of an object can be determined by calculating the area under its acceleration-versus-time graph, and change in velocity = $a_{av}\Delta t$.

9.7.4 Follow through

Players of ball games are often advised to 'follow through'. The force is then applied to the ball by the bat, racquet, club, stick or arm for a larger time interval. The impulse, $F\Delta t$, is larger and the change in momentum, Δp, is therefore larger. Consequently, the change in velocity of the ball as a result of the applied force is greater.

9.7.5 Protecting that frail human body

The human body does not cope very well with sudden blows. The skeleton provides a fairly rigid frame that protects the vital organs inside and, with the help of your muscles, enables you to move. A sudden impact to your body, or part of your body, can:

- push or pull the bones hard enough to break them
- tear or strain the ligaments that hold the bones together
- tear or strain muscles or the tendons that join muscles to bones
- push bones into vital organs like the brain and lungs
- tear, puncture or crush vital organs like the kidneys, liver and spleen.

The damage that is done depends on the magnitude of the net force and the subsequent acceleration to which your body is subjected. In any collision, the net force acting on your body, or part of your body, can be expressed as:

$$F_{net} = \frac{\Delta p}{\Delta t}$$

The symbol Δp represents the change in momentum of the part of your body directly affected by that net force. The magnitude of your change in momentum is usually beyond your control. For example, if you are sitting in a car travelling at 100 km h^{-1} when it hits a solid concrete wall, the magnitude of the change in momentum of your whole body during the collision will be your mass multiplied by your initial speed. When you land on a basketball court after a high jump, the magnitude of the change in momentum of each knee will be its mass multiplied by its speed just as your feet hit the floor. You have no control over your momentum.

FIGURE 9.24 Golfers are advised to 'follow through'. The force is applied to the ball for a longer time, giving it more momentum.

You do, however, have control over the time interval during which the momentum changes. If Δt can be increased, the magnitude of the net force applied to you will be decreased. You can do this by:

- bending your knees when you land after jumping in sports such as netball and basketball. This increases the time interval over which your knees change their momentum, and decreases the likelihood of ligament damage.
- moving your hand back when you catch a fast-moving ball in sports such as cricket. The ball changes its momentum over a longer time interval, reducing the force applied to it by your hand. In turn, the equal and opposite force on your hand is less.
- wearing gloves and padding in sports such as baseball, softball and gridiron. Thick gloves are essential for wicketkeepers in cricket, who catch the solid cricket ball while it is travelling at speeds up to 150 km h^{-1}.
- wearing footwear that increases the time interval during which your feet stop as they hit the ground. This is particularly important for people who run on footpaths and other hard surfaces. Indoor basketball and netball courts have floors that, although hard, bend a little, increasing the period of impact of running feet.

FIGURE 9.25 Gloves make it possible for wicketkeepers to catch a solid cricket ball travelling at high speed without severe pain and bruising.

CASE STUDY: Don't be an egghead

After bicycle helmets became compulsory in Victoria in July 1990, the number of head injuries sustained by cyclists decreased dramatically. Bicycle helmets typically consist of an expanded polystyrene liner about 2 centimetres thick, covered in a thin, hard, polymer shell. They are designed to crush on impact.

In a serious bicycle accident, the head is likely to collide at high speed with the road or another vehicle. Even a simple fall from a bike can result in the head hitting the road at a speed of about 20 km h^{-1}. Without the protection of a helmet, concussion is likely as the skull decelerates and collides with the brain because of the large net force on it. If the net force and subsequent deceleration is large enough, the brain can be severely bruised or torn, resulting in permanent brain damage or death. The effect is not unlike that of dropping a soft-boiled egg onto a hard floor.

Although a helmet does not guarantee survival in a serious bicycle accident, it does reduce the net force applied to the skull, and therefore increases the chances of survival dramatically. The polystyrene liner of the helmet increases the time interval during which the skull changes its momentum.

Helmets used by motorcyclists, as well as in horse riding, motor racing, cricket and many other sports, all serve the same purpose — to increase the time interval over which a change in momentum takes place.

FIGURE 9.26 Helmets save lives and prevent serious injury in many activities.

CASE STUDY: Buckle up

Seatbelts are a relatively new introduction to vehicles, with Victoria being one of the first places in the world to make wearing them compulsory in 1970. Their introduction and the increasing emphasis on vehicle safety over recent decades has drastically reduced the number of fatalities.

In a collision a car comes to a stop rapidly. An occupant not wearing a seatbelt continues at the original speed of the car (as described by Newton's First Law) until acted on by a non-zero net force. An unrestrained occupant therefore moves at speed until:
- colliding with part of the interior of the car, stopping even more rapidly than the car itself, usually over a distance of only several centimetres
- crashing through the stationary, or almost stationary, windscreen into the object collided with, or onto the road
- crashing into another occupant closer to the front of the car.

An occupant properly restrained with a seatbelt stops with the car. In a typical suburban crash, the acceleration takes place over a distance of about 50 centimetres. The rate of change of the momentum of a restrained occupant is much less. Therefore, the net force on the occupant is less.

The addition of airbags to complement seatbelts has further improved occupant safety. Airbags provide an additional means of restraining the occupant and reducing the force experienced.

As well as increasing the time interval over which the occupant comes to a stop, the combination of seatbelts and airbags:
- spreads the force over a larger area of the body
- reduces the likelihood of a collision between the body and the interior of the vehicle
- keeps the occupant in an aligned position, reducing injuries to vulnerable areas such as the neck, as well as stopping them from crashing through the windscreen.

CASE STUDY: Computer crash modelling

Automotive engineers use computer modelling during the design and development of new vehicles to investigate the effectiveness of safety features. This sort of crash modelling takes place long before the first prototype is built and the first physical crash tests take place (see figure 9.27).

FIGURE 9.27 Engineers use computers to model collisions in order to design and develop features that improve the safety characteristics of cars.

Computer crash modelling has resulted in improvements to front and side structural design, and to internal safety features such as seatbelt and airbag systems. Modelling crashes allows the investigation of a wide range of collision types, including full frontal, offset frontal, angled frontal and pole or barrier; collisions between trucks and cars; and rear impacts. The possibilities are endless. The computer models are then verified with the crash testing of real vehicles.

During side-impact modelling, the computer is used to test thousands of combinations of seatbelt, cushioning and airbag designs. For each test, the computer can be set up to calculate the forces acting on occupants, estimate the severity of injuries and compare results with other design solutions.

One aspect of design that can be tested is the sensor that triggers airbags to inflate. Complex calculations and comparisons are performed by a microprocessor within the sensing module before it 'decides' whether or not to trigger the airbags. The crash events that are modelled to develop the airbag sensors include high- and low-speed collisions, full-frontal and angled-frontal impacts, and pole- or tree-type collisions.

CASE STUDY: Cars that crumple

Modern cars are designed to crumple at the front and rear. This provision increases the time interval during which the momentum of the car changes in a collision, further protecting its occupants from death or serious injury.

Even though the front and rear of the car crumple, the passenger compartment is protected by a rigid frame. The engine is also surrounded by rigid structures that prevent it from being pushed into the passenger compartment. The tendency of the roof to crush is currently being reduced by increasing the thickness of the windows, using stronger adhesives and strengthening the structure of the roof panel and pillars.

FIGURE 9.28 A vehicle that has undergone a roof crush test

The inside of the passenger compartment is also designed to protect occupants. Padded dashboards, collapsible steering wheels and airbags are designed to reduce the rate of change of momentum of occupants in the event of a collision. Interior fittings like switches, door knobs and handbrakes are sunk so that the occupants do not collide with them.

9.7 Activities

learn on

9.7 Quick quiz	9.7 Exercise	9.7 Exam questions

9.7 Exercise

1. A 1400-kilogram car travels at 60 km h^{-1} east. Calculate the momentum of the car.
2. Make an estimate to one significant figure of the magnitude of each of the following.
   a. The average net force on a car while it is accelerating from 0 to 40 km h^{-1} in 3.2 seconds
   b. The magnitude of the air resistance on an 80-kilogram skydiver who has reached a terminal velocity of 200 km h^{-1}
   c. The momentum of an Olympic class athlete participating in the 100-metre sprint event
   d. The momentum of a family car travelling at the speed limit along a suburban street
   e. The impulse that causes a 70-kilogram football player who is running at top speed to stop abruptly as he collides with a goal post that he didn't see
   f. The impulse applied to a netball by a goal shooter as she pushes it up towards the goal at a speed of 5 m s^{-1}
   g. The change in momentum of a tennis ball as it is returned to the server in a Wimbledon final

3. A 60-gram tennis ball is bounced vertically onto the ground. After reaching the ground with a downwards velocity of 8.0 m s^{-1}, the ball rebounds with a velocity of 6.0 m s^{-1} vertically upwards.
   a. What is the change in momentum of the tennis ball?
   b. What is the impulse applied by the tennis ball to the ground? Explain how you obtained your answer without any information about the change in momentum of the ground.
   c. Does the ground actually move as a result of the impulse applied by the tennis ball? Explain your answer.
   d. If the tennis ball is in contact with the ground for 2.0×10^{-3} s, what is the average net force on the tennis ball during this interval?
   e. What is the average normal force during this time interval?
4. A 75-kilogram basketballer lands vertically on the court with a speed of 3.2 m s^{-1}.
   a. What total impulse is applied to the basketballer's feet by the ground?
   b. If the basketballer's speed changes from 3.2 m s^{-1} to zero in 0.10 seconds, what total force does the ground apply to his feet?
   c. Estimate the height from which the basketballer fell to the court.
5. A car with a total mass of 1400 kilograms (including occupants) travelling at 60 km h^{-1} hits a large tree and stops in 0.080 seconds.
   a. What impulse is applied to the car by the tree?
   b. What force is exerted by the tree on the car?
   c. What is the magnitude of the deceleration of the 70-kilogram driver of the car if he is wearing a properly fitted seatbelt?
6. Airbags are fitted to the centre of the steering wheel of many new cars. In the event of a sudden deceleration, the airbag inflates rapidly, providing extra protection for a driver restrained by a seatbelt. Explain how airbags reduce the likelihood of serious injury or death.
7. Joggers are advised to run on grass or other soft surfaces rather than concrete paths or bitumen roads to reduce the risk of knee and other leg injuries. Explain why this is so.
8. The following graph shows how the horizontal force on the upper body of each of two occupants of a car changes as a result of a head-on collision. One occupant is wearing a seatbelt while the other is not. Both occupants are stationary 0.10 seconds after the initial impact.

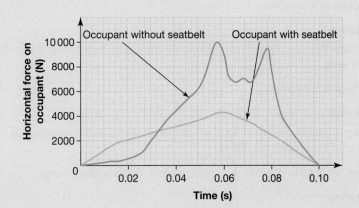

a. What is the horizontal impulse on the occupant wearing the seatbelt?
b. If the mass of the occupant wearing the seatbelt is 60 kilograms, determine the speed of the car just before the initial impact.
c. Is the occupant who is not wearing the seatbelt heavier or lighter than the other (more sensible) occupant?
d. Write a paragraph explaining the difference in shape between the two curves on the graph.

9. The following graph shows how the upward push of the court floor changes as a 60-kilogram basketballer jumps vertically upwards to complete a slam dunk.

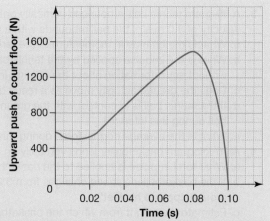

   a. What is the impulse applied to the basketballer by the floor?
   b. With what speed did the basketballer leave the ground?
   c. What was the average force exerted on the basketballer by the floor during the 0.10-second interval?
   d. Explain why the initial upward push of the floor is not zero.

10. A well-known politician makes the suggestion that if cars were completely surrounded by rubber 'bumpers' like those on dodgem cars, they would simply bounce off each other in a collision and passengers would be safer. Discuss the merits of this suggestion in terms of Newton's laws of motion.

9.7 Exam questions

▶ Question 1 (3 marks)

A small test vehicle has a mass of 50 kg. It is travelling at 12 m s^{-1} east when it collides head-on with a barrier and is brought to rest in a time of 0.60 s.
   a. What is the change in momentum of the vehicle? **(2 marks)**
   b. What is the impulse on the barrier? **(1 mark)**

▶ Question 2 (3 marks)

An object, with mass 4.0 kg, is moving in a straight line. It has an initial speed of 6.0 m s^{-1} and then experiences a change in momentum of 20 kg m s^{-1}.

Calculate the final speed of this object.

▶ Question 3 (3 marks)

A force of 10 N west acts for 4.0 s. An unknown force, F_X west, acts for a further 6.0 s. The total impulse of these two forces is 130 N s west.

Calculate the magnitude of the force F_X.

▶ Question 4 (4 marks)

A ball of mass 0.20 kg is travelling north at 30 m s^{-1}. It then strikes a wall and rebounds directly backwards at 20 m s^{-1}. The duration of the impact is 0.025 s.

Calculate the average net force on the ball during the impact. Show your working.

▶ Question 5 (4 marks)

An object has a mass of 2.5 kg. The net force on the object is a constant 5.0 N east for 8.0 s then a constant 3.0 N west for the next 20 s.

Calculate the change in velocity of the object. Show your working.

More exam questions are available in your learnON title.

9.8 Torque

9.8.1 Torque or the turning effect of a force

So far the explanation of forces and motion has treated objects as if they are a single point, or as if the force acts through the middle of the object; that is, its centre of mass. However, reality is more complicated than this. Friction acts at the rim of the front tyre of a bike to make it roll. Other actions involving rotation and a force making an object turn include a billiard cue hitting the bottom edge of a ball to make it spin backwards, the wind blowing over a tree, or pulling on a handle to open a door.

The turning effect of a force is called a torque. The symbol for torque is τ, the Greek letter *tau*. Torque is always measured about a particular point. In some situations, the point about which an object will rotate is obvious (such as in a seesaw), and it may be referred to as a pivot or fulcrum.

FIGURE 9.29 Examples of torque being applied

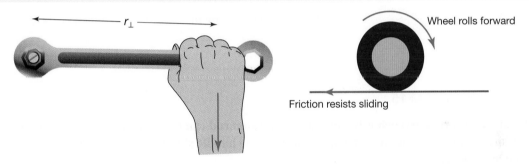

The size of a torque about a point or pivot is determined by the product of two factors:
- The size of the force, F
- The perpendicular distance between the line of action of the force and the point, $r_\perp$.

$$\tau = r_\perp F$$

As a product of force and distance, torque has the units of newton metre (N m). It is also a vector, but because its effect is rotation, the direction of the vector is set by a rule.

The rule is:

If the rotation in the plane of the page is clockwise, the direction of the vector is into the page. If the rotation in the plane of the page is anticlockwise, the direction of the vector is out of the page.

SAMPLE PROBLEM 15 Calculating force using torque and perpendicular distance

A torque wrench is used to tighten nuts onto their bolts to a specific tightness or force. A torque wrench has a handle (black in the photo shown) on one end and a socket that fits over a nut on the other end. In between is a scale that gives a reading in newton metres.

The scale on a torque wrench has a reading of 30 Newton metres. If the hand applying the force is 30 centimetres from the end, what is the size of the force by the hand on the wrench?

THINK	WRITE
1. Recall the formula for calculating torque.	$\tau = r_\perp F$
2. Substitute the torque and perpendicular distance into the equation to find the force applied.	$30 = 0.3 \times F$ $\Rightarrow F = \dfrac{30}{0.3}$ $= 100\,\text{N}$
3. State the solution.	The force by the hand on the wrench is 100 N.

PRACTICE PROBLEM 15

The handle of a torque wrench is hollow so an extension rod can be inserted. If you can exert only 30 N of force, how far along the extension rod from the handle should you place your hand to achieve a torque of 30 N m?

9.8 Activities

learn on

Students, these questions are even better in jacPLUS

Receive immediate feedback and access sample responses

Access additional questions

Track your results and progress

Find all this and MORE in jacPLUS

9.8 Quick quiz on	9.8 Exercise	9.8 Exam questions

9.8 Exercise

1. A mechanic applies a force of 200 N to a wheel nut using a shifter. The perpendicular distance from where they apply the force to the nut is 25 centimetres. What torque are they applying to the nut?
2. A mechanic applies a force of 200 N to a wheel nut using a shifter. The perpendicular distance from where they apply the force to the nut is 25 centimetres, but they are unable to loosen the nut using the torque applied. Suggest two ways in which they could increase the torque in this situation.

3. A lever is used to apply a torque of 20 N m about a pivot point. The perpendicular distance is 0.25 metres between the application of the force and the pivot point. What is the applied force?

4. There are myriad examples of everyday situations where we use devices that have a lever of some form to increase the torque that we apply. Examples include door handles, car steering wheels, electric motors, pushbike pedals, wrenches, wheelbarrows and bottle lids. For one of these examples, or another that you can identify, estimate the force applied and the perpendicular distance to calculate an estimate of the torque involved.

5. Sam is standing at the right-hand end of the seesaw shown in the following figure. He places a bag on the seesaw and then begins walking up the plank to the left. Describe what happens as he walks towards, and then beyond, the fulcrum.

9.8 Exam questions

▶ **Question 1 (2 marks)**

MC Two children are opening a door by pushing at right-angles to the door (from the same side of the door).

One child applies a force of 20 N at the edge of the door, 0.70 m from the hinges. The other child pushes with a force F at a distance of 0.30 m from the hinges. The net torque about the hinges is 26 N m.

Which of the following best gives the magnitude of F?
A. 12 N
B. 14 N
C. 40 N
D. 50 N

▶ **Question 2 (1 mark)**

MC A force F is applied at an angle of 90° to a lever at a distance r from the pivot.

Which of the following changes would decrease the clockwise torque?
A. Increase the distance r
B. Increase the angle between the force and the lever
C. Increase the force F
D. Double the distance r and halve the force F

▶ **Question 3 (1 mark)**

MC Consider a rod pivoted at point P and able to rotate in the vertical plane.

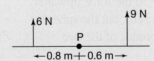

A force of 9.0 N upward is applied on the right-hand side of the bar, 0.6 m from P. A second force of 6.0 N is applied to the left-hand side of the bar, a distance 0.8 m from P.

What is the best estimate of the magnitude of the net torque on the bar about P?
A. 10.2 N m
B. 5.4 N m
C. 4.8 N m
D. 0.60 N m

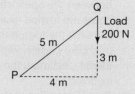
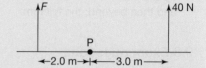
9.9 Equilibrium

9.9.1 Equilibrium or keeping still

Earlier in this topic, 'keeping still' meant not moving. If the net force was zero and the object was at rest, it would stay still. The forces were considered as acting on a single point. However, if the forces act at different points on the object, it is possible to have a net force of zero, but the object can still spin. In figure 9.30 the force upwards equals the force downwards, so the net force is zero, but the sphere rotates. In this case there is a net torque. The torques of the two forces about the centre add together.

FIGURE 9.30 A zero net force can cause rotation.

In cases such as car engines and electric motors, the production of a torque is essential for rotation and movement. But torque, and the rotation and movement it causes, can be detrimental. In bridges and buildings, the torque effect of a force can't be avoided, but needs to be controlled if the structure is to remain standing. Such structures need to be designed so that not only is the net force equal to zero, but the net torque is also zero, and importantly this is true about every point in the structure.

For a structure to be in equilibrium, two conditions need to apply:
1. Translational equilibrium: net force = zero
2. Rotational equilibrium: net torque about any point = zero

9.9.2 Strategy for solving problems involving torque

Questions regarding torque will often involve determining the value of two forces, so the solution will require generating two equations, which can then be solved simultaneously.

1. Draw a diagram with all the forces acting on the structure. Label each force. If its size is given in the question, write the value; for example, 10 N. If the size of the force is unknown, use a symbol such as **F** or **R**.
2. Using translational equilibrium: net force = zero.
   It is easier to break this into two simpler tasks such as:
   a. sum of forces up = sum of forces down
   b. sum of forces left = sum of forces right
3. Using rotational equilibrium: net torque about any point = zero.
   Choose a point about which to calculate the torques.
   Any point is acceptable; however, it can make solving the problem easier if you choose a point through which an unknown force acts. The torque of this force about that point will be zero as its line of action passes through the point, so that unknown will be effectively eliminated from this equation.

   sum of clockwise torques = sum of anticlockwise torques

4. Now you will have two equations with two unknowns: one equation from the net force and one from the net torque. You can then solve the equations simultaneously to determine the unknown quantities.

tlvd 3702

SAMPLE PROBLEM 16 Determining the equilibrium of a seesaw

Where should person 1 sit to balance the seesaw?

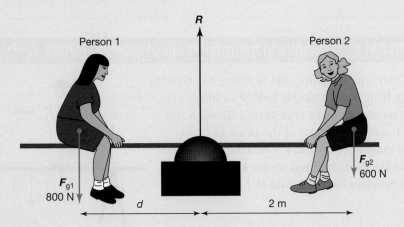

THINK	WRITE
1. To satisfy equilibrium, both the sum of the forces acting on the seesaw and the sum of the torques must equal zero.	$F_{net} = 0$ $\tau_{net} = 0$
2. Consider the net force equilibrium. The sum of the upwards forces must equal the sum of the downwards forces.	$R = 800 + 600$ $= 1400\,\text{N upwards}$

3. Consider the net torque, taking the torques about the fulcrum at the centre.

$\tau_{net} = 0$

$\Rightarrow$ Sum of clockwise torques = sum of anticlockwise torques

$$600 \times 2 = 800 \times d$$

$$\Rightarrow d = \frac{1200}{800}$$

$$= 1.5 \text{ m}$$

4. State the solution.

To balance the seesaw, person 1 must sit 1.5 metres to the left of the fulcrum.

PRACTICE PROBLEM 16

The following seesaw is balanced. Calculate the mass of person 1.

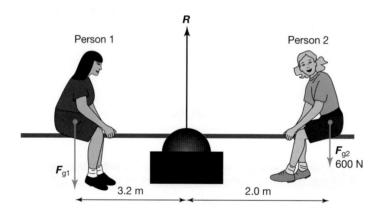

tlvd-0092

SAMPLE PROBLEM 17 Determining the magnitude of reaction forces

Consider the painter's plank supported between two trestles shown. The plank behaves as a simple bridge or beam and the weight of the painter must be transferred through the plank to the two trestles. The mass of the beam is 40 kilograms, the mass of the painter is 60 kilograms and she is a quarter of the distance from trestle 1. What is the magnitude of the reaction forces R_1 and R_2?

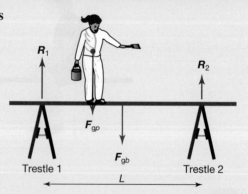

THINK	WRITE
1. For the structure to be stable, the sum of the forces and the sum of the torques must both equal zero.	$F_{net} = 0$ $\tau_{net} = 0$
2. Consider the net force. The sum of the upwards forces must equal the sum of the downwards forces.	$R_1 + R_2 = 40 \times 9.8 + 60 \times 9.8$ $\qquad\qquad = 980 \text{ N}$

3. Consider the net torque, taking the torques about trestle 1.

$\tau_{net} = 0$

$\Rightarrow$ Sum of clockwise torques = sum of anticlockwise torques

$$40 \times 9.8 \times \frac{1}{2}L + 60 \times 9.8 \times \frac{1}{4}L = R_2 \times L$$

$$40 \times 9.8 \times \frac{1}{2} + 60 \times 9.8 \times \frac{1}{4} = R_2$$

$$196 + 147 = R_2$$

$$\Rightarrow R_2 = 343 \text{ N}$$

4. Substitute the value of R_2 into the equation for the net force.

$R_1 + R_2 = 980$

$R_1 + 343 = 980$

$$\Rightarrow R_1 = 980 - 343$$

$$= 637 \text{ N}$$

5. State the solution.

The magnitude of R_1 is 637 N and R_2 is 343 N.

PRACTICE PROBLEM 17

A eucalyptus tree, 15 metres high and with a 200-centimetre diameter, was pulled over until it failed. The applied load was 6.0 kN m about the base of the tree.

a. If the root ball of the tree has an average depth of 0.80 metres, what is the size of the force by the soil on the root ball at the point of failure?
b. If the rope pulling on the tree was attached halfway up the tree, calculate the size of the force at the point of failure:
  i. in the rope (assuming the rope is horizontal)
  ii. by the ground at the base of the tree.

9.9.3 Types of structures: cantilevers

A cantilever is a beam with one end free to move. A diving board, a flagpole and a tree are examples of cantilevers.

The diving board in figure 9.31 is supported by an upward force, R_1, from the bracket. The force due to gravity acts down through the middle of the board at a point further out. If these were the only forces on the diving board, the board would rotate anticlockwise. To prevent this rotation, the other end of the bracket pulls down on the diving board. The board is bolted to each end of the bracket.

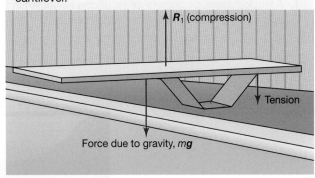

FIGURE 9.31 A diving board is an example of a cantilever.

R_1 (compression)

Tension

Force due to gravity, mg

The tree in figure 9.32 is buffeted by winds from the left. The soil on the right at the base of the tree is compressed and pushes back to the left. The combination of these two forces pushes the roots of the tree to the left, and the soil to the left of the roots pushes back to the right.

FIGURE 9.32 A tree buffeted by winds is a cantilever.

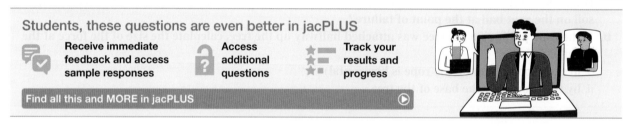

Force on tree by wind

Force on tree by surface

Force on tree (roots) by soil

9.9 Activities

9.9 Quick quiz on	**9.9 Exercise**	**9.9 Exam questions**

9.9 Exercise

1. A truck crosses a concrete girder bridge as shown in the following figure. The bridge spans 20 metres and is supported at each end on concrete abutments.

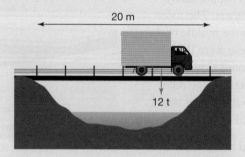

20 m

12 t

a. Describe what happens to the reaction at each abutment as the truck moves across the bridge from left to right.
b. The force due to gravity acting on the truck is 12 kN. Calculate the reaction at each support when the centre of gravity of the truck is 4 metres from the right abutment.

2. A person standing on the outside edge of a cantilevered balcony, shown in the following figure, walks inside.

    a. Explain what forces are necessary to support the balcony.

    b. As the person walks across the balcony, describe what happens to the reaction at the support.

3. The truck crane in the following figure is able to lift a 20-tonne load at a radius of 5 metres.

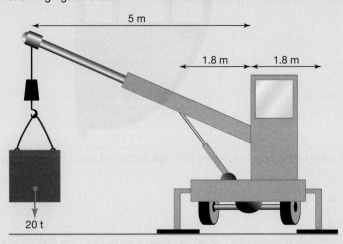

    If the mass of the truck is evenly distributed, how heavy must the truck be if it is not to tip over?

4. The pedestrian bridge spanning the creek in the following figure has a force due to gravity acting at its centre of mass of 2 kN.

    Calculate the reaction at each end when the three people are in the positions shown.

5. The plank in the following figure is 6 m long and extends 2 m beyond the support at the edge of the boat. The forces due to gravity acting at the centre of mass of the plank and Pirate Bill are 800 N and 500 N respectively.

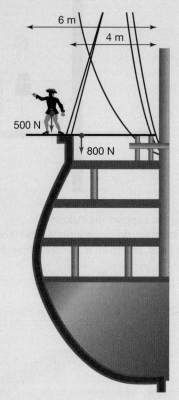

How far beyond the edge of the boat can Pirate Bill walk along the plank before it tips and he falls into the water?

9.9 Exam questions

Question 1 (2 marks)

Two children are trying to balance themselves on either side of a seesaw. The first child sits 0.75 m from the pivot point and has a force due to gravity acting on them of 150 N. The second child has a force due to gravity acting on them of 190 N.

How far should they sit on the other side of the pivot point to balance the seesaw?

Question 2 (1 mark)

MC The two ends of a beam rest on supports P and Q that are 4.0 m apart. The mass of the beam can be neglected.

A load that has a force of gravity acting on it of 80 N is placed 3.0 m from support P, as shown in the following figure.

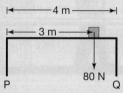

Which of the following best describes the magnitude of the reaction force of support Q on the beam?
A. 60 N
B. 80 N
C. 107 N
D. 120 N

Question 3 (7 marks)

A beam of length 8.0 m has a force due to gravity acting at its centre of mass, C, of 2000 N.

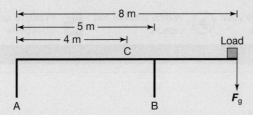

a. A load with a force due to gravity acting on it of F_g stands at the far end of the beam. The pillar B exerts an upward force of 2400 N on the beam.

Calculate the force due to gravity, F_g, acting on the load. **(3 marks)**

b. The load is now removed and replaced with a different load with a force due to gravity acting on it of 2000 N.

Specify fully the force now exerted by the pillar A on the beam. **(4 marks)**

Question 4 (3 marks)

A uniform beam rests at its ends on two supports, A and B, 8.0 m apart. The force due to gravity acting at the centre of mass of the beam is 200 N.

A load of 400 N rests on the beam at a distance 2.0 m from support A, as shown in the following figure.

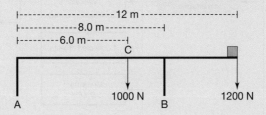

Calculate the magnitude of the force applied to the beam by support A. Show working.

Question 5 (4 marks)

A uniform beam of length 12 m is attached to two supports, A and B. The force due to gravity acting at the centre of mass of the beam is 1000 N. The two supports are 8.0 m apart, so the beam overhangs by 4.0 m. The centre of mass of the beam is at its midpoint. A load of 1200 N acts at the free end of this cantilevered beam.

Calculate the force on the beam applied by the support A. Give magnitude and direction.

More exam questions are available in your learnON title.

9.10 Review

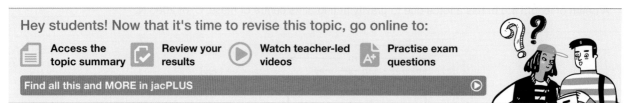

9.10.1 Topic summary

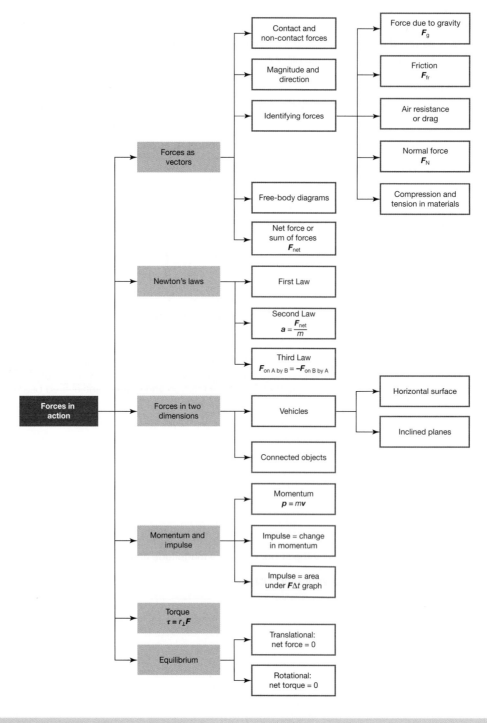

9.10.2 Key ideas summary

9.10.3 Key terms glossary

on Resources

Solutions	Solutions — Topic 9 (sol-0795)
Practical investigation eLogbook	Practical investigation eLogbook — Topic 9 (elog-1578)
Digital documents	Key science skills — VCE Physics Units 1–4 (doc-36950) Key terms glossary — Topic 9 (doc-36967) Key ideas summary — Topic 9 (doc-36968)
Exam question booklet	Exam question booklet — Topic 9 (eqb-0077)

9.10 Activities

9.10 Review questions

1. As part of a practical investigation, a Physics student rolls objects horizontally off the edge of a table and records the motion of their fall onto the floor below. Identify the forces that will be acting on these objects during their fall to the floor. For each force you identify, describe the interaction it represents and the likely direction of the force.

2. A rear-wheel-drive car is accelerating forwards along a horizontal road surface. Draw and label a complete diagram of all external forces acting on the car.

3. A child pulls along a toy on a piece of string at an angle of 30 degrees to the horizontal. They apply a force of 25 N along the string. Determine the horizontal and vertical components of this force.

4. The following forces are acting on ropes used in a four-way tug of war (as viewed from above) as part of a school athletics carnival.

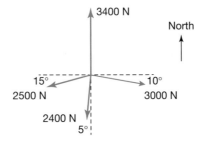

Resolve all of the forces into north–south or east–west components and determine the size and direction of the net force.

5. A sports car of mass 1645 kilograms accelerates at 9.6 m s^{-2}. Determine the net force that would be required to produce this acceleration.

6. The Falcon Heavy rocket produces approximately 2.2×10^6 N of thrust during the first moments of liftoff. If the acceleration of the rocket at this instant is 1.57 m s^{-2}, calculate its mass.

7. A piano is falling through the air near the surface of Earth. It has a mass of 410 kilograms.
   a. At a particular instant during its fall, the force of air resistance acting to oppose its fall is 2400 N. Calculate its acceleration at this instant.
   b. Determine the magnitude of air resistance force that would be required for the piano to reach a constant velocity during its fall.

8. A removalist is pushing two heavy boxes across the floor of their truck. The horizontal forces acting on the boxes are shown in the following diagram.

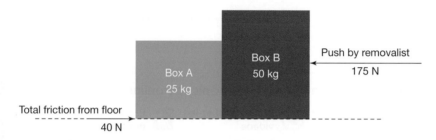

   a. Calculate the acceleration of the boxes.
   b. Calculate the compression force acting between the two boxes.

9. A tennis ball is travelling at 65 m s^{-1} towards a player's racquet an instant before it collides with it. A moment later, the ball leaves the racquet travelling at 33 m s^{-1} in the opposite direction. Assume that the ball has a mass of 57 grams.
   a. Determine the change in momentum that the ball experiences during this collision.
   b. Assuming that the collision duration is 0.0020 seconds, determine the average force applied by the racquet on the ball during the collision.

10. The following graph is a simplified representation of the force applied by a trampoline on an acrobat over the duration of a single rebound.

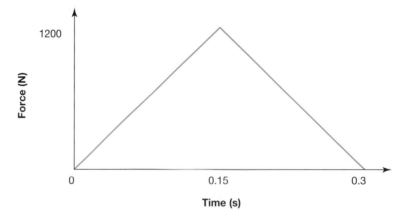

Use this graph to determine the impulse applied to the acrobat.

9.10 Exam questions

In answering the following questions, assume the magnitude of the gravitational field strength near Earth's surface is 9.8 N kg^{-1}.

Question 1

Which of the following is *not* a vector quantity?

A. Force
B. Mass
C. Acceleration
D. Momentum

Question 2

Which of the following forces would be acting on a basketball falling through the air during a game of basketball?

A. Gravity
B. Normal force
C. Net force
D. Air resistance

Question 3

Which of the following is *not* at rest or in a uniform state of motion?

A. A car travelling along a highway at a constant velocity using cruise control
B. A rocket sitting motionless on the launch pad
C. A ladder leaning against a wall
D. A train accelerating uniformly as it departs a station

Question 4

An object of mass 43 kilograms has a net force of 86 N applied to it.

What is its acceleration?

A. 9.8 m s^{-2}
B. 0.50 m s^{-2}
C. 2.0 m s^{-2}
D. 1.6 m s^{-2}

Question 5

Which of the following is *not* a Newton's Third Law pair of forces?

A. The force of a book on a table and the force of the table on the book
B. The normal force on someone sitting in a chair and the force of gravity on them
C. The force of a car tyre on the road and the force of the road on the car tyre
D. The force due to gravity of Earth on you and the force due to gravity of you on Earth

Question 6

A car of mass 2250 kilograms is travelling at 20 m s^{-1}.

What is its momentum?

A. 45 000 kg m s^{-1}

B. 4500 kg m s^{-1}

C. 1125 kg m s^{-1}

D. 112.5 kg m s^{-1}

Question 7

A tennis ball of mass 58 grams experiences a change in velocity of 155 km h^{-1} when struck by a racquet during a serve.

The change in momentum it experiences is

A. 90 kg m s^{-1}.

B. 9 kg m s^{-1}.

C. 25 kg m s^{-1}.

D. 2.5 kg m s^{-1}.

Question 8

A car of mass 1850 kilograms is travelling at 8 m s^{-1} around a carpark when it collides with a parked car. The collision lasts 0.2 seconds, after which the car has come to a complete stop.

What is the magnitude of the average force acting on the car?

A. 9250 N

B. 74 000 N

C. 2960 N

D. 14 800 N

Question 9

Some engineering students are discussing ways to reduce the force acting on a car during a collision.

Based on your understanding of impulse and momentum, which of the following will *not* be effective in reducing the force acting on the car?

A. Reduce the initial speed of the car (using more effective braking or lowered speed limits)

B. Decrease the mass of the car (through better structural design and material selection)

C. Reduce the duration (time taken) for the collision

D. All of the above will be effective.

Question 10

A student sits at the end of a pipe attached to a shifter to apply a force of 735 N at a perpendicular distance of 125 centimetres from a wheel nut that they are struggling to undo.

What is the torque applied to the nut?

A. 93.8 N m

B. 9380 N m

C. 919 N m

D. 91 900 N m

▶ Question 11 (1 mark)

Consider the example of a textbook at rest on a table.

Explain why the normal force acting on the book and the force due to gravity are not a Newton's Third Law pair of forces.

▶ Question 12 (3 marks)

A cycle tourist is towing all of their camping equipment, clothes and food behind their pushbike in a bike trailer. Whilst pedalling at a reasonable pace they produce a driving force of 172 N acting forwards from the rear wheel of their bike. The road friction and air resistance opposing the motion of the bike and trailer is equal to 34 N acting backwards (20 N on the bike and 14 N on the trailer). The total mass of the cyclist and bike is 95 kilograms. The total mass of the trailer and all of its payload is 20 kilograms.

Determine the tension force in the link between the trailer and the bike.

▶ Question 13 (3 marks)

A small rocket of mass 2500 kg is launched up along an inclined ramp at an angle of 42 degrees from the horizontal. During the launch, the rocket engine provides a constant thrust of 18 000 N.

Determine the acceleration of the rocket during its launch. It is reasonable to consider air resistance and friction to be negligible in this situation.

▶ Question 14 (2 marks)

During a standardised car crash test, a vehicle of mass 1980 kilograms is travelling at 64.0 km h^{-1} when it strikes a barrier. The car rebounds and is travelling at 12.0 km h^{-1} in the opposite direction immediately after the collision. The duration of the collision with the barrier is 160 milliseconds.

Determine the average force exerted by the car on the barrier during the collision.

▶ Question 15 (3 marks)

A truck of mass 14 500 kilograms is crossing a bridge over a small river. The bridge span is 40.0 metres between the two supports. The bridge has a total mass of 46 000 kilograms, with its centre of gravity exactly in the middle of its span.

Determine the reaction at each of the supports when the centre of mass of the truck is 9.00 metres from the right-hand support.

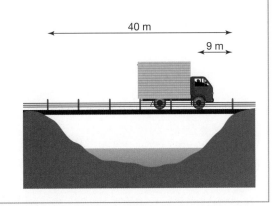

10 Energy and motion

KEY KNOWLEDGE

In this topic, you will:
- analyse impulse in an isolated system (for collisions between objects moving in a straight line): $F\Delta t = m\Delta v$
- investigate and analyse theoretically and practically momentum conservation in one dimension
- apply the concept of work done by a force using:
  - work done = force × displacement: $W = Fs\cos\theta$, where force is constant
  - work done = area under force vs distance graph
- investigate and analyse theoretically and practically Hooke's Law for an ideal spring: $F = -kx$, where x is extension
- analyse and model mechanical energy transfers and transformations using energy conservation:
  - changes in gravitational potential energy near Earth's surface: $E_g = mg\Delta h$
  - strain potential energy in ideal springs: $E_s = \frac{1}{2}kx^2$
  - kinetic energy: $E_k = \frac{1}{2}mv^2$
- calculate the efficiency of an energy transfer system: $\eta = \dfrac{\text{useful energy out}}{\text{total energy in}}$
- analyse rate of energy transfer using power: $P = \dfrac{E}{t}$.

Source: VCE Physics Study Design (2023–2027) extracts © VCAA; reproduced by permission.

Note: Vector quantities are bolded.

PRACTICAL WORK AND INVESTIGATIONS

Practical work is a central component of VCE Physics. Experiments and investigations, supported by a **practical investigation eLogbook** and **teacher-led videos**, are included in this topic to provide opportunities to undertake investigations and communicate findings.

EXAM PREPARATION

▶ Access exam-style questions and their video solutions in every lesson, to ensure you are ready.

10.1 Overview

10.1.1 Introduction

What happens when a ball is dropped from a height of 1 metre? Depending on the type of ball, and the surface it is bounced on, there will be a variety of different answers, but could a ball ever rebound to a greater height than it started at? The answer to this question is based on the conservation of energy, a law of nature that is inherent to the world we live in.

In this topic, we will learn about the different forms of energy and how it can be transferred from one form to another.

FIGURE 10.1 In this collision the kinetic energy of a car is transferred into many other forms of energy. Energy from the car is transformed into the deformation of the materials that the front of the car is made of, and a large amount of energy is transferred to the object it collides with, the road and the particles of air around the car. Some of the energy is transferred into sound and heat.

LEARNING SEQUENCE

 Resources

Solutions	Solutions — Topic 10 (sol-0796)
Practical investigation eLogbook	Practical investigation eLogbook — Topic 10 (elog-1570)
Digital documents	Key science skills — VCE Physics Units 1–4 (doc-36950)
	Key terms glossary — Topic 10 (doc-36969)
	Key ideas summary — Topic 10 (doc-36970)
Exam question booklet	Exam question booklet — Topic 10 (eqb-0078)

10.2 Impulse and momentum

10.2.1 Impulse and momentum in collisions

When two objects, A and B, collide with each other, each object exerts a force on the other. These two forces are an example of Newton's Third Law of Motion. The two forces act on different objects, are equal in magnitude and act in opposite directions. To better understand impulse and momentum, consider an accident between two cars. Momentum is what each vehicle brings to the collision and the impulse is the force dissipated by the vehicle during the collision multiplied by the duration of the impact.

The collision starts at some point in time and finishes at another point in time. The symbol Δt represents the duration of the collision. From Newton's point of view, both objects measure the same duration. Multiplying $F_{\text{on B by A}}$ by Δt gives $F_{\text{on B by A}}\Delta t$, which is the impulse on B by A. This impulse produces a change in momentum, but whose momentum? A or B? Because it is the impulse by A on B, it is B's momentum that is changed.

$$F_{\text{on B by A}}\Delta t = I_{\text{on B by A}} = \Delta p_{\text{B}}$$

Similarly, the force on A by B produces a change in A's momentum.

$$F_{\text{on A by B}}\Delta t = I_{\text{on A by B}} = \Delta p_{\text{A}}$$

Forces are actions by one object on another, but momentum can be said to be a quantity an object has, even if it is a vector. So a force acting for a time changes how much momentum an object has.

In a collision, the two forces are equal in magnitude and opposite in direction. So the changes in momentum of A and B are also equal in magnitude and opposite in direction.

$$\Delta p_{\text{B}} = -\Delta p_{\text{A}}$$

Because momentum is a quantity, this statement can be interpreted as 'the momentum that B gains equals the momentum that A loses'. This is the basis of a conservation principle. Total momentum is conserved.

The 'change' in something is always the 'final' value minus the 'initial' value, or what is added to the 'initial' value to get the 'final'.

$$\Delta p = p_{\text{final}} - p_{\text{initial}} \text{ or } p_{\text{final}} = p_{\text{initial}} + \Delta p$$

For object B: $\Delta p_{\text{B}} = p_{\text{B final}} - p_{\text{B initial}}$

For object A: $\Delta p_{\text{A}} = p_{\text{A final}} - p_{\text{A initial}}$

So, from the above relationship:

$$\Delta p_{\text{B}} = -\Delta p_{\text{A}}$$
$$p_{\text{B final}} - p_{\text{B initial}} = -\left(p_{\text{A final}} - p_{\text{A initial}}\right)$$

Putting the initial momenta together gives:

$$p_{A\ \text{final}} + p_{B\ \text{final}} = p_{A\ \text{initial}} + p_{B\ \text{initial}}$$

Sum of momentum after = sum of the momentum before

Total momentum is conserved.

From this analysis it can be seen that the conservation of momentum is a logical consequence of Newton's Third Law.

The interaction between objects A and B can be summarised as follows:
- The total momentum of the system remains constant.
- The change in momentum of the system is zero.
- The change in momentum of object A is equal and opposite to the change in momentum of object B.

FIGURE 10.2 The net force on this system of two blocks is zero; therefore, momentum is conserved.

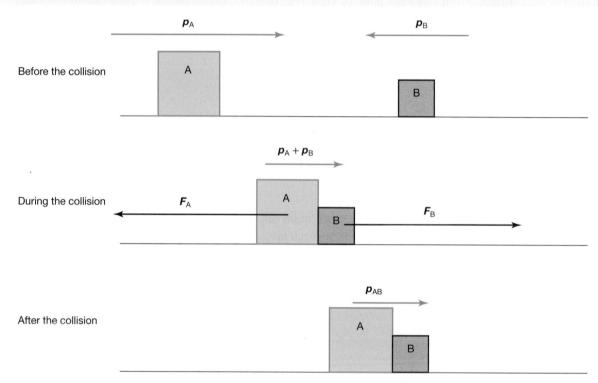

The total momentum of the system p_{AB} after the collision is the same as the total momentum of the system before and during the collision.

tlvd-0093

SAMPLE PROBLEM 1 Determining velocity, impulse and momentum during collisions

Consider the collision illustrated in figure 10.2. Block A has a mass of 5 kilograms, block B has a mass of 3 kilograms and each block has a speed of 4 m s⁻¹ before the collision. After the collision, the blocks move off together. Friction may be ignored.

a. Determine the velocity of the blocks after the collision.

b. What is the change in momentum of each of the blocks?

c. **What is the impulse on block A during the collision?**
d. **Determine the final velocity of block B if, instead of moving off together, block A rebounds to the left with a speed of 0.5 m s⁻¹.**

THINK	WRITE
a. 1. Recall the formula for conservation of momentum.	a. $p_A + p_B = p_{A+B}$
2. Calculate the momentum of each block to determine $p_A + p_B$. Take right as the positive direction.	$\begin{aligned} p_A + p_B &= m_A v_A + m_B v_B \\ &= 5 \times 4 + 3 \times -4 \\ &= 20 - 12 \\ &= 8 \text{ kg m s}^{-1} \text{ to the right} \end{aligned}$
3. Use the formula for conservation of momentum to determine the velocity of the blocks after the collision.	$\begin{aligned} p_{A+B} &= p_A + p_B \\ m_{A+B} v_{A+B} &= 8 \\ (5+3)v &= 8 \\ \Rightarrow v &= 1 \text{ m s}^{-1} \end{aligned}$
4. State the solution.	The velocity of the blocks after the collision is 1 m s^{-1} to the right.
b. 1. Consider block A only.	b. $\begin{aligned} \Delta p_A &= m_A \Delta v_A \\ &= 5(1-4) \\ &= 5 \times -3 \\ &= -15 \text{ kg m s}^{-1} \\ &= 15 \text{ kg m s}^{-1} \text{ to the left} \end{aligned}$
2. Δp_B should be 15 kg m s⁻¹ to the right since the change in momentum of the whole system is zero. To confirm this, consider block B alone.	$\begin{aligned} \Delta p_B &= m_B \Delta v_B \\ &= 3(1-(-4)) \\ &= 3 \times 5 \\ &= 15 \text{ kg m s}^{-1} \text{ to the right} \end{aligned}$
3. State the solution.	The change in momentum of each of the blocks is 15 kg m s⁻¹.
c. Impulse is equal to the change in momentum.	c. The impulse on block A is equal to the change in momentum of block A: 15 kg m s⁻¹ to the left.
d. 1. By conservation of momentum: $p_A + p_B = 8 \text{ kg m s}^{-1}$	d. $\begin{aligned} p_{Af} + p_{Bf} &= 8 \\ 5 \times -0.5 + 3v_{Bf} &= 8 \\ -2.5 + 3v_{Bf} &= 8 \\ 3v_{Bf} &= 10.5 \\ \Rightarrow v_{Bf} &= \frac{10.5}{3} \\ &= 3.5 \text{ m s}^{-1} \text{ to the right} \end{aligned}$
2. State the solution.	The final velocity of block B if, instead of moving off together, block A rebounds to the left with a speed of 0.5 m s⁻¹, is 3.5 m s⁻¹ to the right.

PRACTICE PROBLEM 1

Consider a collision in which a model car of mass 5 kilograms travelling at 2 m s^{-1} in an easterly direction catches up to and collides with an identical model car travelling at 1 m s^{-1} in the same direction. The cars lock together after the collision. Friction can be assumed to be negligible.

a. What was the total momentum of the two-car system before the collision?
b. Calculate the velocity of the model cars as they move off together after the collision.
c. What is the change in momentum of the car that was travelling faster before the collision?
d. What is the change in momentum of the car that was travelling slower before the collision?
e. What was the magnitude of impulse on both cars during the collision?
f. How are the impulses on the two cars different from each other?

10.2.2 Modelling real collisions

The Law of Conservation of Momentum makes it possible to predict the consequences of collisions between two cars or between two people on a sporting field. For example, if a 2000-kilogram delivery van travelling at 30 m s^{-1} (108 km h^{-1}) collided with a small, stationary car of mass 1000 kilograms, the speed of the tangled wreck (the two vehicles locked together) could be predicted. However, you would need to assume that the frictional forces and driving force acting on both cars were zero after the collision. A reasonably good estimate can be made of the speed of the tangled wreck immediately after the collision in this way.

The initial momentum, p_i, of the system is given by:

$$p_{van} + p_{car} = 2000\,\text{kg} \times 30\,\text{m s}^{-1} + 0\,\text{kg m s}^{-1}$$
$$= 60\,000\,\text{kg m s}^{-1}$$

where the initial direction of the van is taken to be positive.

The momentum of the system after the collision, p_f, is the momentum of just one object — the tangled wreck:

$$p_f = 3000\,\text{kg} \times v$$

where v is the velocity of the tangled wreck after the collision.

Since $p_f = p_i$:

$$3000\,\text{kg} \times v = 60\,000\,\text{kg m s}^{-1}$$
$$\Rightarrow v = \frac{60\,000\,\text{kg m s}^{-1}}{3000\,\text{kg}}$$
$$= 20\,\text{m s}^{-1}$$

The speed of the small car changes a lot more than the speed of the large van. However, the change in the momentum of the car is equal and opposite to that of the van.

$$\Delta p_{car} = 1000\,\text{kg} \times 20\,\text{m s}^{-1} - 0\,\text{kg m s}^{-1}$$
$$= 20\,000\,\text{kg m s}^{-1}$$
$$\Delta p_{van} = 2000\,\text{kg} \times 20\,\text{m s}^{-1} - 2000\,\text{kg} \times 30\,\text{m s}^{-1}$$
$$= 40\,000\,\text{kg m s}^{-1} - 60\,000\,\text{kg m s}^{-1}$$
$$= -20\,000\,\text{kg m s}^{-1}$$

If the small car hit the stationary van at a speed of 30 m s⁻¹, and the two vehicles locked together, the speed of the tangled wreck would be less than 20 m s⁻¹. Apply the Law of Conservation of Momentum to predict the speed of the tangled wreck immediately after this collision.

elog-1873

tlvd-0823

on Resources

 **Interactivity** Colliding dodgems (int-6610)

10.2 Activities learn on

| 10.2 Quick quiz **on** | 10.2 Exercise | 10.2 Exam questions |

10.2 Exercise

1. Explain in terms of the Law of Conservation of Momentum how astronauts walking in space can change their speed or direction.
2. A Physics student is experimenting with a low-friction trolley on a smooth horizontal surface. Predict the final velocity of the 2-kilogram trolley in each of the following experiments.
   a. The trolley is travelling at a constant speed of 0.6 m s⁻¹. A suspended 2-kilogram mass is dropped onto it as it passes.
   b. The trolley is loaded with 2 kilograms of sand. As the trolley moves with an initial speed of 0.6 m s⁻¹, the sand is allowed to pour out through a hole behind the rear wheels.
3. Two stationary ice skaters, Catherine and Lauren, are facing each other and use the palms of their hands to push each other in opposite directions. Catherine, with a mass of 50 kilograms, moves off in a straight line with a speed of 1.2 m s⁻¹. Lauren moves off in the opposite direction with a speed of 1.5 m s⁻¹.
   a. What is Lauren's mass?
   b. What is the magnitude of the impulse that results in Catherine's gain in speed?
   c. What is the magnitude of the impulse on Lauren while the girls are pushing each other away?
   d. What is the total momentum of the system of Catherine and Lauren just after they push each other away?
   e. Would it make any difference to their final velocities if they pushed each other harder? Explain.

4. Nick and his brother Luke are keen rollerbladers. Nick approaches his stationary brother at a speed of 2.0 m s^{-1} and bumps into him. As a result of the collision Nick, who has a mass of 60 kilograms, stops moving and Luke, who has a mass of 70 kilograms, moves off in a straight line. The surface on which they are 'blading' is smooth enough that friction can be ignored.
   a. With what speed does Luke move off?
   b. What is the magnitude of the impulse on Nick as a result of the bump?
   c. What is the magnitude of Nick's change in momentum?
   d. What is the magnitude of Luke's change in momentum?
   e. How would the motion of each of the brothers after their interaction be different if they pushed each other instead of just bumping?
   f. If Nick held onto Luke so that they moved off together, what would be their final velocity?
5. An unfortunate driver of mass 50 kilograms travelling on an icy road in her 1200-kilogram car collides with a stationary police car of total mass 1500 kilograms (including occupants). The tangled wreck moves off after the collision with a speed of 7.0 m s^{-1}. The frictional force on both cars can be assumed to be negligible.
   a. At what speed was the unfortunate driver travelling before her car hit the police car?
   b. What was the impulse on the police car due to the collision?
   c. What was the impulse on the driver of the offending car (who was wearing a properly fitted seatbelt) due to the impact with the police car?
   d. If the duration of the collision was 0.10 seconds, what average net force was applied to the police car?
6. The Law of Conservation of Momentum can be written as $\Delta p_B = -\Delta p_A$, which equates the change in momentum of two different objects, A and B.
   Newton's Third Law is often expressed as $F_{\text{on B by A}} = -F_{\text{on A by B}}$. Although this equation looks very similar to a law of conservation, no such law exists for forces. Explain why this is the case.

10.2 Exam questions

Question 1 (2 marks)

A student is riding a skateboard travelling at 3 m s^{-1}. The total mass of the student and skateboard is 70 kg. A friend gives the student a push, causing them to gain 80 kg m s^{-1} of momentum.

Calculate their new speed.

Question 2 (2 marks)

Tram B, of mass 32 000 kg, is travelling at 12 m s^{-1} when it collides with the back of tram C, of mass 40 000 kg, which was travelling at 4 m s^{-1} in the same direction. They stick together and move off together as a tangled mess.

Calculate the combined speed of the trams after the collision.

Question 3 (3 marks)

A frictionless trolley of mass 6.0 kg is moving to the right at a speed of 2.4 m s^{-1} when it collides with a stationary trolley of mass m. The two trolleys stick together and move off to the right with a common speed of 1.8 m s^{-1}.

Calculate the value of the mass, m.

Question 4 (2 marks)

Block P, of mass 40 kg, is sliding east on a frictionless surface at 6.0 m s^{-1}. It collides head-on with a stationary block Q, of mass 20 kg. After collision the two blocks are both moving east. Block P has a speed of 2.0 m s^{-1} and block Q has speed v.

What is the value of v?

Question 5 (3 marks)

Two blocks, R and S, are on a frictionless surface. Block R has mass 2.0 kg and block S has mass 3.0 kg. Moving at 5.0 m s^{-1} east, block R collides with stationary block S. After impact, block S is moving east at 4.0 m s^{-1}.

What is the velocity of block R after the collision?

More exam questions are available in your learnON title.

10.3 Work and energy

10.3.1 The concept of energy

Energy is a concept — an idea — that is used to describe and explain change. The following list of some of the characteristics of energy provides some clues as to what it really is:

- All matter possesses energy.
- Energy takes many different forms. It can therefore be classified. Light, internal energy, kinetic energy, gravitational potential energy, chemical energy and nuclear energy are some of the different forms of energy. The names given to different forms of energy sometimes overlap. For example, light is an example of radiant energy. Gamma radiation could be described as nuclear energy or radiant energy. Sound energy and electrical energy involve kinetic energy of particles.
- Energy cannot be created or destroyed. This statement is known as the Law of Conservation of Energy. The quantity of energy in the universe is a constant. However, nobody knows how much energy there is in the universe.
- Energy can be stored, transferred to other matter or transformed from one form into another.
- Some energy transfers and transformations can be seen, heard, felt, smelt or tasted.
- It is possible to measure the quantity of energy transferred or transformed. However, many energy transfers and transformations are not observable and therefore cannot be measured.

The concept of energy allows us to keep track of the changes that take place in a system. The system could be the universe, Earth, your home, the room you are in, your body or a car.

10.3.2 Getting down to work

Energy can be transferred to or from matter in several different ways. It can be transferred by:

- emission or absorption of electromagnetic radiation or nuclear radiation
- heating or cooling as the result of a temperature difference
- the action of a force on an object resulting in movement.

An interaction that involves the transfer of energy by the action of a force is called a **mechanical interaction**.

When mechanical energy is transferred to or from an object, the amount of mechanical energy transferred is called **work**. Simply put, work is a change in energy.

mechanical interaction an interaction in which energy is transferred from one object to another by the action of a force

work a measure of energy transferred to or from an object by the action of a force; it is a scalar quantity

---

The work (W) done when a force (F) causes a displacement (s) in the direction of the force is defined as:

Work = magnitude of the force × magnitude of displacement in the direction of the force

$$W = Fs = Fs\cos\theta$$

Note: If θ is equal to 90°, then no work is done and no change in energy can take place. Some component of the force must be in the direction of the required displacement, otherwise no movement will take place.

Work is a scalar quantity. The SI unit of work is the joule. One joule of work is done when a force of magnitude of 1 newton (N) causes a displacement of 1 metre in the same direction of the force. That is:

$$1\,J = 1\,N \times 1\,m$$
$$= 1\,N\,m$$

It is important to remember that work is always done *by* a force acting *on* something.

tlvd-0094

SAMPLE PROBLEM 2 Calculating work using displacement and force

A shopper pushes horizontally on a loaded supermarket trolley of mass 30 kilograms with a force of 150 N to move it a distance of 5 metres along a horizontal, straight path. The friction force opposing the motion of the trolley is a constant 120 N. How much work is done on the trolley by:
a. the force applied by the shopper   **b. the net force**
c. the shopper to oppose the friction force?

THINK	WRITE
a. 1. Recall the formula for work done.	a. $W = Fs$ $= 150 \times 5$ $= 750\,J$
2. State the solution.	750 J of work is done on the trolley by the force applied by the shopper.
b. 1. Calculate the net force on the trolley, then use the formula for work done.	b. $F_{net} = 150 - 120$ $= 30\,N$ $W = F_{net}\,s$ $= 30 \times 5$ $= 150\,J$
2. State the solution.	150 J of work is done on the trolley by the new force.
c. 1. The force applied to oppose the friction is equal to the friction force. Use this force to calculate the work done.	c. $W = Fs$ $= 120 \times 5$ $= 600\,J$
2. State the solution.	600 J of work is done on the trolley by the shopper to oppose the friction force.

PRACTICE PROBLEM 2

A warehouse worker pushes a heavy crate a distance of 2 metres across a horizontal concrete floor against a constant friction force of 240 N. He applies a horizontal force of 300 N on the crate. How much work is done on the crate by:
a. the warehouse worker
b. the net force?

10.3.3 Force-versus-distance graphs

The work done by a force can be calculated from a force-versus-displacement, or force-versus-distance, graph. For this to be applicable the force presented in the graph and the displacement must be in the same direction.

> The work done is equal to the area under the force-versus-distance graph.

tlvd-3793

SAMPLE PROBLEM 3 Using force-versus-distance graphs to determine work

A filing cabinet is pushed in a straight line across the floor of an office during some spring cleaning. The force applied in the direction of its motion and its displacement are recorded in the graph provided. Determine the work done.

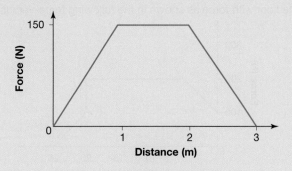

THINK

1. The work done can be calculated by the area under the graph.

2. The area of a trapezium is $\frac{1}{2}(a+b)h$, where a and b are the lengths of the horizontal sides and h is the height.

3. State the solution.

WRITE

$W = \text{area under graph}$

$W = \frac{1}{2}(3+1)\,150$
$ = 300\text{ J}$

300 J of work is done.

PRACTICE PROBLEM 3

A spring is used in an old-fashioned pinball machine to launch the ball from a rest position into the arcade game. The following graph shows the force applied to the ball by the spring during this motion against the distance travelled in the direction of the force. Determine the work done.

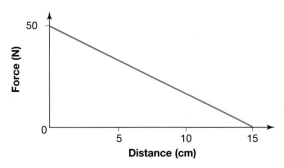

10.3 Quick quiz on	10.3 Exercise	10.3 Exam questions

10.3 Exercise

1. A box is pushed across the floor with force as shown in the following force-versus-displacement graph.

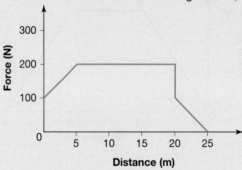

   Calculate the work done to push the box 25 metres.
2. How much work is done on a 4-kilogram brick as it is pushed a distance of 1.5 metres by a net force of 40 N?
3. Imagine that you are trying to push-start a 2000-kilogram truck with its handbrake on. How much work are you doing on the truck?
4. A toddler swings her fluffy toy dog by a string around in circles at a constant speed. How much work does she do on the toy in completing:
   a. one full revolution
   b. half of a full revolution?
5. A gardener uses a rope to drag a heavy log across a flat, horizontal garden path. The log has a mass of 350 kilograms and the rope makes an angle of 35° to the horizontal. The gardener applies a constant force of 275 N and does 1400 J of work on the log. Determine how far they were able to move the log along the path.

10.3 Exam questions

▶ Question 1 (1 mark)

MC A child pulls a toy along using a string inclined at 60° to the horizontal. The tension in the string is 16 N. The toy moves a distance of 12 m horizontally.

What is the work done by this force?
A. 192 J   B. 96 J   C. 48 J   D. 0

▶ Question 2 (1 mark)

MC A block is dragged a horizontal distance of 1.5 m across a rough, horizontal surface. The constant, horizontal pulling force has magnitude 4.0 N. The normal reaction force has magnitude 3.0 N.

What is the work done by the pulling force and what is the work done by the normal reaction force?
A. Pulling force = 6.0 J; normal reaction force = 0
B. Pulling force = 6.0 J; normal reaction force = 4.5 J
C. Pulling force = 0; normal reaction force = 4.5 J
D. Pulling force = 7.5 J; normal reaction force = 0

Question 3 (3 marks)

The graph shows the net force on an object as a function of distance moved.

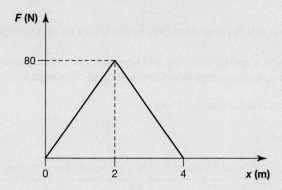

Calculate the work done on the object as it moves 4.0 m.

Question 4 (3 marks)

The graph shows the variation of the net force on an object for a total 5.0 m of displacement.

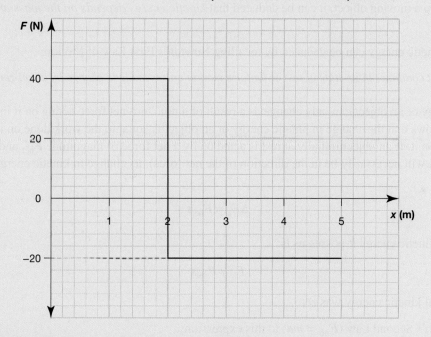

Calculate the net work done on the object. Show your reasoning.

Question 5 (2 marks)

The assisted emergency braking system in a 1200-kilogram car detects a potential collision and applies the brakes, bringing the car to a complete stop. The average retarding force on the four tyres by the road due to friction is 8000 N in total. The work done during the braking was 128 000 J.

Determine how far the car travelled in the time between when the brakes were applied and when the car stopped.

More exam questions are available in your learnON title.

10.4 Energy transfers

10.4.1 Kinetic energy

Kinetic energy is the energy associated with the movement of an object. By imagining how much energy it would take to stop a moving object, it can be deduced that *kinetic energy depends on the mass and speed of the object.*

A formula for kinetic energy can be deduced by recalling Newton's First Law of Motion:

> *Every object continues in its state of rest or uniform motion unless made to change by a non-zero net force.*

The kinetic energy of an object can only change as a result of a non-zero net force acting on it in the direction of motion. It follows that the change in kinetic energy of an object is equal to the work done on it by the net force acting on it. If an object initially at rest is acted on by a net force of magnitude F_{net} and moves a distance s (which will necessarily be in the direction of the net force), its change in kinetic energy, ΔE_k, can be expressed as:

$$\Delta E_k = F_{net}s$$

The quantity of kinetic energy it possesses is:

$$E_k = F_{net}s$$

because the initial kinetic energy was zero.

Applying Newton's Second Law ($F_{net} = ma$) to this expression:

$$E_k = mas$$

where m is the mass of the object.

The movement of the object can also be described in terms of its initial velocity, v, and its final velocity, u. The magnitudes of the quantities a, s, v and u are related to each other by the equation:

> **kinetic energy** the energy associated with the movement of objects; like all forms of energy, it is a scalar quantity

$$v^2 = u^2 + 2as$$

If the object acquires a speed v as a result of the work done by the net force:

$$v^2 = 2as \quad \text{since } u = 0$$
$$\Rightarrow as = \frac{v^2}{2}$$

Substituting this into the expression for kinetic energy:

$$E_k = mas$$

$$\Rightarrow E_k = \frac{mv^2}{2}$$

The kinetic energy of an object of mass m and speed v can therefore be expressed as:

$$E_k = \frac{1}{2}mv^2$$

Note that the momentum ($p = mv$) is a vector quantity, whereas kinetic energy, $E_k = \frac{1}{2}mv^2$, is a scalar quantity.

tlvd-0096

SAMPLE PROBLEM 4 Comparing kinetic energy of objects of various mass and speed

Compare the kinetic energy of an Olympic track athlete running the 100-metre sprint with that of a family car travelling through the suburbs. Assume the athlete is 70 kilograms and travelling at 10 m s⁻¹ and the car is 1500 kilograms and travelling at 60 km h⁻¹.

THINK	WRITE
1. Use the estimated mass and speed to calculate the kinetic energy of the athlete.	$E_k = \frac{1}{2}mv^2$ $= \frac{1}{2} \times 70 \times 10^2$ $= 3500 \text{ J}$
2. Convert the speed into m s⁻¹.	Speed $\approx 60 \text{ km h}^{-1}$ $\approx 17 \text{ m s}^{-1}$
3. Use the estimated mass and speed to calculate the kinetic energy of the family car.	$E_k = \frac{1}{2}mv^2$ $= \frac{1}{2} \times 1500 \times 17^2$ $\approx 217\,000 \text{ J}$
4. Interpret the results.	$\frac{217\,000}{3500} = 62$
5. State the solution.	The family car has approximately 62 times more kinetic energy than the athlete.

PRACTICE PROBLEM 4

a. **Calculate the kinetic energy of a 2000-kilogram elephant charging at a speed of 8.0 m s⁻¹.**
b. **Estimate the kinetic energy of:**
   i. **a cyclist riding to work**
   ii. **a snail crawling across a footpath.**

SAMPLE PROBLEM 5 Calculating the final speed using kinetic energy, displacement, force and mass

A shopper pushes horizontally on a loaded supermarket trolley of mass 30 kilograms with a force of 150 N to move it a distance of 5 metres along a horizontal, straight path. The friction force opposing the motion of the trolley is a constant 120 N. If the trolley starts from rest, what is its final speed?

THINK	WRITE
1. The change in kinetic energy of the trolley is equal to the work done on it by the net force acting on it.	$\Delta E_k = F_{net}s$
2. As the trolley was initially at rest, the change in kinetic energy is equal to the final kinetic energy.	$E_k = F_{net}s$ $= 30 \times 5.0$ $= 150$ J
3. Use the formula for kinetic energy to calculate the speed.	$E_k = \dfrac{1}{2}mv^2$ $150 = \dfrac{1}{2} \times 30 \times v^2$ $v^2 = \dfrac{150}{15}$ $= 10$ $\Rightarrow v = \sqrt{10}$ $\approx 3.2 \text{ m s}^{-1}$
4. State the solution.	The trolley's final speed is 3.2 m s^{-1}.

PRACTICE PROBLEM 5

A gardener pushes a loaded wheelbarrow with a mass of 60 kilograms a distance of 4 metres along a horizontal, straight path against a constant friction force of 120 N. He applies a horizontal force of 150 N on the wheelbarrow. If the wheelbarrow is initially at rest, what is its final speed?

If the net force is in the opposite direction to that in which the object is moving, the object slows down. For example, the work done by the net force to stop a 70-kilogram athlete running at a speed of 10 m s^{-1} is given by:

$$\text{Work done by net force} = \Delta E_k$$
$$= 0 - \frac{1}{2}mv^2$$
$$= -\frac{1}{2} \times 70 \text{ kg} \times \left(10 \text{ m s}^{-1}\right)^2$$
$$= -3500 \text{ J}$$

The negative sign indicates that the direction of the net force is opposite to the direction of the displacement.

EXTENSION: Speed kills

The truth of the slogan 'Speed kills' can be appreciated by comparing the kinetic energy of a 1500-kilogram car travelling at 60 km h^{-1} (16.7 m s^{-1}) with that of the same car travelling at 120 km h^{-1} (33.3 m s^{-1}).

At 60 km h^{-1}, the car's kinetic energy is:

$$E_k = \frac{1}{2}mv^2$$
$$= \frac{1}{2} \times 1500\,\text{kg} \times \left(16.7\,\text{m s}^{-1}\right)^2$$
$$= 2.1 \times 10^5\,\text{J}$$

At 120 km h^{-1}, its kinetic energy is:

$$E_k = \frac{1}{2}mv^2$$
$$= \frac{1}{2} \times 1500\,\text{kg} \times \left(33.3\,\text{m s}^{-1}\right)^2$$
$$= 8.3 \times 10^5\,\text{J}$$

In other words, a 100 per cent increase in speed produces a 400 per cent increase in the kinetic energy, and therefore four times as much work needs to be done on the car to stop it during a crash with a solid object.

10.4.2 Potential energy

Energy that is stored is called potential energy. Objects that have potential energy have the capacity to apply forces and do work. Potential energy takes many forms.

- The food that you eat contains potential energy. Under certain conditions, the energy stored in food can be transformed into other forms of energy. Your body is able to transform the potential energy in food into internal energy so that you can maintain a constant body temperature. Your body transforms some of the food's potential energy into the kinetic energy of blood, muscles and bones so that you can stay alive and move. Some of it is transformed into electrochemical energy to operate your nervous system.
- Batteries contain potential energy.
- An object that is in a position from which it could potentially fall has **gravitational potential energy**. The gravitational potential energy of an object is 'hidden' until the object is allowed to fall. Gravitational potential energy exists because of the gravitational attraction of masses towards each other. All objects with mass near Earth's surface are attracted towards the centre of Earth. The further away from Earth's surface an object is, the more gravitational potential energy it has.
- Energy can be stored in objects by compressing them, stretching them, bending them or twisting them. If the change in shape can be reversed, energy stored in this way is called **strain potential energy**. Strain potential energy can be transformed into other forms of energy by allowing the object to reverse its change in shape.

gravitational potential energy the energy stored in an object as a result of its position relative to another object to which it is attracted by the force of gravity

strain potential energy the energy stored in an object as a result of a reversible change in shape; it is also known as elastic potential energy

10.4.3 Gravitational potential energy

When an object is in free fall, work is done on it by the force of gravity, transforming gravitational potential energy into kinetic energy. When you lift an object, you do work on it by applying an upwards force on it greater than or equal to its weight. Although the gravitational field strength, g, decreases with distance from Earth's surface, it can be assumed to be constant near the surface. The increase in gravitational potential energy, ΔE_{gp}, by an object of mass m lifted through a height Δh can be found by determining how much work is done on it by the force (or forces) opposing the force of gravity.

$$W = Fs$$
$$= mg\Delta h \ \text{(substituting } F = mg \text{ and } s = \Delta h\text{)}$$
$$\Rightarrow \Delta E_g = mg\Delta h$$

This formula only provides a way of calculating *changes* in gravitational potential energy. If the gravitational potential energy of an object is defined to be zero at a reference height, a formula for the quantity of gravitational potential energy can be found for an object at height h above the reference height.

$$\Delta E_g = mg\Delta h$$
$$\Rightarrow E_g - 0 = mg(h - 0)$$
$$\Rightarrow E_g = mgh$$

Usually the reference height is ground or floor level. Sometimes it might be more convenient to choose another reference height. However, it is the change in gravitational potential energy that is most important in investigating energy transformations. Figure 10.3 shows that the gain in gravitational potential energy as a raw egg is lifted from the surface of a table is mgd. When the raw egg is dropped to the table, the result will be the same whether you use the height of the table or ground level as your reference height. The gravitational potential energy gained will be transformed into kinetic energy as work is done on the egg by the force of gravity.

FIGURE 10.3 The choice of reference height does not have any effect on the change in gravitational potential energy.

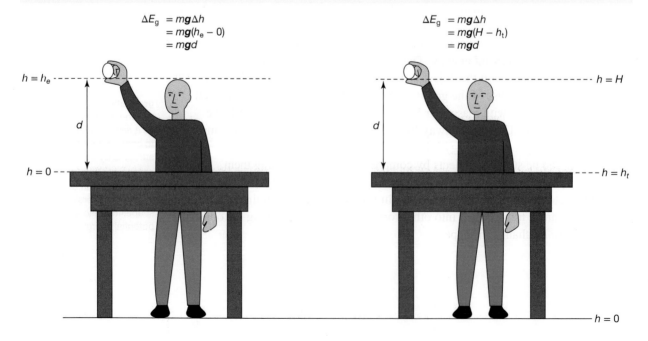

10.4.4 Strain potential energy

Work must be done on an object by a force in order to change its strain potential energy. However, when objects are compressed, stretched, bent or twisted, the force needed to change their shape is not constant. For example, the more you stretch a rubber band, the harder it is to stretch it further. Similarly, the more you compress the sole of a running shoe, the harder it is to compress it further, just like with a spring. The amount of strain potential energy gained by stretching a rubber band or by compressing the sole of a running shoe can be determined by calculating the amount of work done on it.

The amount of work done by a changing force is given by:

$$W = F_{av}s$$

It can be determined by calculating the area under a graph of force versus displacement in the direction of the force. In the case of a simple spring, a rubber band or the sole of a running shoe, the work done (and hence the change in strain potential energy) can be calculated by determining the area under a graph of force versus extension, or force versus compression.

FIGURE 10.5 Graphs showing the force applied by an ideal spring versus **a.** compression and **b.** extension

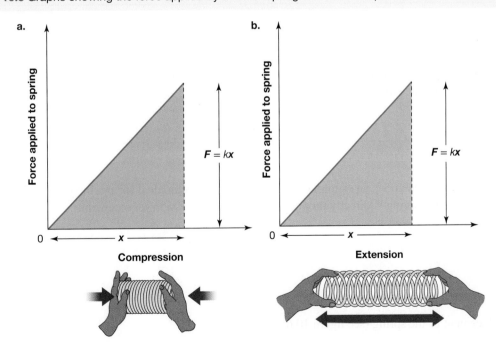

tlvd-3794

SAMPLE PROBLEM 6 Calculating strain potential energy in a spring

The following figure shows how the force required to compress a spring changes as the spring is compressed. How much strain potential energy is stored in the spring when it is compressed by 25 centimetres?

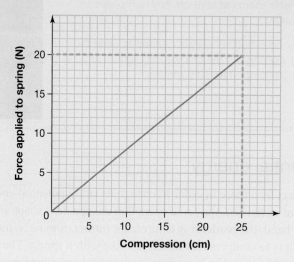

THINK

1. The amount of strain potential energy added to the spring when it is compressed is equal to the amount of work done to compress it.

2. Work is equal to the area under the graph.

3. State the solution.

WRITE

$E_s = W$

$W = $ area under graph

$$= \frac{1}{2} \times 20 \times 0.25$$

$$= 2.5 \text{ J}$$

2.5 J of strain potential energy is stored in the spring when it is compressed by 25 centimetres.

PRACTICE PROBLEM 6

How much strain potential energy is stored in the spring described in sample problem 6 when it is compressed by a distance of:
a. 10 centimetres
b. 20 centimetres?

Hooke's Law

The spring in sample problem 6 is an example of an ideal spring. For an ideal spring, the force required to compress (or extend) the spring is directly proportional to the compression (or extension):

$$F \propto x$$

where:

F is the force exerted by the spring

x is the displacement of the spring (see figure 10.6).

This relationship is expressed fully by Hooke's Law, which states:

$$F = -kx$$

where k is known as the spring constant.

FIGURE 10.6 Displacement for a spring **a.** at equilibrium, **b.** compressed and **c.** elongated

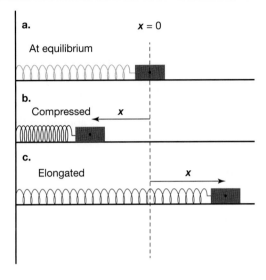

The negative sign in Hooke's Law is necessary because the direction of the force applied by the spring is always opposite to the direction of the spring's displacement. For example, if the spring is compressed, it pushes back in the opposite direction. If the spring is extended, it pulls back in the opposite direction.

Hooke's Law is more conveniently expressed without vector notation as:

$$F = kx$$

where:

F is the magnitude of the force applied by the spring

x is the magnitude of the extension or compression of the spring.

The strain energy stored in a spring that is changed in length by x, whether it is compressed or extended, is equal to the area under the force-versus-compression graph or the force-versus-extension graph. That is:

$$\text{Strain potential energy } (E_s) = \frac{1}{2} \times kx \times x$$
$$= \frac{1}{2} kx^2$$

SAMPLE PROBLEM 7 Using Hooke's Law to calculate work, strain potential energy and force

A wooden block is pushed against an ideal spring of length 30 centimetres until its length is reduced to 20 centimetres. The spring constant of the spring is 50 N m^{-1}.
a. What is the magnitude of the force applied on the wooden block by the compressed spring?
b. How much strain potential energy is stored in the compressed spring?
c. How much work was done on the spring by the wooden block?

THINK	WRITE
a. 1. Recall Hooke's Law.	a. $F = kx$
2. Substitute the values into the equation.	$x = 0.30 - 0.20 = 0.10 \, \text{m}$ $F = 50 \times 0.10$ $= 5.0 \, \text{N}$
3. State the solution.	5 N is applied on the wooden block by the compressed spring.
b. 1. Recall the formula for strain potential energy.	b. $E_s = \frac{1}{2}kx^2$
2. Substitute the values into the equation.	$E_s = \frac{1}{2} \times 50 \times 0.10^2$ $= 0.25 \, \text{J}$
3. State the solution.	0.25 J of strain potential energy is stored in the compressed spring.
c. 1. The work done on the spring is equal to the elastic potential energy.	c. $W = E_s$ $= 0.25 \, \text{J}$
2. State the solution.	0.25 J of work was done on the spring by the wooden block.

PRACTICE PROBLEM 7

a. An object hanging from the end of a spring extends the spring by 20 centimetres. The spring constant is 60 N m^{-1}.
   i. What upwards force is applied to the object by the spring?
   ii. How much strain potential energy is stored in the spring when it is extended by 50 centimetres?
   iii. What is the mass of the object?
b. What is the spring constant of the spring described in sample problem 6?

10.4.5 Conservation of energy

Conservation of energy is a law of the universe. It states that the total energy in an isolated system is constant. Note that in Year 11, you will mainly see this as the total energy being the sum of the kinetic energy, the gravitational potential energy and the strain potential energy. Considering the universe is an isolated system, the total amount of energy is constant, meaning energy cannot be created or destroyed.

Energy can, however, be transferred from one form to another. A single bounce of a tennis ball onto a hard surface involves the following mechanical energy transformations:

- As the ball falls, the force of gravity does work on the ball, transforming gravitational potential energy into kinetic energy.
- As soon as the bottom of the tennis ball touches the ground, the upwards push of the ground does work on the tennis ball, transforming kinetic energy into strain potential energy. A small amount of gravitational potential energy is also transformed into strain potential energy. This continues until the kinetic energy of the ball is zero.
- As the ball begins to rise and remains in contact with the ground, the upwards push of the ground does work on the tennis ball, transforming strain potential energy into kinetic energy and a small amount of gravitational potential energy until the ball loses contact with the ground.
- As the ball gains height, the force of gravity does work on the ball, transforming kinetic energy into gravitational potential energy.

FIGURE 10.7 Energy is transferred into different forms when a tennis ball is bounced.

tlvd-0100

SAMPLE PROBLEM 8 Determining the speed of a skateboarder at the bottom of a ramp

A skateboarder of mass 50 kilograms, starting from rest, rolls from the top of a curved ramp, a vertical drop of 1.5 metres (see the following figure). What is the speed of the skateboarder at the bottom of the ramp? (The frictional force applied to the skateboarder by the ramp is negligible.)

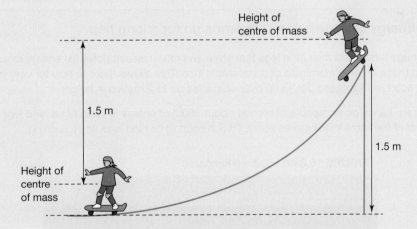

Height of centre of mass

1.5 m

1.5 m

Height of centre of mass

THINK

1. It can be assumed in this case that the total mechanical energy is conserved. The only transformation that takes place is that from gravitational potential energy to kinetic energy. The gain in kinetic energy of the skateboarder is therefore equal to the magnitude of the loss of gravitational potential energy.

WRITE

$$\Delta E_k = \Delta E_{gp}$$
$$\frac{1}{2}mv^2 = mg\Delta h$$

2. Substitute the values into the equation.

$$\Delta E_k = \Delta E_{gp}$$

$$\frac{1}{2}mv^2 = mg\Delta h$$

$$\frac{1}{2} \times 50v^2 = 50 \times 9.8 \times 1.5$$

$$25v^2 = 735$$

$$v^2 = \frac{735}{25}$$

$$\Rightarrow v = \sqrt{\frac{735}{25}}$$

$$v \approx 5.4 \text{ m s}^{-1}$$

3. State the solution.

The speed of the skater at the bottom of the ramp is 5.4 m s^{-1}.

PRACTICE PROBLEM 8

A toy car of mass 0.5 kilograms is pushed against an ideal spring so that the spring is compressed by 0.1 metres. The spring constant of the spring is 80 N m^{-1}.

a. How much strain potential energy is stored in the spring when it is compressed?

b. After the toy car is released, what will be its speed at the instant that the spring returns to its natural length?

EXTENSION: Energy storage helps kangaroos go for a long hop

Kangaroos have huge tendons in their hind legs that store and return elastic potential energy much more efficiently than do those of other mammals of comparable size. This allows them to hop for very large distances without tiring. An adult red kangaroo can jump over obstacles up to 2 metres in height.

A young 50-kilogram kangaroo is capable of storing about 360 J of energy in each of its hind legs. A typical four-legged animal of the same mass stores about 55 J in each of its hind legs while running.

FIGURE 10.8 An adult red kangaroo

10.4 Activities

10.4 Quick quiz on	10.4 Exercise	10.4 Exam questions

10.4 Exercise

1. Use the formulae for work and kinetic energy to show that their units are equivalent.
2. Estimate the kinetic energy of:
   a. a car travelling at 60 km h^{-1} on a suburban street
   b. a 58-gram tennis ball as it is returned to the server in a Wimbledon final.
3. Estimate the amount of work done on a 58-gram tennis ball by the racquet when the ball is served at a speed of 200 km h^{-1}.
4. Estimate the change in gravitational potential energy of:
   a. a skateboarder riding down a half-pipe
   b. a child sliding from the top to the bottom of a playground slide
   c. you at your maximum height as you jump up from rest.
5. A truck driver wants to lift a heavy crate of books with a mass of 20 kilograms onto the back of a truck through a vertical distance of 1 metre. The driver needs to decide whether to lift the crate straight up (figure a), or push it up along a ramp (figure b).

   a. 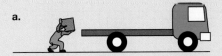   b.

   a. What is the change in gravitational potential energy of the crate of books in each case?
   b. How much work must be done against the force of gravity in each case?
   c. If the ramp is perfectly smooth, how much work must be done by the truck driver to push the crate of books onto the back of the truck?
   d. In view of your answers to parts **b** and **c**, which of the two methods is the best way to get the crate of books onto the back of the truck? Explain your answer.
6. World-class hurdlers raise their centre of mass as little as possible when they jump over the hurdles. Why?
7. Two ideal springs, X and Y, have spring constants of 200 N m^{-1} and 100 N m^{-1} respectively. They are each extended by 20 centimetres by pulling with a hook. For each of the springs, determine:
   a. the magnitude of the force applied to the hook
   b. the strain potential energy.
8. A tourist on an observation tower accidentally drops her 1.2-kilogram camera to the ground 20 metres below.
   a. What kinetic energy does the camera gain before shattering on the ground?
   b. What is the speed of the camera as it hits the ground?

9. The following figure shows part of a roller-coaster track. As a fully loaded roller-coaster car of total mass 450 kilograms approaches point A with a speed of 12 m s^{-1}, the power fails and it rolls freely down the track. The friction force on the car can be assumed to be negligible.

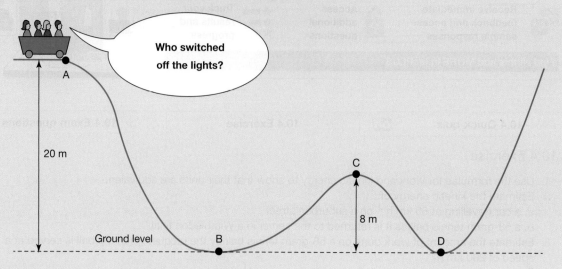

Who switched off the lights?

20 m

C

8 m

Ground level   B   D

a. What is the kinetic energy of the loaded car at point A?
b. Determine the speed of the loaded car at each of points B and C.
c. What maximum height will the car reach after passing point D?

10. The following graph shows how the driving force on a 1200-kilogram car changes as it accelerates from rest over a distance of 1 kilometre on a horizontal road. The average force opposing the motion of the car due to air resistance and road friction is 360 N.

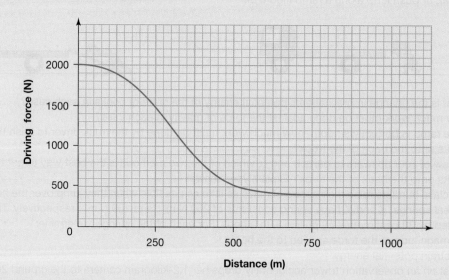

a. How much work has been done by the forward push (the driving force) on the car?
b. How much work has been done on the car to overcome both air resistance and road friction?
c. What is the speed of the car when it has travelled 1 kilometre?

11. A toy truck of mass 0.5 kilograms is pushed against a spring so that it is compressed by 0.1 metres. The spring obeys Hooke's Law and has a spring constant of 50 N m^{-1}. When the toy truck is released, what will be its speed at the instant that the spring returns to its natural length? Assume that there is no frictional force resisting the motion of the toy truck.

12. A pogo stick contains a spring that stores energy when it is compressed. The following graph shows how the upwards force of a pogo stick on a 30-kilogram child jumping on it changes as the spring is compressed. The maximum compression of the spring is 8 centimetres. Assume that all of the energy stored in the spring is transformed to the mechanical energy of the child. The mass of the pogo stick itself can be ignored.

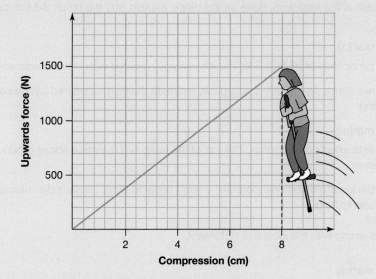

a. How do you know that the spring in the pogo stick is an ideal spring?
b. What is the spring constant of the spring?
c. How much work is done on the child by the pogo stick as the spring expands?
d. What is the kinetic energy of the child at the instant that the compression of the pogo stick spring is zero?
e. How high does the child rise from the ground? Assume that the child leaves the ground at the instant that the compression of the pogo stick spring is zero.

13. Describe the mechanical energy transformations that take place when a child jumps up and down on a trampoline.

14. Discuss the mechanical energy transformations that take place when a diver uses a springboard to dive into the water, from the time that the diver is standing motionless on the springboard until the time she reaches her lowest point in the water. Use a graph or diagram describing the energy transformations in both the springboard and the diver to illustrate your answer.

15. Consider the mechanical energy transformations that take place when a skateboard rider gets airborne off the end of a ramp (see the following figure).

    a. Use a graph to describe the energy transformations that occur during the time interval between starting at one end of the ramp, getting airborne at the other end and returning to the starting point.
    b. Explain, in terms of the energy transformations, how it is possible for the rider's feet to remain in contact with the skateboard while in the air.

10.4 Exam questions

▶ Question 1 (3 marks)

A car is initially moving forwards at 14 m s^{-1}. The driver applies the brakes and the car comes to rest over a distance of 12 m. The mass of the car is 1200 kg.

Calculate the magnitude of the average retarding force on the car.

Question 2 (3 marks)

A student exerted a constant horizontal force of 6.0 N as she pulled a wooden block a distance of 2.0 m across a rough surface. The block moved at constant speed.

a. How much work did the student do on the block? **(1 mark)**

b. Using the concept of the total work done on the block, explain why the block did not gain any kinetic energy. **(2 marks)**

Question 3 (3 marks)

In an explosion, a cannonball of mass 1.5 kg moves vertically upward with initial kinetic energy of 147 J.

Assuming that air resistance on the ball is negligible, estimate the height reached by the cannonball. Show your reasoning.

Question 4 (3 marks)

A marble of mass m is rolled down a ramp. The top of the ramp is 3.0 metres above ground level. The length of the ramp is 5.0 metres.

Jack says the marble loses gravitational potential energy equal to $5mg$. Jill says the marble loses gravitational potential energy equal to $3mg$.

a. Who is correct? **(1 mark)**

b. Explain what is wrong with the incorrect statement. **(2 marks)**

Question 5 (3 marks)

Consider the graph of force versus extension shown for an ideal spring.

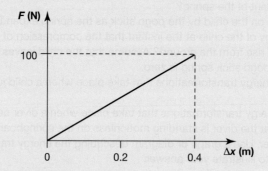

Calculate the work needed to stretch the spring by 0.40 m. Show your working.

More exam questions are available in your learnON title.

10.5 Efficiency and power

KEY KNOWLEDGE

- Calculate the efficiency of an energy transfer system: $\eta = \dfrac{\text{useful energy out}}{\text{total energy in}}$.

- Analyse rate of energy transfer using power: $P = \dfrac{E}{t}$

Source: VCE Physics Study Design (2023–2027) extracts © VCAA; reproduced by permission.

10.5.1 Efficiency

Kinetic energy, gravitational potential energy and strain potential energy are referred to as forms of mechanical energy. Transformation to or from each of these forms of energy requires the action of a force.

If mechanical energy were conserved, a ball would return to the same height from which it was dropped. In fact, mechanical energy is not conserved. During each of the mechanical transformations that occur during a bounce, some of the ball's mechanical energy is transformed to thermal energy of the air, ground and ball, resulting in a small temperature increase. Some mechanical energy is also transformed into sound and permanent deformation of the ball.

Mechanical energy transformations to thermal energy, sound and so on are largely permanent. It is very difficult to convert this energy back into mechanical energy and so it is not considered useful.

> The efficiency, η, of an energy transfer is calculated from the ratio:
>
> $$\eta = \frac{\text{useful energy out}}{\text{total energy in}}$$
>
> where η is the Greek letter 'eta'.
>
> Note that you have to multiply the ratio 'useful energy out to total energy' in by 100 to express the efficiency as a percentage.

tlvd-3795

SAMPLE PROBLEM 9 Calculating the efficiency of a rebound

A ball dropped from 1.5 metres rebounds to 1.2 metres. What is the efficiency?

THINK	WRITE
1. Recall the formula for efficiency.	$\eta = \dfrac{\text{useful energy out}}{\text{total energy in}}$
2. Calculate the total energy in.	The 'total energy in' is the initial gravitational potential energy of the ball. $E_g = mgh$ $= mg \times 1.5$
3. Calculate the useful energy out.	The 'useful energy out' is the gravitational potential energy of the ball at its rebound height of 1.2 metres. $E_g = mgh$ $= mg \times 1.2$
4. Substitute these values into the formula to determine the efficiency.	$\eta = \dfrac{\text{useful energy out}}{\text{total energy in}}$ $= \dfrac{1.2mg}{1.5mg}$ $= 0.8$ $= 80\%$
5. State the solution.	The efficiency is 80%.

PRACTICE PROBLEM 9

A basketball is pumped up to give an efficiency of 80% when dropped. If this basketball is dropped from a height of 2 metres, to what height does it rebound after the fourth bounce?

10.5.2 Power

Power is the rate at which energy is transferred or transformed.

$$P = \frac{E}{t}$$

where:

E = the energy transformed, in joules

t = the time taken, in seconds.

In the case of conversions to or from mechanical energy or between different forms of mechanical energy, power, P, can be defined as the rate at which work is done:

$$P = \frac{W}{\Delta t}$$

where:

W = the work done, in joules

Δt = the time interval during which the work is done, in seconds.

The SI unit of power is the watt (W), which is defined as $1\ \text{J s}^{-1}$.

The power delivered when a force, F, is applied to an object can also be expressed in terms of the object's velocity, v.

power the rate of doing work, or the rate at which energy is transformed from one form to another

$$P = \frac{W}{\Delta t} = \frac{Fx}{\Delta t}$$

$$= F \times \frac{x}{\Delta t}$$

$$= Fv$$

$$P = Fv$$

where P (in W) is the power, delivered by a force F (in N) to an object with velocity v (in m s^{-1}).

tlvd-0102

SAMPLE PROBLEM 10 Calculating power output and work while climbing stairs

A student of mass 40 kilograms walks briskly up a flight of stairs to climb four floors of a building, a vertical distance of 12 metres in a time interval of 40 seconds.

a. At what rate is the student doing work against the force of gravity?

b. If energy is transformed by the leg muscles of the student at the rate of 30 kJ every minute, what is the student's power output?

THINK	WRITE
a. 1. The work done by the student against the force of gravity is equal to the gain in gravitational potential energy.	**a.** $W = mg\Delta h$ $= 40 \times 9.8 \times 12$ $= 4704$ J
2. Recall the formula for power.	$P = \dfrac{W}{\Delta t}$ $= \dfrac{4704}{40}$ ≈ 118 W
3. State the solution.	The student is doing work against the force of gravity at a rate of 118 W.
b. 1. Recall the formula for power.	**b.** $P = \dfrac{\text{energy transferred}}{\text{time taken}}$ $= 30$ kJ min^{-1} $= \dfrac{30\,000 \text{ J}}{60 \text{ s}}$ $= 500$ W
2. State the solution.	If energy is transformed by the leg muscles of the student at the rate of 30 kJ every minute, the student's power output is 500 W.

PRACTICE PROBLEM 10

a. If all of the 720 J of energy stored in the hind legs of a young 50-kilogram kangaroo were used to jump vertically, how high could it jump?

b. What is the kangaroo's power output if the 720 J of stored energy is transformed into kinetic energy during a 1.2 second interval?

---

elog-1838

INVESTIGATION 10.2 — on line only

Climbing to the top

Aim

To investigate the difference that extra load makes to the work done against gravity and the power developed

---

EXTENSION: Which is easier — riding a bike or running?

A normal bicycle being ridden at a constant speed of 4 m s^{-1} on a horizontal road is subjected to a rolling friction force of about 7 N and air resistance of about 6 N. The forward force applied to the bicycle by the ground must therefore be about 13 N. The mechanical power output required to push the bicycle along at this speed is:

$$P = Fv$$
$$= 13 \text{ N} \times 4 \text{ m s}^{-1}$$
$$= 52 \text{ W}$$

Running at a speed of 4 m s^{-1} requires a mechanical power output of about 300 W. Even walking at a speed of 2 m s^{-1} requires a mechanical power output of about 75 W.

Riding a bicycle on a horizontal surface is less tiring than walking or running for a number of reasons.
- Less mechanical energy is needed. The body of the rider does not rise and fall as it does while walking or running, eliminating the changes in gravitational potential energy.
- Because the rider is seated, the muscles need to transform much less chemical energy to support body weight. The strongest muscles in the body can be used almost exclusively to turn the pedals.
- The system of pedals, cogs, chain and wheels transforms a smaller force applied by the rider on the pedal into a larger force of the tyre on the road. The addition of gears also allows the rider to select the most suitable level of pedalling effort for a given situation.

Once you start riding uphill or against the wind, the mechanical power requirement increases significantly. For example, in riding along an incline that rose 1 metre for every 10 metres of road distance covered, the additional power needed by a 50-kilogram rider travelling at 4 m s^{-1} would be:

$$P = \frac{\Delta E_{gp}}{\Delta t}$$
$$= \frac{mg\Delta h}{\Delta t}$$

In a time interval of 1 second, the vertical climb is $\frac{1}{10}$ of 4 m = 0.4 m.

$$\Rightarrow P = \frac{50 \text{ kg} \times 10 \text{ N kg}^{-1} \times 0.4\text{m}}{1 \text{ s}}$$
$$= 200 \text{ W}$$

10.5 Activities

 10.5 Quick quiz on   **10.5 Exercise**   **10.5 Exam questions**

10.5 Exercise

1. Human muscle has an efficiency of about 20%. Take a heavy mass, about 1–2 kilograms, in your hand. With your hand at your shoulder, raise and lower the mass 10 times as fast as you can. Measure the mass, your arm extension and the time taken, and calculate the amount of energy expended and your power output.
2. A pile driver has an efficiency of 80%. The hammer has a mass of 500 kilograms and the pile a mass of 200 kilograms. The hammer falls through a distance of 5 metres and drives the pile 50 millimetres into the ground. Calculate the average resistance force exerted by the ground.
3. Estimate the average power delivered to a 58-gram tennis ball by a racquet when the ball is served at a speed of 200 km h^{-1} and the ball is in contact with the racquet for 4 milliseconds.
4. A small car travelling at a constant speed of 20 m s^{-1} on a horizontal road is subjected to air resistance of 570 N and road friction of 150 N. What power provided by the engine of a car is used to keep it in motion at this speed?
5. While a 60-kilogram man is walking at a speed of 2 m s^{-1}, his centre of mass rises and falls 3 centimetres with each stride. At what rate is he doing work against the force of gravity if his stride length is 1 metre?

10.5 Exam questions

▶ Question 1 (1 mark)

MC Which of the following machines would have the highest useful energy output?
A. 12 kW input at 80% efficiency
B. 200 kW input at 6% efficiency
C. 9 kW input at 95% efficiency
D. 25 kW input at 49% efficiency

▶ Question 2 (3 marks)

A tractor engine has a power output of 80 kW. The tractor is able to travel to the top of a 500-metre-high hill in 4 minutes and 30 seconds. The mass of the tractor is 2.2 tonnes.

What is the efficiency of the engine?

▶ Question 3 (2 marks)

At what average rate is work done on a 4-kilogram brick as it is lifted through a vertical distance of 1.5 metres in 1.2 seconds?

▶ Question 4 (3 marks)

In the sport of weightlifting, the clean-and-jerk involves bending down to grasp the barbell, lifting it to the shoulders while squatting and then jerking it above the head while straightening to a standing position. In 1983, Bulgarian weightlifter Stefan Topurov became the first man to clean and jerk three times his own body mass when he lifted 180 kilograms. Assume that he raised the barbell through a distance of 1.8 metres in a time of 3.0 seconds.
a. How much work did Stefan do in overcoming the force of gravity acting on the barbell?   **(1 mark)**
b. How much power was supplied to the barbell to raise it against the force of gravity?   **(1 mark)**
c. How much work did Stefan do on the barbell while he was holding it stationary above his head?   **(1 mark)**

▶ Question 5 (7 marks)

A bicycle is subjected to a rolling friction force of 6.5 N and an air resistance of 5.7 N. The total mass of the bicycle and its rider is 75 kilograms. Its mechanical power output while being ridden at a constant speed along a horizontal road is 56 W.
a. At what speed is it being ridden?   **(3 marks)**
b. If the bicycle was ridden at the same speed up a slope inclined at 30° to the horizontal, what additional mechanical power would need to be supplied to maintain the same speed? Assume that the rolling friction and air resistance are the same as on the horizontal road.   **(4 marks)**

More exam questions are available in your learnON title.

10.6 Review

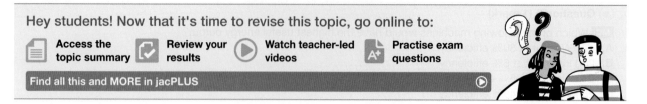
10.6.1 Topic summary

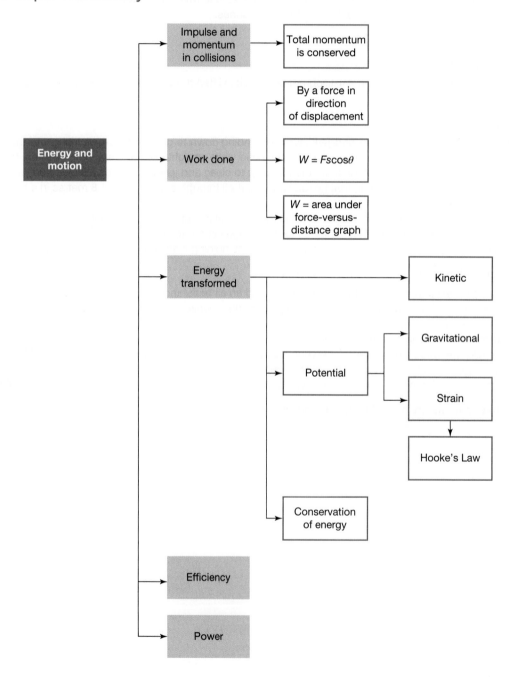

10.6.2 Key ideas summary online only

10.6.3 Key terms glossary online only

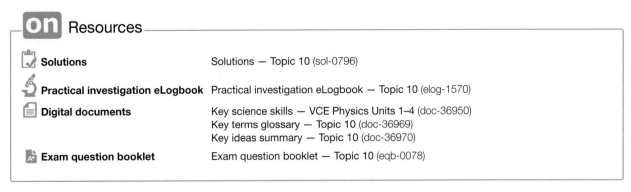
10.6 Activities

learnon

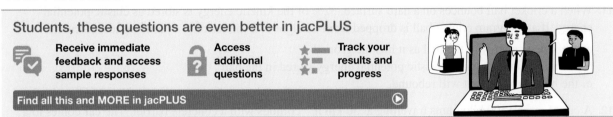
10.6 Review questions

1. John is practising for a tennis tournament by hitting a ball against a wall. The ball strikes the wall at 27 m s^{-1} and rebounds in the opposite direction at 19 m s^{-1}. The ball has a mass of 58.5 grams.

   a. Calculate the change in momentum of the ball.
   b. Calculate the average force exerted by the wall on the ball during this collision if it lasts 50 ms.

2. A large car of mass 1980 kilograms travelling at 11 m s^{-1} collides head on with a small car of mass 970 kilograms. Immediately after they collide, the cars are stuck together and have come to a complete stop. Calculate the speed of the smaller car immediately before the collision.

3. A teacher applies a constant vertical force of 40 N to lift a 4-kilogram box of Physics exam papers 0.8 metres directly upwards from the floor onto a bench. Assume that the motion is at a constant speed. Calculate the work done by the teacher on the box of exam papers.

4. An Airbus A380 of mass 560 000 kilograms lands on a runway at a speed of 250 km h^{-1}. Determine its kinetic energy at this instant.

5. As part of a practical investigation a Physics student hangs various objects on an ideal spring and measures its extension. The spring constant is 90 N m^{-1}. One of the objects that the student hangs on the spring causes it to extend by 12 centimetres.

   a. What upwards force is applied to the object by the spring?
   b. What is the mass of the object?
   c. How much strain potential energy is stored in the spring?

6. A rock climber climbs a cliff, gaining a vertical height of 185 metres above where they started by the end of their climb. Assuming that their mass is 72 kilograms, how much gravitational potential energy have they gained as a result of their climb?

7. A vase of flowers is at rest 175 metres vertically above the surface of Earth. It falls freely, smashing onto the ground below. Assume that it has a mass of 420 grams and air resistance is negligible. Considering the instant immediately before it smashes on the ground, determine the vase's:

   a. kinetic energy
   b. speed.

8. A car of mass 1500 kilograms is travelling along a level road at a constant speed of 62 km h^{-1}. The driver applies the accelerator and the force on the driven wheels does 550 000 J of work to accelerate the car to a higher speed. Determine what the new speed of the vehicle is.

9. A softball of mass 180 grams strikes a hard stationary surface while travelling at a speed of 23 m s^{-1}. It rebounds in the opposite direction at 10 m s^{-1}.

   a. Calculate the kinetic energy of the ball before and after this collision.
   b. Determine the efficiency of the collision.

10. A car travelling at a constant speed of 30 m s^{-1} on a horizontal stretch of highway experiences a combined resistance force of 1150 N from the air and the road. Determine the power that the engine of the car must deliver to the driven wheels to keep it in motion at this constant speed.

11. When a cricket ball bounces on a hard surface, 32% of the kinetic energy is stored as elastic potential energy. If a 160-gram cricket ball is dropped from a height of 2 metres onto a hard surface, calculate:

    a. the kinetic energy of the ball as it hits the ground
    b. the maximum amount of elastic potential energy stored in the ball
    c. the height to which it will rebound.

12. A car of mass 1500 kilograms travelling at 50 km h^{-1} collides with a concrete barrier. The car comes to a stop over a distance of 60 centimetres as the front end crumples.

    a. What is the average net force on the car as it stops?
    b. What is the average acceleration of the car and its occupants? Assume that the occupants are wearing properly fitted seatbelts.
    c. What would be the average acceleration of properly restrained passengers in a very old car with no crumple zone if it stopped over a distance of only 10 centimetres? (The maximum magnitude of acceleration that humans can survive is about 600 m s^{-2}.)
    d. Explain, in terms of mechanical energy transformations, why the front and rear ends of cars are designed to crumple.
    e. In a collision with a rigid barrier, would you feel safer in a large or a small car? Use some sample calculations to illustrate your answer.

13. A girl of mass 50 kilograms rollerblades freely from rest down a path inclined at 30° to the horizontal. The following graph shows how the magnitude of the net force on the girl changes as she progresses down the path.

    a. What is the kinetic energy of the girl after rolling a distance of 8 metres?
    b. What is the sum of the friction force and air resistance on the girl over the first 8 metres?
    c. What is the kinetic energy of the girl at the end of her 20-metre roll?
    d. How much gravitational potential energy has been lost by the girl during her 20-metre roll?
    e. Account for the difference between your answers to parts c and d.

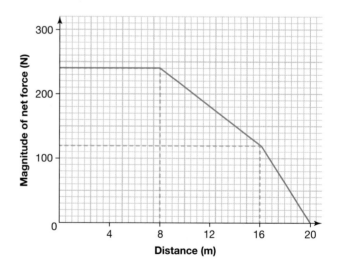

14. The following graph shows the results of a roof crush test conducted in the laboratories of the Department of Civil Engineering at Monash University.

    a. How much work has been done on the roof when the ram has reached its maximum displacement?

    b. If the car has a mass of 1400 kilograms, from what height would it need to be dropped on its roof to crush it by 127 millimetres?

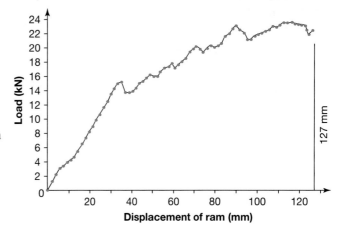

15. Jo and Bill are conducting an experimental investigation into the bounce of a basketball. Bill drops the ball from various heights and Jo measures the rebound height. They also use an electronic timer with thin and very light wires attached to the ball and to alfoil on the floor to measure the impact time. A top-loading balance measures the mass of the ball. What physical quantities can they calculate using these four measurements?

10.6 Exam questions

▶ **Question 1**

Two identical toy cars are travelling directly towards each other at 5 m s^{-1}. They have a head-on collision. They are stuck together after the collision. Assume friction is negligible.

What is their combined speed after the collision?

A. 0 m s^{-1}

B. 2.5 m s^{-1}

C. 5 m s^{-1}

D. 10 m s^{-1}

▶ **Question 2**

Two Physics students, Steve and Terri, are standing at rest next to each other on skateboards. They push against each other and move off in opposite directions. The friction on the floor is negligible. Steve has a mass of 75 kilograms and was moving at a speed of 5 m s^{-1} immediately after he lost contact with Terri. Terri has a mass of 60 kilograms.

What is the impulse that Steve exerted on Terri?

A. 15 N s

B. 300 N s

C. 375 N s

D. 6.25 N s

Question 3

Trish applies a constant horizontal force of 125 N to push a 95-kilogram rock a distance of 5 metres horizontally across her lawn.

What is the work done?

A. 11 875 J
B. 625 J
C. 59 375 J
D. 475 J

Question 4

Emily kicks a football high into the air before she is tackled. The ball strikes the ground at a speed of 25 m s^{-1}.

Assuming that the ball is of regulation size and has a mass of 500 grams, what is its kinetic energy the instant before it strikes the ground?

A. 12.5 J
B. 156 J
C. 313 J
D. 6.25 J

Question 5

Kate creates a toy for her cats to play with by hanging a fluffy object at the bottom of a spring. The spring constant is 5 N m^{-1}. One of the cats is able to stretch the spring by 10 centimetres.

How much energy has the cat stored in the spring?

A. 0.5 J
B. 0.05 J
C. 0.25 J
D. 0.025 J

Question 6

Rohan rides his pushbike home from work. At the end of his journey he has gained 25 metres in altitude.

Assuming that his mass is 80 kilograms, how much gravitational potential energy has he gained over this journey?

A. 2000 J
B. 19 600 J
C. 245 J
D. 784 J

Question 7

Louise stretches a spring, causing it to extend in length by 17 centimetres. The spring constant is 125 N m^{-1}.

What is the force that Louise has applied to the spring at this extension?

A. 21.25 N
B. 2125 N
C. 7.36 N
D. 736 N

Question 8

To assist them in their studies of motion in Physics, Grover and Bailey are riding on a roller coaster. The combined mass of the two students and the carriage is 369 kilograms. As part of the ride they stop momentarily at the highest point in the track. The cart then accelerates down a steep slope, dropping 40 metres in vertical height at the end of the slope.

What is the total kinetic energy of the cart and students at the end of the slope?

A. 200 295 J

B. 295 200 J

C. 648 144 J

D. 144 648 J

Question 9

Scott is building a small robot that uses a number of electrical engines to run its wheels and arms. One of the engines is specified to be 90% efficient.

If the engine is provided with 42 J of energy to perform a manoeuvre, how much useful energy will it output?

A. 34 J

B. 37.8 J

C. 42 J

D. 46.7 J

Question 10

Matt is building a wall out of stone. This requires a lot of stone to be lifted into place. Matt lifts a 3.7-kilogram stone 47 centimetres in 0.8 seconds.

What is the average rate that he is doing work on the stone during this motion?

A. 2130 W

B. 1704 W

C. 21.3 W

D. 17 W

Section B — Short answer questions

Question 11 (2 marks)

As part of a practical investigation, Physics students are measuring the energy transformations of objects sliding down a ramp that they have constructed. A heavy box is allowed to slide down the ramp. The students find that the kinetic energy gained is less than the change in gravitational potential energy.

Explain why this is the case. Assume that there are no significant errors in the students' methodology or measurement.

Question 12 (5 marks)

A runaway tram of mass 26 000 kilograms is travelling at a speed of 54 km h^{-1} when it collides with a stationary tram of mass 35 000 kilograms. After an impact of duration 0.40 seconds, the two trams are stuck together and continue to move along the track.

a. Calculate the speed of the trams after the collision. **(3 marks)**

b. Determine the average force acting on the runaway tram during the collision. **(2 marks)**

Question 13 (2 marks)

A Physics student is waiting tables at a restaurant. They have been doing this work for some time and have perfected the ability to carry a tray of drinks across the room in a purely horizontal motion at a constant speed.

Is the student doing work on the tray of drinks during this constant speed motion? Refer to relevant physics principles in your response.

Question 14 (3 marks)

A new energy storage system is proposed as an alternative to using chemical batteries. It stores energy as gravitational potential energy instead. A prototype of this system uses an old mine shaft with a 450-metre vertical drop to reach the bottom. A 2500-kilogram mass is attached to an electrical motor/generator. The mass is lowered to the bottom of the shaft. When there is electricity available it is used to run the motor to raise the mass up. When electricity is needed, the mass is dropped and the motor is used to generate electricity. At peak capacity, the system takes 520 seconds for the mass to travel the entire 450-metre drop.

Calculate the power output of the system.

Question 15 (6 marks)

A Physics student is investigating the energy transformations involved in the workings of pinball machines. In particular, they decide to focus on the launch of the ball. The following is a diagram they made of the key details of the machine.

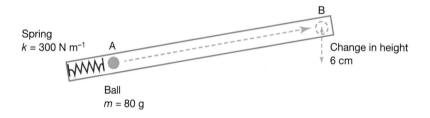

At the start of the game the spring is compressed by 10 centimetres by the machine and the ball sits motionless against it at point A. When the spring is released it pushes on the ball, transferring its energy to the ball, which travels up the machine to point B. At point B the ball has gained 6 centimetres in height.

Calculate the speed of the ball when it reaches point B. Assume that friction is negligible.

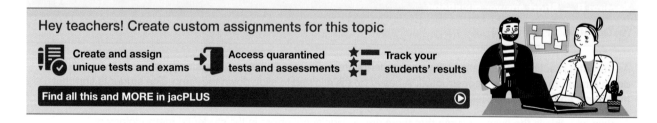

AREA OF STUDY 1 How is motion understood?

OUTCOME 1

Investigate, analyse, mathematically model and apply force, energy and motion.

PRACTICE EXAMINATION

STRUCTURE OF PRACTICE EXAMINATION		
Section	**Number of questions**	**Number of marks**
A	20	20
B	4	20
	Total	**40**

Duration: 50 minutes

Information:
- This practice examination consists of two parts. You must answer all question sections.
- Pens, pencils, highlighters, erasers, rulers and a scientific calculator are permitted.
- You may use the VCAA Physics formula sheet for this task.

 Resources

 Weblink VCAA Physics formula sheet

SECTION A — Multiple choice questions

All correct answers are worth 1 mark each; an incorrect answer is worth 0.

Use the following information to answer questions 1 and 2.

A body travels 25.0 m north in a time of 8.0 seconds. It stops for 2.0 s, then moves 7.0 m south in a time of 2.0 s.

1. Which of the following is the best estimate of its average velocity?
   A. 2.7 m s^{-1}
   B. 1.5 m s^{-1}
   C. 3.3 m s^{-1}
   D. 3.5 m s^{-1}

2. Which of the following is the best estimate of its average speed?
   A. 2.7 m s^{-1}
   B. 1.5 m s^{-1}
   C. 3.3 m s^{-1}
   D. 3.5 m s^{-1}

3. A dodgem car was travelling at 5.0 m s^{-1} south when it collided head-on with another dodgem car, and its velocity changed to 3.0 m s^{-1} north over a time interval of 0.80 s.
   What was the average acceleration of the first dodgem car?
   A. 2.5 m s^{-2} north
   B. 2.5 m s^{-2} south
   C. 10.0 m s^{-2} north
   D. 10.0 m s^{-2} south

Use the following information to answer questions 4 and 5.

The following is a velocity-versus-time graph of a car moving away from a traffic stop.

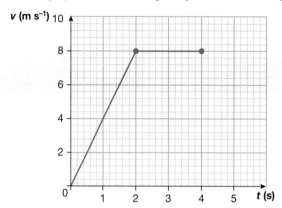

4. What is the average acceleration of the car in the first two seconds?
   A. 2 m s^{-2}
   B. 4 m s^{-2}
   C. 6 m s^{-2}
   D. 8 m s^{-2}

5. What is the distance covered by the car in the first four seconds?
   A. 24 m
   B. 32 m
   C. 40 m
   D. 48 m

6. Consider the following four descriptions for the motion of a body.
   I. Velocity is positive and acceleration is positive.
   II. Velocity is positive and acceleration is negative.
   III. Velocity is negative and acceleration is positive.
   IV. Velocity is negative and acceleration is negative.
   Which of the following best describes the two scenarios in which the body is *slowing down*?
   A. I and II
   B. II and III
   C. I and III
   D. I and IV

Use the following information to answer questions 7 and 8.

A truck travelling on a road changes its velocity from 3.0 m s^{-1} to 9.0 m s^{-1} and covers a distance of 24 m during this period of acceleration.

7. What is the magnitude of the truck's acceleration?
   A. 0.25 m s^{-2}
   B. 0.50 m s^{-2}
   C. 1.5 m s^{-2}
   D. 4.0 m s^{-2}

8. Which of the following is the best estimate for the time taken to accelerate the truck?

   A. 2 s

   B. 4 s

   C. 6 s

   D. 8 s

9. A passenger is not wearing a seatbelt in a car moving forward. The car runs head-on into a barrier and the passenger hits her head against the windscreen in front of her.
   Which of the following is the best explanation for the passenger's motion immediately after the car hit the barrier?

   A. She experienced a force that pushed her forwards towards the windscreen.

   B. She experienced no force on her and kept moving forwards.

   C. She experienced a force from the barrier that pushed her towards the windscreen.

   D. She experienced no force on her by the barrier but one from the car, which pushed her forwards.

10. A man is in a lift travelling upwards at constant speed. He stands on a spring balance that reads 750 N.
    What is the magnitude of the force exerted on the man by the floor of the lift?

    A. Zero

    B. Slightly less than 750 N

    C. 750 N

    D. Slightly more than 750 N

11. A small car of mass 800 kg is accelerating at 2.5 m s^{-2}.
    What is the magnitude of the net force on the car?

    A. 320 N

    B. 1200 N

    C. 1600 N

    D. 2000 N

12. A truck of mass 3600 kg is carrying a load of mass M. A net driving force of 9600 N causes an acceleration of 2.0 m s^{-2}.
    What is the mass M?

    A. 6000 kg

    B. 4800 kg

    C. 2400 kg

    D. 1200 kg

13. A body, J, travelling to the right at a high velocity, collides with a stationary body, K.
    Which of the following best describes the forces in the interaction between J and K?

    A. $F_{\text{on } J \text{ by } K} = F_{\text{on } K \text{ by } J}$

    B. $-F_{\text{on } J \text{ by } K} = -F_{\text{on } K \text{ by } J}$

    C. $-F_{\text{on } J \text{ by } K} = F_{\text{on } K \text{ by } J}$

    D. $F_{\text{on } J \text{ by } K} - F_{\text{on } K \text{ by } J} = 0$

14. A window-washer with a mass of 60 kg is standing on a platform that is accelerating upwards at a constant rate of 1.2 m s^{-2}.
    What is the force exerted on the platform by the window-washer?

    A. 660 N down

    B. 660 N up

    C. 588 N up

    D. 588 N down

Use the following information to answer questions 15 and 16.

A box is at rest on a ramp inclined at 40° to the horizontal, as shown in the following diagram. The force due to gravity is 98 N.

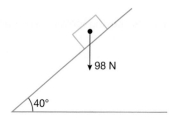

15. What is the magnitude of the normal force that the surface of the ramp is exerting on the box?
  A. 63 N
  B. 75 N
  C. 82 N
  D. 98 N
16. What is the magnitude of the friction force that the surface of the ramp is exerting on the box?
  A. 63 N
  B. 75 N
  C. 82 N
  D. 98 N

Use the following information to answer questions 17 and 18.

A jogger with a mass of 80 kg is moving east at 7 m s^{-1}. He then slows down to a velocity of 2 m s^{-1} east over a period of 2 s.

17. Which of the following best describes the change in momentum of the jogger?
  A. 560 kg m s^{-1} east
  B. 560 kg m s^{-1} west
  C. 400 kg m s^{-1} east
  D. 400 kg m s^{-1} west

18. What is the magnitude of the average force exerted on the jogger to slow him down?
  A. 280 N
  B. 200 N
  C. 160 N
  D. 80 N

19. A ball of mass 0.42 kg is at rest on a field. It is kicked and reaches a maximum speed of 13.5 m s^{-1}. What was the work done on the ball when it was kicked?
  A. 12 J
  B. 24 J
  C. 38 J
  D. 77 J

20. A spring has an unstretched length of 0.7 m. After a mass of 2.5 kg is attached to it, its length is 1.2 m. What is the spring constant of this spring?

A. 20 N m^{-1}

B. 35 N m^{-1}

C. 49 N m^{-1}

D. 98 N m^{-1}

SECTION B — Short answer questions

Question 21 (4 marks)

A cyclist is pedalling along a straight stretch of road at a constant velocity of 6.0 m s^{-1}. She then accelerates at a constant rate for 4.0 s, reaching a final velocity of 14 m s^{-1}.

a. What is the magnitude of the cyclist's acceleration? **(1 mark)**

b. How far did she travel during the period that she was accelerating? **(1 mark)**

While travelling at 14 m s^{-1} the cyclist applied constant braking, coming to a stop over a distance of 24.5 m.

c. What was the magnitude of the cyclist's acceleration while she was braking? **(1 mark)**

d. How long did the bicycle slow down to a stop? **(1 mark)**

Question 22 (6 marks)

A van with mass 150 kg tows a small trailer with mass of 50 kg. They are connected by a towing rod that can be considered to be massless. The driving force is entirely provided by the electric motor on the van. The total resistive force on the van is 30 N, while the total resistive force on the trailer is 20 N. The van and trailer are accelerating at 1.9 m s^{-2}.

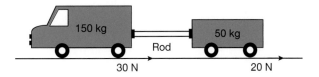

a. Calculate the net force acting on the van and trailer. **(1 mark)**

b. Determine the size of the driving force provided by the electric motor on the van. **(2 marks)**

c. What is the size of the tension force of the rod on the trailer? **(3 marks)**

Question 23 (4 marks)

Suzy, an ice skater with a mass of 55 kg, is skating at a constant velocity of 4.8 m s^{-1} east. She collides with Kai, another ice skater, who is stationary. They move off together at a constant velocity of 3.0 m s^{-1} east. The collision may be considered as isolated.

a. What was Suzy's initial momentum before the collision? **(1 mark)**

b. Considering that the collision is isolated, what is the mass of Kai? **(3 marks)**

A stationary cart with mass 600 kg is at the top of a sloping ramp at point A. It is released and rolls down the ramp, and attains its maximum speed of 14.4 m s^{-1} at point B.

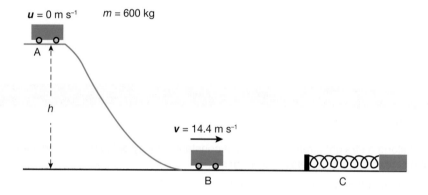

a. Calculate the kinetic energy of the cart at point B. (2 marks)

b. Assuming that friction is negligible and there is no energy lost, what is the height, h, of the ramp? (2 marks)

c. The cart hits a spring at point C, which compresses and slows the cart down to a stop. The spring compresses a distance of 7.40 m.
What is the spring constant of the spring? (2 marks)

PRACTICE SCHOOL-ASSESSED COURSEWORK

ASSESSMENT TASK — DATA ANALYSIS

In this task, you will be required to investigate, analyse and mathematically model the motion of particles and bodies.

- A scientific calculator is permitted. You will need graph paper for this task.

Total time: 50 minutes (5 minutes reading and 45 minutes writing)

Total marks: 34 marks

TRACKING THE POSITION OF ATHLETES

The motion of an athlete can be studied using a radar system that can track their position on a course. The data obtained (shown in table 1 below) gives the position of the runner down a straight athletics track in half-second intervals for the first three seconds of the race and one-second intervals thereafter. The athlete is training for a 100-metre race. At time $t = 0$, the athlete is in the starting blocks and the gun goes off.

USEFUL FORMULAE FOR THIS TASK

Average speed: $v = \dfrac{\Delta x}{\Delta t}$

Average acceleration: $a = \dfrac{\Delta v}{\Delta t}$

Momentum: $p = mv$

Kinetic energy: $E_k = \dfrac{1}{2}mv^2$

1. Use the data in table 1 to plot a distance-versus-time graph.

TABLE 1 Distance versus time

Time (s)	Distance (m)
0.00	0.00
0.50	0.60
1.00	2.40
1.50	5.40
2.00	9.60
2.50	15.0
3.00	21.0
4.00	32.9
5.00	44.7
6.00	56.4
7.00	67.8
8.00	79.0
9.00	89.9
10.0	100.7

2. By taking pairs of data, complete a table for the average speed of the runner for each pair of data points in table 1.

   The first data point, in which the average speed of the athlete in the first interval is 1.2 m s⁻¹ (determined through dividing the change in distance by the change in time), has been done for you. It occurs at the midpoint in time of the first frame. This will be at $t = 0.25$ s.

TABLE 2 Average speed versus time

Time (s)	Average speed (m s⁻¹)
0.25	1.20
0.75	
1.25	

3. Using your data from table 2, plot a graph of the athlete's average speed versus time.

Now we are going to find the acceleration of the athlete during the course of the race.

4. Complete table 3 for acceleration versus time. Again, your time should be the midpoint between each set of data points in table 2.

TABLE 3 Acceleration versus time

Time (s)	Average acceleration (m s⁻²)
0.50	

5. Provide a definition of the term 'acceleration' and explain, using an example from table 3, how you used this definition to determine the average acceleration between two time frames.
6. Plot a graph of the athlete's acceleration versus time.
7. Give a detailed summary of the athlete's motion, making reference to their speed and acceleration during the race.
8. Use your velocity-versus-time graph to estimate the distance run after 7.0 s, showing all workings. How does your estimate compare to the data in table 1?
9. By appealing to Newton's Third Law of Motion, explain why the athlete is able to accelerate out of the starting blocks.
10. The mass of the athlete is 75 kg. Draw a free-body force diagram to illustrate the three significant forces acting on the athlete during the time they are accelerating. (*Hint:* There is one vertical force, one horizontal force and one force acting on the athlete that has both horizontal and vertical components.)
11. During the period of acceleration, what is the magnitude of the horizontal component of the net force acting on the athlete?
12. From your graph of acceleration versus time, approximately when is the net force acting on the athlete zero?
13. Use your results from one of your graphs to estimate both the momentum and the kinetic energy of the athlete at $t = 4$.

 Resources

Digital document School-assessed coursework (doc-38061)

11 Socio-scientific issues

KEY KNOWLEDGE

In this topic, you will investigate and apply physics knowledge to develop and communicate
an informed response to a contemporary societal issue or application related to a selected
option among the 18 options available.

COMMUNICATING PHYSICS

- Evaluate validity of sources of information
- Apply physics concepts specific to the investigation; definitions of key terms; and use
  of appropriate scientific terminology, conventions and representations
- Apply the use of data representations, models and theories in organising and
  explaining observed phenomena and physics concepts, and discuss the limitations
  of the explanations
- Discuss the influence of sociocultural, economic, legal and political factors relevant to
  the selected issue or application
- Apply physics understanding to justify a stance, opinion or solution to the selected
  issue or application

Source: VCE Physics Study Design (2023–2027) extracts © VCAA; reproduced by permission.

This topic is available online at **www. jacplus.com.au**.

How does physics inform contemporary issues and applications in society?

Develop a deeper understanding of an area of interest within diverse areas of physics.

11.1 How does physics explain climate change?

11.2 How do fusion and fission compare as viable nuclear energy power sources?

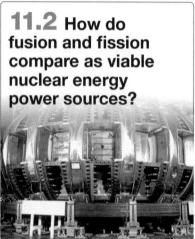

11.3 How do heavy things fly?

11.4 How do forces act on structures and materials?

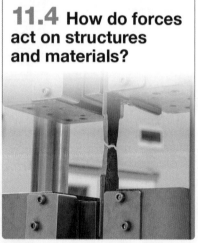

11.5 How do forces act on the human body?

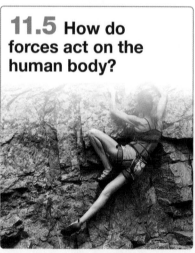

11.6 How is radiation used to maintain human health?

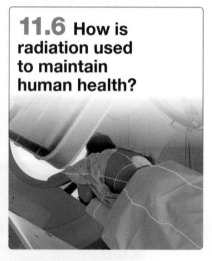

11.7 How does the human body use electricity?

11.8 How can human vision be enhanced?

11.9 How is physics used in photography?

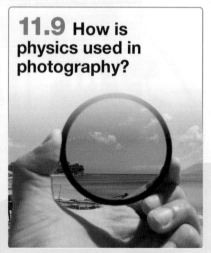

- [ ] **Select one option** from eighteen observations of the physical world.

- [ ] **Access the online content** and resources for background information, starting points and suggestions for your investigation.

- [ ] **Explore the related physics** and use this physics to form a stance, opinion or solution to the contemporary societal issue or application.

11.10 How do instruments make music?

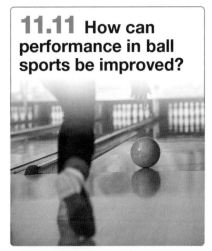

11.11 How can performance in ball sports be improved?

11.12 How can AC electricity charge a DC device?

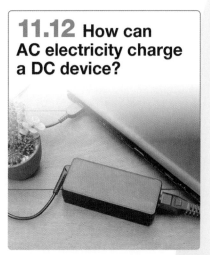

11.13 How do astrophysicists investigate stars and black holes?

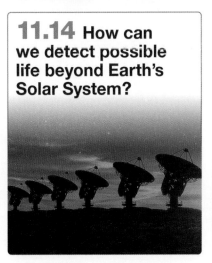

11.14 How can we detect possible life beyond Earth's Solar System?

11.15 How can physics explain traditional artefacts, knowledge and techniques?

11.16 How do particle accelerators work?

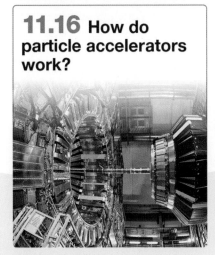

11.17 How does physics explain the origins of matter?

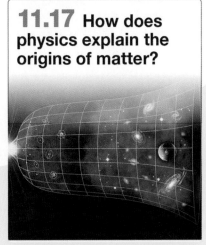

11.18 How is contemporary physics research being conducted in our region?

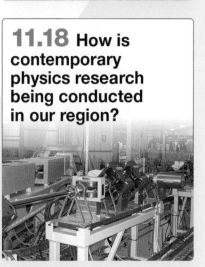

12 Scientific investigations

KEY KNOWLEDGE

In this topic, you will:

Investigation design

- apply the physics concepts specific to the selected investigation and explain their significance, including definitions of key terms, and physics representations
- evaluate the characteristics of the scientific methodology relevant to the investigation, selected from: experiment, fieldwork, classification and identification, modelling, simulation, and the development of a product, process or system
- apply techniques of primary qualitative and quantitative data generation relevant to the investigation
- identify and apply concepts of accuracy, precision, repeatability, reproducibility, resolution, and validity of data in relation to the investigation
- identify and apply health, safety and ethical guidelines relevant to the selected scientific investigation

Scientific evidence

- distinguish between an aim, a hypothesis, a model, a theory and a law
- identify and explain observations and experiments that are consistent with, or challenge, current models or theories
- describe the characteristics of primary data
- evaluate methods of organising, analysing and evaluating primary data to identify patterns and relationships including scientific error, causes of uncertainty, and limitations of data, methodologies and methods
- model the scientific practice of using a logbook to authenticate generated primary data

Science communication

- apply the conventions of scientific report writing including scientific terminology and representations, standard abbreviations, units of measurement, significant figures and acknowledgement of references
- apply the key findings of the selected investigation and their relationship to key physics concepts.

Source: VCE Physics Study Design (2023–2027) extracts © VCAA; reproduced by permission.

This topic is available online at **www.jacplus.com.au**.

APPENDIX 1
Periodic table

	Group 1	Group 2		Group 3	Group 4	Group 5	Group 6	Group 7	Group 8	Group 9
Period 1	1 Hydrogen **H** 1.0									
Period 2	3 Lithium **Li** 6.9	4 Beryllium **Be** 9.0								
Period 3	11 Sodium **Na** 23.0	12 Magnesium **Mg** 24.3								
Period 4	19 Potassium **K** 39.1	20 Calcium **Ca** 40.1		21 Scandium **Sc** 45.0	22 Titanium **Ti** 47.9	23 Vanadium **V** 50.9	24 Chromium **Cr** 52.0	25 Manganese **Mn** 54.9	26 Iron **Fe** 55.8	27 Cobalt **Co** 58.9
Period 5	37 Rubidium **Rb** 85.5	38 Strontium **Sr** 87.6		39 Yttrium **Y** 88.9	40 Zirconium **Zr** 91.2	41 Niobium **Nb** 92.9	42 Molybdenum **Mo** 96.0	43 Technetium **Tc** (98)	44 Ruthenium **Ru** 101.1	45 Rhodium **Rh** 102.9
Period 6	55 Caesium **Cs** 132.9	56 Barium **Ba** 137.3		57–71 Lanthanoids	72 Hafnium **Hf** 178.5	73 Tantalum **Ta** 180.9	74 Tungsten **W** 183.8	75 Rhenium **Re** 186.2	76 Osmium **Os** 190.2	77 Iridium **Ir** 192.2
Period 7	87 Francium **Fr** (223)	88 Radium **Ra** (226)		89–103 Actinoids	104 Rutherfordium **Rf** (261)	105 Dubnium **Db** (262)	106 Seaborgium **Sg** (266)	107 Bohrium **Bh** (264)	108 Hassium **Hs** (267)	109 Meitnerium **Mt** (268)

Key

Period 1	1 Hydrogen **H** 1.0	2 Helium **He** 4.0

← Atomic number
← Name
← Symbol
← Relative atomic mass

- Alkali metal
- Alkaline earth metal
- Transition metal
- Lanthanoids
- Actinoids
- Unknown chemical properties
- Post-transition metal
- Metalloid
- Reactive non-metal
- Halide
- Noble gas

Lanthanoids

57 Lanthanum **La** 138.9	58 Cerium **Ce** 140.1	59 Praseodymium **Pr** 140.9	60 Neodymium **Nd** 144.2	61 Promethium **Pm** (145)	62 Samarium **Sm** 150.4	63 Europium **Eu** 152.0

Actinoids

89 Actinium **Ac** (227)	90 Thorium **Th** 232.0	91 Protactinium **Pa** 231.0	92 Uranium **U** 238.0	93 Neptunium **Np** (237)	94 Plutonium **Pu** (244)	95 Americium **Am** (243)

Group 10	Group 11	Group 12	Group 13	Group 14	Group 15	Group 16	Group 17	Group 18
								2 Helium **He** 4.0
			5 Boron **B** 10.8	6 Carbon **C** 12.0	7 Nitrogen **N** 14.0	8 Oxygen **O** 16.0	9 Fluorine **F** 19.0	10 Neon **Ne** 20.2
			13 Aluminium **Al** 27.0	14 Silicon **Si** 28.1	15 Phosphorus **P** 31.0	16 Sulfur **S** 32.1	17 Chlorine **Cl** 35.5	18 Argon **Ar** 39.9
28 Nickel **Ni** 58.7	29 Copper **Cu** 63.5	30 Zinc **Zn** 66.4	31 Gallium **Ga** 69.7	32 Germanium **Ge** 72.6	33 Arsenic **As** 74.9	34 Selenium **Se** 79.0	35 Bromine **Br** 79.9	36 Krypton **Kr** 83.8
46 Palladium **Pd** 106.4	47 Silver **Ag** 107.9	48 Cadmium **Cd** 112.4	49 Indium **In** 114.8	50 Tin **Sn** 118.7	51 Antimony **Sb** 121.8	52 Tellurium **Te** 127.6	53 Iodine **I** 126.9	54 Xenon **Xe** 131.3
78 Platinum **Pt** 195.1	79 Gold **Au** 197.0	80 Mercury **Hg** 200.6	81 Thallium **Tl** 204.4	82 Lead **Pb** 207.2	83 Bismuth **Bi** 209.0	84 Polonium **Po** (210)	85 Astatine **At** (210)	86 Radon **Rn** (222)
110 Darmstadtium **Ds** (271)	111 Roentgenium **Rg** (272)	112 Copernicium **Cn** (285)	113 Nihonium **Nh** (280)	114 Flerovium **Fl** (289)	115 Moscovium **Mc** (289)	116 Livermorium **Lv** (292)	117 Tennessine **Ts** (294)	118 Oganesson **Og** (294)

64 Gadolinium **Gd** 157.3	65 Terbium **Tb** 158.9	66 Dysprosium **Dy** 162.5	67 Holmium **Ho** 164.9	68 Erbium **Er** 167.3	69 Thulium **Tm** 168.9	70 Ytterbium **Yb** 173.1	71 Lutetium **Lu** 175.0

96 Curium **Cm** (247)	97 Berkelium **Bk** (247)	98 Californium **Cf** (251)	99 Einsteinium **Es** (252)	100 Fermium **Fm** (257)	101 Mendelevium **Md** (258)	102 Nobelium **No** (259)	103 Lawrencium **Lr** (262)

APPENDIX 2
Astronomical data

	Mean radius of orbit		Orbital period		Equatorial radius (m)	Mass (kg)
	(AU)	(m)	(years)	(seconds)		
Sun					6.96×10^8	1.99×10^{30}
Mercury	0.387	5.79×10^{10}	0.241	7.60×10^6	2.44×10^6	3.30×10^{23}
Venus	0.723	1.08×10^{11}	0.615	1.94×10^7	6.05×10^6	4.87×10^{24}
Earth	1.00	1.50×10^{11}	1.00	3.16×10^7	6.37×10^6	5.98×10^{24}
Moon	2.57×10^{-3}	3.84×10^8	27.3 days*	$2.36 \times 10^{6*}$	1.74×10^6	7.35×10^{22}
Mars	1.52	2.28×10^{11}	1.88	5.94×10^7	3.39×10^6	6.42×10^{23}
Jupiter	5.20	7.78×10^{11}	11.9	3.74×10^8	6.99×10^7	1.90×10^{27}
Saturn	9.58	1.43×10^{12}	29.5	9.30×10^8	5.82×10^7	5.68×10^{26}
Titan	8.20×10^{-3}	1.22×10^9	15.9 days*	$1.37 \times 10^{6*}$	2.57×10^6	1.35×10^{23}
Uranus	19.2	2.87×10^{12}	84.0	2.65×10^9	2.54×10^7	8.68×10^{25}
Neptune	30.1	4.50×10^{12}	165	5.21×10^9	2.46×10^7	1.02×10^{26}
Pluto	39.48	5.91×10^{12}	248	7.82×10^9	1.15×10^6	1.31×10^{22}

*The orbital period for the Moon and Titan is the time it takes to complete one orbit around Earth and Saturn respectively. All other listed measurements for the orbital period shows the time to orbit the Sun.

The Milky Way	1.50×10^5 light-years across
Alpha Centauri	4.37 light-years away
Andromeda	2.25×10^6 light-years away
Edge of observable universe	4.65×10^{10} light-years away

Source: Data derived from https://solarsystem.nasa.gov.

Answers

1 Electromagnetic radiation

1.2 Explaining waves as the transmission of energy

1.2 Exercise

1. A periodic wave is a disturbance that repeats itself at regular intervals; a single pulse is, as the name suggests, a disturbance without repeats.
2. For transverse waves, the disturbance is at right angles to the direction the waves travel (such as a pulse on a rope or ripple on the surface of water), whereas for longitudinal waves, the disturbance is parallel to the direction the waves are travelling.
3. $332 \, \text{m s}^{-1}$
4. As a transverse wave passes through a medium, particles in the medium will vibrate/oscillate up and down at right angles to the passage of the wave.
5. As a longitudinal wave passes through a medium, particles in the medium will vibrate/oscillate back and forth in the same direction as the passage of the wave.

1.2 Exam questions

1. Waves transmit **energy** without the net transfer of **matter**.
2. A medium
3. D
4. B
5. a. A longitudinal wave is characterised by the disturbance being parallel to the direction of wave propagation.
   b. Answers may vary. One example is sound waves.
   c. Answers may vary. One example is light waves.

1.3 Properties of waves

Sample problem 1
$340 \, \text{m s}^{-1}$

Practice problem 1
$340 \, \text{m s}^{-1}$

Sample problem 2
$0.609 \, \text{m}$

Practice problem 2
$355 \, \text{m s}^{-1}$

1.3 Exercise

1. One wavelength
2. Periodic sound waves travel at the same speed from a stationary source independent of how loud they are. The loudness is related to the amplitude of the longitudinal waves, which is a measure of the difference in air pressure between the compressions and rarefactions of the wave.
3. The marching band appears out of step with the music because the speed of light is significantly different to the speed of sound. Optical information arrives at an observer before audio information does.
4. Yes, you will, because the speed of sound is the same for low- and high-frequency sound waves produced by a stationary source, such as a group of musicians at a concert.
5. $0.780 \, \text{s}$
6. $1.02 \, \text{m}$
7. $330 \, \text{m s}^{-1}$
8. a. $1.33 \, \text{m}$
   b. $5.86 \, \text{m}$
9. In this instance, the source produces sound with a source frequency of 100 Hz. An observer moving towards the source will intercept more waves per second and so will measure a higher frequency; the observer moving towards the source measures 110 Hz. Likewise, an observer moving away from the source will intercept fewer waves per second and will measure a lower frequency; the observer moving away from the source measures 90 Hz.

1.3 Exam questions

1. C
2. $2.92 \, \text{m s}^{-1}$
3. $1.7 \times 10^{-15} \, \text{s}$
4. $0.22 \, \text{Hz}$
5. $3.82 \times 10^{-3} \, \text{s}$

1.4 Energy from the Sun

Sample problem 3
a. 44 times   b. 248 °C

Practice problem 3
$1.413 \times 10^{27} \, \text{W}$

Sample problem 4
a. $2.64 \times 10^{-7} \, \text{m}$ b. Ultraviolet

Practice problem 4
6444 K

1.4 Exercise

1. Absorption, emission and reflection
2. The shiny surfaces on the thermos would minimise the emission of infrared radiation (heat) from the thermos, thus minimising the heat loss from any liquid it contains.
3. a. Blue, $5000 \times 10^{-10} \, \text{m}$ (500 nm)
   b. Blue. As the temperature of a theoretical blackbody increases, the emission spectrum becomes bluer.
4. As the temperature of an object increases:
   - the *wavelength* of emitted radiation will decrease
   - the *frequency* of emitted radiation will increase.

5. As the temperature of the metal filament increases, the peak wavelength of emitted radiation will decrease. Thus, when the filament is relatively cool, it will glow a dull red colour. As its temperature increases, the intensity of emission of shorter wavelengths of visible radiation (yellow, green, blue) will increase. This will progressively change the colour of the filament to yellow (when the peak emission is yellow) to white (when significant amounts of all visible wavelengths are being emitted) as the filament becomes hotter.

6. 1.00×10^{-5} m

7. 5686 K

1.4 Exam questions

1. D

2. The shiny white surfaces (that is, with an albedo close to 1.0) reflect high proportions of incoming sunlight back into space. Examples include clouds and snow/ice (glaciers and icecaps).

3. The colour of each star is related to its surface temperature. The blue-white colour of Rigel suggests that it is much hotter (~10 000 K) compared to Betelgeuse (~4000 K).

4. A

5. a. 12.8 times
   b. 1066 °C

1.5 The electromagnetic spectrum

Sample problem 5

a. 1.8×10^{-15} s
b. 5.4×10^2 nm

Practice problem 5

$f = 6.7 \times 10^{14}$ Hz

$T = 1.5 \times 10^{-15}$ s

1.5 Exercise

1. 2.1×10^{-15} seconds

2. 3.0 cm to 0.30 mm (fractions of a millimetre to a few centimetres)

3. $f = 1.1 \times 10^{19}$ Hz

   $T = 9.0 \times 10^{-20}$ s

4. a. 50 Hz
   b. 6.0×10^6 m

5. a. 4.6×10^{-7} m
   b. 3.1×10^{-7} m

1.5 Exam questions

1. See the figure at the foot of the page*

2. C

3. B

4. 3×10^8 m s^{-1}

5. Approximately 750 nm

1.6 Review

1.6 Review questions

1. Without a medium, there is nothing for the wave to distort, so it can't travel anywhere

2. a. 0.50 s
   b. 1.3 m
   c. The wavelength will decrease. Speed is unchanged.

3. a. 1.08×10^{-3} s
   b. 0.370 m

4. 20 Hz: 17 m; 20 kHz: 0.017 m

5. Blue light: 4.6×10^{-7} m
   Yellow light: 5.8×10^{-7} m

6. The pulse of light is 46 cm long

1.6 Exam questions

Section A — Multiple choice questions

1. C or D
2. B
3. B
4. B
5. B
6. B
7. D
8. C
9. A
10. C

Section B — Short answer questions

11. a. A longitudinal wave consists of the vibration of a medium parallel to the direction of propagation, whereas for a transverse wave the vibration of the medium is perpendicular to the direction of propagation of the wave. An example of a longitudinal wave is a sound wave, and an example of a transverse wave is a vibrating string such as the string on a guitar.

    b. Both types of waves transfer energy from one place to another via a medium. Both types of waves travel at a constant speed dependent on the properties of the medium. And both types of waves have the same parameters, namely frequency, amplitude and wavelength.

12. a. 4.50 cm
    b. 8.4 Hz
    c. 3.8 cm to the right

*1.

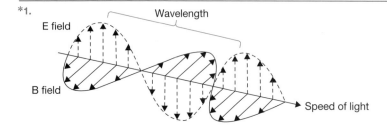

13. a. i. 7250 K

   ii. 11 600 K

   b. i. 193 nm

   ii. 522 nm

14. 10.45 m

15. a. Electromagnetic waves refer to the waves that form from the interactions between an electric and a magnetic field. An electromagnetic wave is a wave that propagates at a unique speed in uniform media.

   b.

	Wavelength (m)	Electromagnetic wave type
i.	1×10^3	Radio waves
ii.	1×10^{-2}	Microwaves
iii.	1×10^{-12}	γ-rays
iv.	1×10^{-8}	UV light

2 Investigating light

2.2 Refraction using Snell's Law

Sample problem 1

8 minutes 20 seconds

Practice problem 1

2.5 s

Sample problem 2

9.5×10^{12} km

Practice problem 2

2.0×10^8 m s^{-1}

This speed is less than the speed of light in a vacuum.

Sample problem 3

20°

Practice problem 3

1.29

Sample problem 4

a. 2.0×10^8 m s^{-1} b. 2.3×10^8 m s^{-1}

Practice problem 4

a. 1.24×10^8 m s^{-1} b. 1.84×10^8 m s^{-1}

2.2 Exercise

1. a. Longest time: 21 minutes
      Shortest time: 4 minutes and 20 seconds

   b. Longest time: 4 hours and 4 minutes
      Shortest time: 4 hours, 1 minute and 40 seconds

2. a. 65°   b. 25°

3. 45°

4. The angle of refraction increases by 6°.

5. 1.5

6. a. 36°   b. 65°

7. 23.8°

8. a. 2.3×10^8 m s^{-1}

   b. 2.0×10^8 m s^{-1}

   c. 1.2×10^8 m s^{-1}

9. a. For acetone, the angle of refraction is 18.2°; for glycerol, it is 16.7°; for carbon tetrachloride, it is 16.8°; and for glass, it is 16.0°. The light ray will emerge from the glass into the air at the angle it left the air: 25°.

   b. All the angles will be the same.

10. 0.85 cm

2.2 Exam questions

1. D

2. C

3. 11°

4. 1.2

5. Possible correct answers:
   - The refractive index of the two mediums must be the same.
   - The speed of light in the two materials needs to be the same.
   - The angle of incidence must be zero.

2.3 Total internal reflection and critical angle

Sample problem 5

50°

Practice problem 5

a. 24.4°

b. The critical angle would increase.

c. 33.3°

Sample problem 6

a. If the refractive index of the cladding was larger than the refractive index of the core, the light would be refracted *towards the normal* and internal reflection would not be possible. When the refractive index of the cladding is less than that of the core, the light is refracted *away from the normal* so that total internal reflection is possible if the angle of incidence is larger than the critical angle.

b. 64.0°

c. 1.90×10^8 m s^{-1}

Practice problem 6

a. As the refractive index for the core is smaller, the critical angle will be larger.

b. 65.5°

c. 1.92×10^8 m s^{-1}
   The speed of light in the new optical fibre is larger.

2.3 Exercise

1. a. 24°   b. 32.1°

2. a. The reflective index is 1.4, and $n = 1.4$ is the minimum value for total internal reflection to occur for an angle of incidence of 45°.

b. The rays would be inverted and swap positions (this is often used in binoculars to correct the normally inverted image). The initial ray of light would move down and refract off the surface shown, bouncing to the left at a right angle. The other ray of light would enter the block and refract at a right angle as well.

c. Both rays of light would be expected to emerge at the same time.

3. 1.49

4. a. 7.00×10^{-8} m

b. The speed is 3.0×10^{-8} m s^{-1} and the time difference is 2.3×10^{-16} s. This could be a problem as it adds up to 3.28×10^{-8} s per km of optical fibre, which would limit the upper frequency of light pulses sent down the optical fibre; if the frequency was too high, the pulses would overlap. The problem can be overcome by using a narrower optical fibre or using an optical fibre whose refractive index gets smaller when the distance gets greater from the centre

5. 1.25

6. 24.6°

2.3 Exam questions

1. B

2. a. 5.3×10^{14} Hz   b. 60.3°   c. 1.80×10^8 m s^{-1}

3. a. 44°

b.

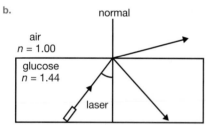

c. All light rays from the laser are totally internally reflected at the surface. Hence, no light rays reach the observer above the surface, so the laser cannot be seen at X.

4. A

5. A

2.4 Dispersion

2.4 Exercise

1. a.

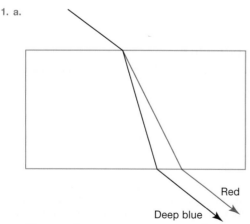

The emerging coloured rays are parallel to each other.

b. In a triangle, the rays emerge at different angles and spread further apart the further they travel, so on a distant wall the colours are seen separately. With the rectangle, the colours stay the same distance apart regardless of how far they travel.

2. Red light travels faster through crown glass. The speed difference is 1.96×10^6 m s^{-1}.

3. Violet light

4. a.

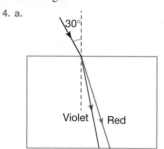

b. Red: 19.28° and violet: 19.07°

c. 0.21°

5. This result indicates that red light travels faster in the prism material, as it is refracted less.

2.4 Exam questions

1. a. The observed effect is known as dispersion. The refractive index of the glass is different for different wavelengths. Different wavelengths refract at different angles. Therefore, the different colours exit the prism at different angles.

b. Colour visible at point X: Red light
   Colour visible at point Y: Deep blue/violet light

2. a. 25.4°   b. 25.0°
   c. 0.4°   d. 1.97×10^8 m s^{-1}

3. The amount of refraction is determined with the frequency of the light, thus supporting the wave model of light.

4. The diamond has a higher refractive index, n, therefore the light is dispersed more.

5. The majority of the light will be parallel to the incident light. There will be a slight red and violet streak at the edges of the refracted light due to their differences in wavelength.

2.5 Optical phenomena

2.5 Exercise

1. In the morning the Sun rises in the east, and you should have your back to it; thus, you should face towards the west. The water droplets should be in front of you, thus the rain should come from the west.

2. A

3. Rainbows involve the internal reflection of light within the water droplet. The spectrum from a prism is observed from the one source, whereas a rainbow is made up from light from numerous water droplets.

4. C

5. This phenomenon is a mirage, occurring above a road (hotter than the air above), and is due to total internal reflection caused by atmospheric refraction. This is caused by the fact that the refractive index of air increases when the temperature decreases. Thus, as a ray of light moves from the cooler top layer of air into hotter air near the ground, it

bends away from the normal. After successive refractions, the angle of incidence exceeds the critical angle for air at that temperature and the ray is totally internally reflected. What is observed is a refraction light from the sky, giving the impression that there is water on the road.

2.5 Exam questions

1. Emma is correct. The white light entering the water droplets will bend the different wavelengths different amounts. The red light that Emma sees in her rainbow will be from different water droplets.

2. White light consists of all colours (frequencies) of light across the visible spectrum, from red to violet. The different colours travel at slightly different speeds outside a vacuum. When a beam of white light strikes a material boundary, the colours are refracted by different amounts due to their differing speeds in the material. This leads to the production of a rainbow of colours, each emerging at slightly different angles.

3. • The Sun needs to be behind the person observing the rainbow.
   • Some water droplets (rain, fog) need to be airborne in front of the observer.
   • The angle between the observer, the Sun and the raindrops needs to be approximately 42°.

4. It is morning, as in order for her to see a rainbow the Sun has to be behind her. She is looking west, so the Sun is in the east.

5. C

2.6 Review

2.6 Review questions

1.
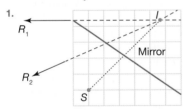

2. a. Blue light: 4.6×10^{-7} m
   Yellow light: 5.8×10^{-7} m

   b. Blue light in the glass: 3.1×10^{-7} m
   Yellow light in the glass: 3.8×10^{-7} m

3. a. 1.36

   b. 23°

   c. 31°
   When light passes from air into perspex (refractive index is greater than 1), it bends towards the normal; the angle of refraction is thus less than the angle of incidence. However, when light passes from perspex to air, it bends away from the normal; in this case, the angle of refraction is greater than the angle of incidence.

4. a. 1.84×10^8 m s^{-1}

   b. The speed of light in saltwater is greater than the speed of light in carbon disulfide.

   c. The light will bend away from the normal.

   d. 57°

5. a. The term *disperses* describes the splitting of white light into its constituent colours. This occurs when refraction into a medium has a refractive index that is dependent on colour, or frequency. Such media are known as dispersive media. In dispersive media, light of different colours will travel at different speeds and have different angles of refraction for the same angle of incidence.

   b. 0.24°

   c. 0.46°

6.
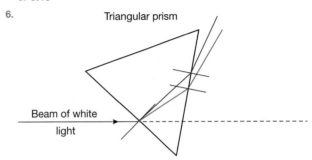

Triangular prism

Beam of white light

7. a. Glass type Z has the smallest critical angle of the three materials.

   b. 39.6°

   c. 77.0°

2.6 Exam questions

Section A — Multiple choice questions

1. A
2. D
3. A
4. B
5. C
6. D
7. B
8. C
9. A
10. B

Section B — Exam questions

11. $n = 1.56$

12. a. 51°

    b. Yes, some light will be transmitted to the cladding.

13. a.

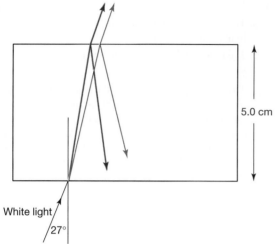

5.0 cm

White light
27°

b. 17.3°

c. The blue light takes slightly longer.

14. a. Since $n_{cladding} < n_{core}$, there exists a critical angle. Ideally, the refractive index of the cladding is only a little smaller than the core so that the critical angle is as large as possible.

b. $n_{cladding} = 1.57$

c. $c_{core} = 1.9 \times 10^8 \, \text{m s}^{-1}$

d. 253 nm

15. 1.39

3 Thermal energy and its interaction with electromagnetic radiation

3.2 Explaining heat using the kinetic theory

Sample problem 1

273 K

Practice problem 1

−270 °C

3.2 Exercise

1. The Celsius scale is commonly used because it is more convenient as it is based on the properties of water (the freezing of ice and the boiling of water with 100 divisions between) and is easy to reproduce anywhere in the world.

2. The Kelvin scale is an absolute by which everything in the universe can be referenced.

3. **a.** Approximately 310 K
   b. Approximately 275 K
   c. Approximately 298 K
   d. Approximately 288 K
   e. 373 K

4. The large thermometer is also surrounded by the air in the school laboratory. The liquid in the test tube might be cold, but the thermometer is also measuring the surrounding air, giving a falsely high reading.

5. 194.5 K

6. −17 °C and −107 °C

7. The particles of the air are in constant motion; this enables the molecules that make up the aroma of the food to travel from the kitchen to the front door.

8. Translational kinetic energy describes the displacement of particles from one position to another. Other types of internal energy involve motion within the particles (movement within bonds of a molecule) or potential energy from atomic and subatomic interactions.

9. The particles in the hot tea are moving rapidly and experiencing many collisions with each other and the sides of the cup. The collisions with the cup transfer the kinetic energy to the cup particles and they start to move faster. The temperature of the tea drops and the temperature of the cup rises. The air particles surrounding the cup also experience similar collisions, thus kinetic energy is transferred to the air. Hence, the overall translational kinetic energy of the cup of tea is transferred to the air and the tea cools down.

10. As the translational kinetic energy of the soup particles is transferred to its surroundings, the particles slow down, thus exerting much less pressure on the cling film. The cling film is pushed into the bowl from the external air pressure.

11. Examples may include a meteorite crashing into Earth, a car hitting a wall, rubbing your hands together, a spoon stirring a cup of tea, an eraser on a piece of paper and water tumbling over a waterfall.

12. A red-hot pin has less internal energy, less kinetic energy and fewer molecules to vibrate than a nail of the same material at the same temperature. There is just less matter to vibrate, so there is less energy to transfer to the water.

13. While you are swimming in the sea, your body maintains an even core temperature of 37.6 °C. The sea around Melbourne is a lower temperature than this. The water particles do not vibrate as quickly as your molecules. Your molecules transfer their greater kinetic energy to the water. Temperature is a measure of the average random kinetic energy of the particles. Internal energy is the total energy of the particles, including the potential energy due to intermolecular forces. You have less internal energy than the sea simply because you have fewer particles than the sea.

3.2 Exam questions

1. D

2. **a.** The Celsius scale is based on the melting point (0 °C) and boiling point (100 °C) of water. The Kelvin scale is based on the kinetic theory of matter, where 0 K corresponds to the lowest possible temperature such that particles completely stop moving.
   b. Both scales use the same scale increments. 1 °C ≡ 1 K.
   c. An absolute scale starts at 0. There are no negative values.
   d. Kelvin does not use a degree symbol.

3. A

4. The atoms in a solid are tightly packed, with strong attractive forces holding them in a rigid structure. In liquids, molecules are free to move around within the volume. Some of the attractive forces holding the atoms together in the solid must be broken in order for the molecules to be free to move

around as a liquid. This requires an increase in energy to break the bonds, and this energy comes from an increase in temperature.

5. When a tennis ball is hit by a racquet the same number of air molecules occupy a smaller volume. This increases the air pressure, causing the air molecules to collide more frequently and increase in speed. The increased speed of the air molecules corresponds to an increase in temperature.

3.3 Transferring heat

3.3 Exercise

1. Some of the kinetic energy of the particles in the hot region is transferred to the particles in the cooler region. Thus, the temperature in the cooler region increases.

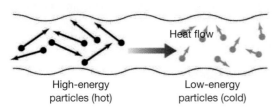

●→ Particle moving slowly

●——→ Particle moving faster

Heat flow

High-energy particles (hot)   Low-energy particles (cold)

2. Conduction is the transfer of heat through a substance as a result of collisions between vibrating neighbouring particles. Liquids and gases have particles that are much further apart than those in solids. The particles have to travel further to transfer kinetic energy, so the process of conduction is generally slower in gases and liquids than in solids.

3. Heat transfer through water occurs mainly by convection. That is, warmer water moves upwards as it is displaced by the denser, cooler water. Warm water does not generally move down a test tube because it is less dense than cooler water.

4. Convection is the transfer of heat through a substance as a result of the movement of particles between regions with different temperatures. In solids, the particles are tightly bound together and unable to move to regions of different temperatures.

5. Radiant energy moves through space at 3.0×10^8 m s^{-1}. This is significant as it is the same value as the speed of light.

6. a. The water on the surface of a still body of water becomes warmer than the water deeper down due to it absorbing radiated energy from the Sun and heat transfer from the warm air near the surface. Although there is a temperature difference between the warm air and water surface, the water is a poor conductor, so the energy is not transferred very far into the water over a short time period. There will be no convection currents as the cooler water is under the warm water.

   b. If a wind is blowing, chopping up the surface, the cold water underneath mixes with the warm water on the top. The heat transfer from the surface water reduces its temperature. The wind also evaporates some of the surface water. In doing so, it uses energy from the water to change the state from liquid to vapour. This reduces the surface temperature further.

7. a. The feeling of warmth is caused by the concrete wall radiating infrared energy due to its temperature being higher than its surroundings.

   b. The walls of the building would have become warm due to heat transfer by radiation from the Sun warming them and by conduction from the warm air surrounding the walls.

8. Aluminium is a good conductor. It will conduct energy from the coffee to the surrounding air. The coffee will cool quickly. The aluminium will also transfer energy to the hand holding the cup, making this an unpleasant, painful experience before it cools below 50 °C.

9. Air within a room circulates by convection, with the warm air rising up to the ceiling. Floor ducts used for heating rely on the warm air leaving the duct and rising up through the room. Ceiling ducts need more powerful blower fans to counteract the tendency of the warm air to stay at ceiling level. It needs to be forced down to mix with the cold air to warm it by conduction and convection.

3.3 Exam questions

1. A

2. D

3. The greater the temperature difference, the greater the difference in kinetic energy between colliding particles. More energy will be transferred in each collision, and so the rate of heat transfer is faster.

4. The higher-energy water at the bottom expands and becomes less dense and rises. The cooler, denser water at the top of the pot falls to the bottom. This motion causes a movement of material and transfers energy; that is, a convection current.

5. The hotter, faster vibrating particles collide with the slower vibrating particles. Heat is transferred from hotter to colder particles. This continues through the solids until both reach the same temperature; that is, thermal equilibrium is achieved.

3.4 Specific heat capacity

Sample problem 2

   a. 2.4×10^3 kJ   b. 37 °C

Practice problem 2

227.5 kJ

Sample problem 3

   a. 5.5×10^3 J   b. 660.3 °C

Practice problem 3

3.45×10^4 J

3.4 Exercise

1. a. 400 kJ   b. 1200 kJ

2. a. This is because the human body is mostly water, and the specific heat capacity of water is quite high compared to that of other substances.

   b. This is because deserts are dry, while fertile soils contain a lot of moisture, and water has a high specific heat capacity.

c. The steel saucepan

d. Metals have low heat capacities; water and things containing water have high specific heat capacities.

3. 1.2×10^6 J

4. a. 230 °C

b. The energy was used to increase the potential energy of the particles as they changed state from solid to liquid.

c. Liquid

d. $160 \, \text{kJ} \, \text{kg}^{-1}$

e. The specific heat capacity of solid candle wax is higher. More energy is required to raise the temperature of the solid wax by 20 °C than the liquid wax.

f. Over the interval DE, the temperature remains the same. The energy put into the wax does not result in an increase in the kinetic energy of the particles; it just breaks the attractive bonds between the particles as the substance changes state from liquid to gas.

5. 8.2 kg

6. A simmering saucepan loses heat by convection and evaporation. By keeping the lid on, the water is prevented from evaporating because the air inside the pan is saturated. The evaporation of the water would tend to cool the thing you are heating. In addition, escaping steam would be replaced by cooler air, meaning that more energy would be needed to sustain the temperature of the water.

7. The evaporation of water requires energy for the water particles to change from their liquid state to the vapour state. The energy comes from the body of the liquid. Therefore, the net result is a reduction in the kinetic energy of the water particles and the temperature of the body of the liquid.

3.4 Exam questions

1. C

2. B

3. 127 kJ

4. 82.0 °C

5. 1.00 MJ

3.5 Understanding climate change and global warming

3.5 Exercise

1. Some of the radiation that is emitted by the Sun is absorbed by various gases in the atmosphere before it reaches Earth.

2. Some gases naturally present in small proportions in Earth's atmosphere — such as water vapour, carbon dioxide, nitrous oxide, ozone and methane — absorb infrared light and re-emit it in all directions. The radiation re-emitted towards Earth's surface increases its temperature.

3. The enhanced greenhouse effect is the disruption to Earth's climate equilibrium. It is caused by the release of greenhouse gases, mainly carbon dioxide and methane, in the atmosphere due to human activities such as burning fossil fuels. This leads to an increase in global average surface temperatures.

4. Properties of water include high specific heat capacity, convective properties and the fact that it can exist in solid, liquid and gaseous states. These properties have different impacts on the climate, with ice and snow having a high albedo, water vapour being a greenhouse gas, and clouds reflecting light, for example.

5. Many species can survive only within specific conditions (such as temperatures and gas concentrations). The greenhouse effect may change these conditions so that they are not compatible with life on Earth.

3.5 Exam questions

1. B

2. A

3. D

4. Water vapour (H_2O), carbon dioxide (CO_2), methane (CH_4), nitrous oxide (N_2O) and ozone (O_3)

5. Sunlight heats Earth's surface, which then re-radiates infrared radiation. Some of this infrared radiation is absorbed by small gas molecules in the atmosphere. By trapping this energy, these molecules heat up and so the atmospheric temperature increases.

3.6 Review

3.6 Review questions

1. Testing your own temperature is a subjective exercise.

2. Absolute temperature is measured using the Kelvin scale. On the Kelvin scale 28 °C is not twice as hot as 14 °C (301 K and 287 K, respectively).

3. The warm, less dense liquid rises, while the cooler, denser liquid sinks.

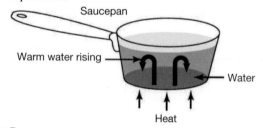

4. B

5. C

6. Conventional ovens have heating elements at the bottom to allow air to become warmed, and then rise. This sets up convection currents, which move around the food. Food is then warmed by conduction and convection as well as by radiation from the element. The advantage of having an oven with a fan is that the inside of the oven is kept at an even temperature.

7. a. The water evaporates by gaining energy from the skin to change state. Taking energy from the surface of the skin results in a reduction in the temperature of the skin.

b. You feel cooler when the wind is blowing because the rate of evaporation of moisture from your skin is increased, lowering your skin temperature.

c. Sweating in humid weather does not cool the body because the sweat does not evaporate.

8. Conduction is the transfer of thermal energy through a substance as a result of collisions between neighbouring vibrating particles. Particles in solids are more tightly bound and closer together in solids than in liquids, and in liquids

than in gases. Thus, solids are better conductors of heat than liquids, and liquids are better conductors of heat than gases.

9. The vaporisation of water is an example of positive feedback. The accelerated formation of clouds following a temperature increase is an example of negative feedback.

10. Ice has a high albedo; it is highly reflective. Thus, the more ice there is, the more solar energy is reflected back to space, cooling Earth, and a cooler Earth means a greater ice cover.

3.6 Exam questions

Section A — Multiple choice questions

1. D
2. A
3. D
4. C
5. B
6. C
7. B
8. D
9. A
10. D

Section B — Short answer questions

11. 68 °C
12. 6.1 MJ
13. In an oven at 300 °C, your hand is bombarded by only a limited number of fast-moving air particles, and the energy transferred is not too detrimental. Touching the metal tray at the same temperature brings your hand into contact with more fast-moving particles (in a solid, the molecules are closer together), which transfers more energy to your hand, thus causing injury.
14. a. Sand heats in one-quarter of the time it takes to heat the same mass of water. Sand cools four times faster than the same mass of water.

    b. The temperature of dry sand increases rapidly during the day due to its low specific heat capacity. The temperature of the water increases very slowly due to its high specific heat capacity. In addition, there is a much greater mass of water to heat than sand.

    c. The cooling effect of evaporation from bodies of water, the modifying influence of the sea breezes, and the coolness of the air above the oceans due to the high specific heat capacity of water all contribute to a lower air temperature above beach sand than desert sand.

15. The temperature of Earth and the atmosphere as a whole balances itself out; the differential warming of the atmosphere and ocean causes heat transfer from the tropics to the poles by convection in the form of winds and ocean currents.

4 Radiation from the nucleus and nuclear energy

4.2 Nuclear stability and nuclear radiation

Sample problem 1

Thorium-234 (^{234}Th)

Practice problem 1

Iron-56 (^{56}Fe)

Sample problem 2

There is an attractive force called the strong nuclear force that acts between protons and neutrons. The strong nuclear force can only hold a nucleus together when it is stronger than the electromagnetic force pushing the protons apart. The force holding protons and neutrons together to form nuclei must be an attractive force that works between neutrons and protons and is stronger than the electromagnetic force at close range.

Practice problem 2

The strong nuclear force acts between the two protons and each proton and the neutron. It is a small nucleus, so the strong nuclear force extends to all nucleons from each of the other nucleons. The electromagnetic force also acts to repel the two protons but not as strongly as the strong nuclear force attracts them.

Sample problem 3

a. One-quarter would be left after 12 hours.

b. $\dfrac{1}{256}$ would be left after 48 hours.

Practice problem 3

The cobalt-60 source will need to be replaced in just over 21 years.

4.2 Exercise

1. Sodium-23
2. a. One proton and one neutron
   b. 95 protons and 146 neutrons
   c. 63 protons and 101 neutrons
3. The force between a proton and neutron is the strong nuclear force, but it only acts if the proton and neutron are very close together (within about 2×10^{-15} m).
4. $\dfrac{1}{32}$ of the initial sample remains.
5. A stable isotope is one that never decays.

6. A nucleus can be unstable because the electromagnetic force becomes greater than the strong nuclear force at only a very small separation distance. Therefore, the sum of all of the repulsive electromagnetic forces can total more than the sum of the strong nuclear forces in large nuclei.

7. 45 minutes

8.

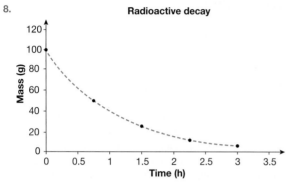

Radioactive decay

9. Nuclei decay at random. It is impossible to predict when any individual nucleus will decay but if a large sample of nuclei are decaying, it is possible to determine when a proportion (such as a half) of them will decay.

10. After each half-life, half of the sample remaining decays. Only half of the nuclei are left to decay so the number of decays that will happen per second will halve.

11. 5120 grams or 5.12 kilograms

4.2 Exam questions

1. C

2. A

3. B

4. 160 000

5. 8 hours

4.3 Types of nuclear radiation

Sample problem 4

a. $^{234}_{92}\text{U} \rightarrow ^{230}_{90}\text{Th} + ^{4}_{2}\text{He} + \text{energy}$

b. $^{210}_{82}\text{Pb} \rightarrow ^{210}_{83}\text{B} + ^{0}_{-1}\text{e} + \text{energy}$

c. $^{11}_{6}\text{C} \rightarrow ^{11}_{5}\text{B} + ^{0}_{+1}\text{e} + \text{energy}$

Practice problem 4

a. $^{241}_{95}\text{Am} \rightarrow ^{237}_{93}\text{Np} + ^{4}_{2}\text{He} + \text{energy}$

b. $^{197}_{78}\text{Pt} \rightarrow ^{197}_{79}\text{Au} + ^{0}_{-1}\text{e} + \bar{v} + \text{energy}$

c. $^{23}_{12}\text{Mg} \rightarrow ^{23}_{11}\text{Na} + ^{0}_{+1}\text{e} + \bar{v} + \text{energy}$

Sample problem 5

In a decay chain, a nucleus may decay to a less stable nucleus. This, however, is a step towards increased stability as the final nucleus is always stable.

Practice problem 5

a. Uranium-238 is the most stable nucleus as it has the longest half-life.

b. If the decay chain started with an unstable nucleus with a shorter half-life, it would have all decayed by now. As uranium-238 has such a long half-life — one similar to the age of Earth — much of it still remains.

4.3 Exercise

1. $^{222}_{86}\text{Rn} \rightarrow ^{218}_{84}\text{Po} + ^{4}_{2}\text{He} + \text{energy}$

2. $^{214}_{82}\text{Pb} \rightarrow ^{214}_{83}\text{Bi} + ^{0}_{-1}\text{e} + \bar{v} + \text{energy}$

3. $^{60}_{20}\text{Co}^* \rightarrow ^{60}_{20}\text{Co} + \gamma$

4. a. 2

   b. 4

   c. An alpha particle

5. Both the neutrino and the gamma ray have a mass number and atomic number of zero.

6. Kinetic energy of the emitted particles

7. Beta particles have atomic numbers of −1 or +1; however, they are only electrons or positrons. There are no protons. In these cases the atomic number refers to the charge of the particle.

4.3 Exam questions

1. B

2. A

3. Yes, both nuclei could be members of the same decay series

4. The atomic number of the final isotope is 88.

5. $^{222}_{86}\text{Rn} \rightarrow ^{4}_{2}\text{He} + ^{222}_{86}\text{Rn}$

4.4 Radiation and the human body

Sample problem 6

a. The absorbed dose is 9×10^{-4} Gy.

b. The equivalent dose if the energy was delivered by γ rays is 0.9 mSv.

c. The equivalent dose if the energy was delivered by α particles is 18 mSv.

d. The equivalent dose delivered by the α particles would cause about 20 times more damage than that delivered by the γ rays.

Practice problem 6

Each astronaut (estimated mass of 80 kg) absorbed approximately 96 joules of energy.

Sample problem 7

a. 0.3 mSv   b. 73 days

Practice problem 7

0.7 mSv

Sample problem 8

a. 39 hours   b. 5 mg

Practice problem 8

a. 16 mg   b. 0.0625 mg

Sample problem 9

a. 10 minutes   b. 30 minutes

Practice problem 9

a. 32 MBq   b. 2 MBq   c. 40 days

4.4 Exercise

1. 9×10^{-2} J
2. 1.5 mGy
3. 3 mSv
4. 54 mSv
5. α particles cause a lot of localised damage. They give much of their energy quickly to nearby atoms.
6. Equivalent dose takes into account the different types of damage caused by different forms of ionising radiation.
7. In its early stages a foetus consists of only a few cells, which divide to form the rest of the baby. If these cells are damaged, the foetus may not be able to form properly.
8. 24 000 each second
9. Cancer cells grow more rapidly than normal tissue cells. Cells are more vulnerable to radiation damage when they are dividing. This means that more cancer cells are killed than normal cells.

4.4 Exam questions

1. C
2. 13
3. B
4. 2.1 mJ
5. a. 1.2×10^{-5} Gy   b. 2.4 μSv

4.5 Energy from mass

Sample problem 10

3.9 MeV (to one decimal place)

Practice problem 10

$7.487\ 177 \times 10^{-30}$ kg

4.5 Exercise

1. The equation $E = mc^2$ expresses the fact that energy and mass are equivalent. Mass and energy cannot be considered independently.
2. 9×10^{15} J
3. 1.638×10^{-13} J
4. 2.8×10^{25} eV
5. 5.0×10^{-11} kg

4.5 Exam questions

1. C
2. C
3. B
4. 3.1×10^{-29} J (to two significant figures)
5. 2.1×10^{-12} J (to two significant figures)

4.6 Energy from the nucleus

Sample problem 11

a. 160.001 MeV
b. $2.857\ 62 \times 10^{-28}$ kg
c. $2.568\ 301 \times 10^{-11}$ J or 160.300 789 MeV

Practice problem 11

a. 166.744 812 MeV
b. $2.977\ 62 \times 10^{-28}$ kg
c. 167.032 MeV

Sample problem 12

a. 12.566 671 MeV
b. 1.863×10^{-29} kg
c. 1.674×10^{-12} J or 10.45 MeV

Practice problem 12

a. 5.639 928 MeV
b. 1.0541×10^{-29} kg
c. $9.473\ 778 \times 10^{-13}$ J or 5.913 070 MeV

4.6 Exercise

1. Fission is the name given to the process in which a very large nucleus splits into two smaller nuclei. Fusion is the process of two small nuclei combining to form a single nucleus.
2. Fusion
3. Energy is released in a fission or fusion reaction if the binding energy per nucleon (the energy required to remove a nucleon from the nucleus) increases in the process. When uranium-235 nuclei are split into two roughly equal-sized nuclei, the binding energy per nucleon for each of these new nuclei is greater than that in uranium-235. When fusion of two hydrogen atoms occurs, the binding energy per nucleon also increases.
4. The total energy of the fission products is 1984.8 MeV, which is 199 MeV greater than the uranium-235 nucleus. This agrees well with the measured value of 200 MeV.
5. When two small nuclei are fused together to form a larger one, energy is released because the new nucleus exists in a lower energy state than the two nuclei that formed it. Energy is conserved, so the difference in energy needs to be released in some form.
6. For light elements (mass number < 56), the graph of binding energy per nucleon is much steeper than for heavy elements.
7. Fusion occurs when the combining of two light nuclei results in an increase in binding energy per nucleon. This means that energy has been released to bind the nuclei more tightly together. The release of energy is equivalent to a release of mass, so the nucleus after fusion is lighter than the two nuclei that it was fused from.
8. The fission of uranium-236 results in two nuclei with higher binding energy per nucleon. This involves energy release, so the sum of the masses of all of the particles after the fission is less than the mass of the nucleus that underwent fission.

4.6 Exam questions

1. A
2. C
3. B
4. 22.3722 MeV
5. $2.850\ 84 \times 10^{-11}$ J

4.7 Fission chain reactions

Sample problem 13

Barium

Practice problem 13

a. 134

b. Caesium

c. 37, therefore it is rubidium

4.7 Exercise

1. A flat mass has a much lower volume-to-surface-area ratio than a sphere. That is, for the same volume and mass as a sphere, more of the flat sheet is exposed to the air. This allows many more free neutrons to escape rather than sustain a chain reaction.

2. Kinetic energy of free neutrons and product nuclei, and some radiation

3. Neutrons are good at initiating nuclear reactions because they have no charge and can easily enter a nucleus.

4. A chain reaction occurs when more than one of the free neutrons produced in a fission reaction triggers another fission reaction.

5. Fusion has fewer waste problems and the potential to use widely available sources of fuel, such as seawater. The disadvantage is that, after about 60 years of research, a sustainable fusion reaction has still not been achieved.

4.7 Exam questions

1. B
2. B
3. C
4. In a fission chain reaction the neutrons emitted from one fission reaction go on to initiate fission in surrounding atoms.
5. Control rods are made of materials, such as cadmium or boron, that absorb neutrons. Movement of the control rods into the reactor interrupts the chain of fission reactions, allowing the rate of reaction to be controlled.

4.8 Review

4.8 Review questions

1. a. 30 protons and 36 neutrons

   b. 90 protons and 140 neutrons

   c. 20 protons and 25 neutrons

   d. 14 protons and 17 neutrons

2. a. $^{4}_{2}\text{He}$ b. $^{13}_{7}\text{N}$ c. $^{234}_{91}\text{Pa}$

3. a. β^- particle   b. β^- particle   c. α particle

4. a. i. $^{226}_{88}\text{Ra} \rightarrow\ ^{222}_{86}\text{Rn} +\ ^{4}_{2}\alpha + \text{energy}$

      ii. $^{214}_{84}\text{Po} \rightarrow\ ^{210}_{82}\text{Pb} +\ ^{4}_{2}\alpha + \text{energy}$

      iii. $^{241}_{95}\text{Am} \rightarrow\ ^{237}_{93}\text{Np} +\ ^{4}_{2}\alpha + \text{energy}$

   b. i. $^{60}_{27}\text{Co} \rightarrow\ ^{60}_{28}\text{Ni} +\ ^{0}_{-1}\beta + \bar{v} + \text{energy}$

      ii. $^{90}_{38}\text{Sr} \rightarrow\ ^{90}_{39}\text{Y} +\ ^{0}_{-1}\beta + \bar{v} + \text{energy}$

      iii. $^{32}_{15}\text{P} \rightarrow\ ^{32}_{16}\text{S} +\ ^{0}_{-1}\beta + \bar{v} + \text{energy}$

5. Approximately 2 hours

6. 17 190 years

7. The absorbed dose is the energy absorbed by each kilogram of the tissue being irradiated.

8. Extremely high temperatures and pressures — such as exist in the centre of the Sun. These conditions are required to bring the protons close enough together for the strong nuclear force to exceed the electromagnetic repulsion.

9. The critical mass of a fissionable substance is the smallest mass that will sustain an uncontrolled chain reaction.

10. a. 67.212 434 MeV

    b. 1.198×10^{-28} kg

    c. 67.202 898 MeV

4.8 Exam questions

Section A — Multiple choice questions

1. C
2. B
3. A
4. D
5. B
6. D
7. C
8. C
9. D
10. B

Section B — Short answer questions

11. a. 173.334 717 MeV

    b. $3.089\ 62 \times 10^{-28}$ kg

    c. $2.776\ 812 \times 10^{-11}$ J or 173.315 040 MeV

12. a. 9.07×10^{-4} Gy

    b. 0.907 mSv

    c. 18 mSv

    d. The α particles would cause more damage.

13. a. Twelve hours   b. 4.375 mg

14. a. A neutron

    b. Nuclei stability follows a curve of neutron number versus proton number. Adding a neutron takes the nucleus away from that stability curve. Of the decay options α, β^+ and β^-, and γ, only β^- decay moves the nucleus back towards the line of stability.

c. Alpha emission relies on the bonding between a group of nucleons being stronger than the bonding of that group to the entire nucleus. As the strong nuclear force is stronger than the electromagnetic force for adjacent nucleons, this will only happen when the combined electromagnetic repulsion from a large nucleus affects protons that are less attached by the strong nuclear force.

15. a. 186.411 188 MeV

   b. 3.323×10^{-28} kg

   c. $2.986\,563 \times 10^{-11}$ J or 186.406 674 MeV

   d. 0.789 859 MeV

   e. 0.008 484%

5 Concepts used to model electricity

5.2 Static and current electricity

5.2 Exercise

1. Rubbing the plastic pen with wool left the pen with a net negative charge and the wool a net positive charge. When the charged plastic pen is brought close to the neutral paper the charges in the paper rearrange themselves and the positively charged paper is attracted to the negatively charged pen.

2. Your hair has transferred electrons to the balloon, giving it a net negative charge. Your hair now has a net positive charge.

3. Each balloon will have a net negative charge, which means they will repel each other.

4. Your socks rub electrons off the carpet, leaving you with a net negative charge. You get an electric shock when electrons jump from you to the metal doorknob.

5. A substance that is resistant to current flow

5.2 Exam questions

1. D

2. D

3. Electrons in metal can flow freely and carry electric current with little resistance.

4. Rubbing the balloon on her hair caused it to become negatively charged. The electrons in the can moved away from the balloon and caused the can to roll.

5. a. A conductor is a material that contains charge carriers. Examples include metals and salt solutions.

   b. Conductors always allow for charged particles to pass through, whereas semiconductors allow these particles to move under certain conditions.

5.3 Electric charge and current

Sample problem 1

4×10^{19} electrons

Practice problem 1

The answer is not a whole number of electrons so the charge cannot exist.

Sample problem 2

2.0 A

Practice problem 2

5.0 A

Sample problem 3

9.6×10^2 C

Practice problem 3

3.0 seconds

Sample problem 4

1 C to the right

Practice problem 4

4 C of charge moving downwards

Sample problem 5

0.450 A

Practice problem 5

0.280 A

Sample problem 6

25 000 s

Practice problem 6

16 000 s

5.3 Exercise

1. Conventional current is the direction that positive charges would flow if they were free to do so. Electron current is the direction that electrons flow.

2. Direct current: the charge carriers move in only one direction
   Alternating current: the charge carriers periodically change direction

3. 37.5 C

4. 0.045 A

5. 0.23 mA

6. No. Electrons cannot be destroyed. Current is *not* used up in a light globe.

7. 4.2×10^3 C

8. 8.0×10^{-5} m s^{-1}

5.3 Exam questions

1. D

2. A

3. 4.5×10^{-4} A

4. 0.30 A

5. 6.0 A

5.4 Electric potential difference

Sample problem 7
a. 0.75 J
b. 2.4×10^{-19} J

Practice problem 7
18 800 J (or 18.8 kJ)

5.4 Exercise
1. 1.5 V
2. a. 3.3 V   b. 6.0 V   c. 31.5 J
   d. 1.02 J   e. 2.7 C   f. 31 C
3. 60 μC or 6.0×10^{-5} C
4. a. 0.0073 V   b. 2.3 kV   c. 0.42 mV
   d. 932 000 V   e. 0.000 002 7 V
5. A torch does not need as much energy to produce light compared to a car, so can use a battery that supplies less energy.
6. 1.5 V

5.4 Exam questions
1. The total energy is the sum of the electrical energy and light energy.
12 V
2. 3.4×10^{-5} J
3. D
4. A 1.5-V battery provides each charge with less energy than a 9-V battery. The 1.5-V battery would need to run for six times as long as the 9-V battery to provide the same energy to an appliance.
5. 9.0 V

5.5 Electric energy and power

Sample problem 8
240 V

Practice problem 8
12 V

Sample problem 9
1.2 kW

Practice problem 9
0.60 W

Sample problem 10
16 kJ

Practice problem 10
0.175 A (or 175 mA)

Sample problem 11
30 W

Practice problem 11
12 W

5.5 Exercise
1. a. 0.25 A   b. 3.3 A   c. 1.1 A   d. 5.00 A
2. 90 s
3. 2.4 kW
4. 40 s
5. No. Mark did not use the correct units in his calculation. He needed to convert kW to W and hours to seconds in order to get the correct answer.
6. a. Each coulomb of charge passing through the heater will transform 240 J of energy.
   b. 6.0×10^3 J

5.5 Exam questions
1. D
2. C
3. The sum of the power in each device is equal to the power, thus $P_R = P_{total} - P_Q - P_P$.
Total power – power in P and Q = 8.0 W
4. 2.0 A
5. a. 4.3×10^4 J   b. 24 W

5.6 Electrical resistance

Sample problem 12
a. $27 \times 10^3 \, \Omega \pm 1350 \, \Omega$ b. $1.0 \times 10^3 \, \Omega \pm 100 \, \Omega$

Practice problem 12
a. $39 \, \Omega \pm 2 \, \Omega$
b. $56 \, 000 \, \Omega \pm 5600 \, \Omega$
c. $750 \, k\Omega \pm 37.5 \, k\Omega$

Sample problem 13
$20 \, \Omega$

Practice problem 13
$40 \, \Omega$

Sample problem 14
2.0 W

Practice problem 14
1200 W (or 1.2 kW)

Sample problem 15
a. 3.33 A   b. $72 \, \Omega$

Practice problem 15
a. 2.5 A   b. $96 \, \Omega$

5.6 Exercise
1. 1.08 kJ
2. a. 32 V   b. 48 V   c. 2.0 A
   d. 3.0 mA   e. $1.5 \, \Omega$ f. $33 \, \Omega$
3. a. Non-ohmic. The graph of current versus voltage is non-linear.
   b. 0 A

c. ∞

d. 0.65 V

   32.5 Ω

4. 0.45 W

5. a. 960 Ω   **b.** 5.7 Ω   **c.** 3.6 Ω

6. 1.2 kW

5.6 Exam questions

1. A

2. Bill is correct.
   Ben is incorrect.
   Ben would have to say '$V = IR$ where R is a constant'.

3. Yes
   Resistance is the same (25 Ω) at 12.5 V and at 30 V.

4. 270 J

5. 0.20 A

5.7 Review

5.7 Review questions

1. 2×10^8 V

2. a. 11 A

   b. 405 kJ

3. a. 110 V

   b. The current would also increase.

4. a. 0.2 W

   b.

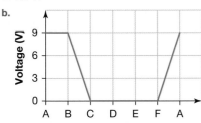

   c.

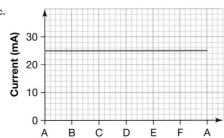

   d. 360 Ω

5. a. 0.25 A   **b.** 12 Ω   **c.** 0.75 W

6. a. 6×10^2 kg m s^{-1} **b.** 32 kJ

7. a. 15 Ω

   b. For voltages up to 4.5 V, there is a constant relationship between the voltage across the globe and the current flowing through it. In this range, the globe exhibits ohmic characteristics. However, for higher voltages the relationship is not constant and therefore non-ohmic.

8. a. A device that has a varying resistance

   b. 20 mA

   c. 2.5 Ω

5.7 Exam questions

Section A — Multiple choice questions

1. A

2. C

3. D

4. B

5. C

6. D

7. D

8. C

9. D

10. B

Section B — Short answer questions

11. a. i. 4.8×10^{-19} J   **ii.** 3.0 J

   b. 2.0×10^{-3} W

12. Disagree. When a light switch is turned on the light comes on almost instantaneously. Similarly, Luke should feel the pull on the rope almost immediately. The energy transfer is quick and he does not need to wait for the same segment of rope that was pulled on by the teacher. Although electrons carry the electric charge in a circuit, they move more slowly than the energy that is transferred to the loads in the circuit.

13. a. 0.8 A   **b.** 4 Ω

14. a. 3.30×10^5 C   **b.** 11.5 A

15. a. Non-ohmic; the resistance is not constant

   b. 10 mA

   c. Approximately 25 Ω

   d. 2.0×10^4 W

6 Circuit electricity

6.2 BACKGROUND KNOWLEDGE Electrical circuit rules

Sample problem 1

The unknown current flowing out of the junction is 1.2 A.

Practice problem 1

The unknown current flowing out of the junction is 1.5 A.

Sample problem 2

$a = 7.4$ mA, $b = 15.3$ mA, $c = 13.2$ mA, $d = 6.7$ mA, $e = 2.1$ mA, $f = 15.3$ mA

Practice problem 2

$a = 11.1$ A
$b = 2.4$ A

Sample problem 3

3.8 V

Practice problem 3

7.2 V

6.2 Exercise

1. **a.** A connection between two or more conducting paths
   b. The flow of electrons through the circuit
   c. The amount of electrical potential energy converted in a load for every coulomb of charge passing through it
   d. The wire connecting the elements in an electric circuit

2. **a.** 4.5 A
   The current flows towards the junction.
   b. 1.1 A
   Ic = 0. The current flows towards the junction.

3. Both voltages are equal to 16 V.

4. **a.** 12 V   **b.** 12 V   **c.** 12 V   **d.** 0.30 A   **e.** 40 Ω

6.2 Exam questions

1. **a.**

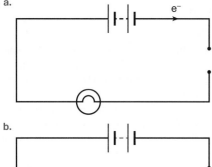

   b.

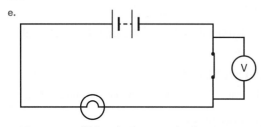

   c. 3.0 A
   d. 0
   e.

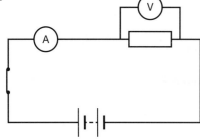

   f. No current will flow in the open circuit.

2. **a.**

   b. 5.0 A

3. Pushing the button creates a closed circuit, which is needed for electric current to flow and operate the bell.

4. **a.** 10 mA
   b. 10 V across the first person, 15 V across the second person
   c. 25 mA in the first person, 17 mA in the second person
   d. 25 V across each person

5. Short circuiting the resistor branch will result in the current flowing through the short circuit and not through the resistor. Under these conditions the voltage drop across the short circuit will be zero. Substituting in $V = IR$, where $V = 0$, R must be equal to 0.

6.3 Series circuits

Sample problem 4
74 Ω

Practice problem 4
13.9 kΩ

Sample problem 5
a. 1 A   **b.** 60 V   **c.** 40 V   **d.** 40 Ω

Practice problem 5
a. 0.6 A   **b.** 9 V   **c.** 15 V   **d.** 25 Ω

Sample problem 6
4.4 kΩ

Practice problem 6
730 Ω

6.3 Exercise

1. **a.** 62 Ω   **b.** 15.4 kΩ
2. 9.0 V
3. 10 Ω
4. $I_a = 1$ A
   $V_1 = 4$ V
   $V_2 = 2$ V
   $R_2 = 2.0$ Ω
5. B
6. **a.** 4.0 V
   b. 3.0 V
7. 4.5 V
8. The equivalent resistance will increase, hence the current will decrease. Hence, V_1 will decrease (R_1 is fixed) and hence, V_2 will increase (V_{in} is fixed).
9. **a.** 3.0 V
   b. 6.0 V

6.3 Exam questions

1. C
2. A
3. C
4. D
5. **a.** 2 A   **b.** 60 V   **c.** 10 Ω   **d.** 40 Ω   **e.** 80 V

6.4 Parallel circuits

Sample problem 7
2.9 Ω

Practice problem 7

$2\,\Omega$

Sample problem 8

a. 9 V   b. 0.9 A   c. 0.45 A
d. 20 Ω e. 6.7 Ω

Practice problem 8

a. 24 V   b. 0.6 A   c. 0.2 A   d. 120 Ω e. 30 Ω

Sample problem 9

$9.99\,\Omega$

Practice problem 9

$960\,\Omega$

Sample problem 10

$0\,\Omega$

Practice problem 10

$330\,\Omega$

Sample problem 11

a. 6.67 Ω b. 0.5 A   c. 3.5 V

Practice problem 11

a. 5 Ω b. 0.8 A

6.4 Exercise

1. a. 3.0 Ω b. 10 Ω

2. a. 5 Ω b. 3.0 A   c. 1.5 A

3. a. 10 Ω
   b. 9.0 A
   c. 1.5 A, 3.0 A and 4.5 A for 60 Ω, 30 Ω and 20 Ω
      respectively

4. $I_a = 1.5$ A
   $V_1 = 6$ V
   $V_2 = 6$ V
   $I_1 = 1.0$ A
   $I_2 = 0.5$ A
   $R_2 = 12\,\Omega$

5. a. 6.0 A through R_1, 2.0 A through R_2 and 4.0 A through R_3
   b. 12 A
   c. 3.0 Ω

6. a. 2 Ω b. 3 A   c. 2.4 V

6.4 Exam questions

1. a. G_1, G_2, G_3 b. G_1, G_3 c. G_1
2. 3 A
3. a. 10 Ω b. 2.0 A   c. 6.0 Ω
4. 4.0 Ω
5. D

6.5 Non-ohmic devices in series and parallel

Sample problem 12

a. 10 V   b. 3 A

Practice problem 12

a. 6 V   b. 5 mA

Sample problem 13

750 Ω

Practice problem 13

2.5 kΩ

Sample problem 14

R must increase.

Practice problem 14

Connect the switch across a 2.5-kΩ fixed resistor.

Sample problem 15

2.3×10^{-2} A

Practice problem 15

1.8 V

6.5 Exercise

1. a. 6 mA   b. 140 V   c. 8.8 kΩ

2. a. 30 V   b. 6.0 mA   c. 100 V   d. 130 V

3. a. 100 V   b. 100 V   c. 6 mA   d. 26 mA

4. Turning the dial on a light dimmer moves the slider along a
   potentiometer. This has the effect of changing the resistance
   of the arm of the voltage divider circuit that is connected
   to the light. Changing the resistance of this arm results in
   a changing voltage experienced by the light. If the voltage
   across the light decreases, the light will become dimmer.

5. a. 1.82×10^{-2} W
   b. 0.73 W
   c. Using LEDs vastly decreases the amount of power needed
      to run the traffic lights. This makes traffic lights cheaper to
      run and decreases their environmental impact.

6.5 Exam questions

1. A
2. D
3. 3.0 kΩ
4. R must decrease.
5. a. 0 A. The diode is reverse biased so no current will flow.
   b. Connect the diode in forward bias. The current would be
      400 μA.
   c. This would damage the diode instead of increasing its
      power output.

6.6 Power in circuits

Sample problem 16

8.91 A

Practice problem 16

5.04 A

6.6 Exercise

1. a. 0.20 A

  b. 5.0 V across the 25-Ω resistor, 3.0 V across the 15-Ω resistor, and 2.0 V across the 10-Ω resistor

  c. 1.0 W for the 25-Ω resistor, 0.6 W for the 15-Ω resistor, and 0.4 W for the 10-Ω resistor

  d. 2.0 W

2. a. 2.1 A

  b. 10 V

  c. 4.0 W for the 25-Ω resistor, 6.7 W for the 15-Ω resistor, and 10 W for the 10-Ω resistor

  d. 21.7 W

6.6 Exam questions

1. 600 Ω

2. 11.7 A

3. a. 4.6 kW

  b. Two appliances

4. 108 mW

5. 0.48 W

6.7 Review

6.7 Review questions

1.

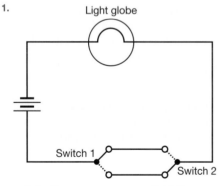

2. a. 3.3 kΩ   **b.** 2.5 kΩ

3. a. 0.30 A

  b. 5.4 W in the 60-Ω resistor, and 1.8 W in the 20-Ω resistor

  c. 1.2 A though the 20-Ω resistor, 0 A through the 60-Ω resistor

  d. 28.8 W for the 20-Ω resistor

4. a. 14.8 Ω   **b.** 20.3 V   **c.** 1.22 A

5. a. The lamp will dim.

  b. The power is inversely proportional to the resistance, thus the power will decrease.

  c. The brightness of the lamp will remain the same.

  d. The power consumed will increase.

6. a. 20 Ω   **b.** 5 Ω

7.

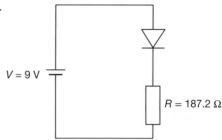

8. a. 4.4 Ω

  b. No. The 24-W lamp will be brighter.

9. a. 10.8 Ω   **b.** 0.22 A   **c.** 3.3 V

10. 1.22 A

6.7 Exam questions

Section A — Multiple choice questions

1. C

2. A

3. C

4. B

5. A

6. D

7. C

8. C

9. C

10. C

Section B — Short answer questions

11. a. 3 mA

  b. 4.5 V for the first person, 7.5 V for the second person

  c. In parallel the voltage drop across each person will be 12 V.

  d. 8 mA for the first person, 48 mA for the second person

12. a. 625 Ω

  b. 14.4 mW

  c. V_{ab} will also increase.

13. a. 400 Ω   **b.** 5000 Ω

14. a. 1.5 kΩ   **b.** 0.7 V

15.

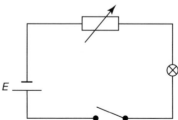

The circuit consumes more power when the light is bright since the current flowing through the globe is greater, and for the circuit $P = VI$.

7 Using electricity and electrical safety

7.2 Household electricity and usage

Sample problem 1
a. 6.09 A b. 37.8 Ω

Practice problem 1
a. 65 mA b. 3.5 kΩ

Sample problem 2
3.6 MJ

Practice problem 2
28.8 MJ

Sample problem 3
a. 85 W b. 12 kW h c. $1.97

Practice problem 3
a. 10 mA b. 1.45 MJ c. $48.38

7.2 Exercise
1. • It is an alternating current.
   • The voltage varies between +325 V and −325 V.
   • The voltage oscillates 50 times per second.
   • The voltage provides heating effects equivalent to a DC voltage of 230 V.
2. Drawing too much current can cause wires to melt or materials surrounding the wires to catch fire.
3. Active: brown
   Neutral: blue
   Earth: green-and-yellow striped
4. See the figure at the foot of the page*
5. The neutral wire is used to provide an insulated path for the electricity to return to the generator. The earth wire connects the metal case of the appliance to earth so that, if a live active wire accidentally contacts the outer case, a low-resistance path is provided for the electricity to return to earth.
6. The earth wire is used in household lighting circuits when the light fittings have a metal case. Plastic and glass fittings are good insulators and, unlike metal cases, do not need to be earthed.
7. $411.02

8. a.

Product	Energy (kW h)
Laptop computer	127.0
Modem	29.8
Cordless phone equipment	32.4
DVD player	21.0
Television	54.3

b. 381.9 kg

7.2 Exam questions
1. B
2. 20 ms
3. The neutral wire is earthed, or connected to the ground, at the switchboard. Hence, there is zero volts difference between the neutral wire and a grounded person.
4. 300 W
5. 60 kWh

7.3 Electrical safety

Sample problem 4
a. The total current is 13.8 A, so the fuse will not 'blow'.
b. The total current is 24.2 A, so the fuse will 'blow'.

Practice problem 4
a. 760 mA
b. 1.7 A
c. 3.4 A

7.3 Exercise
1. An electric shock is a violent disturbance of the nervous system caused by an electrical discharge or current through the body. Electrocution is death resulting from an electric shock.
2. Breaks or cuts in the skin, and water, oil and other fluids reduce the resistance of human skin.
3. Nerve impulses are electrical in nature. The size of the current will influence the size and type of muscle contraction.
4. Fibrillation is the disorganised rapid contraction of separate parts of the heart so that it pumps no blood.
5. Your muscles would contract and you could grip onto the victim and not be able to let go.
6. The longer the time of exposure, the more severe the shock.
7. Double insulation is a way to protect the user of handheld appliances. There are two separate layers of insulation between the functional parts of the appliance and the user.

*4.

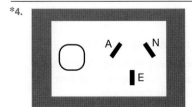 

8. Voltage doesn't flow; current does. The paper should have reported that there was a voltage drop across his body of 50 000 V. The human body is not resistant to current flow at large voltage. However, it is the size of the current that determines the amount of injury; it is not the voltage that kills.

7.3 Exam questions

1. B

2. Wet hands and feet are more likely in the bathroom; this will reduce skin resistance and hence, larger current will flow through the body.

3. Current is more important as it directly affects or disrupts the actions of nerves and muscles. Voltage is mainly important in that higher voltage means larger current.

4. The earth wire acts only when the metal casing comes into contact with a live wire and so becomes live itself. It allows a large current to flow to earth and so 'blows' the fuse. This disconnects the appliance from the live wire and removes the hazard of shock to the user.

5. A residual current device acts when the current in the active wire is greater than that in the neutral wire. It disconnects the active wire:
   - quickly enough to prevent dangerous shock
   - at a value of current small enough to not be dangerous.

7.4 Review

7.4 Review questions

1. a. 3942 kW h
   b. $1104.00

2. a. They should be connected in parallel, so that if one fails, the other can still function. If they are connected in series, then if one fails, there is a total loss of sound.
   b. 4.6 kW
   c. $8.83
   d. 10 A (the fuse should fail at, or just above, the operating current of the device)

3. a. 4.35 A
   b. 52.9 Ω
   c. $0.85

4. a. The metal casing on the appliance is connected to the earth wire and not the live wiring that provides energy to the appliance. This keeps the voltage of the casing at zero.
   b. A shock could occur if the casing has a non-zero voltage, which could happen if the insulation between the casing and the functional parts of the appliance broke down.

5. a. An electric shock occurs when electric current passes through the body. Electrocution is death caused by an electric shock.
   b. An electric shock is more likely to be fatal if there is a large amount of current flowing through the body, the pathway of the current is through the trunk of the body, or there is an extended time of exposure to the current.

6. Both fuses and circuit breakers provide overload protection in circuits. As such, each has the potential to protect

electrical equipment and reduce the likelihood of fire from overheating.

A fuse comprises a metal wire that melts when its current limit is exceeded. Fuses can be used in a variety of places, including plugs, appliances and household circuits. They are also readily used in DC circuits. After the source of an electrical problem is resolved a melted fuse must be replaced, which also means that unless the electrical problem causing the problem is also resolved, replacement fuses will continue to 'blow'.

Circuit breakers comprise resettable switches activated by one of two methods — either by using the heat generated by an excessive current, or by the electromagnetic effect in a circuit to activate a switch. They are initially more expensive to install but are readily reset. They also allow the isolation of circuits within a system of circuits as found in households.

7. a. 2.63 A
   b. No

8. As the magnitude of the current passing through the body increases it takes less time to cause fibrillation. It is therefore important to safely remove a person in shock from the source.

9. Dry skin has a higher resistance than wet hands. Wet hands provide an easier pathway to earth and therefore attract a larger current that can cause damage when travelling through the body.

10. Whereas a circuit breaker provides overload protection in a circuit by detecting increasing current, a residual current device effectively detects a leakage of current, as might happen when there is a short circuit. Residual current devices rely on the magnetic properties of electric currents to detect when there is a difference in the backwards and forwards movement of AC current to then activate a switch. Residual current devices can be activated quickly and at low currents, providing good safety protection as long as the leakage of current is to earth. However, a residual current device is ineffective when a person acts as a conductor between the active and neutral wires of a circuit.

7.4 Exam questions

Section A — Multiple choice questions

1. C
2. B
3. A
4. B
5. C
6. B
7. D
8. C
9. B
10. B

Section B — Short answer questions

11. Factors to decrease the severity of the shock include:
    - not having wet hands or standing on a wet floor
    - wearing shoes with an insulating sole

- not touching an easier path to earth, such as a kitchen tap
- letting go of the appliance as soon as possible (turning off the electricity and removing the person from the source if the person cannot let go).

12. a. See the figure at the foot of the page*

b. Any two of the following are acceptable answers:
- The earth wire in the main circuit provides a low-resistance conducting path for current should there be a short circuit.
- Insulation between the appliance and the case insulates the metal case from the live circuit.
- Separately earthing the metal case provides a low-resistance conducting path for current to flow to earth if the insulation 'barrier' breaks down.
- A fuse, circuit breaker or residual current device on the active wire between the power supply and the appliance breaks the circuit if the current to the device is excessive, or in the case of a residual current device, there is a current leakage.

13. See the figure at the foot of the page*

14. a.

Product	Power (W)	Power (kW h per year)
Laptop computer	14.5	127.0
Microwave	4.2	36.8
Laser printer	8.5	74.5
Set-top box	11.2	98.1
Television	6.2	54.3

b. 564.2 kg

c. $195.35

15. a. The person will experience a shock but, provided the person does not continue to hold the appliance, it should not be dangerous.

b. 30 kV

c. The current will be 76.7 mA. With this current, there is a risk that the person's muscles will contract and they will be unable to let go of the appliance. Fibrillation is likely and the person must be safely removed from the source of electricity.

8 Analysing motion

8.2 Describing movement

Sample problem 1

a. 100 metres north
b. 220 metres
c. 100 metres
d. 60 metres south

Practice problem 1

a. 16 km

b. 5.7 km

c. 0 metres

Sample problem 2

2700 kilometres

Practice problem 2

720 kilometres

Sample problem 3

a. 11 m s^{-1}

b. 5 m s^{-1} north

Practice problem 3

a. 3.5 m s^{-1}

b. 1.2 m s^{-1} east

---

*12. a.

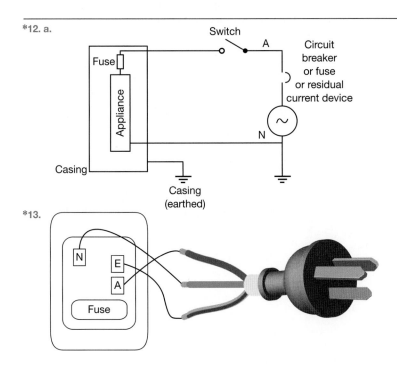

*13.

Sample problem 4

a. 2.5 m s^{-2} towards the letterbox

b. 5 m s^{-2} away from the letterbox

c. 4 m s^{-2} towards the letterbox

d. Braking at the letterbox and accelerating back towards home from the letterbox

e. A decreasing speed in the negative direction is a positive change in velocity, and hence a positive acceleration.

Practice problem 4

a. 15 m s^{-2}

b. i. 70 km h^{-1} s^{-1} ii. 19.4 m s^{-2}

8.2 Exercise

1. a. Scalar b. Vector c. Scalar
   d. Vector e. Vector

2. 27.8 m s^{-1}

3. 5.4 km h^{-1}

4. a. 88 km h^{-1} b. 24 m s^{-1}

5. a. 800 metres b. 400 metres c. 0 metres

6. a.

Event (m)	Average speed (m s^{-1})
100	9.53
200	9.37
400	8.40
800	7.06
1500	6.52
3000	6.17
5000	5.91
10 000	5.74

   b. The acceleration from rest to the maximum speed takes place over a significant fraction of the time taken for the 100-metre event.

   c. 1 h 47 min 52 s

   d. Only Florence Griffith Joyner. Her event is the only one that involves straight-line motion.

7. a. i. 10 km h^{-1} ii. 2.8 m s^{-1}
   b. 12 s

8. a. 15.30 m s^{-1}
   b. 2 h 44 min
   c. 1 h 53 min
   d. i. 75.5 km h^{-1} ii. 0 km h^{-1}

9. Approximately 3 hours 37 minutes

10. 3.3 m s^{-1}

11. 89 km h^{-1}

12. a. i. −40 km^{-1} ii. 40 km h^{-1} south
    b. i. −20 m s^{-1}
       ii. −20 m s^{-1} in original direction
    c. i. +5 m s^{-1}
       ii. −55 m^{-1} in original direction

13. Yes, there is an acceleration. Even though the speed has not changed, the velocity has changed. The acceleration is therefore not 0.

14. Answers will vary.
    For a car that accelerates from rest to 60 km h^{-1} (17 m s^{-1}) in 5 seconds: $a = 3.4$ m s^{-2}

15. Answers will vary.
    v_{av} for a 100-metre sprint in say 10.49 s = 9.5 m s^{-1}.
    Estimate $v_{max} = 12$ m s^{-1} is reached after 2 seconds:
    $a = 6$ m s^{-2}

8.2 Exam questions

1. B

2. −2 m s^{-1}

3. 2400 m north-west

4. −2.0 m s^{-2}

5. Yes, the car is accelerating. A change in direction (or change in magnitude) of velocity requires an acceleration.

8.3 Analysing motion graphically

Sample problem 5

Approximately 4.7 seconds

Practice problem 5

a. 30 m b. 28 m

8.3 Exercise

1. a. B and C b. B and D c. A and E
   d. A and E e. D

2. a. A: Constant negative acceleration with an initial positive velocity
   B: Constant positive acceleration from rest
   C: Constant positive velocity, followed by an interval of constant negative acceleration until a negative velocity equal in magnitude to the initial velocity is reached. The velocity then remains constant.

   b.

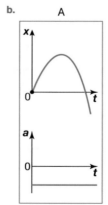
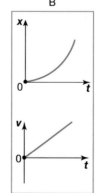
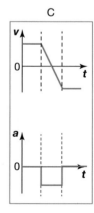

3. a.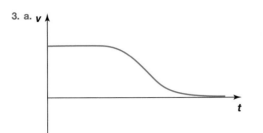

b.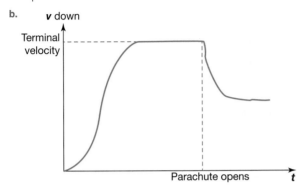

c.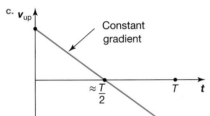

4. a.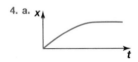

b.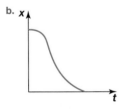

c.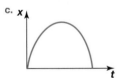

5. a. B

b. A, D, E

c. 40 seconds

d. 20 metres north

e. 260 metres

f. D

g. E

h. 3.3 m s^{-1}

i. 6.0 m s^{-1} south

j. 3 m s^{-1}

6. a. B, D, F

b. +20 m

c. 0.25 m s^{-1}

d. 30 seconds

e. It didn't

f. C, G

g. The first half of interval C, the first half of interval E and all of interval G

h. A negative acceleration doesn't always decrease the speed and a positive acceleration doesn't always increase the speed. A negative acceleration increases the speed if the velocity is negative and decreases the speed if the velocity is positive. Similarly, a positive acceleration decreases the speed if the velocity is negative and increases the speed if the velocity is positive.

i. 0.20 m s^{-2}

j. 0.050 m s^{-2}

k. First 10 seconds: The toy robot started from rest and increased its speed at a constant rate until reaching a speed of 1.0 m s^{-1} after 10 seconds.
10 s to 20 s: It maintained a constant speed of 1.0 m s^{-1}.
20 s to 30 s: It slowed down at a constant rate. It was at rest for an instant, 30 seconds after starting.
30 s to 40 s: It increased its speed at the same constant rate as the first interval, but in the opposite direction, to reach a maximum speed of 1.0 m s^{-1}.
40 s to 50 s: It maintained a constant speed of 1.0 m s^{-1}.
50 s to 55 s: It decelerated to rest at a constant rate.
55 s to 60 s: It increased its speed at a constant rate in the original direction. The acceleration was twice that of the first interval.
60 s to 70 s: It maintained a constant speed of 1.0 m s^{-1}.
70 s to 80 s: It decelerated to rest at a constant rate.

7. a. The acceleration of the jet ski becomes zero first, after 8 seconds.

b. i. 21 m s^{-1}

ii. 33 m s^{-1}

c.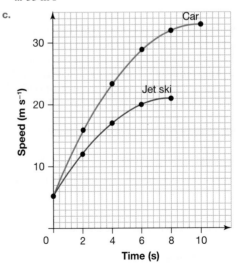

8.3 Exam questions

1. D
2. A
3. +2 m
4. a. -4 m s^{-1}
   b. In the interval 4 to 5 s (just after $t = 4.5$ s)
5. a. 0 m
   b. 5 m s^{-1}

8.4 Equations for constant acceleration

Sample problem 6

a. 29 m s^{-1}
b. 44 metres

Practice problem 6

a. 6.0 seconds
b. 36 metres

Sample problem 7

a. 12 m s^{-1}
b. -6 m s^{-2}

Practice problem 7

a. 18 metres
b. -16 m s^{-2}

8.4 Exercise

1. a. 1.8 s
   b. 5.0 s
2. a. 2.7 s
   b. 27 m s^{-1}
3. a. 12 m s^{-1}
   b. -6 m s^2
4. a. -8.0 m s^{-2}
   b. 3.5 s
   c. The reaction time of the driver needs to be known to determine the distance travelled between the instant that the branch is seen and the instant that the brakes are applied. An estimate of 0.2 seconds would be reasonable for the reaction time. At a constant speed of 100 km h^{-1} (27.8 m s^{-1}), the car would travel a distance of:
   27.8 m s^{-1} × 0.2 s = 5.6 m
   The total distance required to stop is therefore:
   5.6 m + 48 m = 53.6 m
   The car would not stop in time.
5. The leap would take 0.8 seconds. The leap is not possible.
6. 793 metres
7. a. 32 m s^{-1}
   b. The balls meet 50 metres from the ground.

8.4 Exam questions

1. C
2. A
3. -5 m s^{-2}
4. 19.6 m
5. a. 2.0 m s^{-2}
   b. 40 m
   c. 10 m s^{-1}
   d. -4.0 m s^{-2}
   e. 3.5 s

8.5 Review

8.5 Review questions

1. Distance = 31 m; displacement = 3 m north
2. 185 m s^{-1}
3. 660 m s^{-2} away from the racquet
4. 4 m s^{-1}
5. 117.5 m
6. 10 m s^{-2}
7. -1.875 m s^{-1}
8. 2.5 m s^{-1}
9. 458.3 m
10. $u = 43.2$ km h^{-1}
    Yes, the driver was travelling faster than the speed limit when they braked.

8.5 Exam questions

Section A — Multiple choice questions

1. C   2. A
3. C   4. D
5. B   6. A
7. C   8. D
9. B   10. C

Section B — Short answer questions

11.

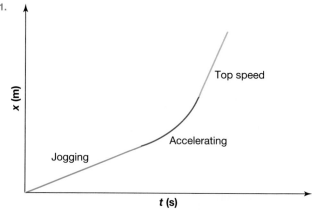

12. In this instance, the magnitude of the velocity (the speed) has not changed, but the direction has. Therefore, there has been a change in velocity, and thus an acceleration.
13. a. 8.5 metres
    b. Assume that the baton change starts at the 3-second mark and is just inside the interchange area at that point. From the calculations for the previous question, Bo starts 7 metres before the interchange, so therefore complies with the 10-metre maximum rule.

The exchange happens at a constant speed of 7 m s^{-1} and takes 2 seconds, so by inspection Alex and Bo will travel 14 metres together during the interchange.
Therefore, they also comply with the 20-metre interchange area rule.

14. 4.23 seconds

15. a. 3.2 seconds
    b. 2.5 m s^{-2}
    c. 10 seconds
    d. 80 metres

9 Forces in action

9.2 Forces as vectors

Sample problem 1

a. 490 N downwards
b. 80 N downwards

Practice problem 1

a. At the North Pole: 688 N
   At the equator: 685 N
b. Denver

Sample problem 2

a.
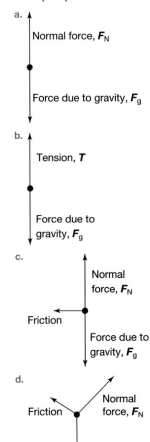

b.

c.

d.

Practice problem 2

a.
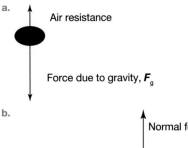

b.
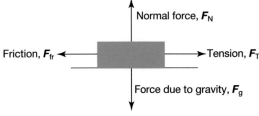

c.
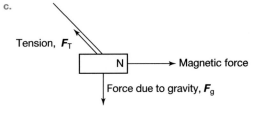

Sample problem 3

a. 20 N   b. 15 N   c. 21.2 N

Practice problem 3

89 N

9.2 Exercise

1. Vector quantities have magnitude and direction. Scalar quantities have magnitude only.

2. II and III

3. a. 13 720 N   b. 5040 N   c. 1400 kg

4. a. 0.98 N (for an apple with mass = 100 g)
   b. 9.8 N (for a textbook with mass = 1 kg)
   c. 784 N (for a teacher with mass = 80 kg)

5. a. 588 N (for a student with mass = 60 kg)
   b. 216 N (for a student with mass = 60 kg)
   c. Mass is unchanged

6. a.
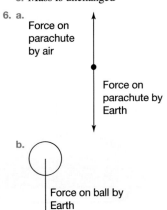

b.

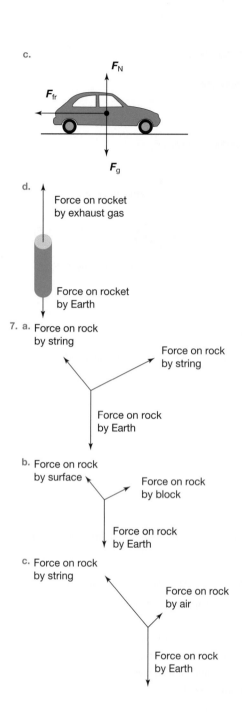

c.

F_N
F_{fr}
F_g

d. Force on rocket by exhaust gas

Force on rocket by Earth

7. a. Force on rock by string

Force on rock by string

Force on rock by Earth

b. Force on rock by surface

Force on rock by block

Force on rock by Earth

c. Force on rock by string

Force on rock by air

Force on rock by Earth

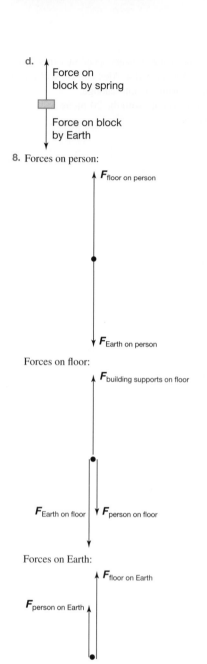

d. Force on block by spring

Force on block by Earth

8. Forces on person:

$F_{\text{floor on person}}$

$F_{\text{Earth on person}}$

Forces on floor:

$F_{\text{building supports on floor}}$

$F_{\text{Earth on floor}}$ $F_{\text{person on floor}}$

Forces on Earth:

$F_{\text{floor on Earth}}$

$F_{\text{person on Earth}}$

9. a. 3 N to the east
   b. 141.4 N to the east
10. a. 346 N to the east
    b. 53.6 N to the east
11. See the figure at the foot of the page*

---

*11.

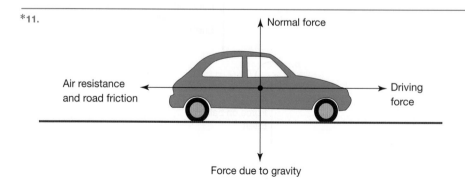

Normal force

Air resistance and road friction

Driving force

Force due to gravity

12. a. 181 N

   b. 100 N

   c. 0 N

9.2 Exam questions

1. B

2. 7.1 N north-east

3. 4.5 N

4. 784 N

5. a. 25 kg

   b. The person's mass is still 50 kg but the force due to gravity acting on them will be halved, due to the reduction in g. Since the calibration of the dial is unchanged, the reading for mass will be only half of the correct value.

9.3 Newton's First Law of Motion

9.3 Exercise

1. For a smartphone sliding across a table to maintain a constant speed there must be no unbalanced forces on the phone. Though the vertical forces cancel out, there will always be a horizontal friction force opposing the direction of motion. These horizontal forces act in the same direction, so there is a non-zero net force acting on the phone. According to Newton's First Law, the net force will cause the motion of the phone to change, slowing until it comes to a stop.

2. The vehicle experiences a non-zero net force that slows it down. No such force acts on you. The net force on you is zero. Therefore, you continue in your state of constant velocity.

3. There is an unbalanced force on the bike and so its velocity changes. Your inertia keeps you moving forward as there is no unbalanced force to change your motion (apart from gravity).

4. 125 N

5. It changed due to the air resistance slowing it down in the horizontal direction.

9.3 Exam questions

1. Air resistance and friction act to cause the ball to slow down eventually.
   These two forces acting to oppose the horizontal motion of the ball result in a non-zero net force upon the ball. In accordance with Newton's First Law, this will change the state of motion of the ball, which in this case means eventually slowing it to a stop.

2. 600 N upwards

3. An object remains in the same state of motion if the net force acting on it is zero. The net force on a space probe in deep space is close to zero, while an aircraft experiences significant air resistance opposing its motion, thus it would slow down if its engines stopped producing thrust.

4. 900 N

5. The passenger and car were initially moving forward at speed. The obstacle exerted a large net force on the car, which stopped very quickly. With no net force acting on the passenger, he kept moving forward at the original speed. The passenger keeps moving forward until he strikes something that exerts a net force backwards, to slow him down.

9.4 Newton's Second Law of Motion

Sample problem 4

$9.3 \, \text{m s}^{-1}$

Practice problem 4

a. 696 N

b. 0.8 kg

Sample problem 5

a. $5.2 \, \text{m s}^{-2}$

b. No, she wouldn't stop before hitting the water

Practice problem 5

8259 N

Sample problem 6

a. 160 N north

b. $3 \, \text{m s}^{-2}$ north

c. 80 N south

Practice problem 6

a. $2 \, \text{m s}^{-2}$

b. 320 N

Sample problem 7

a. −15 N

b. 4.5

Practice problem 7

a. 400 N

b. 1800 N

9.4 Exercise

1. Idealisations can be made to allow the use of a simple mathematical model to solve a physical problem. For example, in order to use simple equations to analyse the motion of a falling ball, the idealisation can be made that the air resistance is insignificant and the ball does not spin.

2. a. 6.6×10^6 N

   b. 2.8×10^7 N

3. a. Both the bowling ball and gold bar have a very small air-resistance-to-mass ratio, so they will experience a very similar acceleration. Therefore, they will fall at the same rate, hitting the ground simultaneously.

   b. The air resistance on the doormat is significant when compared with the force due to gravity acting on it. Therefore, the net force on the doormat is less than that of the bowling ball (which has the same mass as the doormat) and the acceleration of the doormat $\dfrac{F_{\text{net}}}{m}$ is smaller than that of the bowling ball (and the gold bar).

4. a. 686 N downwards

   b. i. The upwards force is greater when the jumper is decelerating downwards or, in other words, accelerating upwards.

      ii. The force due to gravity is greater than the upwards pull when the jumper is accelerating downwards.

c. The tension in the bungee cord must be equal in magnitude to the force due to gravity acting on the jumper in order for the speed to be constant; that is, 686 N. This occurs only for an instant during the fall. At this instant, the cord is extending and the tension is increasing.

5. a. 7500 N
   b. 6.250 m s^{-2}
   c. 31.25 m s^{-1}
   d. 78.13 m
6. 6.944×10^6 N
7. 100 N
8. a. 700 N
   b. 557 N
   c. 840 N
9. a. 0 m s^{-1}
   b. 9.8 m s^{-2}
   c. 4.9 N downwards

9.4 Exam questions

1. 1.8 m s^{-2}
2. 1.5 N
3. 95 N
4. The student overlooked the friction.
   1 N
5. 1300 N

9.5 Newton's Third Law of Motion

Sample problem 8

$F_{\text{on Jack by Jill}}$

Practice problem 8

a.

$F_{\text{on boy by girl}}$ $F_{\text{on girl by boy}}$

b.

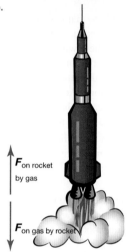
$F_{\text{on rocket by gas}}$

$F_{\text{on gas by rocket}}$

c.

$F_{\text{on Earth by apple}}$

$F_{\text{on apple by Earth}}$

d.

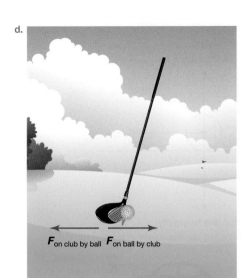

$F_{\text{on club by ball}}$ $F_{\text{on ball by club}}$

9.5 Exercise

1.

Force 1	Pair of force 1
You push on a wall with the palm of your hand.	The wall pushes on your palm in the opposite direction.
Your foot pushes down on a bicycle pedal.	The bicycle pedal pushes up on your foot.
The ground pushes up on your feet while you are standing.	You push down on the ground when you are standing.
Earth pulls down on your body.	Your body pulls up on Earth.
You push on a broken-down car to try to get it moving.	The broken-down car pushes on you in the opposite direction.
A hammer pushes down on a nail.	The nail pushes up on the hammer.

2. a.

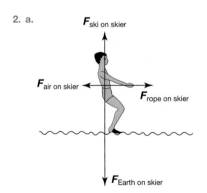

$F_{\text{ski on skier}}$

$F_{\text{air on skier}}$ $F_{\text{rope on skier}}$

$F_{\text{Earth on skier}}$

b.

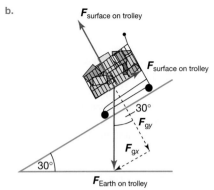

$F_{\text{surface on trolley}}$

$F_{\text{surface on trolley}}$

$30°$

F_{gy}

F_{gx}

$30°$

$F_{\text{Earth on trolley}}$

3. The dinghy:

Force	Action–reaction pair
Resistance forces	$F_{\text{dinghy on air and water}}$, $F_{\text{air and water on dinghy}}$
Gravity	$F_{\text{Earth on dinghy}}$, $F_{\text{dinghy on Earth}}$
Tension	$F_{\text{rope on dinghy}}$, $F_{\text{dinghy on rope}}$
Normal force	$F_{\text{dinghy on water}}$, $F_{\text{water on dinghy}}$

The boat:

Force	Action–reaction pair
Resistance forces	$F_{\text{boat on air and water}}$, $F_{\text{air and water on boat}}$
Gravity	$F_{\text{Earth on boat}}$, $F_{\text{boat on Earth}}$
Tension	$F_{\text{boat on rope}}$, $F_{\text{rope on boat}}$
Normal force	$F_{\text{boat on water}}$, $F_{\text{water on boat}}$
Driving force	$F_{\text{boat on water}}$, $F_{\text{water on boat}}$

4. All swimmers move forwards in the water because the water pushes them forwards. This is the unbalanced force that provides the acceleration during each stroke (Newton's First Law). The size of the forward force is equal to, and opposite in direction to, the force that the swimmer applies to the water (Newton's Third Law). A freestyle stroke pushes water back with a greater force than a breaststroke stroke. Therefore, the forward force is greater for a freestyle swimmer.

5. The student is correct but needs to make it clear that the two friction forces are not an action–reaction pair. The best way to do this is to identify the two action–reaction pairs involved. If the car is a front-wheel drive, the friction on each of the rear tyres is a reaction to the backwards push of the front tyres on the road. The rear tyres are being pulled forward. So the friction on the rear tyres is a reaction to the forward push of the rear tyres on the road.

9.5 Exam questions

1. Any two of the following or equivalent:
- $F_{\text{on car by air (drag)}} = -F_{\text{on air by car}}$
- $F_{\text{on car by road (friction)}} = -F_{\text{on road by car}}$
- $F_{\text{on car by road (normal)}} = -F_{\text{on road by car}}$
- $F_{\text{on car by Earth}} = -F_{\text{on Earth by car}}$

2. B

3. C

4. 1.5 m s^{-2}

5. • To cause motion, the tyres push backwards on the road and, by Newton's Third Law, the road friction pushes forward on the tyres.
  • However, the muddy road surface can provide very little friction force on the tyres.
  • Hence, with little force forward, the acceleration of the wheels (and car) is negligible.

9.6 Forces in two dimensions

Sample problem 9

4058 N

Practice problem 9

a. **i.** 16 759 N
  ii. 46 045 N

b. **i.** 0 N
  ii. 15 146 N

Sample problem 10

a. 3.6 m s^{-1}

b. 16 m

Practice problem 10

a. 35 N down the slope

b. 0.35 m s^{-2} down the slope

c. 4.2 m s^{-1}

Sample problem 11

a. 2 m s^{-2} to the right

b. 1800 N

Practice problem 11

a. 2 m s^{-2}

b. 200 N

c. 300 N

9.6 Exercise

1. a.

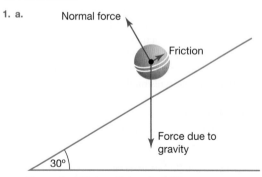

b. Down the hill (parallel to the hill)

c. The force due to gravity

d. The ball slows to a stop because of the effect of the friction acting on the ball. On a horizontal surface, the normal reaction is equal in magnitude to the force due to

gravity, so the friction force is the net force, which causes the ball to decelerate.

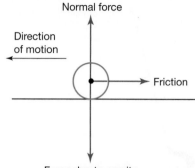

2. The friction between the stationary tyres and the road

3.

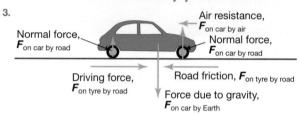

4. a. Zero

b. 392 N

c. 402 N

d. 679 N

5. a. Down the slope

b.

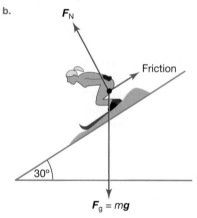

c. 294 N

d. 286 N

6. 174 N

7. 5.85 m s^{-2}

8. a. 42 N

b. 19°

9. a. Traction (friction) on your blades

b. The component of gravity down the slope and the force of the ground on your poles if you use them

c. The tension in the rope attached to the handle that you are holding

d. The traction (friction) as you push back on the ground

e. The force exerted by the water on your hands, arms, legs and feet as you push back with your hands and kick

f. The force exerted by the water on the oar as the oar pushes back on the water

10. The driving force, the forward force applied to the tyres by the road, is a reaction to the force applied backwards to the road by the tyres. The size of the driving force is, therefore, controlled by the driver's use of the accelerator. The driving force acts on the front wheels of a front-wheel-drive vehicle, whereas it acts only on the rear wheels of a rear-wheel-drive car. The front wheels of a rear-wheel-drive car are pushed forward as a result of the driving force on the rear wheels. So the force applied to the front wheels cannot be controlled directly by the driver.

11. a. A sample spreadsheet is shown here.

	A	B	C	D	E
1	Speed (km h^{-1})	Driving force (N)	Friction (N)	Force due to air resistance (N)	Net force (N)
2	20	1800	−300	−240	1260
3	25	1800	−300	−375	1125
4	30	1800	−300	−540	960
5	35	1800	−300	−735	765
6	40	1800	−300	−960	540
7	45	1800	−300	−1215	285
8	50	1800	−300	−1500	0

b.

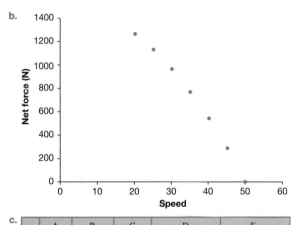

c.

	A	B	C	D	E
1	Speed (km h^{-1})	Driving force (N)	Friction (N)	Force due to air resistance (N)	Net force (N)
2	20	1800	−300	−240	3302.124
3	25	1800	−300	−375	3167.124
4	30	1800	−300	−540	3002.124
5	35	1800	−300	−735	2807.124
6	40	1800	−300	−960	2582.124
7	45	1800	−300	−1215	2327.124
8	50	1800	−300	−1500	2042.124

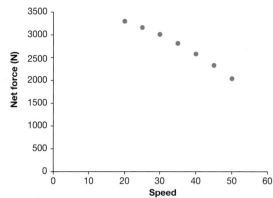

12. 0.9 m s^{-2}

13. The friction on the rear wheels is the driving force, which pushes the bicycle forward. The friction on the front wheel opposes the motion of the bicycle. When the driving force is greater than the friction on the front wheel, the bicycle will accelerate.

14. a. 2.0 m s^{-2}

    b. 6.0 N

    c. 8.0 N

    d. 3.5 m s^{-2}

15. a. 4.0 m s^{-2} to the right

    b. 160 N to the right

    c. 240 N to the left

    d. 240 N to the right

    e. Reversing the two blocks would not make it easier.

9.6 Exam questions

1. 0.2 m s^{-2}

2. C

3. C

4. B

5. C

9.7 Momentum and impulse

Sample problem 12

1.2×10^8 kg m s^{-1} north

Practice problem 12

a. 18 000 kg m s^{-1} east

b. 25 200 kg m s^{-1} east

Sample problem 13

a. 0.81 kg m s^{-1} away from the wall

b. 0.81 N s away from the wall

c. 540 N

Practice problem 13

a. 28 000 kg m s^{-1}

b. 28 000 kg m s^{-1}

c. 20 000 N

Sample problem 14

$12.25 \, \text{m s}^{-1}$

Practice problem 14

a. $5.5 \, \text{m s}^{-1}$

b. $5.0 \, \text{m s}^{-2}$

c. 245 N

9.7 Exercise

1. $23\,000 \, \text{kg m s}^{-1}$ east
2. a. $3 \times 10^3 \, \text{N}$
   b. $8 \times 10^2 \, \text{N}$
   c. $8 \times 10^2 \, \text{kg m s}^{-1}$
   d. $2 \times 10^4 \, \text{kg m s}^{-1}$
   e. $6 \times 10^2 \, \text{kg m s}^{-1}$
   f. 2 N s
   g. $3 \, \text{kg m s}^{-1}$
3. a. $0.84 \, \text{kg m s}^{-1}$ up
   b. $0.84 \, \text{kg m s}^{-1}$ down, as the impulse applied by the tennis ball to the ground is equal and opposite to the impulse applied by the ground to the tennis ball
   c. No, Earth is so massive that this impulse effect on the velocity is negligible
   d. $4.2 \times 10^2 \, \text{N}$ up
   e. $4.2 \times 10^2 \, \text{N}$ up
4. a. 240 N s upwards
   b. $3.1 \times 10^3 \, \text{N}$ upwards
   c. 0.52 m
5. a. $2.3 \times 10^4 \, \text{N s}$ opposite to the initial direction of the car
   b. $2.9 \times 10^5 \, \text{N}$ opposite to the initial direction of the car
   c. $2.1 \times 10^2 \, \text{m s}^{-2}$
6. The airbags allow the change in momentum (impulse) of the driver's head to take place over a longer time interval than would be the case if it collided directly with the steering wheel. The average net force on (and the magnitude of the acceleration of) the driver's head is therefore less.
7. The change in momentum (impulse) on the legs takes place over a longer interval, reducing the force exerted by the ground on the knee joints and muscles, tendons and ligaments in the legs.
8. a. $2.2 \times 10^2 \, \text{N s}$
   b. $3.7 \, \text{m s}^{-1}$
   c. It can be estimated that the unbelted occupant is approximately 1.6 times heavier than the belted occupant.
   d. The graph describing the force on the occupant with the seatbelt shows that the force is applied immediately and is applied for a relatively large amount of time compared with the force applied to the occupant without the seatbelt. The occupant without the seatbelt experiences no immediate force as she or he continues to move forward at the same speed as the car was moving before impact. The force applied to this occupant increases rapidly to a magnitude greater than the force applied to the occupant with the seatbelt. The multiple peaks in force on the second occupant can be explained by multiple impacts with the dashboard or other parts of the car.

9. a. 95 N s upwards
   b. $0.58 \, \text{m s}^{-1}$
   c. $9.5 \times 10^2 \, \text{N}$ upwards
   d. The normal force is present as the basketballer is initially pushing down on the floor with a force equal to the basketballer's weight (force due to gravity).
10. Bouncing off during collision results in a greater change in momentum of the cars in a similar or smaller time interval. The rate of change in momentum of the cars and the resulting net force on the passengers, would therefore be greater ($F\Delta t = m\Delta v$). In low-speed collisions with small vehicles (such as dodgem cars), this is not a problem. However, in real cars at typical road speeds, more injuries would occur.

9.7 Exam questions

1. a. $600 \, \text{kg m s}^{-1}$ west
   b. 600 N s east
2. $11 \, \text{m s}^{-1}$ (change of momentum in the direction of initial motion)
   OR
   1.0m s^{-1} (change of momentum opposite the direction of initial motion)
4. 15 N
5. 400 N south
6. $8 \, \text{m s}^{-1}$ west

9.8 Torque

Sample problem 15

100 N

Practice problem 15

1 m along the handle

9.8 Exercise

1. 50 N m
2. They can either increase the force they are applying, or increase the distance from the nut at which they are applying the force.
3. 80 N
4. The solution will depend on the chosen example and estimates made. Calculate using $\tau = r_\perp F$ with estimated quantities in appropriate units (m and N). For example, turning a steering wheel in a car might require a force of 15 N acting over a distance of 0.2 metres.
5. As Sam moves beyond the left of the fulcrum, the seesaw will rotate anticlockwise. This will happen when the torque from Sam about the fulcrum is larger than the torque of the bag.

9.8 Exam questions

1. C
2. B
3. D
4. 800 N m clockwise
5. 80 N

9.9 Equilibrium

Sample problem 16
1.5 m to the left of the fulcrum

Practice problem 16
38 kg

Sample problem 17
Magnitude of R_1: 637 N

Magnitude of R_2: 343 N

Practice problem 17
a. 7500 N

b. i. 800 N

   ii. 8300 N

9.9 Exercise
1. a. The reaction from the left abutment decreases and the reaction from the right abutment increases.

   b. R_L = 2.4 kN and R_R = 12 kN

2. a. The wall resists the loads from the balcony and the person with an upwards force and an anticlockwise torque.

   b. As the person moves towards the wall, the total reaction remains constant. However, the torque on the wall decreases.

3. 36 tonnes

4. Left: 1767 N
   Right: 1833 N

5. 1.6 m

9.9 Exam questions
1. 0.59 m from the pivot point

2. A

3. a. 500 N

   b. 800 N down

4. 400 N

5. 350 N downwards

9.10 Review

9.10 Review questions
1. • Gravity: the force due to gravity of Earth on the object
   • Air resistance: the force of the air on the object; opposite direction to the motion

2.
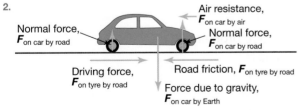

3. Horizontal component = 21.7 N
   Vertical component = 12.5 N

4. The net force has a magnitude of 366 N and acts in the direction 115.7° true.

5. 15 792 N

6. 1.4×10^6 kg

7. a. $3.9 \ \text{m s}^{-2}$

   b. 4018 N

8. a. $1.8 \ \text{m s}^{-2}$

   b. 85 N

9. a. $-5.6 \ \text{kg m s}^{-1}$

   b. -2793 N

10. 180 N s

9.10 Exam questions

Section A — Multiple choice questions
1. B

2. A and D

3. D

4. C

5. B

6. A

7. D

8. B

9. C

10. C

Section B — Short answer questions
11. For two forces to be a Newton's Third Law pair, they must be equal and opposite in magnitude and act on different objects, such that $F_{\text{on A by B}} = -F_{\text{on B by A}}$.
    In the given scenario, while the forces are equal and opposite, they are both acting on the book.

12. 38 N

13. $0.64 \ \text{m s}^{-2}$

14. $-261\,250$ N

15. Left support: 257 372.5 N
    Right support: 335 527.5 N

10 Energy and motion

10.2 Impulse and momentum

Sample problem 1
a. $1 \ \text{m s}^{-1}$ to the right

b. $15 \ \text{kg m s}^{-1}$

c. $15 \ \text{kg m s}^{-1}$ to the left

d. $3.5 \ \text{m s}^{-1}$ to the right

Practice problem 1
a. $15 \ \text{kg m s}^{-1}$ east

b. $1.5 \ \text{m s}^{-1}$ east

c. $2.5 \ \text{kg m s}^{-1}$ west

d. $2.5 \ \text{kg m s}^{-1}$ east

e. $2.5 \ \text{kg m s}^{-1}$

f. The impulses on the two cars are equal in magnitude, but opposite in direction.

10.2 Exercise

1. By releasing a high-pressure propellant, the astronaut gains momentum in one direction while the propellant gains the same amount of momentum in the opposite direction. The total change in momentum of the astronaut and the contents of their backpack is zero.

2. a. $0.3 \, \mathrm{m\,s^{-1}}$
   b. $0.6 \, \mathrm{m\,s^{-1}}$

3. a. $40 \, \mathrm{kg}$
   b. $60 \, \mathrm{N\,s}$
   c. $60 \, \mathrm{N\,s}$
   d. Zero. As there is no external net force acting on the system of the two girls, momentum is conserved.
   e. If they pushed each other harder they would have greater speeds but still in the same ratio as before. The total momentum would remain zero as there are no external horizontal forces acting on the girls.

4. a. $1.7 \, \mathrm{m\,s^{-1}}$
   b. $120 \, \mathrm{N\,s}$
   c. $120 \, \mathrm{kg\,m\,s^{-1}}$
   d. $120 \, \mathrm{kg\,m\,s^{-1}}$
   e. They would have different speeds but the total momentum would be conserved as there are no external horizontal forces acting on the boys.
   f. $0.92 \, \mathrm{m\,s^{-1}}$

5. a. $15.4 \, \mathrm{m\,s^{-1}}$
   b. $10\,500 \, \mathrm{N\,s}$ in the initial direction of motion of the car
   c. $420 \, \mathrm{N\,s}$ opposite to the initial direction of motion of the car
   d. $105\,000 \, \mathrm{N}$ in the initial direction of motion of the car

6. The two forces act on different objects. Forces are what objects do, rather than something that objects have, so force cannot be conserved.

10.2 Exam questions

1. $4.1 \, \mathrm{m\,s^{-1}}$
2. $7.6 \, \mathrm{m\,s^{-1}}$
3. $2.0 \, \mathrm{kg}$
4. $8 \, \mathrm{m\,s^{-1}}$
5. $1 \, \mathrm{m\,s^{-1}}$ west

10.3 Work and energy

Sample problem 2

a. $750 \, \mathrm{J}$
b. $150 \, \mathrm{J}$
c. $600 \, \mathrm{J}$

Practice problem 2

a. $600 \, \mathrm{J}$
b. $120 \, \mathrm{J}$

Sample problem 3

$300 \, \mathrm{J}$

Practice problem 3

$3.75 \, \mathrm{J}$

10.3 Exercise

1. $4000 \, \mathrm{J}$
2. $60 \, \mathrm{J}$
3. None. As there is no displacement in the direction of the force, the work done is zero.
4. a. Zero
   b. Zero
5. $6.2 \, \mathrm{m}$

10.3 Exam questions

1. B
2. A
3. $160 \, \mathrm{J}$
4. $20 \, \mathrm{J}$
5. $16 \, \mathrm{m}$

10.4 Energy transfers

Sample problem 4

The family car has approximately 62 times more kinetic energy than the athlete.

Practice problem 4

a. $64\,000 \, \mathrm{J}$
b. i. $1000 \, \mathrm{J}$
   ii. $2 \times 10^{-8} \, \mathrm{J}$

Sample problem 5

$3.2 \, \mathrm{m\,s^{-1}}$

Practice problem 5

$2 \, \mathrm{m\,s^{-1}}$

Sample problem 6

$2.5 \, \mathrm{J}$

Practice problem 6

a. $0.40 \, \mathrm{J}$
b. $1.6 \, \mathrm{J}$

Sample problem 7

a. $5 \, \mathrm{N}$
b. $0.25 \, \mathrm{J}$
c. $0.25 \, \mathrm{J}$

Practice problem 7

a. i. $12 \, \mathrm{N}$
   ii. $7.5 \, \mathrm{J}$
   iii. $1.2 \, \mathrm{kg}$
b. $80 \, \mathrm{N\,m^{-1}}$

Sample problem 8

$5.4 \, \mathrm{m\,s^{-1}}$

Practice problem 8

a. 0.4 J

b. 1.3 m s^{-1}

10.4 Exercise

1. Work = Fs

   Unit = N m

   But 1 N = 1 kg m s^{-2}

   $\Rightarrow$ Unit = kg m s^{-2} × m

   $= $ kg m^2 s^{-2}

   Kinetic energy $= \dfrac{1}{2} mv^2$

   Unit = kg (m s^{-1})^2

   $= $ kg m^2 s^{-2}

   Therefore, both kinetic energy and work are the same unit.

2. a. 100 000 J

   b. 23 J

3. 90 J

4. a. 1715 J

   b. 588 J

   c. 510 J

5. a. 196 J

   b. 196 J

   c. 196 J

   d. It is better to use the ramp. Although the amount of work needed to move the crate is the same, the force that needs to be applied to the crate is less if the ramp is used.

6. So that as little of their kinetic energy as possible is transferred to gravitational potential energy. Subsequently, a greater proportion of their kinetic energy is available to cover the horizontal distance as fast as possible.

7. a. Spring X: 40 N

   Spring Y: 20 N

   b. Spring X: 4.0 J

   Spring Y: 2.0 J

8. a. 235 J

   b. 20 m s^{-1}

9. a. 32 400 J

   b. Point B: 23 m s^{-1}; point C: 19.5 m s^{-1}

   c. 27 m

10. a. 900 000 J

    b. 360 000 J

    c. 30 m s^{-1}

11. 1.0 m s^{-1}

12. a. The force applied by the spring is proportional to its compression.

    b. 18 750 N m^{-1}

    c. 60 J

    d. 60 J

    e. 0.2 m

13. As the child falls through the air from maximum height, gravitational potential energy is transformed into kinetic energy.

    After the child touches the trampoline after falling through the air, kinetic energy and gravitational potential energy are transformed into strain potential energy until the trampoline is at maximum extension.

    After maximum extension, strain potential energy is transformed into gravitational potential energy and kinetic energy until contact is lost with the trampoline.

    After contact is lost, kinetic energy is transformed into gravitational potential energy until maximum height is achieved.

14.

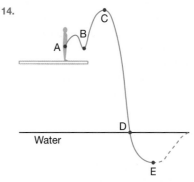

The energy transformations include the following.

A–B:	Chemical energy is transformed into strain potential energy of muscles, tendons and ligaments.
B–C:	Kinetic energy and some gravitational potential energy are transformed into strain potential energy of the springboard until the springboard reaches its maximum deflection.
C–D:	Gravitational potential energy of the diver is transformed into kinetic energy until the diver strikes the water.
D–E:	Kinetic energy of the diver is transferred to the water (as kinetic energy) and eventually transformed into thermal energy of the water particles.

15.

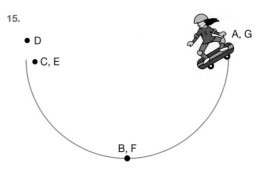

a. The energy transformations can be displayed with a graph of energy versus time or energy versus position, as follows.

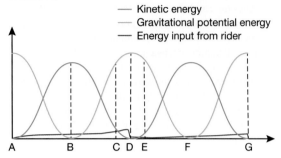

— Kinetic energy
— Gravitational potential energy
— Energy input from rider

As the rider moves down and up the slope, gravitational potential energy is transformed to kinetic energy and back again to gravitational kinetic energy. However, the total mechanical energy is not quite conserved and the rider needs to provide some additional energy 'input' to reach the top of the slope and point C. Further energy input is needed from the rider in order to gain the gravitational potential energy required at point D. Gravitational potential energy is then transformed into kinetic energy as the rider returns to point F and transformed into gravitational potential energy at point G. At points A, D and G, the rider's kinetic energy is zero.

b. The horizontal components of the speed of both the rider and the skateboard are also the same, as long as air resistance is negligible. The rider therefore needs to make little effort to remain in contact with the skateboard. There is some skill involved in ensuring that the frictional forces made possible by the contact are used to turn the skateboard so that it lands on the ramp before the feet or any other part of the rider's body.

10.4 Exam questions

1. 9800 N
2. **a.** 12 J
   b. The rough surface exerted an opposing force that did negative work = −12 J. Hence, the net work on the block is zero and so, there is no change in kinetic energy.
3. 10 m
4. **a.** Jill is correct.
   b. The force due to gravity is vertically down. To calculate the work done by the gravitational force, Jack must use the displacement parallel to this force; that is, 3.0 m, not the sloping distance of 5.0 m.
5. 20 J

10.5 Efficiency and power

Sample problem 9
80%

Practice problem 9
0.82 m

Sample problem 10
a. 118 W
b. 500 W

Practice problem 10
a. 1.5 m
b. 600 W

10.5 Exercise
1. Answers will depend on individual measurements. For a 2-kg mass, a 60-cm arm length and a 30-s time interval, the total work done is 1176 J and the total power output is 392 W.
2. 4.0 kN
3. 2.0×10^4 W
4. 1.4×10^4 W
5. 36 W

10.5 Exam questions
1. D
2. 50%
3. 49 W
4. **a.** 3.2×10^3 J
   b. 1.1×10^3 W
   c. None
5. **a.** 4.6 m s^{-1}
   b. 1.7×10^3 W

10.6 Review

10.6 Review questions
1. **a.** −2.7 kg m s^{-1} (away from the wall)
   b. 54 N
2. −22 m s^{-1} (the smaller car was travelling in the opposite direction to the larger car)
3. 32 J
4. 1.35×10^9 J
5. **a.** 10.8 N
   b. 1.1 kg
   c. 0.65 J
6. 130 000 J
7. **a.** 7.2×10^2 J
   b. 59 m s^{-1}
8. 32 m s^{-1}
9. **a.** $E_{\text{K before}} = 48$ J
   $E_{\text{k after}} = 9.0$ J
   b. 19%
10. 35×10^3 W
11. **a.** 3.1 J
    b. 1.0 J
    c. 0.64 m
12. **a.** 2.4×10^5 N
    b. 1.6×10^2 m s^{-2}
    c. 10^3 m s^{-2}

d. The kinetic energy of the car is transformed into potential energy of the materials in the crumple zone, which undergo a permanent change in shape. This leaves a smaller amount of kinetic energy to be transferred to the passengers.

e. One could argue that a large car is safer. For a given force applied by an obstacle or another vehicle, the deceleration of a large car is less than that of a small car. Therefore, the deceleration of the occupants inside is less.

13. a. 1920 J

b. 5 N

c. 3600 J

d. 4900 J

e. Some of the gravitational potential energy is transformed into thermal energy and sound, due to the frictional force and air resistance.

14. a. 2079 J

b. 0.15 m

15. Speed and momentum before impact, speed and momentum after impact, change in momentum on impact, average force during impact, loss of gravitational potential energy on falling, gain in kinetic energy on falling, gain of gravitational potential energy on rebounding, loss of kinetic energy on rebounding, percentage of energy lost.

10.6 Exam questions

Section A — Multiple choice questions

1. A

2. C

3. B

4. B

5. D

6. B

7. A

8. D

9. B

10. C

Section B — Short answer questions

11. Total energy is conserved; however, not all of the gravitational potential energy is converted into kinetic energy. The most likely cause for the discrepancy is the work done on the object by friction acting to oppose its motion down the ramp. This will decrease the amount of gravitational potential energy that is transformed into kinetic energy.

12. a. $6.4 \, \text{m s}^{-1}$

b. $-559 \, \text{kN}$ (in the opposite direction to the initial motion)

13. The student is not doing any work on the tray during the motion described. For work to be done there must be a force applied and a displacement in the direction of that force. In this instance the student would be exerting an upwards force on the tray to balance the downwards force due to gravity on the tray. This is perpendicular to the horizontal motion of the tray.

14. 21.2 kW

15. $6 \, \text{m s}^{-1}$

GLOSSARY

absolute refractive index the relative refractive index for light travelling from a vacuum into a substance, commonly referred to as the refractive index

absolute temperature the temperature of an object taken in the scale using absolute zero

absolute zero the lowest temperature that is physically possible, equal to 0 K or approximately −273 °C; at this temperature, particles cease to vibrate

absorbed dose a quantity describing the amount of energy absorbed by each kilogram of tissue that is irradiated

acceleration the rate at which the velocity of an object changes; it is a vector quantity

active wire the wire connected to the 230 V_{RMS} supply at the switchboard

air resistance the force applied to an object, opposite to its direction of motion, by the air through which it is moving

albedo the proportion of solar radiation reflected by a surface

α particle a relatively slow-moving decay product consisting of two protons and two neutrons, equivalent to a helium nucleus and carrying a positive charge

ampere the unit of current

amplitude a periodic disturbance that is the maximum variation from zero

angle of incidence the angle between an incident ray and the normal

angle of reflection the angle between a reflected ray and the normal

angle of refraction the angle between a refracted ray and the normal

atomic number the number of protons in a nucleus

binding energy the energy required to split a nucleus into individual nucleons

blackbody an object that absorbs all radiation that falls on it

blackbody radiation the characteristic radiation emitted by a blackbody when heated

centre of mass the point at which all of the mass of an object can be considered to be situated

chain reaction a reaction occurring when neutrons, emitted from the decay of one atom, are free to initiate fission in surrounding nuclei

charge carrier a charged particle moving in a conductor

circuit breaker a device that breaks a circuit when the current through it exceeds a certain value, carrying out the same function as a fuse

components parts of a vector; any vector can be resolved into a number of components, and when all the components are added together, the result is the original vector

compression a region of increased pressure in a medium during the transmission of a wave

conduction the transfer of heat through a substance as a result of collisions between neighbouring vibrating particles

conductor a material that contains charge carriers

control rods rods of a neutron-absorbing material used in a nuclear reactor to regulate the rate of nuclear fission

convection the transfer of heat in a fluid (a liquid or gas) as a result of the movement of particles within the fluid

convection current the movement of particles during the transfer of heat through a substance

cosmic radiation very energetic charged particles, mainly protons, originating from beyond the solar system

coulomb the unit of electric charge

critical angle the angle of incidence for which the angle of refraction is 90°; the critical angle exists only when light passes from one substance into a second substance with a lower refractive index

critical mass the smallest spherical mass of a fissionable substance that will sustain an uncontrolled chain reaction

current electricity electricity in which electrons move though electrical conductors

daughter nucleus the nucleus remaining after an atom undergoes radioactive decay; it is more stable than the original nucleus

decay chain the sequence of stages that a radioisotope passes through to become more stable; at each stage, a more stable isotope forms, and the chain ends when a stable isotope forms; also known as the decay series

decay curve a graph of the number of nuclei remaining in a substance versus the time elapsed; the half-life of a substance can be determined by looking at the time that corresponds to half of the substance remaining

decay equation a representation of a decay reaction; it shows the changes occurring in nuclei and lists the products of the decay reaction

diffuse reflection reflection from a rough or irregular surface

diode a device that allows current to pass through it in one direction only; its resistance changes with voltage

dispersion the separation of light into different colours as a result of refraction

displacement a measure of the change in position of an object; it is a vector quantity

distance a measure of the length of the path taken by an object; it is a scalar quantity

disturbance the movement of particles due to an energy wave passing through them

earth wire the wire connecting the case of an appliance to the earth, as a safety device

eccentricity a measure of how elliptical an object's orbit is

effective dose the sum of the tissue-equivalent doses weighted by the weighting factors

electric charge the basic property of matter; it occurs in two states: positive (+) charge and negative (−) charge

electric circuit a closed loop of moving electric charge

electric conductor a material that contains charge carriers; that is, charged particles can move and travel freely through the material

electric current the movement of charged particles from one place to another

electric insulator a material in which the electrons are bound tightly to the nucleus and are not free to travel through the material

electric shock a violent disturbance of the nervous system caused by an electrical discharge or current through the body

electrocution death brought about by an electrical shock

electromagnetic radiation an electromagnetic wave or radiation that includes visible light, radio waves, gamma rays and x-rays

electromotive force (emf) a measure of the energy supplied to a circuit for each coulomb of charge passing through the power supply

electron volt a unit of energy, where 1 electron volt (eV) is equal to the energy of electrons

electrostatic force the force between two stationary charged objects

element a substance that consists only of atoms of the same name

enhanced greenhouse effect the disruption to Earth's climate equilibrium caused by the release of greenhouse gases in the atmosphere due to human activities, which leads to an increase in global average surface temperatures

enrichment the process of increasing the percentage of uranium-235 in a sample of uranium to enable it to sustain a chain reaction

equivalent dose a quantity taking into account the type of radiation absorbed and representing the biological effect of radiation that has been absorbed by living tissue

excited nucleus a nucleus that does not have an ideal arrangement of protons and neutrons within it; an excited nucleus emits γ radiation to become more stable

feedback refers to when a system's input is fed by its previous output

fibrillation the disorganised, rapid contraction of separate parts of the heart that prevent it from pumping blood; death may follow

fission fragments the products from a nucleus that undergoes fission; they are smaller than the original nucleus

force an interaction between two objects that can cause a change; commonly a push, pull or twist applied to one object by another, measured in newtons (N); it is a vector quantity

force due to gravity the force applied to an object due to gravitational attraction

forward biased refers to when a diode's positive side is connected to the positive output of a power source

free radical an uncharged fragment of a molecule resulting from a covalent bond being broken

frequency a measure of how many times per second an event happens, such as the number of times a wave repeats itself every second

friction the force applied to the surface of an object when it is pushed or pulled against the surface of another object

fuse a short length of conducting wire or strip of metal that melts when the current through it reaches a certain value, breaking the circuit

γ ray a packet of electromagnetic energy released when a nucleus remains unstable after α or β decay; γ rays travel at the speed of light and carry no charge

gravitational field strength (g) the force of gravity on a unit of mass

gravitational potential energy the energy stored in an object as a result of its position relative to another object to which it is attracted by the force of gravity

half-life the time taken for half of a group of unstable nuclei to decay

heat the transfer of energy from one body to another due to a temperature difference

idealisation assuming ideal conditions that don't exactly match the real situation to make modelling a phenomenon or event easier

impulse the product of the force and the time interval over which it acts; it is a vector quantity

incandescent refers to luminous objects that produce light because they are hot; the higher the temperature, the brighter the light, and the colour also changes

inertia the resistance of any physical object to a change in its speed or direction of motion

instantaneous speed the speed at a particular instant of time

instantaneous velocity the velocity at a particular instant of time

insulator a substance that does not readily allow the passage of heat

ionising radiation high-energy radiation that can change atoms by removing electrons, thus giving the atom an overall charge

ion a charged particle

isotope an atom containing the same number of protons but a different numbers of neutrons

joule the SI unit of work or energy; 1 joule is the energy expended when a force of 1 newton acts through a distance of 1 metre

kilowatt-hour (kW h) the amount of energy transformed by a 1000 W appliance when used for one hour

kinetic energy the energy associated with the movement of objects; like all forms of energy, it is a scalar quantity

latent heat the heat added to a substance undergoing a change of state that does not increase the temperature

light dependent resistor (LDR) a device that has a resistance that varies with the amount of light falling on it

light-emitting diode (LED) a small semiconductor diode that emits light when a current passes through it

light ray an infinitely narrow beam of light, represented as a straight line

load a device in which electrical energy is converted into other forms to perform tasks, such as heating or lighting

longitudinal wave a wave for which the disturbance is parallel to the direction of propagation

luminosity the amount of radiated electromagnetic energy emitted by a light-emitting or luminous object

luminous refers to objects seen because they give off their own light

magnitude the size or quantity of an object or variable

mass defect the difference in the mass of the products and reactants in a nuclear reaction; also known as the mass deficit

mass number the total number of nucleons in an atom

mechanical interaction an interaction in which energy is transferred from one object to another by the action of a force

model a representation of ideas, phenomena or scientific processes; can be a physical model, mathematical model or conceptual model

moderator a material that slows down the speed of a neutron

momentum the product of the mass of an object and its velocity; it is a vector quantity

negative feedback refers to when the response to the feedback is in the opposite direction to the input

negligible a quantity so small that it can be ignored when modelling a phenomenon or an event

net force the vector sum of the forces acting on an object

neutral an object that carries an equal amount of positive and negative charge

neutral wire the wire connected to the neutral link at the switchboard, which is connected to the earth

neutralise when two opposite charges of the same magnitude cancel each other out

non-ohmic device a device for which the resistance varies with changing physical conditions, including voltage, light level or temperature, but is constant for all currents passing through it under each set of physical conditions

non-uniform motion motion in which the velocity changes over time

normal a line that is perpendicular to a surface or a boundary between two surfaces

normal force the force that acts perpendicularly to a surface as a result of an object applying a force to the surface

nuclear fission the process of splitting a large nucleus to form two smaller, more stable nuclei

nucleon a particle found in the nucleus — proton or neutron

nucleus the solid centre of an atom where most of its mass is concentrated

obliquity a measure of the angle tilt of a planet against its plane of orbit

ohmic device a device for which the resistance is constant for all currents under a range of changing physical conditions, including changing voltage, light level and temperature

optical fibre a thin tube of transparent material that allows light to pass through without being refracted into the air or another external medium

parallel refers to devices connected so that one end of each device is joined at a common point, and the other end of each device is joined at another common point

period the amount of time, measured in seconds, that one cycle or event takes, such as the time taken for an object moving in a circular path and at a constant speed to complete one revolution

periodic wave a disturbance that repeats itself at regular intervals

photosynthesis a chemical reaction, converting light energy into chemical energy (sugars), that takes place in the chloroplasts of a plant cell consuming carbon dioxide and releasing oxygen as a byproduct

positron a positively charged particle with the same mass as an electron, formed when a proton disintegrates to form a neutron and a positron

potential difference the amount of electrical potential energy, in joules, lost by each coulomb of charge in a given part of a circuit

potentiometer a variable voltage divider, with a fixed resistor that can slide up and down

power the rate of doing work, or the rate at which energy is transformed from one form to another

power radiated by a blackbody the total energy radiated by a blackbody every second

precession a change in direction of the rotational axis of a spinning object

pulse a wave of short duration

radiation heat transfer without the presence of particles

radioactive decay the process by which an unstable atomic nucleus loses energy by radiation

radioisotope an unstable isotope

rarefraction a region of reduced pressure in a medium during the transmission of a sound wave

refraction the bending of light as it passes from one medium into another

regular reflection reflection from a smooth surface; also referred to as specular reflection

relative refractive index a measure of how much light bends when it travels from any one substance into any other substance

residual current device a current-activated circuit breaker; it operates by making use of the magnetic effects of a current to break a circuit in the event of an electrical fault

resistance the ratio of voltage drop, V, across a material or device to the current, I, flowing through it

resistor a device used to control the current flowing through, and the voltage drop across, parts of a circuit

reverse biased refers to when a diode's positive side is connected to the negative output of a power source

rotational kinetic energy the energy due to the rotational motion of objects

scalar a quantity that specifies magnitude (size) but not direction

semiconductor a material that allows electrons to pass through in certain conditions

series refers to devices joined together one after the other

short circuit a malfunction or fail that can occur when frayed electrical cords or faulty appliances allow the current to flow from one conductor to another with little or no resistance, resulting in a rapid increase of the current and potentially causing the wires to get hot and start a fire

specific latent heat of fusion the quantity of energy required to change 1 kilogram of a substance from a solid to a liquid without a change in temperature

specific latent heat of vaporisation the quantity of energy required to change 1 kilogram of a substance from a liquid to a gas without a change in temperature

speed the rate at which distance travelled changes over time; it is a scalar quantity

speed of a periodic wave the product of the wavelength of the wave multiplied by its frequency

static electricity electricity produced as the result of an imbalance between negative and positive charges in objects containing charges that are usually static (unmoving)

strain potential energy the energy stored in an object as a result of a reversible change in shape; it is also known as elastic potential energy

strong nuclear force the attractive force that holds nucleons together in a nucleus of an atom, acting over only very short distances

switch a device that stops or allows the flow of electricity through a circuit

switch-on voltage the voltage needed to allow current to flow freely through a diode

temperature a measure of how hot or how cold something is; a measure of the average translational kinetic energy of particles

tension a pulling force along the length of an object (such as a rope or cable) that is being stretched

thermal equilibrium the state obtained when the temperature of two regions in contact is uniform

thermistor a device that has a resistance that changes with a change in temperature

tissue weighting factor a dimensionless factor reflecting the variable radiosensitivity of specific tissues and organs that is used to calculate the equivalent dose

total internal reflection the total reflection of light from a boundary between two substances; it occurs when the angle of incidence is greater than the critical angle

transducer a device that converts energy from one form to another form

translational kinetic energy the energy due to the motion of an object from one location to another

transverse wave a wave for which the disturbance is at right angles to the direction of propagation

uniform motion motion in which the velocity remains constant over time

vector a quantity that specifies direction as well as magnitude (size)

velocity the rate at which displacement changes over time, or the rate of change in position; it is a vector quantity

vibrational kinetic energy the energy due to the vibrational motion of objects

voltage divider used to reduce, or divide, a voltage to a value needed for a part of a circuit

voltage drop another term used to describe potential difference

wave the transfer of energy through a medium without any net movement of matter

wavelength the distance between successive corresponding parts of a periodic wave

weak nuclear force the force that explains the transformation of neutrons into protons, and vice versa

work a measure of energy transferred to or from an object by the action of a force; it is a scalar quantity

x-rays electromagnetic waves of very high frequency and very short wavelength

INDEX

normal   42
normal force   374, 400
nuclear fission   158–62
nuclear fusion   158, 162–5
nuclear radiation   125–32
  alpha (α) decay   133–5
  atoms and isotopes   125–9
  beta (β) decay   135–8
  decay series   139–42
  gamma (γ) decay   138–9
  half-life   129–31
  types of   132–43
nuclear stability   129
nuclear transformations   141
nuclei   157
nucleons   125
nucleus   125, 127–8, 135
  energy from   157–66

O

obliquity   104
ohmic devices   218
Ohm's Law   217–8
optical fibres   53
  internal reflection in   53–6
  refraction and critical
      angles in   55
optical phenomena   65–8
  mirages   66–7
  rainbows   65–6
ozone layer   105

P

parallel circuits   242, 250–61
  current in   250
  modelling resistors in   251–5
  short circuits   255–6
  voltage drop in   250–1
particle radiation, electromagnetic
      radiation and   143–4
paying for electricity   289–92
period   8, 285
periodic table   126
periodic waves   5
phlogiston   79
photosynthesis   104
plug-in fuse   297
polygraph   213
position-versus-time graphs
      335–8, 346
positrons   135, 163
potential difference
  analogies and models for   205–6
  bicycle chain model of   206
  definition of   203–4
  hydraulic model of   205–6

potential energy   455–6
  gravitational   455–7
  strain   455, 457–60
potentiometers   267
power   207–13, 220–3, 271
      468–70
  in circuits   271–3
  definition of   208
  delivered by circuit   208–11
power circuit   287
power radiated by blackbody   20
power ratings   288–9
  of device   291
  using voltage drop and
      current   209
precession   104
propagation of light   39–41
pulse   5

R

radiation   88, 91–2
  absorbed dose   145
  α, β and γ radiation on
      humans   144–8
  background radiation   147
  detecting   137
  in diagnosis and treatment   148–51
  effective dose   146–8
  electromagnetic radiation   143–4
  equivalent dose   145–6
  and human body   143–52
  ionising radiation   143
  particle radiation   143–44
  quality factors for   145
radioactive decay   103
radioactive elements   124
radioactivity   124
  as a diagnostic tool   148–9
  in smoke detectors   134
radioisotopes   128, 149
  in body   150
  used in medical diagnosis
      149, 150
rainbows   65–6
rarefactions   10
ray model for light   41–2
reaction   396
reaction forces, magnitude of   426
real collisions, modelling   444–5
reflected ray   42
reflection
  angle of   42
  regular and diffuse   42–3
refraction   44
  early history of   45
  using Snell's Law   39–52

refractive index   48
  critical angle for water using   53
  values for different coloured
      light   60
regular reflection   42
relative refractive index   46
residual current device   300
resistance   213–4, 220–3, 243
  using coloured bands   215
  of different resistors   215
  forces   389
  of human body   294
  with short circuit   256
  using voltage and current   217
resistors   214–7, 235
  in combinations of series and
      parallel   257–8
  in parallel   251–5
  in series   243–4
  in voltage divider   246
resultant force   377
reverse biased diode   265
road friction   401
rotational kinetic energy   81

S

safety in household circuits   296–8
salt solutions   194
scalars   323
scattering of light   61–2
scientific principles,
      climate   108–9
semiconductor   194
sensors   262–4
series circuits   242–50
  current in   242
  resistors in   243–4
  voltage divider   245–6
  voltage drop in   242–3
shape, into triangles and rectangles
      340–3
short circuits   255–6, 296
  resistance with   256
sight   38
Snell's Law   45–8
sodium   125
solids   82–3
sound wave
  speed of   10
  wavelength of   11
specific heat capacity   95–7
specific latent heat
  of fusion   98
  of vaporisation   98–100
speed   325–6
  converting units of   326–7